P9-CDM-370

Diffusional Mass Transfer

Diffusional Mass Transfer

A. H. P. SKELLAND
Georgia Institute of Technology

Krieger Publishing Company
Malabar, Florida

Original Edition 1974
Reprint Edition 1985

Printed and Published by
ROBERT E. KRIEGER PUBLISHING COMPANY, INC.
KRIEGER DRIVE
MALABAR, FL 32950

Copyright © 1974 By John Wiley & Sons, Inc.
Reprinted by Arrangement

All rights reserved. No part of this book may be reproduced in any form or by any electronic or mechanical means including information storage and retrieval systems without permission in writing from the publisher.

No liability is assumed with respect to the use of the information contained herein.

Printed in the United States of America

Library of Congress Cataloging in Publication Data

Skelland, A. H. P.
Diffusional mass transfer.

Reprint. Originally published: New York: Wiley, 1974.
Includes index.
1. Mass transfer. 2. Diffusion. 3. Molecules.
I. Title.
QC318.M3S54 1985 531'.1137 84-14403
ISBN 0-89874-792-9

10 9 8 7

This book is dedicated to

the late Professor F. H. Garner, O.B.E.

Preface

The occurrence of mass-transfer processes throughout the biological, chemical, physical, and engineering fields is extremely widespread. Biological involvements include respiration mechanisms and the oxygenation of blood, kidney functions, and food and drug assimilation. A few engineering examples are the ablative cooling of space vehicles during reentry to the atmosphere, the transpiration and film cooling of rocket and jet-engine exhaust nozzles, and the separation of ores and isotopes. Chemical-engineering applications arise in such processes as distillation, gas absorption, stripping, liquid and solid extraction, adsorption, crystallization, air conditioning, water cooling, drying, ion exchange, sublimation, and chromatography.

This book describes a representative selection of topics, many of which are common to a wide variety of applications. Nevertheless, any text with less than several thousand pages must inevitably omit more than it covers in the enormous field of mass transfer. Indeed, entire books have been written on some single process from among those listed above. It would therefore be a simple matter to prepare an extensive list of "omissions," although one in particular should be mentioned. The field of mass transfer in chemically reacting systems is not treated in this text because good presentations of the subject are available in two recent books. They are *Mass Transfer with Chemical Reaction* by G. Astarita (Elsevier, Amsterdam, London, and New York, 1967), and *Gas-Liquid Reactions* by P. V. Danckwerts (McGraw-Hill, New York, 1970).

The approach throughout is from the diffusional or rate-process point of view. The nature and diversity of mass-transfer processes are indicated in Chapter 1, and Chapter 2 describes steady and unsteady-state molecular diffusion under conditions often encountered in practice. The prediction of molecular diffusivities in gases, liquids, and solids is discussed in Chapter 3. The concepts of individual and overall transfer coefficients are introduced in Chapter 4, where some of the more prominent theories on transfer mechanisms are also presented. Chapters 5 and 6 consider mass transfer under laminar and turbulent flow conditions, mostly with known velocity fields, and for a variety of external and internal flow systems. The last three chapters are concerned with column and tower designs for several gas-liquid and liquid-liquid processes where the details of the velocity fields are unknown. Chapter 7 deals with continuous columns; the first half of the chapter considers the location of the operating line in various cases, for use in evaluation of the NTU relationships derived later in the chapter. A provisional attempt is made in Chapter 8 to integrate some of the many and diverse studies on droplet phenomena into a coherent design procedure for perforated-plate extraction columns. The approach is clearly amenable to refinement after further research. Rate equations are applied in Chapter 9 to the design of cooling towers. Basic concepts from the earlier parts of the book influence the formulation of the last three chapters in ways which are outlined in Chapter 1. Examples showing numerical computations appear throughout, and the unworked problems at the end of each chapter—for solution by the reader—are intended to consolidate and extend material in the text.

Results of digital-computer solutions to certain boundary-value problems are incorporated at appropriate places in the book (for example, when considering mass transfer during laminar flow through tubes and between parallel plates). However, in common with E. R. G. Eckert and R. M. Drake (*Analysis of Heat and Mass Transfer*, Preface, McGraw-Hill, New York, 1972), I assume that the student is already familiar with the computer programming techniques needed to obtain results from the relevant equations, so that such programs have generally not been included. An exception is the rather extensive computer program for perforated-plate column design in Chapter 8. This facilitates the use of a complex design procedure and would perhaps pose unusual difficulties if left to the student.

The book would be suitable for use by either senior or graduate students and should prove helpful to the practicing engineer. To this end the derivation of important or representative relationships has been presented in sufficient detail to enable a clear understanding and, I hope, an avoidance of misapplication.

A. H. P. SKELLAND

Lexington, Kentucky
June 1973

Contents

Diffusional Mass Transfer

1

Introduction

Diffusional mass transfer involves the migration of one substance through another under the influence of a concentration gradient. The occurrence of mass-transfer processes throughout the biological, chemical, physical, and engineering fields is extremely widespread. Biological involvements include respiration mechanisms and the oxygenation of blood, kidney functions, and food and drug assimilation. Engineering examples are found in the ablative cooling of space vehicles during reentry into the atmosphere, in the transpiration and film cooling of rocket and jet-engine exhaust nozzles, and in the separation of ores and isotopes. Chemical-engineering applications of mass transfer in separation processes involve the diffusional transport of some component within a single phase or between two immiscible phases which have been brought into contact to enable the transfer of the component from one phase to the other. Components may migrate from the bulk of one phase to the interface between phases and remain there, as in adsorption or crystallization. Alternatively, penetration of the interface may occur, followed by diffusion into the bulk of the other phase, as in distillation, gas absorption, and liquid-liquid extraction.

The design of equipment for the diffusional separation of mixtures is determined by two major considerations, namely, the distribution of components between phases in a state of thermodynamic equilibrium, and the rate at which mass transfer occurs under conditions prevailing in the equipment.

In *distillation* intimate contact is promoted between saturated-vapor and boiling-liquid phases to facilitate the transfer of a less volatile component from the vapor to the liquid and of a more volatile component in the opposite direction. *Gas absorption* refers to the transfer of a soluble component from a gas phase into a nonvolatile liquid absorbent. The reverse process is called *desorption* or *stripping*. Transfer of solute between two immiscible or partially miscible liquid phases occurs in *liquid-liquid extraction*, whereas in *solid-liquid extraction* a liquid solvent is used to dissolve a soluble solid component from its mixture with another insoluble solid. In *adsorption* operations a gaseous or liquid mixture is separated by preferential adsorption of some component on the surface of a solid. Subsequent recovery of the adsorbed material is often effected by heating or steaming. *Crystallization* is used to separate a crystalline solid from its solution by inducing supersaturation. *Air humidification* and some forms of *air conditioning* and *water cooling* involve the transport of water vapor through an air stream which has been contacted with water. *Drying* operations depend on the transport of both liquid and vapor within the solid and of vapor in the drying gas. Clearly, many other examples could be cited, such as ion exchange, sublimation, chromatography, and reverse osmosis, all of which are linked by their common dependence upon rate-process mass transfer.

A great diversity of equipment has evolved for carrying out these various separation operations. In cases involving the contacting of a gas and a liquid phase or of two liquid phases the equipment may generally be classified as either a continuous or a stagewise contactor.

Continuous contactors usually consist of vertical columns, frequently filled with some sort of packing. The two phases generally flow countercurrently through the interstices in the packing, which is provided to promote good contact for mass transfer between the phases. The necessary column height to achieve a specified separation is a major design objective in such equipment.

Stagewise contactors provide intermittent, rather than continuous, contact between the phases. The stages often take the form of horizontal plates or trays of varied design, arranged vertically above each other in a column. The two phases usually flow countercurrently, mix together to allow interphase mass transfer on a given stage, and then separate and flow, respectively, up and down to the next stages in the series. The design of such contactors involves a determination of the number of stages needed to effect a given separation of components in a phase, or of the separation obtainable from an existing column with a fixed number of stages.

The choice between continuous and stagewise contactors in a particular situation is determined by such factors as the attainable stage efficiency, capacity or scale of operation, corrosion problems, tolerable pressure drop, availability of performance data, and, of course, relative cost. Both types of contactor are considered here in terms of rate-process mass transfer. Other forms of phase-contacting equipment, including dryers, crystallizers, mixer-settlers, and the like, are available in great variety. In all cases, however, their performance is substantially dependent on relationships to be presented here.

Transfer in solids may occur by mechanisms other than diffusion. Thus in particulate solids with an extensive structure of large and varying pores, liquid transport may take place under the influence of capillary forces in a manner that is not directly proportional to the concentration gradient. In fluids at rest, mass is transferred by purely molecular diffusion when a concentration gradient is present. When convective movement exists in the fluid, however, transfer occurs both by molecular diffusion and by bulk motion of the whole mixture. The latter contribution to the mass-transfer process depends on the details of the flow pattern within the fluid. Knowledge of the relevant fluid dynamics is therefore essential to the solution of any convective-mass-transfer problem. To keep the length of the treatment within bounds, fluid-dynamical developments are here confined to those aspects necessary to the case under consideration.

Forced-convection problems are those in which the flow field is imposed by some device such as a pump or fan, or by the propulsion of a body through a fluid. Natural or free convection arises, for example, under the influence of a gravitational field acting on density differences associated with variations in solute concentration or temperature.

It will be found that certain mass-, heat-, and momentum-transfer processes are in some ways analogous when solute concentrations and transfer rates are low. Under conditions of high mass flux, however, the flow field is modified by velocity components associated with the mass transfer, and this introduces significant differences between mass transfer on the one hand and heat and momentum transfer processes without mass transfer on the other. Mass-transfer coefficients that are restricted to low concentrations and low transfer rates are usually marked with an asterisk in this book.

Any text with fewer than several thousand pages must inevitably omit more than it covers in the enormous field of mass transfer. Indeed, entire books have been written on some single process from among those described earlier. An encyclopedic survey of mass transfer is therefore not feasible in one volume, although one omission in particular should be mentioned. The study of mass transfer in chemically reacting systems is not treated in this text, because good presentations of the subject are already available in two recent books (Astarita, 1967; Danckwerts, 1970).

Design applications in the final three chapters are confined to the gas-liquid and liquid-liquid processes of distillation, gas absorption, stripping or desorption, liquid-liquid extraction, humidification, and gas and liquid cooling. Solid-fluid operations, such as adsorption, chromatography, ion exchange, drying, and crystallization, are excluded, although some related topics are dealt with in earlier parts of the book. These include unsteady-state diffusion in bodies of several geometries and *j* factors in packed and fluidized beds. The approach throughout is from the diffusional or rate-process point of view. Design procedures in terms of the nondiffusional aspects of material and energy conservation, subject to the constraints of phase equilibrium, are to be found in texts by Smith (1963), Henley and Staffin (1963), and Brian (1972).

The last three chapters contrast with situations in Chapters 2, 5, and 6 in that the details of the hydrodynamics involved are much less well known. Basic relationships and concepts from earlier chapters nevertheless contribute significantly to the treatment in these areas, as outlined in the following three paragraphs.

In Chapter 7, the four rate equations from Chapter 4, the two-film theory (Chapter 4), and the respective mechanisms of equimolal counterdiffusion and unimolal unidirectional diffusion (Chapter 2) are all used in formulating the expressions for the number of transfer units (NTU) in distillation on the one hand and in absorption, stripping or desorption, and extraction on the other. The concept of the additivity of resistances to mass transfer between phases (Chapter 4) is then applied in obtaining the relationships between overall and individual heights of transfer units (HTUs) for the two different transport mechanisms. Evaluation of the individual HTUs requires a knowledge of the molecular diffusivities, predictable according to Chapter 3.

Chapter 8 again uses rate equations and the concept of additive resistances (Chapter 4) to formulate overall mass-transfer coefficients from individual coefficients describing extraction in perforated plate columns. Various theoretical expressions for coefficients during droplet formation, rise, and coalescence have been developed using the penetration theory (Chapter 4), and the disperse-phase coefficient during the free rise of stagnant droplets is derived from the treatment of unsteady-state diffusion in a sphere given in Chapter 2. The corresponding continuous phase coefficient is obtained from relationships for spheres and spheroids provided in Chapter 6. The rate expressions during all stages of the process utilize diffusivities obtainable from Chapter 3.

In Chapter 9, the wet-bulb temperature analysis involves individual coefficients (Chapter 4), *j* factors (Chapter 6), and diffusivities (Chapter 3); both individual and overall coefficients (Chapter 4) are used in formulating the NTU and HTU expressions. Limitations on the use of constant overall coefficients in Chapters 7 and 9 are as prescribed in Chapter 4. In both Chapters 7 and 9 the preference for design in terms of transfer units rather than coefficients follows from demonstrations in earlier chapters (4, 5, and 6) of the substantial variations in coefficients with flow rate and, in some cases, with concentration.

REFERENCES

Astarita, G., *Mass Transfer with Chemical Reaction*, Elsevier, Amsterdam, (1967).

Brian, P. L. T., *Staged Cascades in Chemical Processing*, Prentice-Hall, Englewood Cliffs, N. J., (1972).

Danckwerts, P. V., *Gas-Liquid Reactions*, McGraw-Hill, New York, (1970).

Henley, E. J., and H. K. Staffin, *Stagewise Process Design*, Wiley, New York, (1963).

Smith, Buford D., *Design of Equilibrium Stage Processes*, McGraw-Hill, New York, (1963).

2

Molecular Diffusion

When the composition of a fluid mixture varies from one point to another, each component has a tendency to flow in the direction that will reduce the local differences in concentration. If the bulk fluid is either stationary or in laminar flow in a direction normal to the concentration gradient, the mass transfer reducing the concentration difference occurs by a process of molecular diffusion. This mechanism, characterized by random movement of individual molecules, contrasts with the bulk transport by eddies which occurs in a turbulent fluid.

Consideration is first given to some of the ways in which concentration and flux are defined.

Flux Definitions

A wide variety of methods for expressing the composition of multicomponent systems is in use, including mole or mass fraction, moles or mass per unit volume, and moles or mass of component A per mole or unit mass of non-A. The mass concentration or mass of component A per unit volume of solution is denoted by ρ_A, and the mass fraction, ρ_A/ρ, by w_A. The molar concentration (the number of moles of component A per unit volume of solution) is written as c_A, and the mole fraction, c_A/c, as x_A. It may be noted that "molal" and "molar" refer to different definitions of concentration in classical chemistry. However, both terms are also widely used in a broader sense in the engineering literature to denote quantities and processes relating to moles. Among those using "molal" in this way are Rohsenow and Choi (1961), Bennett and Myers (1962), Kay (1963), Sherwood and Pigford (1952), and McCabe and Smith (1967), whereas those preferring "molar" include Bird, Stewart, and Lightfoot (1960), Welty, Wicks, and Wilson (1969), Oliver (1966), and Foust et al. (1960). The term "molal" is generally used in this text to mean "pertaining to moles," in accordance with the *International Dictionary of Physics and Electronics*, 2nd ed., Van Nostrand, Princeton, N. J., (1961), p. 761.

Attention is now directed to a nonuniform multicomponent fluid mixture that is undergoing bulk motion and within which the various components move with different velocities because of diffusional activity. Some procedures are considered by which the component velocities may be averaged to provide different definitions of the average fluid velocity. Detailed developments of such relationships and the attendant expressions for flux have been presented in the engineering literature by Bird (1956), Bird, Stewart, and Lightfoot (1960), Rohsenow and Choi (1961), and Bennett and Myers (1962).

The statistical mean velocity of component i in the x direction with respect to stationary coordinates is written as u_i, so that the mass flux of component i through a stationary surface normal to u_i is $\rho_i u_i$. For an n-component system the *mass-average velocity in the x direction* is then defined by

$$u=\frac{1}{\rho}\sum_{i=1}^{n}\rho_i u_i \tag{2.1}$$

Another form of mean velocity for the mixture is the *molal-average velocity in the x direction*, given by the expression

$$U=\frac{1}{c}\sum_{i=1}^{n}c_i u_i \tag{2.2}$$

Evidently u and U are approximately equal at low solute concentrations in binary systems—a situation which has received extensive theoretical and experimental study. The velocities u and U are also the same in nonuniform mixtures of compounds having the same molecular weight. Another case in which the mass and molal average velocities are equal is the bulk flow of a mixture with uniform composition throughout, regardless of the relative molecular weights of the components.

The velocity of component i may clearly be defined in three frames of reference. In relation to stationary coordinates it is u_i, in relation to the mass-average velocity it is u_i-u, and in relation to the molal-average velocity it is $u_i - U$. These various velocities lead to corresponding definitions of *mass fluxes* in the x direction for component i as follows:

$$\text{Relative to stationary coordinates,} \quad n_{ix} = \rho_i u_i \tag{2.3}$$

$$\text{Relative to the mass-average velocity,} \quad i_{ix} = \rho_i (u_i - u) \tag{2.4}$$

$$\text{Relative to the molal-average velocity,} \quad j_{ix} = \rho_i (u_i - U) \tag{2.5}$$

Similar definitions of *molal fluxes* in the x direction can be written for component i. Thus:

$$\text{Relative to stationary coordinates,} \quad N_{ix} = c_i u_i \tag{2.6}$$

$$\text{Relative to the mass average velocity,} \quad I_{ix} = c_i (u_i - u) \tag{2.7}$$

$$\text{Relative to the molal average velocity,} \quad J_{ix} = c_i (u_i - U) \tag{2.8}$$

These expressions enable ready development of the relationships between the various mass and molal fluxes. For example, to relate mass flux i_{ix} to mass flux n_{ix}, consider equations 2.1, 2.3, and 2.4:

$$i_{ix} = \rho_i u_i - \rho_i u = n_{ix} - \frac{\rho_i}{\rho} \sum_{i=1}^{n} \rho_i u_i$$

$$= n_{ix} - w_i \sum_{i=1}^{n} n_{ix} \tag{2.9}$$

and for a binary system of components A and B,

$$i_{Ax} = n_{Ax} - w_A (n_{Ax} + n_{Bx}) \tag{2.10}$$

Summing equation 2.9 for all components gives

$$\sum_{i=1}^{n} i_{ix} = 0 \tag{2.11}$$

or, for the binary system,

$$i_{Ax} + i_{Bx} = 0 \tag{2.12}$$

To relate the mass flux j_{ix} to the mass flux n_{ix}, consider equations 2.2, 2.3, and 2.5:

$$j_{ix}=\rho_i u_i-\rho_i U=n_{ix}-\frac{\rho_i}{c}\sum_{i=1}^{n} c_i u_i$$

$$=n_{ix}-\frac{c_i M_i}{c}\sum_{i=1}^{n} N_{ix} \tag{2.13}$$

and for a binary system,

$$j_{Ax}=n_{Ax}-\frac{c_A M_A}{c}(N_{Ax}+N_{Bx})=n_{Ax}-x_A\left(n_{Ax}+\frac{M_A}{M_B}n_{Bx}\right) \tag{2.14}$$

since $N_{Ax}M_A=n_{Ax}$; $N_{Bx}M_B=n_{Bx}$. Summing equation 2.13 for all components,

$$\sum_{i=1}^{n} j_{ix}=\rho(u-U) \tag{2.15}$$

or, for the binary system,

$$j_{Ax}+j_{Bx}=\rho(u-U) \tag{2.16}$$

From equations 2.1 and 2.3 for the binary system, $n_{Ax}+n_{Bx}=\rho u$.

The development of the corresponding relationships between the *molal* fluxes is summarized by equations 2.17 to 2.24 in Table 2.1. Other relationships between the fluxes may be developed by analogous procedures and for the coordinate directions y and z.

Now consider a binary mixture of nonreacting components A and B. Suppose that the total mixture is flowing steadily with mass- and molal-average velocities u and U in the x direction. If the composition is nonuniform, molecular diffusion occurs within the mixture in accordance with Fick's first law. For steady one-dimensional transfer this diffusive flux may be written as follows:

$$i_{Ax}=-D_{AB}\frac{d\rho_A}{dx} \tag{2.25}$$

which is shown below to require constant density ρ. More generally,

$$i_{Ax}=-\rho D_{AB}\frac{dw_A}{dx} \tag{2.26}$$

Table 2.1. Development of some relationships between molal fluxes N_{ix}, I_{ix}, and J_{ix}.

	$I_{ix}=f(N_{ix})$	Eq. No.	$J_{ix}=f(N_{ix})$	Eq. No.
Relevant equations	2.1, 2.6, and 2.7		2.2, 2.6, and 2.8	
Flux	$I_{ix}=c_iu_i-c_iu$		$J_{ix}=c_iu_i-c_iU$	
	$=N_{ix}-\frac{c_i}{\rho}\sum_{i=1}^{n}\rho_iu_i$		$=N_{ix}-\frac{c_i}{c}\sum_{i=1}^{n}c_iu_i$	
	$=N_{ix}-\frac{w_i}{M_i}\sum_{i=1}^{n}n_{ix}$	2.17	$=N_{ix}-x_i\sum_{i=1}^{n}N_{ix}$	2.21
For binary systems	$I_{Ax}=N_{Ax}-w_A\left(N_{Ax}+\frac{M_B}{M_A}N_{Bx}\right)$	2.18	$J_{Ax}=N_{Ax}-x_A(N_{Ax}+N_{Bx})$	2.22
Sum for all components	$\sum_{i=1}^{n}I_{ix}=c(U-u)$	2.19	$\sum_{i=1}^{n}J_{ix}=0$	2.23
For binary systems	$I_{Ax}+I_{Bx}=c(U-u)$	2.20	$J_{Ax}+J_{Bx}=0$	2.24

From equations 2.2 and 2.6: $\sum_{i=1}^{n}N_{ix}=cU$, and for a binary system, $N_{Ax}+N_{Bx}=cU$

which will be shown not to require constancy of ρ. In molal terms,

$$J_{Ax}=-D_{AB}\frac{dc_A}{dx} \tag{2.27}$$

for which, it can be proved, constant total molar concentration c is required. More generally,

$$J_{Ax}=-cD_{AB}\frac{dx_A}{dx} \tag{2.28}$$

for which variation in c is permissible. In these expressions, $\rho=\rho_A+\rho_B$, $c=c_A+c_B$, and $D_{AB}=D_{BA}$ is the molecular diffusivity in the binary system.

Whether ρ and c need to be constant is now considered using expressions due to Mikic (1970). Equation 2.1 may be written for the binary mixture as

$$u\rho=u\rho_A+u\rho_B=u_A\rho_A+u_B\rho_B \tag{2.29}$$

Combining equations 2.4 and 2.25,

$$u_A\rho_A = u\rho_A - D_{AB}\frac{d\rho_A}{dx} \tag{2.30}$$

and for component B,

$$u_B\rho_B = u\rho_B - D_{BA}\frac{d\rho_B}{dx} \tag{2.31}$$

Inserting equations 2.30 and 2.31 into equation 2.29 and dividing throughout by $D_{AB} = D_{BA}$ yields

$$\frac{d\rho_A}{dx} + \frac{d\rho_B}{dx} = \frac{d\rho}{dx} = 0 \tag{2.32}$$

showing that equation 2.25 is restricted to constant density ρ. Equation 2.26 may be expanded to

$$i_{Ax} = -\rho D_{AB}\frac{dw_A}{dx} = -D_{AB}\frac{d\rho_A}{dx} + \frac{D_{AB}}{\rho}\rho_A\frac{d\rho}{dx} \tag{2.33}$$

so that equations 2.30 and 2.31 become

$$u_A\rho_A = u\rho_A - D_{AB}\frac{d\rho_A}{dx} + \frac{D_{AB}}{\rho}\rho_A\frac{d\rho}{dx} \tag{2.34}$$

$$u_B\rho_B = u\rho_B - D_{BA}\frac{d\rho_B}{dx} + \frac{D_{BA}}{\rho}\rho_B\frac{d\rho}{dx} \tag{2.35}$$

The combination of equations 2.29, 2.34, and 2.35 gives

$$\frac{d\rho_A}{dx} + \frac{d\rho_B}{dx} = \frac{d\rho}{dx} \tag{2.36}$$

The validity of this result indicates that equation 2.26 does not require constant density.

An entirely analogous treatment may be performed, beginning with equations 2.2 and 2.8, to show that equation 2.27 is confined to constant c, whereas equation 2.28 is not.

It may be noted that for dilute mixtures of A in B the quantities ρ and c are effectively constant throughout. In this case equations 2.26 and 2.28 simplify, respectively, to equations 2.25 and 2.27.

STEADY-STATE MOLECULAR DIFFUSION

Under steady-state conditions the concentration at a given point is constant with time. Attention is here confined to nonreacting systems of two components A and B, for which Fick's first law of molecular diffusion may be written for steady one-dimensional transfer with constant c as

$$J_{Az} = -D_{AB}\frac{dc_A}{dz} \tag{2.37}$$

$$J_{Bz} = -D_{BA}\frac{dc_B}{dz} \tag{2.38}$$

where J_{Az} and J_{Bz} are the molal fluxes of A and B in the z direction relative to the molal average velocity of the whole mixture, the latter being with respect to stationary coordinates; z is the distance in the direction of diffusion; c_A and c_B are the molar concentrations of A and B; and D_{AB} and D_{BA} are the molecular diffusivities of A in B and of B in A, respectively. Now for a perfect gas,

$$c_A = \frac{p_A}{RT}, \qquad c_B = \frac{p_B}{RT} \tag{2.39}$$

so that equations 2.37 and 2.38 become

$$J_{Az} = -\frac{D_{AB}}{RT}\frac{dp_A}{dz} \tag{2.40}$$

$$J_{Bz} = -\frac{D_{BA}}{RT}\frac{dp_B}{dz} \tag{2.41}$$

Consider first the general case in which a steady total or bulk flow is imposed upon the fluid mixture in the direction in which component A is diffusing. The magnitude of this molal flux of the whole mixture relative to stationary coordinates is $N_{Az} + N_{Bz}$. The fluxes of components A and B relative to stationary coordinates are now each the resultant of two vectors, namely the flux caused by the bulk flow and the flux caused by molecular diffusion. Whereas these two vectors are in the same direction for component A, they are clearly in opposite directions for component B. The total flux of component A relative to stationary coordinates, then, is the sum of that resulting from bulk flow and that due to molecular diffusion; for a gaseous mixture this is

$$N_{Az} = (N_{Az} + N_{Bz})\frac{p_A}{P} - \frac{D_{AB}}{RT}\frac{dp_A}{dz} \tag{2.42}$$

This relationship is another expression of equation 2.22 given earlier in Table 2.1. Assuming constant D_{AB},

$$\frac{D_{AB}}{RT}\int_{p_{A1}}^{p_{A2}} \frac{dp_A}{N_{Az} - \left(\dfrac{N_{Az}+N_{Bz}}{P}\right)p_A} = -\int_{z_1}^{z_2} dz \tag{2.43}$$

Integrating for constant N_{Az}, N_{Bz},

$$N_{Az} = \frac{D_{AB}P}{RTz}\left(\frac{1}{1+\gamma}\right)\ln\left[\frac{1-(1+\gamma)(p_{A2}/P)}{1-(1+\gamma)(p_{A1}/P)}\right] \tag{2.44}$$

where $\gamma = N_{Bz}/N_{Az}$. Equation 2.44 reduces to two special cases of molecular diffusion which are customarily considered. In *equimolal counterdiffusion*, component A diffuses through component B, which is diffusing at the same molal rate as A relative to stationary coordinates, but in the opposite direction. This process is often approximated in the distillation of a binary system. In *unimolal unidirectional diffusion*, only one molecular species—component A—diffuses through component B, which is motionless relative to stationary coordinates. This type of transfer is frequently approximated in the operations of gas absorption, stripping or desorption, liquid-liquid extraction, and adsorption.

Steady-State Equimolal Counterdiffusion in Gases

In this case the total molal flux with respect to stationary coordinates is zero, so that $N_{Az} = -N_{Bz}$. Then from equations 2.40, 2.41, and 2.42,

$$N_{Az} = J_{Az} = -N_{Bz} = -J_{Bz} \tag{2.45}$$

but

$$p_A + p_B = P = \text{constant} \tag{2.46}$$

Therefore

$$\frac{dp_A}{dz} = -\frac{dp_B}{dz} \tag{2.47}$$

and from equations 2.40, 2.41, and 2.45,

$$D_{AB} = D_{BA} = D \tag{2.48}$$

At steady state N_{Az} and N_{Bz} are constants, so that equations 2.40, 2.45, and 2.48 may be combined and integrated for constant D to give

$$N_{Az}=\frac{D}{RTz}(p_{A1}-p_{A2}) \tag{2.49}$$

where z is z_2-z_1; p_{A1} and p_{A2} are the partial pressures of A at z_1 and z_2, respectively. Equation 2.49 is alternatively obtainable from equation 2.44 after applying L'Hôpital's rule for $\gamma=-1$.

Equations 2.40, 2.45, and 2.49 demonstrate that the partial-pressure distribution is linear in the case of steady-state equimolal counterdiffusion.

Steady-State Unimolal Unidirectional Diffusion in Gases

In this case the flux of component B in one direction because of the bulk flow is equal to the flux of B in the opposite direction because of molecular diffusion. Component B is therefore motionless in relation to stationary coordinates, and N_{Bz} equals zero. Setting γ equal to zero in equation 2.44 and recalling that $P-p_A=p_B$,

$$N_{Az}=\frac{DP}{RTz}\ln\frac{p_{B2}}{p_{B1}} \tag{2.50}$$

which may be written as

$$N_{Az}=\frac{DP}{RTz}\left(\frac{p_{B2}-p_{B1}}{p_{BLM}}\right)=\frac{D}{RTz}\left(\frac{P}{p_{BLM}}\right)(p_{A1}-p_{A2}) \tag{2.51}$$

where

$$p_{BLM}=\frac{p_{B2}-p_{B1}}{\ln(p_{B2}/p_{B1})} \tag{2.52}$$

The increase in transfer—by the factor P/p_{BLM}—due to bulk flow in the direction of diffusion of A is indicated by a comparison between equations 2.49 and 2.51.

Equation 2.50 demonstrates that the partial-pressure distribution is nonlinear in the case of steady-state unimolal unidirectional diffusion.

Illustration 2.1.

Two large vessels are connected by a truncated conical duct which is 2 ft in length and has internal diameters of 8 and 4 in., respectively, at its larger and smaller ends. One vessel contains a uniform mixture of 80 mole percent nitrogen and 20 mole percent oxygen, and the other a uniform mixture of 30 mole percent nitrogen and 70 mole percent oxygen. The pressure throughout the system is 1 atm, and the temperature is 32°F. The diffusivity for the nitrogen-oxygen system under these conditions is 0.702 ft^2/hr. Determine the rate of transfer of nitrogen between the two vessels in the early stages of the process, assuming the complete absence of convection and that nitrogen diffuses along the duct in the direction of decreasing diameter.

Compare the result with that which would be obtained if the conical duct were replaced with a circular tube of diameter equal to the average of the terminal diameters of the truncated cone—that is, 6 in.

SOLUTION. Steady-state transfer is assumed in view of the large capacities of the reservoirs on each side of the connecting conduit. Equimolal counterdiffusion then takes place in the duct, such that the rate of transfer at a given cross section is in accordance with equations 2.40, 2.45, and 2.48:

$$N_{Az} = \frac{q_{Az}}{A} = \frac{-D}{RT}\frac{dp_A}{dz}$$

where q_{Az} is the rate of transfer of nitrogen (component A) in the z direction, in moles per unit time. In general the terminal diameters of the truncated cone may be denoted by d_1 and d_2, located at axial points z_1 and z_2, where $d_1 > d_2$. Then the cross section at distance z from z_1 is

$$A = \frac{\pi d^2}{4} = \frac{\pi}{4}\left[d_1 - \left(\frac{d_1 - d_2}{z_2 - z_1}\right)z\right]^2$$

Substitution in the foregoing expression for N_{Az} gives the following result after integration:

$$q_{Az} = \frac{\pi D}{4RT}\left(\frac{d_1 - d_2}{z_2 - z_1}\right)\frac{p_{A1} - p_{A2}}{\dfrac{1}{d_1 - \left(\dfrac{d_1 - d_2}{z_2 - z_1}\right)z_2} - \dfrac{1}{d_1 - \left(\dfrac{d_1 - d_2}{z_2 - z_1}\right)z_1}}$$

For the conditions of this problem, $p_{A1}=0.8(1\ \text{atm})=0.8$ atm; $p_{A2}=0.3(1\ \text{atm})=0.3$ atm; $T=492°$R; $z_1=0$; $z_2=2$ ft; $R=0.7302$ (atm)(ft^3)/(lb-mole)(°R).

$$q_{Az}=\frac{\pi(0.702)}{4(0.7302)(492)}\left(\frac{0.67-0.33}{2}\right)\frac{0.8-0.3}{\dfrac{1}{0.67-[(0.67-0.33)/2]\times 2}-\dfrac{1}{0.67}}$$

$=8.69\times10^{-5}$ lb-mole/hr.

When $d_1=d_2=d$ the foregoing relationship gives $q_{Az}=0/0$, but application of L'Hôpital's rule for this case leads to

$$q_{Az}=\frac{D(p_{A1}-p_{A2})}{RT(z_2-z_1)}\frac{\pi d^2}{4}$$

This expression corresponds to equation 2.49 multiplied by the cross-sectional area of the circular tube. For the present conditions,

$$q_{Az}=\frac{0.702(0.8-0.3)}{0.7302(492)2}\frac{\pi(0.5)^2}{4}=9.6\times10^{-5}\ \text{lb-mole/hr.}$$

Illustration 2.2.

The diffusivity of the binary gas system cyclohexane-nitrogen is to be determined from the measured rate of evaporation of liquid cyclohexane, which partially fills a vertical, narrow, glass tube. A steady stream of pure nitrogen is blown across the top of the tube, which has an internal diameter of 1 cm. The system is at 15°C, the total pressure is 1 atm, and the liquid level is located 10 cm below the top of the tube at the start of the experiment. If the drop in level is 0.21 cm after 20 hr of continuous operation, calculate the diffusivity under these conditions.

The gas constant R is 82.06 atm cm^3/gm-mole °K, and the density and vapor pressure of cyclohexane at 15°C are 0.779 gm/cm^3 and 0.082 atm, respectively.

SOLUTION. This process represents an example of unimolal unidirectional diffusion in which cyclohexane is transferred through stationary nitrogen. The molal flux in equation 2.51 is determined by the rate of fall of the liquid level, so that

$$N_{Az}=\frac{\rho_{AL}}{M_A}\frac{dz}{dt}=\frac{D}{RTz}\left(\frac{P}{p_{BLM}}\right)(p_{A1}-p_{A2})$$

where ρ_{AL} is the density of the liquid and z is the diffusion distance in the gas. Integrating over the range $0 \leqslant t \leqslant t$, $z_0 \leqslant z \leqslant z_t$,

$$D = \frac{\rho_{AL} R T p_{BLM} (z_t^2 - z_0^2)}{2 M_A t P (p_{A1} - p_{A2})}$$

The calculation is conveniently performed in cgs units.

$$p_{BLM} = \frac{1 - (1 - 0.082)}{\ln[1/(1 - 0.082)]} = 0.957 \text{ atm}$$

$$D = \frac{0.779(82.06)(288)(0.957)\left[(10.21)^2 - (10)^2\right]}{2(84.16)(20)(3600)(1)(0.082 - 0)} = 0.076 \text{ cm}^2/\text{sec}$$

Conversion to units of ft^2/hr is effected by multiplying the value in cm^2/sec by 3.88; thus

$$D = 0.076(3.88) = 0.295 \text{ ft}^2/\text{hr}$$

Molecular Diffusion in Liquids

In the absence of a fully developed kinetic theory for liquids the relationships for molecular diffusion are usually assumed to parallel those for gases, although diffusivities are often more substantially dependent on concentration of the diffusing components. In the case of equimolal counterdiffusion, the expression analogous to equation 2.49 is

$$N_{Az} = \frac{D}{z}(c_{A1} - c_{A2}) \tag{2.53}$$

If $c_A + c_B = c$, then $c_A = x_A c$ and $c_B = x_B c$, where x_A, x_B are the mole fractions of A and B, and

$$N_{Az} = \frac{Dc}{z}(x_{A1} - x_{A2}) \tag{2.54}$$

For unimolal unidirectional diffusion the liquid-phase analog of equation 2.51 is

$$N_{Az} = \frac{D}{z}\left(\frac{c}{c_{BLM}}\right)(c_{A1} - c_{A2}) \tag{2.55}$$

$$= \frac{Dc}{z}\frac{(x_{A1} - x_{A2})}{x_{BLM}} \tag{2.56}$$

where

$$c_{BLM} = \frac{c_{B2} - c_{B1}}{\ln(c_{B2}/c_{B1})} \tag{2.57}$$

$$x_{BLM} = \frac{x_{B2} - x_{B1}}{\ln(x_{B2}/x_{B1})} \tag{2.58}$$

In addition to the variation in D, the total molal concentration c also varies, and a mean value of $(c_1 + c_2)/2$ is used when variations are not excessive.

Molal and Volumetric Diffusivities

An expression that may be thought of as intermediate between equations 2.55 and 2.56 is often seen:

$$N_{Az} = \frac{D_m}{z}\left(\frac{c}{c_{BLM}}\right)(x_{A1} - x_{A2}) \tag{2.59}$$

Rearrangement of equation 2.55 shows the dimensions of D to be $(\text{length})^2/\text{time}$, whereas from equation 2.59, D_m has the dimensions of mole/(length)(time). D_m is called the molal diffusivity, while D is called the volumetric diffusivity. Experimentally measured values are more commonly reported as volumetric diffusivities, which is why the preceding derivations have been in terms of this quantity.

Steady-State Diffusion from a Sphere

Consider a sphere of radius r_s located inside a concentric spherical shell of radius r_0. The surface of the sphere is maintained at the constant partial pressure p_{As} with respect to component A. The spherical shell contains a stagnant gas in which the diffusivity of component A is constant. The boundary of the spherical shell is held at another uniform partial pressure p_{Ab}, where $p_{Ab} < p_{As}$, and this boundary constitutes a "sink" for component A. It follows that for steady-state diffusion the material between the surface of the sphere and the boundary of the shell cannot act as a sink. The equation for steady-state diffusion from the surface of the sphere is as follows:

$$4\pi r_s^2 N_{Ar_s} = 4\pi r^2 N_{Ar} = -\frac{4\pi r^2 DP}{RT(P - p_A)}\frac{dp_A}{dr} = \text{constant} \tag{2.60}$$

where N_{Ar}, the radial flux at r, has been replaced by equation 2.42 with $N_{Br}=0$. Integrating,

$$4\pi r^2 N_{Ar} = \frac{4\pi DP}{RT}\left(\frac{r_s r_0}{r_0 - r_s}\right)\ln\frac{P-p_{Ab}}{P-p_{As}} \tag{2.61}$$

Now a mass transfer coefficient k_G is defined as

$$4\pi r^2 N_{Ar} = k_G \pi d_s^2 (p_{As} - p_{Ab}) \tag{2.62}$$

where $d_s = 2r_s$. Combining equations 2.61 and 2.62 gives the following relationship for the Sherwood number:

$$N_{\text{Sh}} = \frac{k_G p_{\text{BLM}} R T d_s}{PD} = \frac{2r_0}{r_0 - r_s} \tag{2.63}$$

(Relations between the quantity $k_G p_{BLM} RT/P$ and a variety of mass transfer coefficients are given later in Problem 4.3 of Chapter 4.) Equation 2.63 shows, for example, that when the radius ratio r_0/r_s assumes values of 2, 5, 10, 50, and infinity, the corresponding Sherwood numbers are 4, 2.5, 2.22, 2.04, and 2.0. The latter value is often regarded as the lower limit of the Sherwood number for a sphere. Cornish (1965), however, has pointed out that values of N_{Sh} much lower than 2 may be obtained when the sphere is located within an array of spheres, as in the case of some packed-bed studies.

Consider a limiting case of two equal-size spheres with the same constant surface concentration, located in an infinite stagnant medium. If the spheres are brought closer together there will be a reduction in the steepness of the concentration gradient in the vicinity of parts of the sphere surfaces. This reduces the rate of mass transfer and consequently the Sherwood number. The dependence of Sherwood number on distance between sphere centers is shown in Table 2.2, after Cornish.

The mean Sherwood number, based on $p_{As} - p_{Ab}$, decreases as the number of spheres increases, approaching a minimum theoretical value of zero as the array of spheres approaches an infinite extent. It may be noted that if the continuous medium flows through the array of spheres, it may then act as a sink for component A. Even in this case, however, N_{Sh} values less than 2 may arise for fluid velocities in the vicinity of zero.

A different situation is often found in experimental work, where a sphere undergoing mass (or heat) transfer is located within an array or bed of inert, dummy spheres. This case may be regarded as one in which the continuous medium and the embedded dummy spheres constitute a composite medium with properties varying with position. It is then necessary to

Table 2.2. Dependence of N_{Sh} on the distance between two equal spheres in an infinite stationary medium.

Distance between centers of spheres	Sherwood number N_{Sh}
∞	2
$100r_s$	≈ 1.98
$4r_s$	≈ 1.6
$2r_s$ (surfaces touching)	1.386

estimate the proper diffusivity for this composite medium. The analogous heat-transfer case for spheres in a cubic array has been considered by Rayleigh (1892) and by Maxwell (1904) for small spheres separated by large distances. De Vries (1952) reviews several such analyses. When the composite medium surrounding the active sphere is infinite, the Sherwood number is again 2, provided that the effective diffusivity appropriate to the composite medium is used. However, if the Sherwood number is calculated using the diffusivity of A in the continuous medium between the dummy spheres, a value either substantially above or substantially below 2 may be obtained, depending on how closely the dummy spheres are packed and on the value and sign of the difference between the diffusivities of A in the dummy spheres and in the continuous medium.

Illustration 2.3

A water droplet with an initial diameter of 0.1 in. is suspended on a thin wire in a large volume of stationary air at 80°F, containing water vapor that exerts a partial pressure of 0.01036 atm. Estimate the time required for complete evaporation of the droplet if the total pressure is 1 atm.

SOLUTION. Equation 2.61 may be written as follows (because in this case $r_0 \gg r_s$):

$$-\frac{dm_A}{dt} = 4\pi r^2 N_{Ar} = \frac{4\pi D P r_s}{RT} \ln\left(\frac{P - p_{Ab}}{P - p_{As}}\right)$$

where $-dm_A/dt$ is the instantaneous evaporation rate in lb-mole/hr and

$$m_A = \frac{4\pi r_s^3 \rho_A}{3M_A}; \qquad r_s = \left(\frac{3M_A m_A}{4\pi\rho_A}\right)^{1/3}$$

Substituting for r_s and integrating,

$$t = \frac{\rho_A r_s^2 RT}{2M_A DP \ln[(P-p_{Ab})/(P-p_{As})]}$$

Evaporation of the droplet causes its temperature to fall to a steady-state value called the wet-bulb temperature, T_w. For the present conditions Illustration 9.1 shows T_w to be 60°F, corresponding to a water vapor pressure of 0.01743 atm. The increase in vapor pressure with surface curvature is less than 4 percent down to drop diameters of 0.1 μ in the case of water (Perry, 1963, p. **18**-61). This effect is, therefore, neglected, as are any convective effects caused by the difference in density between the drop surface and the bulk of the air. The diffusivity of water vapor in air at the average temperature of 70°F is taken to be 1.01 ft^2/hr, so that

$$t = \frac{62.43(0.05/12)^2(0.73)(530)}{2(18)(1.01)(1)2.303\log[(1-0.01036)/(1-0.01743)]} = 1.612 \text{ hr}$$

Additional effects due to natural convection are discussed later near equation 6.141.

UNSTEADY-STATE MOLECULAR DIFFUSION

Unsteady-state mass-transfer processes are those in which the concentration at a given point varies with time. The mathematical solution of the differential equations for unsteady diffusion is complicated and has been performed only for transfer through bodies of simple geometry, such as slabs, cylinders, and spheres, subject to particular sets of boundary conditions.

Unsteady-State Diffusion in a Sphere

A major fraction of all engineering mass-transfer operations involves transfer between two phases, one of which is dispersed as droplets or bubbles in the other. Various attempts at theoretical analysis of such processes have assumed that the droplets or bubbles of the disperse phase may be regarded as spheres, in which mass transfer occurs by unsteady-state molecular diffusion. These considerations, plus the occurrence of some drying problems for this geometry, justify a detailed presentation of the solution for the case of the sphere. The following assumptions are made (see Figure 2.1):

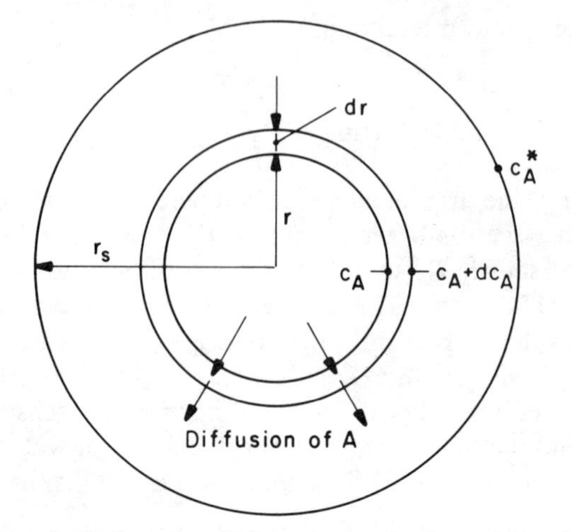

Figure 2.1. Section of a sphere in which mass transfer is occurring by unsteady-state molecular diffusion.

(a) The concentration of solute (component A) is uniform at c_{A0} throughout the sphere at the start of diffusion ($t=0$).

(b) The resistance to transfer in the medium surrounding the sphere is negligible, so that the surface concentration of the sphere is constant at c_A^*, in equilibrium with the entire continuous phase—the latter having constant composition.

(c) Diffusion is radial, there being no variation in concentration with angular position, and physical properties are constant.

The origin of coordinates is at the center of the sphere; the concentration at the spherical surface of radius r will be c_A, and at the spherical surface of radius $r+dr$ it will be c_A+dc_A. A control volume is defined as bounded by these two surfaces at r and $r+dr$.

The rate of flow of solute into the control volume is

$$-D(4\pi r^2)\frac{\partial c_A}{\partial r} \tag{2.64}$$

and the rate of flow out of the control volume is

$$-D\left[4\pi(r+dr)^2\right]\left[\frac{\partial c_A}{\partial r}+\frac{\partial}{\partial r}\left(\frac{\partial c_A}{\partial r}\right)dr\right] \tag{2.65}$$

The net flow rate of solute into the control volume is obtained by subtracting equation 2.65 from equation 2.64, and neglecting second- and third-order differentials, the result is

$$4\pi D\left(r^2\frac{\partial^2 c_A}{\partial r^2}\,dr+2r\frac{\partial c_A}{\partial r}\,dr\right) \tag{2.66}$$

The rate of accumulation of solute in the control volume is also given by

$$(4\pi r^2\,dr)\frac{\partial c_A}{\partial t} \tag{2.67}$$

The expressions 2.66 and 2.67 may therefore be equated and solved for $\partial c_A/\partial t$ to obtain

$$\frac{\partial c_A}{\partial t}=D\left(\frac{\partial^2 c_A}{\partial r^2}+\frac{2}{r}\frac{\partial c_A}{\partial r}\right) \tag{2.68}$$

The boundary conditions follow from the initial assumptions:

$$c_A(r,0)=c_{A0}$$

$$c_A(r_s,t)=c_A^*$$

$$\lim_{r\to 0} c_A(r,t)=\text{bounded}$$

where r_s is the radius of the sphere. Let

$$y'=c_A-c_A^*$$

Then

$$\frac{\partial y'}{\partial t}=D\left(\frac{\partial^2 y'}{\partial r^2}+\frac{2}{r}\frac{\partial y'}{\partial r}\right) \tag{2.69}$$

and the boundary conditions become

$$y'(r,0)=c_{A0}-c_A^*$$

$$y'(r_s,t)=0$$

$$\lim_{r\to 0} y'(r,t)=\text{bounded}$$

Following the conventional procedure, assume a solution of the following form in order to separate the variables:

$$y'(r,t)=R(r)T(t) \tag{2.70}$$

where R is some function of r only, and T is some function of t only. Then clearly

$$\frac{\partial y'}{\partial t} = R\frac{dT}{dt}, \qquad \frac{\partial y'}{\partial r} = T\frac{dR}{dr}, \qquad \frac{\partial^2 y'}{\partial r^2} = T\frac{d^2R}{dr^2}$$

and equation 2.69 becomes

$$\frac{1}{DT}\frac{dT}{dt} = \frac{1}{R}\left(\frac{d^2R}{dr^2} + \frac{2}{r}\frac{dR}{dr}\right) = -\lambda^2 \tag{2.71}$$

The definitions of R and T show that the left side of equation 2.71 is independent of r, while the right side is independent of t. Both sides are therefore equal to a constant, which can be $+\lambda^2$, 0, or $-\lambda^2$. The constant $-\lambda^2$, however, can be shown to give a nontrivial result. Solving the first differential equation,

$$T = C_1 \exp(-\lambda^2 Dt) \tag{2.72}$$

The second differential equation is

$$\frac{d^2R}{dr^2} + \frac{2}{r}\frac{dR}{dr} + \lambda^2 R = 0 \tag{2.73}$$

Let $rR = \beta$, then

$$\frac{dR}{dr} = -\frac{\beta}{r^2} + \frac{1}{r}\frac{d\beta}{dr} \tag{2.74}$$

$$\frac{d^2R}{dr^2} = \frac{2\beta}{r^3} - \frac{2}{r^2}\frac{d\beta}{dr} + \frac{1}{r}\frac{d^2\beta}{dr^2} \tag{2.75}$$

Substituting equations 2.74 and 2.75 into equation 2.73,

$$\frac{d^2\beta}{dr^2} + \lambda^2\beta = 0$$

This equation has the solution (Li, 1960, p. 310)

$$\beta = C_2 \sin\lambda r + C_3 \cos\lambda r$$

from which

$$R = \frac{C_2}{r}\sin\lambda r + \frac{C_3}{r}\cos\lambda r$$

for R to be finite at $r=0$, $C_3=0$. Also, from the boundary condition $y'(r_s,t)=0$ and equation 2.70,

$$R(r=r_s)=0=\frac{C_2}{r_s}\sin\lambda r_s$$

Therefore

$$\lambda=\frac{n\pi}{r_s} \tag{2.76}$$

where $n=1,2,3,\ldots$.

$$R=\frac{C_2}{r}\sin\left(\frac{n\pi r}{r_s}\right) \tag{2.77}$$

Combining equations 2.70, 2.72, 2.76, and 2.77,

$$y'=\sum_{n=1}^{\infty}A_n\frac{1}{r}\sin\left(\frac{n\pi r}{r_s}\right)\exp\left(\frac{-Dn^2\pi^2t}{r_s^2}\right) \tag{2.78}$$

When $t=0$,

$$r(c_{A0}-c_A^*)=\sum_{n=1}^{\infty}A_n\sin\left(\frac{n\pi r}{r_s}\right)$$

This is a Fourier sine series for $f(r)=r(c_{A0}-c_A^*)$, so that A_n may be evaluated as follows (Li, 1960, Chapter 5):

$$A_n=\frac{2}{r_s}\int_0^{r_s}r(c_{A0}-c_A^*)\sin\left(\frac{n\pi r}{r_s}\right)dr$$

$$=\frac{2r_s}{n\pi}(c_{A0}-c_A^*)(-1)^{n+1} \tag{2.79}$$

Then equation 2.78 becomes

$$c_A=c_A^*+\frac{2r_s}{\pi}(c_{A0}-c_A^*)\sum_{n=1}^{\infty}\frac{(-1)^{n+1}}{n}\frac{1}{r}\sin\left(\frac{n\pi r}{r_s}\right)\exp\left(\frac{-Dn^2\pi^2t}{r_s^2}\right) \tag{2.80}$$

The evaluation of local $c_A(r,t)$ is facilitated by adaptation of the Gurney-Lurie heat conduction charts in Perry (1963). For this purpose temperatures in the ordinates (Y) of the charts are replaced by the corresponding concentrations, the thermal diffusivity $k/\rho c_p$ in the abscissa (X) is replaced by D, and k/h_T in the parameter m is replaced by $D/k_{c,\text{ surroundings}}$. The rate of transfer at time t across the surface of the sphere is

$$4\pi r_s^2 N_{Ar}(t) = -4\pi r_s^2 D\left(\frac{\partial c_A}{\partial r}\right)_{r=r_s} \tag{2.81}$$

and evaluating $\partial c_A/\partial r$ at $r=r_s$ from equation 2.80,

$$4\pi r_s^2 N_{Ar}(t) = 8\pi r_s D(c_{A0}-c_A^*)\sum_{n=1}^{\infty}\exp\left(\frac{-Dn^2\pi^2 t}{r_s^2}\right) \tag{2.82}$$

The total transfer up to time t is N_A', where

$$N_A' = 4\pi r_s^2\int_0^t N_{Ar}(t)\,dt$$

$$= \frac{8r_s^3}{\pi}(c_{A0}-c_A^*)\sum_{n=1}^{\infty}\frac{1}{n^2}\left[1-\exp\left(\frac{-Dn^2\pi^2 t}{r_s^2}\right)\right] \tag{2.83}$$

The total transfer per unit surface up to time t is

$$\frac{N_A'}{4\pi r_s^2} = (c_{A0}-c_A^*)\frac{r_s}{3}\left[\frac{6}{\pi^2}\sum_{n=1}^{\infty}\frac{1}{n^2}-\frac{6}{\pi^2}\sum_{n=1}^{\infty}\frac{1}{n^2}\exp\left(\frac{-Dn^2\pi^2 t}{r_s^2}\right)\right] \tag{2.84}$$

where it can be shown (Li, 1960, p. 349) that

$$\sum_{n=1}^{\infty}\frac{1}{n^2} = \frac{\pi^2}{6}$$

A material balance on the transfer up to time t is

$$(c_{A0}-\bar{c}_A)\times\tfrac{4}{3}\pi r_s^3 = N'_A \tag{2.85}$$

in which $\bar{c}_A$ is the average concentration throughout the sphere at t. The fractional extraction from the sphere at time t may be defined as follows and combined with equation 2.84:

$$\frac{c_{A0}-\bar{c}_A}{c_{A0}-c_A^*}=\frac{3N'_A}{4\pi r_s^3(c_{A0}-c_A^*)}=1-\frac{6}{\pi^2}\sum_{n=1}^{\infty}\frac{1}{n^2}\exp\left(\frac{-Dn^2\pi^2 t}{r_s^2}\right) \tag{2.86}$$

It may be noted that the series in equations 2.80 and 2.86 converge rapidly only for large times or large values of Dt/r_s^2. Alternative solutions, useful for small times, are obtainable by use of the Laplace transform. The results are in terms of infinite series of error functions and associated functions as follows (Crank, 1956, pp. 86–87):

$$c_A=c_{A0}+\frac{r_s}{r}(c_A^*-c_{A0})\sum_{n=0}^{\infty}\left[\operatorname{erfc}\frac{(2n+1)r_s-r}{2\sqrt{Dt}}-\operatorname{erfc}\frac{(2n+1)r_s+r}{2\sqrt{Dt}}\right] \tag{2.87}$$

and

$$\frac{c_{A0}-\bar{c}_A}{c_{A0}-c_A^*}=6\sqrt{\frac{Dt}{r_s^2}}\left[\frac{1}{\sqrt{\pi}}+2\sum_{n=1}^{\infty}\operatorname{ierfc}\frac{nr_s}{\sqrt{Dt}}\right]-3\frac{Dt}{r_s^2} \tag{2.88}$$

where

$$\operatorname{erf}\alpha=\frac{2}{\sqrt{\pi}}\int_0^{\alpha}\exp(-\theta^2)\,d\theta$$

in which θ is any "dummy" variable, used merely to describe the function to be integrated; θ is eliminated by the limits of integration.

$$\operatorname{erfc}\alpha=\frac{2}{\sqrt{\pi}}\int_{\alpha}^{\infty}\exp(-\theta^2)\,d\theta=1-\operatorname{erf}\alpha$$

$$\operatorname{ierfc}\alpha=\int_{\alpha}^{\infty}\operatorname{erfc}\theta\,d\theta=\frac{1}{\sqrt{\pi}}\exp(-\alpha^2)-\alpha\operatorname{erfc}\alpha$$

Tabulated values of these three functions are readily available (e.g., Crank, 1956, p. 326).

Equation 2.86 agrees with the expression stated without derivation by Treybal (1963) and by Newman (1931). For ready solution of numerical

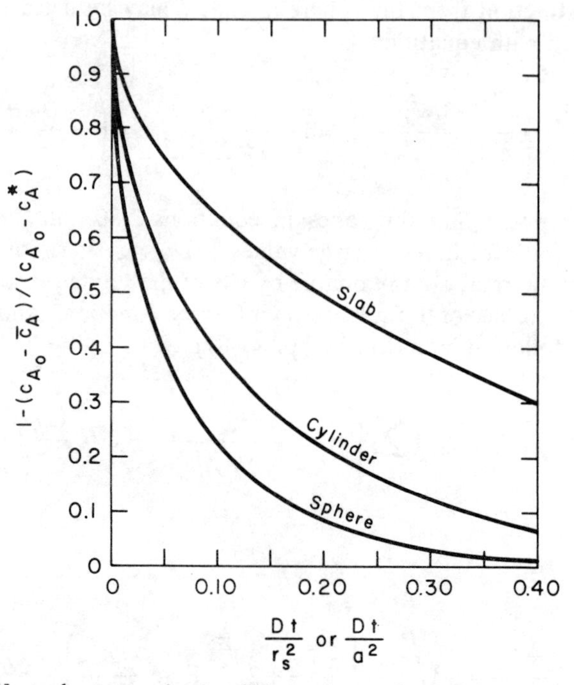

Figure 2.2. Unsteady-state molecular diffusion in a sphere, slab, or cylinder (Newman, 1931).

problems Newman (1931) presents a graphical representation of equation 2.86 in the form of a plot of $1-(c_{A0}-\bar{c}_A)/(c_{A0}-c_A^*)$ against the dimensionless quantity Dt/r_s^2, as shown in Figure 2.2.

Unsteady-State Diffusion in a Slab

The problem of unsteady-state diffusion in a slab or plate is of importance, for example, in some drying operations on certain colloidal or gel-like materials, where it may be necessary to know the distribution of moisture in the slab as a function of position and time, or the relation between the average moisture content of the slab and the duration of drying. For purposes of analysis it will be supposed that the edges of the slab are sealed against transfer. Alternatively, the slab or sheet may be regarded as thin enough for edge effects to be neglected. Diffusion through two opposite surfaces and then through a single surface are considered in turn.

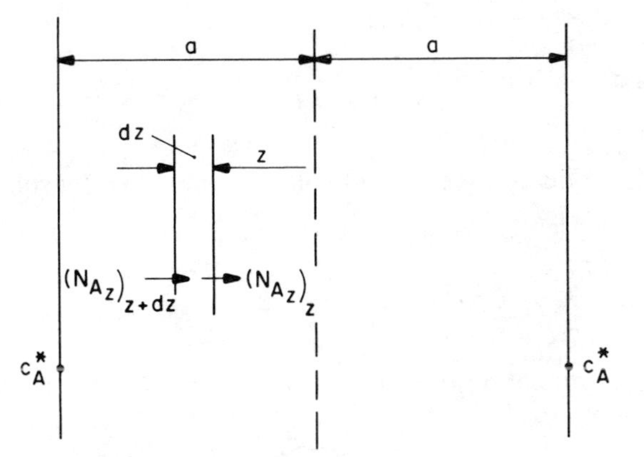

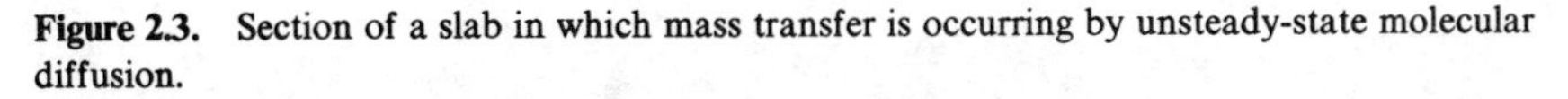
Figure 2.3. Section of a slab in which mass transfer is occurring by unsteady-state molecular diffusion.

Diffusion Through Two Opposite Surfaces of a Slab

The assumptions made are analogous to those used for the sphere and are as follows, with reference to Figure 2.3: uniform concentration c_{A0} throughout the slab at $t=0$; constant concentration c_A^* at the two large surfaces; diffusion confined to a direction normal to the two large surfaces of the slab, which are both permeable to solute (A); and constant physical properties.

The origin of coordinates is at the midplane of the slab, which has area A normal to z.

Consider a control volume defined by the element of slab at z, having thickness dz as shown. The concentration gradient at $z+dz$ at some given instant is

$$\frac{\partial c_A}{\partial z}+\frac{\partial}{\partial z}\left(\frac{\partial c_A}{\partial z}\right)dz$$

and the flow rate of solute into the control volume is then

$$(N_{Az})_{z+dz}A=DA\left[\frac{\partial c_A}{\partial z}+\frac{\partial}{\partial z}\left(\frac{\partial c_A}{\partial z}\right)dz\right] \tag{2.89}$$

The flow rate of solute out of the control volume at the same instant occurs across the face at z and is

$$(N_{Az})_z A = DA \frac{\partial c_A}{\partial z} \tag{2.90}$$

The net flow rate into the control volume is obtained by subtracting 2.90 from 2.89 to obtain

$$DA \frac{\partial^2 c_A}{\partial z^2} dz \tag{2.91}$$

The rate of accumulation of solute in the control volume is also given by

$$A\, dz \frac{\partial c_A}{\partial t} \tag{2.92}$$

The expressions 2.91 and 2.92 may therefore be equated and solved for $\partial c_A / \partial t$ to obtain Fick's second law of molecular diffusion, namely

$$\frac{\partial c_A}{\partial t} = D \frac{\partial^2 c_A}{\partial z^2} \tag{2.93}$$

It may be noted that for general three-dimensional diffusion with constant D the expression corresponding to equation 2.93 would be obtained in an analogous manner by considering the rates at which solute enters, leaves, and accumulates in a cubic control volume of differential dimensions $dx\, dy\, dz$. The resulting expression is then

$$\frac{\partial c_A}{\partial t} = D\left(\frac{\partial^2 c_A}{\partial x^2} + \frac{\partial^2 c_A}{\partial y^2} + \frac{\partial^2 c_A}{\partial z^2}\right) = D\nabla^2 c_A \tag{2.94}$$

For the present problem, however, the terms $\partial^2 c_A / \partial x^2$ and $\partial^2 c_A / \partial y^2$ are zero, and the boundary conditions follow from the initial assumptions as

$$c_A(a,t) = c_A^*$$

$$c_A(z,0) = c_{A0}$$

$$\frac{\partial c_A}{\partial z}(0,t) = 0$$

Let

$$y' = c_A - c_A^*, \qquad y' = y'(z,t)$$

Then

$$\frac{\partial y'}{\partial t} = D\frac{\partial^2 y'}{\partial z^2} \tag{2.95}$$

and the boundary conditions become

$$y'(a,t)=0$$

$$y'(z,0)=c_{A0}-c_A^*=y_0'$$

$$\frac{\partial y'}{\partial z}(0,t)=0$$

Equation 2.95 could be solved by the method of separation of variables in a manner analogous to that already demonstrated for the sphere. The results are useful for large diffusion times, because the series converges rapidly under such conditions. An alternative method of solution is provided by use of the Laplace transform. This procedure is presented here, to complement the separation-of-variables method already given for the sphere. It yields results suitable for small times of diffusion.

In general the Laplace transform $\bar{f}(s)$ of a function $f(t)$ is defined as

$$\bar{f}(s)=L[f(t)]=\int_0^\infty e^{-st}f(t)\,dt \tag{2.96}$$

where s is a number large enough to assure the convergence of the integral. The inverse Laplace transform is expressed as

$$f(t)=L^{-1}[\bar{f}(s)] \tag{2.97}$$

For a function $f(x,t)$ of two independent variables x and t, the (partial) Laplace transform of $f(x,t)$ with respect to t is defined by

$$\bar{f}(x,s)=L_t[f(x,t)]=\int_0^\infty e^{-st}f(x,t)\,dt \tag{2.98}$$

where the subscript t denotes transformation with respect to t. Properties of Laplace transforms of interest here include those relating to partial derivatives; thus

$$L_t\left[\frac{\partial f(x,t)}{\partial t}\right] = s\bar{f}(x,s) - f(x,0) \tag{2.99}$$

$$L_t\left[\frac{\partial f(x,t)}{\partial x}\right] = \frac{\partial \bar{f}(x,s)}{\partial x} \tag{2.100}$$

assuming that the order of integration and of differentiation with respect to x may be interchanged. More extended information on Laplace transforms may be found, for example, in the text by Holl, Maple, and Vinograde (1959).

Taking Laplace transforms of both sides of equation 2.95 with respect to t,

$$L_t\left[\frac{\partial y'}{\partial t}\right] = s\bar{y} - y_0' = DL_t\left[\frac{\partial^2 y'}{\partial z^2}\right]$$

where

$$\bar{y} = \bar{y}(z,s) = L_t[y'(z,t)]$$

$$\frac{d^2\bar{y}}{dz^2} - \frac{s\bar{y}}{D} = -\frac{y_0'}{D}$$

with s regarded as a parameter. This equation has the solution

$$\bar{y} = c_1 e^{-\sqrt{s/D}\,z} + c_2 e^{\sqrt{s/D}\,z} + \frac{y_0'}{s} \tag{2.101}$$

$$\frac{d\bar{y}}{dz} = -c_1\sqrt{\frac{s}{D}}\; e^{-\sqrt{s/D}\,z} + c_2\sqrt{\frac{s}{D}}\; e^{\sqrt{s/D}\,z}$$

and from the boundary condition at $z=0$,

$$\frac{d\bar{y}(0,t)}{dz} = 0 = -c_1\sqrt{\frac{s}{D}} + c_2\sqrt{\frac{s}{D}}$$

from which $c_1 = c_2$, then

$$\bar{y} = c_1\left[e^{-\sqrt{s/D}\,z} + e^{\sqrt{s/D}\,z}\right] + \frac{y_0'}{s}$$

Next, from the boundary condition $\bar{y}=0$ at $z=a$,

$$c_1 = \frac{-y_0'}{s\left[e^{-\sqrt{s/D}\,a} + e^{\sqrt{s/D}\,a}\right]}$$

so that

$$\bar{y} = -\frac{y_0'}{s}\left[\frac{e^{-\sqrt{s/D}\,z} + e^{\sqrt{s/D}\,z}}{e^{-\sqrt{s/D}\,a} + e^{\sqrt{s/D}\,a}}\right] + \frac{y_0'}{s}$$

$$= -\frac{y_0'}{s}\left[\frac{e^{-\sqrt{s/D}\,(a+z)} + e^{-\sqrt{s/D}\,(a-z)}}{e^{-2\sqrt{s/D}\,a} + 1}\right] + \frac{y_0'}{s}$$

Application of the binomial theorem to the denominator of the bracketed term gives

$$\left(1 + e^{-2\sqrt{s/D}\,a}\right)^{-1} = \sum_{n=0}^{\infty} (-1)^n e^{-2n\sqrt{s/D}\,a}$$

Therefore

$$\bar{y} = y_0'\left\{\frac{1}{s} - \sum_{n=0}^{\infty} \frac{(-1)^n}{s} e^{-\sqrt{s/D}\,[(2n+1)a - z]} - \sum_{n=0}^{\infty} \frac{(-1)^n}{s} e^{-\sqrt{s/D}\,[(2n+1)a + z]}\right\} \qquad (2.102)$$

The inverse transform of each term in these two series may be found in tables; for example, as item 8 in the table of Laplace transforms given by

Crank (1956, p. 327). The result can be written as

$$c_A = c_{A0} + (c_A^* - c_{A0})\left[\sum_{n=0}^{\infty} (-1)^n \operatorname{erfc} \frac{(2n+1)a - z}{2\sqrt{Dt}} + \sum_{n=0}^{\infty} (-1)^n \operatorname{erfc} \frac{(2n+1)a + z}{2\sqrt{Dt}}\right] \tag{2.103}$$

The corresponding expression for the average concentration $\bar{c}_A$ throughout the slab at time t is

$$\frac{c_{A0} - \bar{c}_A}{c_{A0} - c_A^*} = 2\sqrt{\frac{Dt}{a^2}}\left[\frac{1}{\sqrt{\pi}} + 2\sum_{n=1}^{\infty} (-1)^n \operatorname{ierfc} \frac{na}{\sqrt{Dt}}\right] \tag{2.104}$$

It should be pointed out that these series converge rapidly for small times, so that these results are of most value when diffusion times are short.

For large times a more rapidly converging series solution is obtained by the separation-of-variables method. The results are

$$c_A = c_A^* + \frac{4}{\pi}(c_{A0} - c_A^*)\sum_{n=0}^{\infty} \frac{(-1)^n}{2n+1} \cos\left[\frac{(2n+1)\pi z}{2a}\right] \exp\left[\frac{-D(2n+1)^2\pi^2 t}{4a^2}\right] \tag{2.105}$$

Evaluation of local $c_A(z,t)$ is facilitated by the Gurney-Lurie charts given in Perry (1963), adapted as described below equation 2.80. For the average concentration $\bar{c}_A$ throughout the slab at time t,

$$\frac{c_{A0} - \bar{c}_A}{c_{A0} - c_A^*} = 1 - \frac{\bar{c}_A - c_A^*}{c_{A0} - c_A^*} = 1 - \frac{8}{\pi^2}\sum_{n=0}^{\infty} \frac{1}{(2n+1)^2} \exp\left[\frac{-D(2n+1)^2\pi^2 t}{4a^2}\right] \tag{2.106}$$

Numerical application is facilitated by the plot of equation 2.106 in the form of $1-(c_{A0}-\bar{c}_A)/(c_{A0}-c_A^*)$ versus Dt/a^2 in Figure 2.2, after Newman (1931).

One sometimes encounters diffusion into a slab or sheet in which the two large opposite faces are maintained at the respective constant but unequal concentrations c^*_{A1} and c^*_{A2}. The interior of the slab is initially at the uniform concentration c_{A0} when $t=0$. This is the case, for example, in diffusion through a membrane. The boundary conditions are

$$c_A(0,t)=c^*_{A1}$$

$$c_A(2a,t)=c^*_{A2}$$

$$c_A(z,0)=c_{A0}$$

Crank (1956, p. 47) shows that for this situation equation 2.106 takes the form

$$\frac{c_{A0}-\bar{c}_A}{c_{A0}-0.5(c^*_{A1}+c^*_{A2})}=1-\frac{8}{\pi^2}\sum_{n=0}^{\infty}\frac{1}{(2n+1)^2}\exp\left[\frac{-D(2n+1)^2\pi^2 t}{4a^2}\right] \tag{2.107}$$

where $2a$ is again the slab thickness. Figure 2.2 may also be used for ready application of equation 2.107, if the ordinate is set equal to one minus the left side of this equation. Eventually, as steady-state conditions are approached ($t\to\infty$), the concentration distribution of course becomes linear in the slab.

Diffusion Through a Single Surface of a Slab

Diffusion may take place across only one large surface of a slab, the opposite face being impermeable to transfer. Such a situation arises, for example, in tray dryers for treating certain colloidal or gel-like materials. The concentration gradient $\partial c_A/\partial z$ is zero at an impermeable surface, which therefore coincides with the midplane of the slab considered earlier, for which diffusion occurred symmetrically through the two large opposite faces, each of which was at the identical concentration c^*_A. Thus placing an impermeable surface down the midplane and removing that half of the slab beyond the midplane will have no effect on the previous solutions, while yielding the case now under consideration. The solutions for the case of symmetrical diffusion therefore apply here, with the permeable surface at $z=a$ and the impermeable surface at $z=0$.

Unsteady-State Diffusion in a Cylinder

The plane ends of the cylinder will be sealed against mass transfer, so that diffusion occurs only in the radial direction; also constant physical properties are assumed. The following boundary-value problem is obtained in a manner analogous to that given earlier for the sphere and the slab. This involves a material balance on component A entering, leaving, and accumulating in an annular, cylindrical volume element:

$$\frac{\partial c_A}{\partial t} = D\left[\frac{\partial^2 c_A}{\partial r^2} + \frac{1}{r}\frac{\partial c_A}{\partial r}\right] \tag{2.108}$$

with the boundary conditions

$$c_A(r,0) = c_{A0}$$

$$c_A(a,t) = c_A^*$$

$$\lim_{r\to 0} c_A(r,t) = \text{bounded}$$

Attention is now directed to a situation considered in Chapter 5, where a detailed analysis is made of diffusion into a fluid in plug flow through a cylindrical tube, the walls of which are at a constant solute concentration. The relevant differential equation 5.103 is reproduced below in terms of c_A.

$$\frac{\partial c_A}{\partial x} = \frac{D}{V}\left[\frac{\partial^2 c_A}{\partial r^2} + \frac{1}{r}\frac{\partial c_A}{\partial r}\right] \tag{5.103}$$

where V is the velocity of plug flow in the axial or x direction along the tube. Suppose that equation 5.103 is multiplied throughout by V, and the quantity x/V then replaced by t. The result is equation 2.108. The solutions derived for equation 5.103 can accordingly be adapted to provide the solutions to equation 2.108 as follows.

The concentration distribution for unsteady-state diffusion is given by equation 5.113, adapted to the present situation, as

$$c_A = c_A^* + \frac{2}{a}(c_{A0} - c_A^*)\sum_{n=1}^{\infty}\frac{J_0(b_n r)}{b_n J_1(b_n a)}\exp(-Db_n^2 t) \tag{2.109}$$

where the b_n's are roots of $J_0(b_n a) = 0$, and $J_0(br)$ is the Bessel function of the first kind of order zero. Evaluation of local $c_A(r,t)$ is facilitated by the Gurney-Lurie charts given by Perry (1963), adapted as described below equation 2.80. The average solute concentration is also obtainable as a function of time from the plug-flow analysis in Chapter 5. Thus equation 5.117 is combined with the equalities noted below equation 5.113, and x/V is replaced by t to give

$$\frac{c_{A0}-\bar{c}_A}{c_{A0}-c_A^*}=1-\frac{\bar{c}_A-c_A^*}{c_{A0}-c_A^*}=1-\frac{4}{a^2}\sum_{n=1}^{\infty}\frac{1}{b_n^2}\exp\left(-Db_n^2t\right) \qquad (2.110)$$

Although solutions useful for small times and corresponding to equations 2.109 and 2.110 are available (Crank, 1956, p. 66), they are more limited both in convenience and in range of application than in the case of the sphere (equations 2.87 and 2.88) or the slab (equations 2.103 and 2.104).

Numerical application is facilitated by the plot of equation 2.110 in the form of $1-(c_{A0}-\bar{c}_A)/(c_{A0}-c_A^*)$ versus Dt/a^2 in Figure 2.2, provided by Newman (1931).

It must be appreciated that the detailed solutions presented here for the sphere, the slab, and the cylinder under particular boundary conditions are merely a sampling of the more important cases which may be encountered. Other solutions for a variety of geometries and boundary conditions are given by Crank (1956).

Illustration 2.4

A stream of air at 20°C and with low, constant humidity is directed over the two large surfaces of a slab of clay that initially contained 15 mass percent of water, distributed uniformly throughout the clay. The slab has a thickness of 2 in. and its edges are sealed against mass transfer. If the rate of drying is controlled by the process of moisture diffusion within the slab, determine the effective diffusivity when the average moisture content falls to 10 mass percent in 375 min. Suppose that the equilibrium moisture content for these conditions of air temperature and humidity is 2 mass percent.

For the same initial moisture content determine the final average percentage of moisture contained in the following bodies made from the same clay and dried for 25 hr under the same conditions:

(a) A solid sphere with a diameter of 6 in.

(b) A solid cylinder that has both ends sealed against mass transfer, a length of 10 in., and a diameter of 8 in.

(c) A solid cylinder as in (b), with one end sealed and the other exposed to transfer.
(d) A solid cylinder as in (b), with both ends exposed.
(e) A rectangular bar with both ends sealed, of thickness 4 in. and width 8 in.
(f) A brick-shaped body, measuring 4×6×8 in.

SOLUTION. The ordinate of Figure 2.2 represents the fraction of solute unremoved and is denoted by E. Concentrations may be converted from terms of c_A to those commonly used in drying, namely the mass of moisture per unit mass of dry solid. Since the weight of dry solid per unit volume is constant throughout the drying process when shrinkage is negligible, it is possible to write

$$E=\frac{\bar{c}_A-c_A^*}{c_{A0}-c_A^*}=\frac{\overline{X}_A-X_A^*}{X_{A0}-X_A^*}$$

where X_A represents the moisture content expressed in pounds of water per pound of dry solid.

Items (c) to (f) of this problem involve relationships not developed in the text; the relevant expressions have been presented by Newman (1931) as follows (see Figure 2.2):

1. For unsteady-state diffusion in a cylinder with one end sealed and one end exposed, when the cylinder radius and length are a and $2a_1$, respectively:

$$E=f\left(\frac{Dt}{a^2}\right)f\left(\frac{Dt}{(2a_1)^2}\right)=E_{\text{cyl}}E_{2a_1}$$

where E_{cyl} is read as the ordinate of the curve for the cylinder corresponding to an abscissa of Dt/a^2. E_{2a_1} is the ordinate of the curve for the slab when the abscissa is $Dt/(2a_1)^2$.

2. For unsteady-state diffusion in a cylinder with both ends exposed, when the cylinder radius and length are a and $2a_1$:

$$E=f\left(\frac{Dt}{a^2}\right)f\left(\frac{Dt}{a_1^2}\right)=E_{\text{cyl}}E_{a_1}$$

where E_{cyl} is as in (1) above and E_{a_1} is the ordinate of the curve for the slab corresponding to an abscissa of Dt/a_1^2.

3. For unsteady-state diffusion in a rectangular bar with both ends sealed, when the bar thickness is $2a_1$ and width is $2a_2$:

$$E = f\left(\frac{Dt}{a_1^2}\right) f\left(\frac{Dt}{a_2^2}\right) = E_{a_1} E_{a_2}$$

where E_{a_1} and E_{a_2} are the ordinates of the curve for the slab corresponding to abscissas of Dt/a_1^2 and Dt/a_2^2.

4. For unsteady-state diffusion in a brick-shaped body of thickness $2a_1$, width $2a_2$, and length $2a_3$:

$$E = f\left(\frac{Dt}{a_1^2}\right) f\left(\frac{Dt}{a_2^2}\right) f\left(\frac{Dt}{a_3^2}\right) = E_{a_1} E_{a_2} E_{a_3}$$

where E_{a_1}, E_{a_2}, and E_{a_3} are ordinates of the curve for the slab for abscissas of Dt/a_1^2, Dt/a_2^2, and Dt/a_3^2.

The effective diffusivity is assumed to be independent of concentration and direction and is calculated as follows:

$$X_{A0} = \frac{15}{85} = 0.1765 \text{ lb moisture/lb dry solid}$$

$$\overline{X}_A = \frac{10}{90} = 0.1110 \text{ lb moisture/lb dry solid}$$

$$X_A^* = \frac{2}{98} = 0.0204 \text{ lb moisture/lb dry solid}$$

$$E = \frac{0.1110 - 0.0204}{0.1765 - 0.0204} = 0.58$$

This ordinate corresponds to the following abscissa of the curve for the slab in Figure 2.2:

$$\frac{Dt}{a^2} = 0.135$$

so that

$$D = 0.135\left(\frac{1}{12}\right)^2 \frac{60}{375} = 1.5 \times 10^{-4} \text{ ft}^2/\text{hr}.$$

SOLUTION (a). For the solid sphere,

$$\frac{Dt}{r_s^2}=\frac{(0.00015)25}{(0.25)^2}=\frac{0.00375}{(0.25)^2}=0.06$$

$$E=0.345 \qquad \text{(from Figure 2.2)}$$

$$\bar{X}_A=0.345(0.1765-0.0204)+0.0204=0.345(0.1561)+0.0204=0.0743$$

or

$$\frac{0.0743}{1.0743}(100)=6.91 \text{ mass percent final average moisture content.}$$

SOLUTION (b). For the solid cylinder, both ends sealed,

$$\frac{Dt}{a^2}=\frac{0.00375}{(0.33)^2}=0.0338$$

$$E_{\text{cyl}}=0.615 \text{ (from Figure 2.2)}$$

$$\bar{X}_A=0.615(0.1561)+0.0204=0.1164$$

or

$$\frac{0.1164}{1.1164}(100)=10.43 \text{ mass percent final average moisture content.}$$

SOLUTION (c). For the solid cylinder, one end exposed,

$$\frac{Dt}{a^2}=\frac{0.00375}{(0.33)^2}=0.0338; \qquad \frac{Dt}{(2a_1)^2}=\frac{0.00375}{(0.833)^2}=0.0054$$

$$E_{\text{cyl}}=0.615 \text{ (from Figure 2.2, cylinder)}$$

$$E_{2a_1}=0.92 \text{ (from Figure 2.2, slab)}$$

$$E=E_{\text{cyl}}E_{2a_1}=0.615(0.92)=0.566$$

$$\bar{X}_A=0.566(0.1561)+0.0204=0.1089$$

or

$$\frac{0.1089}{1.1089}(100)=9.81 \text{ mass percent final average moisture content.}$$

SOLUTION (d). For the solid cylinder, both ends exposed,

$$\frac{Dt}{a^2}=\frac{0.00375}{(0.33)^2}=0.0338;\quad \frac{Dt}{a_1^2}=\frac{0.00375}{(0.4165)^2}=0.0216$$

$$E_{\text{cyl}}=0.615 \text{ (from Figure 2.2, cylinder)}$$

$$E_{a_1}=0.83 \text{ (from Figure 2.2, slab)}$$

$$E=E_{\text{cyl}}E_{a_1}=0.615(0.83)=0.51$$

$$\overline{X}_A=0.51(0.1561)+0.0204=0.0999$$

or

$$\frac{0.0999}{1.0999}(100)=9.09 \text{ mass percent final average moisture content.}$$

SOLUTION (e). For the rectangular bar, both ends sealed,

$$\frac{Dt}{a_1^2}=\frac{0.00375}{(0.166)^2}=0.136;\quad \frac{Dt}{a_2^2}=\frac{0.00375}{(0.33)^2}=0.0338$$

$$E_{a_1}=0.58;\quad E_{a_2}=0.79 \quad \text{(both from Figure 2.2, slab)}$$

$$E=E_{a_1}E_{a_2}=0.58(0.79)=0.458$$

$$\overline{X}_A=0.458(0.1561)+0.0204=0.0919$$

or

$$\frac{0.0919}{1.0919}(100)=8.4 \text{ mass percent final average moisture content.}$$

SOLUTION (f). For the brick-shaped body, all surfaces exposed,

$$\frac{Dt}{a_1^2}=\frac{0.00375}{(0.166)^2}=0.136;\quad \frac{Dt}{a_2^2}=\frac{0.00375}{(0.25)^2}=0.06;\quad \frac{Dt}{a_3^2}=\frac{0.00375}{(0.33)^2}=0.0338$$

$$E_{a_1}=0.58;\quad E_{a_2}=0.72;\quad E_{a_3}=0.79 \quad \text{(all from Figure 2.2, slab)}$$

$$E=E_{a_1}E_{a_2}E_{a_3}=0.58(0.72)(0.79)=0.33$$

$$\overline{X}_A=0.33(0.1561)+0.0204=0.0719$$

or

$$\frac{0.0719}{1.0719}(100) = 6.7 \text{ mass percent final average moisture content.}$$

NOMENCLATURE

A	Component A.
A	Area, ft^2.
A_n	Coefficient in the nth term of an infinite series.
a	Half thickness of a slab; radius of a cylinder, ft.
B	Component B.
b_n	The nth root of $J_0(b_n a) = 0$.
C_1, C_2, C_3	Constants.
c	Total concentration, lb-mole/ft^3.
c_A, c_B	Concentrations of components A and B, lb-mole/ft^3.
c_A^*	Equilibrium concentration at the surface, lb-mole/ft^3.
$\bar{c}_A$	Average concentration throughout the phase, lb-mole/ft^3.
c_{A0}	Uniform concentration of solute throughout at the start of diffusion ($t = 0$), lb-mole/ft^3.
c_{BLM}	Logarithmic-mean concentration of component B between points 1 and 2, distance z apart, lb-mole/ft^3.
D, D_{AB}, D_{BA}	(Volumetric) molecular diffusivity; of A in B; of B in A; ft^2/hr.
D_m	Molal diffusivity, lb-mole/(ft)(hr).
d_s	Diameter of a sphere, ft.
I_{ix}	Molal flux of component i relative to the mass-average velocity, both in the x direction, lb-mole/(ft^2)(hr).
i_{ix}	Mass flux of component i relative to the mass-average velocity, both in the x direction, lb-mass/(ft^2)(hr).
J_{ix}, J_{iz}	Molal flux of component i relative to the molal-average velocity, both in the x and z directions, respectively, lb-mole/(ft^2)(hr).
j_{ix}	Mass flux of component i relative to the molal-average velocity, both in the x direction, lb-mass/(ft^2)(hr).
J_0, J_1	Bessel functions of the first kind of order zero and one.

$k_{c,\text{surroundings}}$	Coefficient of mass transfer between the surroundings and the surface of the body, ft/hr.
k_G	Individual mass-transfer coefficient based on Δp_A, lb-mole/(hr)(ft^2)(lb-force/ft^2).
M_A, M_B	Molecular weights of components A and B.
N'_A	Total transfer up to time t, lb-mole.
N_{Ar}, N_{Ax}, N_{Az}, N_{Bx}, N_{Bz}	Molal fluxes of A and B in the respective directions r, x, and z relative to stationary coordinates, lb-mole/(ft^2)(hr).
N_{ix}	Molal flux of component i in the x direction relative to stationary coordinates, lb-mole/(ft^2)(hr).
N_{Sh}	Sherwood number.
n	An integer.
n_{ix}	Mass flux of component i in the x direction relative to stationary coordinates, lb-mass/(ft^2)(hr).
P	Total pressure, lb-force/ft^2.
p_A, p_B	Partial pressures of components A and B, lb-force/ft^2.
p_{Ab}	Constant partial pressure of component A at boundary of surrounding spherical shell, lb-force/ft^2.
p_{As}	Constant partial pressure of component A at surface of a sphere, lb-force/ft^2.
$p_{B\text{LM}}$	Logarithmic-mean partial pressure of component B between points 1 and 2, distance z (or $r_0 - r_s$) apart, lb-force/ft^2.
R	Gas constant, 1545(ft)(lb-force)/(lb-mole)(°R).
R	Some function of r only—equation 2.70.
r, r_0, r_s	Radius; of a spherical shell; of a sphere, ft.
T	Absolute temperature, °R.
T	Some function of t only, equation 2.70.
t	Time, hr.
U	Molal-average velocity in the x direction, equation 2.2, ft/hr.
u	Mass-average velocity in the x direction, equation 2.1, ft/hr.
u_i	Statistical-mean velocity in the x direction of component i relative to stationary coordinates, ft/hr.
V	Plug-flow velocity along a tube, ft/hr.
w_A	Mass fraction of component A, ρ_A/ρ.

x, y, z	Coordinate directions and distances, ft.
x_A	Mole fraction of component A, c_A/c.
x_{BLM}	Logarithmic-mean mole fraction of component B between points 1 and 2, distance z apart.
y'	$c_A - c_A^*$.
β	rR.
γ	N_{Bz}/N_{Az}.
λ	A constant in equation 2.71.
ρ, ρ_A, ρ_i	Total mass density; mass concentration or mass of components A and i, respectively, per unit volume of solution, lb/ft^3.

Subscripts:

$A, B, i,$	Components A, B, i.
LM	Logarithmic mean.
1, 2	Points 1 and 2.

PROBLEMS

2.1 Derive the following flux relationships for the nonuniform, single-phase, binary fluid $A+B$:

$$j_{Ax} = J_{Ax} M_A = i_{Ax} + w_A(n_{Ax} + n_{Bx}) - x_A\left(n_{Ax} + \frac{M_A}{M_B} n_{Bx}\right)$$

$$I_{Ax} = \frac{i_{Ax}}{M_A} = J_{Ax} + x_A(N_{Ax} + N_{Bx}) - w_A\left(N_{Ax} + \frac{M_B}{M_A} N_{Bx}\right)$$

$i_{Ax} = j_{Ax}$ and $I_{Ax} = J_{Ax}$ when $M_A = M_B$.

2.2 Suppose that the direction of the truncated conical duct is reversed in Illustration 2.1—in other words, the nitrogen now diffuses along the duct in the direction of increasing diameter. Determine the rate of transfer of nitrogen in the early stages of the process under these new conditions and compare with the previous results.

√ **2.3** A narrow, cylindrical vessel is 2 ft high and is filled to a depth of 18 in. with toluene at a temperature of 18.4°C. If the vessel is not closed, how long will it take to lose 5 percent of the toluene by evaporation and escape to the surroundings when the total pressure is 1 atm? The air within the vessel is motionless, but air currents across the open top ensure zero concentration of toluene there. Under these conditions the vapor pressure of toluene is 20 mm Hg, its density is 54.1 lb-mass/ft^3, and the diffusivity of the air-toluene vapor system is 0.326 ft^2/hr.

2.4 A spherical container has a diameter of 2 ft and is open to the atmosphere through a 3-in. diameter hole at the top. If the vessel is half full of liquid toluene, what is the instantaneous rate of toluene loss to the surroundings by evaporation? Assume that conditions and relevant physical properties are as in Problem 2.3.

2.5 Estimate the time required for complete evaporation of a naphthalene sphere having an initial diameter of 0.2 in. when suspended in an effectively infinite amount of still air at 113°F. The naphthalene surface temperature is taken to be 113°F (see Illustration 5.1), so that its vapor pressure is 1.546 lb-force/ft^2 and its density is 71.4 lb/ft^3. For the present purposes neglect any variation in vapor pressure with changing curvature of the surface. For these conditions D is 0.2665 ft^2/hr.

2.6 Equation 2.94 in the x,y,z coordinate system may be expressed in spherical polar coordinates r, θ, and ϕ by use of the transformation equations

$$x = r\sin\theta\cos\phi$$

$$y = r\sin\theta\sin\phi$$

$$z = r\cos\theta$$

For spherically symmetrical diffusion ($\partial c_A/\partial\theta = 0, \partial^2 c_A/\partial\phi^2 = 0$) show that the result is equation 2.68.

The transformation of equation 2.94 to cylindrical coordinates r, θ, and z is performed with the equations

$$x = r\cos\theta$$

$$y = r\sin\theta$$

For radially symmetrical diffusion with negligible end effects ($\partial c_A/\partial\theta = 0$, $\partial^2 c_A/\partial z^2 = 0$), prove that the result is equation 2.108.

2.7 Radial steady-state diffusion is occurring through the rigid walls of a hollow sphere and a hollow cylinder, both bodies having inner and outer radii r_1 and r_2. If the concentrations of solute A at r_1 and r_2 are constant at

c_{A1} and c_{A2}, show that the concentration distributions in the solid shells are given by the respective expressions

$$c_{A\,\text{sphere}} = c_{A1} - (c_{A1} - c_{A2})\frac{r_2(r-r_1)}{r(r_2-r_1)}$$

$$c_{A\text{cyl}} = \frac{c_{A1}\ln(r_2/r) + c_{A2}\ln(r/r_1)}{\ln(r_2/r_1)}$$

Evaluate the radial concentration gradients at $r = r_2$ and obtain the flux at the outer surfaces of the two solid bodies.

2.8 A small amount m of substance A is deposited in a very thin, disk-like film (at $z=0$, $t=0$) through a solid cylinder of B. The cylinder has unit cross section and effectively infinite length on each side of the film. The substance A spreads axially on either side of $z=0$ by diffusion in accordance with equation 2.93. Show by differentiation and by application of the boundary conditions

$$z \to \pm\infty,\ t > 0, \qquad c_A \to 0$$

$$z = 0,\ t = 0, \qquad c_A = \infty$$

$$z \neq 0,\ t = 0, \qquad c_A = 0$$

and with

$$m = \int_{-\infty}^{\infty} c_A\, dz$$

that the solution describing the unsteady-state concentration distribution is

$$c_A(z,t) = \frac{m}{2\sqrt{\pi D t}} \exp\left(\frac{-z^2}{4Dt}\right)$$

where z is the axial distance in either direction normal to the original film of A. How might this process be used to evaluate D in metals?

2.9 Apply the method of separation of variables to the problem of unsteady-state molecular diffusion in a slab, on the assumptions already used in the text for this geometry. Take the origin of coordinates at one large surface of the slab, so that the other large surface is at $z' = 2a$ and z is replaced by z' in equation 2.95. Show that the local concentration of solute (A) is then

$$c_A(z',t)=c_A^*+\frac{4}{\pi}(c_{A0}-c_A^*)\sum_{n=0}^{\infty}\frac{1}{2n+1}\sin\left[\frac{(2n+1)\pi z'}{2a}\right]\exp\left[\frac{-D(2n+1)^2\pi^2 t}{4a^2}\right]$$

Compare this result with equation 2.105, which corresponds to a different coordinate system. Integrate the above expression over $0\leqslant z'\leqslant 2a$ and obtain equation 2.106.

2.10 Consider the unsteady-state molecular diffusion of component A in a semi-infinite medium, $0\leqslant z\leqslant\infty$. The process is described by equation 2.93, and the boundary conditions are $c_A(0,t)=c_A^*$, $c_A(z,0)=0$. Use the Laplace transform with respect to time to obtain the solution giving c_A at any z and t as

$$c_A(z,t)=c_A^*\left[1-\frac{2}{\sqrt{\pi}}\int_0^{z/2\sqrt{Dt}}e^{-\theta^2}\,d\theta\right]$$

[Note: This problem has been solved frequently throughout the literature —see, e.g., Perry (1963), pp.**2**-44 and **10**-6.]

2.11 A wet slab of wood measuring $6\times8\times\frac{3}{4}$ in. is dried in a stream of air with low, constant humidity. The edges of the slab are sealed, and drying occurs by evaporation at the two large surfaces, which are supplied with liquid moisture by diffusion from within the slab. If the equilibrium moisture content under these conditions is 5 mass percent and the moisture content falls from an initial uniform value of 35 mass percent to an average of 20 percent in 8 hr, determine the effective diffusivity of moisture in the wood. It will be assumed that moisture diffusion is the rate-controlling process, that diffusivity is independent of direction and concentration, and that shrinkage can be ignored.

For the same initial and final average moisture contents, how much drying time would be required by the following bodies made from the same wood when dried in a similar manner?

(a) A brick-shaped body measuring $3\times2\times\frac{1}{2}$ in., with one of its smallest faces sealed against transfer.

(b) A solid sphere with a diameter of 1 in.

(c) A cylindrical rod with a diameter of 1 in. and a length of 12 in. with one end sealed.

2.12 Plot the moisture concentration profile in the sphere and in the slab of Problem 2.11 after 8 hr of drying:

(a) Using equations 2.80 and 2.105.
(b) Using Figures 10.3 and 10.4 in Perry (1963), p. **10**-6.

REFERENCES

Bennett, C. O., and J. E. Myers, *Momentum, Heat, and Mass Transfer*, McGraw-Hill, New York, (1962), pp. 438-454.

Bird, R. B., in *Advances in Chemical Engineering*, Vol. 1., T. B. Drew and J. W. Hoopes, Eds., Academic Press, New York, pp. 170–173, (1956).

Bird, R. B., W. E. Stewart, and E. N. Lightfoot, *Transport Phenomena*, Wiley, New York, (1960), pp. 496–501.

Cornish, A. R. H., *Trans. Inst. Chem. Eng.*, (Lond.) **43**, T332–T333, (1965).

Crank, J., *The Mathematics of Diffusion*, Clarendon Press, Oxford, (1956).

de Vries, D. A., *Bull. Inst. Refrig.* (Louvain Meeting) **1**, Annexe, 113 (1952).

Foust, A. S., L. A. Wenzel, C. W. Clump, L. Maus, and L. B. Andersen, *Principles of Unit Operations*, Wiley, New York, (1960).

Holl, D. L., C. G. Maple, and B. Vinograde, *Introduction to the Laplace Transform*, Appleton-Century-Crofts, New York, (1959).

Kay, J. M., *An Introduction to Fluid Mechanics and Heat Transfer*, 2nd ed., Cambridge University Press, Cambridge, (1963).

Li, Wen Hsiung, *Engineering Analysis*, Prentice-Hall, Englewood Cliffs, N. J., (1960).

Maxwell, J. C., *Electricity and Magnetism*, Vol. 1, 3rd ed., Clarendon Press, Oxford, (1904), p. 440.

McCabe, W. L., and J. C. Smith, *Unit Operations of Chemical Engineering*, 2nd ed., McGraw-Hill, New York, (1967).

Mikic, Bora, privately communicated by W. M. Rohsenow, February 27, 1970.

Newman, A. B., Trans. *A. I. Ch. E.*, **27**, 310–333, (1931).

Oliver, E. D., *Diffusional Separation Processes*, Wiley, New York, (1966).

Perry, J. H., *Chemical Engineers' Handbook*, 4th ed., R. H. Perry, C. H. Chilton, and S. D. Kirkpatrick, Eds., p. **10**-5-7, McGraw-Hill, New York, (1963).

Rayleigh, W. R., *Phil. Mag.*, **34**, 481, (1892).

Rohsenow, W. M., and H. Y. Choi, *Heat, Mass, and Momentum Transfer*, Prentice-Hall, Englewood Cliffs, N. J., (1961), pp. 380–382.

Sherwood, T. K., and R. L. Pigford, *Absorption and Extraction*, 2nd ed., McGraw-Hill, New York, (1952).

Treybal, R. E., *Liquid Extraction*, 2nd ed., McGraw-Hill, New York, (1963), p. 186.

Welty, J. R., C. E. Wicks, and R. E. Wilson, *Fundamentals of Momentum, Heat, and Mass Transfer*, Wiley, New York, (1969).

3

Molecular Diffusivities

The theory of molecular diffusion has been the subject of extensive investigation because of its close relationship to the kinetic theory of gases. Detailed reviews are available (Crank, 1956; Jost, 1960; Bird, 1956; Hirschfelder et al., 1954; Barrer, 1941), and Reid and Sherwood (1966) have provided a valuable critical comparison of the various correlations which have been presented for the prediction of diffusivities in gases and liquids, including electrolytes and nonelectrolytes under a variety of conditions.

The purpose here is not to review molecular-diffusivity theory, but instead to assemble predictive correlations necessary for the application of relationships given in other chapters.

DIFFUSIVITIES IN GASES

The kinetic theory of gases, in which molecules are regarded as rigid spheres experiencing elastic collisions, has resulted in several theoretical expressions of the following form for the binary system $A+B$:

$$D_{AB}=\frac{bT^{3/2}}{Pd^2}\sqrt{\frac{1}{M_A}+\frac{1}{M_B}} \tag{3.1}$$

where M_A and M_B are the molecular weights of A and B, T is in ^{0}K, P is the total pressure in atmospheres, and d is the distance in centimeters between the centers of unlike molecules on impact.

Various theoretical values have been assigned to the constant b, for example by Maxwell (1890), Jeans (1921), Chapman (1918), and Sutherland (1894). After comparing the available correlations, Reid and Sherwood (1966) recommend the following expression at pressures below 20 atm:

$$D_{AB}=\frac{0.0018583T^{3/2}}{P(\sigma_{AB})^2\Omega_{D,AB}}\sqrt{\frac{1}{M_A}+\frac{1}{M_B}} \tag{3.2}$$

This equation originates from the Chapman-Enskog kinetic theory and attempts to allow for attractive and repulsive forces between the molecules. Here T is in ^{0}K, P in atmospheres, D_{AB} in cm^2/sec, and σ_{AB} in Angstrom units. The Lennard-Jones potential function is frequently used to approximate the intermolecular potential field for a molecule of A and a molecule of B. The "collision integral" $\Omega_{D,AB}$ is then determined by the temperature and by k_BT/ϵ_{AB}, where k_B is Boltzmann's constant and the Lennard-Jones "force constants" ϵ_{AB} and σ_{AB} are estimated by the following combining rules:

$$\epsilon_{AB}=\sqrt{\epsilon_A\epsilon_B} \quad \text{or} \quad \frac{\epsilon_{AB}}{k_B}=\sqrt{\frac{\epsilon_A}{k_B}\frac{\epsilon_B}{k_B}} \tag{3.3}$$

$$\sigma_{AB}=\tfrac{1}{2}(\sigma_A+\sigma_B) \tag{3.4}$$

The quantities ϵ_A/k_B, ϵ_B/k_B, σ_A, and σ_B may be obtained from Table 1 in the Appendix. (Those that are not listed may be estimated by means of the relations at the foot of the table.) Substitution in equations 3.3 and 3.4 gives ϵ_{AB}/k_B and σ_{AB}; $\Omega_{D,AB}$ is next found as the value corresponding to k_BT/ϵ_{AB} in Table 2 of the Appendix. Insertion in equation 3.2 then yields D_{AB}. The average error between diffusivities calculated from equation 3.2 and 114 measured values in 65 binary systems at various temperatures was 7.5% (Reid and Sherwood, 1966).

Three semiempirical relationships for estimating diffusivity appear in Table 3.1. Equation 3.5 contains atomic diffusion volume increments, v, which are to be summed for each component using values listed in Table 3 of the Appendix. The units used in these equations must be as given in the table of nomenclature at the end of this chapter. It is recommended that

Table 3.1. Semiempirical relationships for diffusivity in binary gas mixtures at low pressures.[a]

Equation number	Equation	Average error[b]	Reference
3.5[c]	$D_{AB}=\dfrac{0.00100T^{7/4}}{P[(\Sigma v)_A^{1/3}+(\Sigma v)_B^{1/3}]^2}\sqrt{\dfrac{1}{M_A}+\dfrac{1}{M_B}}$	6.9%	Fuller et al. (1966)
3.6	$D_{AB}=\dfrac{0.0150T^{1.81}}{P(T_{cA}T_{cB})^{0.1405}(V_{cA}^{0.4}+V_{cB}^{0.4})^2}\sqrt{\dfrac{1}{M_A}+\dfrac{1}{M_B}}$	8.5%	Chen and Othmer (1962)
3.7	$D_{AB}=(2.52\times 10^7)\mu_{\text{air}}^{2.74}\left[\dfrac{\sqrt{\dfrac{1}{M_A}+\dfrac{1}{M_B}}}{(V_{cA}^{0.4}+V_{cB}^{0.4})^2}\right]^{1.23}$	12.6%	Othmer and Chen (1962)

[a] Units as given in the table of nomenclature must be used.
[b] From comparisons by Reid and Sherwood (1966) with the same 114 measurements used to test equation 3.2.
[c] Atomic diffusion volume increments, v, to be summed for each component, are listed in Table 3 of the Appendix, after Fuller et al. (1966).

estimation of D_{AB} be made either by equation 3.2 or 3.5, although more extensive tabulation of v is needed to enhance the range of applicability of the latter expression.

If an experimental value of diffusivity at a temperature T_1 is available, the diffusivity for the same system at T_2 may be estimated by means of a relationship that follows from equation 3.2:

$$D_{T_2}=D_{T_1}\left(\frac{T_2}{T_1}\right)^{3/2}\frac{(\Omega_{D,AB})_{T_1}}{(\Omega_{D,AB})_{T_2}} \tag{3.8}$$

Illustration 3.1

Estimate the binary diffusivity for naphthalene vapor–air at a temperature of 0°C and a total pressure of 1 atm. Compare predictions from equations 3.2, 3.5, 3.6, and 3.7 with the experimental value of 0.0513 cm^2/sec (Perry, 1963, **14**-23).

Use equation 3.8 to convert the experimental value to that corresponding to a temperature of 45°C.

SOLUTION. Call naphthalene component A, and air component B. Consider first the evaluation of equations 3.2 to 3.4.

The critical constants of naphthalene are as follows:

$$T_{cA}=748.4°\text{K}; \quad V_{cA}=408 \text{ cm}^3/\text{gm-mole}$$

(Reid and Sherwood, 1966, p. 576). Table 1 of the Appendix gives $\epsilon_A/k_B=0.75\, T_{cA}=0.75(748.4°\text{K})=561.5°\text{K}$ and $\epsilon_B/k_B=78.6°\text{K}$. Then from equation 3.3,

$$\frac{k_B T}{\epsilon_{AB}}=\frac{273}{\sqrt{(561.5)(78.6)}}=\frac{273}{210}=1.3$$

The corresponding value of $\Omega_{D,AB}$ is obtained from Table 2 of the Appendix as 1.273.

Table 1 of the Appendix shows that $\sigma_A=\frac{5}{6}V_{cA}^{1/3}=\frac{5}{6}(408)^{1/3}=6.18$ Å and $\sigma_B=3.711$Å. Then from equation 3.4, $\sigma_{AB}=\frac{1}{2}(6.18+3.711)=4.945$ Å. Substituting in equation 3.2,

$$D_{AB}=\frac{0.0018583(273)^{3/2}}{(1)(4.945)^2(1.273)}\sqrt{\frac{1}{128.16}+\frac{1}{29}}$$

$$=0.0553 \text{ cm}^2/\text{sec}$$

The evaluation of equation 3.5 requires the following diffusion volume increments, taken from Table 3 of the Appendix:

For naphthalene $(A)=C_{10}H_8$,

Carbon:	10×16.5	$=$ 165
Hydrogen:	8×1.98	$=$ 15.84
Aromatic rings:	$2\times(-20.2)$	$=$ -40.4
	$(\Sigma v)_A$	$=$ 140.44

For air (B), $(\Sigma v)_B=20.1$.

Substituting in equation 3.5,

$$D_{AB}=\frac{0.00100(273)^{7/4}}{(1)\left[(140.44)^{1/3}+(20.1)^{1/3}\right]^2}\sqrt{\frac{1}{128.16}+\frac{1}{29}}$$

$$=0.0605 \text{ cm}^2/\text{sec}$$

Insertion of appropriate values in equation 3.6 gives

$$D_{AB}=\frac{0.0150(273)^{1.81}}{(1)[(748.4)(132.5)]^{0.1405}\left[(408)^{0.4}+(90.52)^{0.4}\right]^2}\sqrt{\frac{1}{128.16}+\frac{1}{29}}$$

$$=0.0538\ \text{cm}^2/\text{sec}.$$

The viscosity of air at 0°C and 1 atm is 0.017 cP, so that equation 3.7 is written for this case as

$$D_{AB}=(2.52\times10^7)(0.017)^{2.74}\left[\frac{\sqrt{\frac{1}{128.16}+\frac{1}{29}}}{\left[(408)^{0.4}+(90.52)^{0.4}\right]^2}\right]^{1.23}$$

$$=0.0469\ \text{cm}^2/\text{sec}$$

The error between these respective predictions and the experimental value of 0.0513 cm^2/sec is defined as

$$\text{percentage error}=\frac{(D_{AB})_{\text{predicted}}-(D_{AB})_{\text{experimental}}}{(D_{AB})_{\text{experimental}}}\times100$$

The resulting errors are +7.8, +17.9, +4.9, and −8.6 percent for equations 3.2, 3.5, 3.6, and 3.7, respectively

Estimation of the diffusivity at 45°C requires evaluation of the collision integral $\Omega_{D,AB}$ at this temperature.

$$\frac{k_BT}{\epsilon_{AB}}=\frac{273+45}{210}=1.515$$

The corresponding $\Omega_{D,AB}$ is found in Table 2 of the Appendix to be 1.193. From equation 3.8,

$$(D_{AB})_{45^\circ\text{C}}=0.0513\left(\frac{318}{273}\right)^{3/2}\frac{1.273}{1.193}=0.0687\ \text{cm}^2/\text{sec}$$

An alternative allowance for the influence of temperature upon diffusivity is provided by equation 3.6 as

$$(D_{AB})_{T_2} = (D_{AB})_{T_1}\left(\frac{T_2}{T_1}\right)^{1.81}$$

so that

$$(D_{AB})_{45^\circ\mathrm{C}} = 0.0513\left(\frac{318}{273}\right)^{1.81} = 0.0675\ \mathrm{cm^2/sec}$$

DIFFUSIVITIES IN LIQUIDS

Different theoretical approaches to the description of diffusion in liquids have been made, depending upon whether the systems are electrolytic or nonelectrolytic. The available prediction procedures must therefore be divided into those suitable for nonelectrolytes and those suitable for electrolytes; for the latter, relationships are unfortunately scarce. Most studies have been devoted to the estimation of diffusivities in very dilute solution, although some progress has been made towards allowance for the substantial variations that occur with increasing concentration of the diffusing solute.

The experimental methods available for the measurement of molecular diffusivity are well reviewed by Johnson and Babb (1956), Tyrell (1961), Jost (1960), and, more briefly, by Nienow (1965).

Nonelectrolytes

Several correlations for dilute solutions are available; a choice may be made on the basis of the accuracy needed and the availability of relevant physical data.

Dilute Solutions

The kinetic theory of liquids is of course much less advanced than that of gases, and this has hampered fundamental developments. There are two well-known theoretical approaches to diffusional theory for nonelectrolytes. Eyring's theory of absolute reaction rates treats the molecules of the liquid as being in a quasicrystalline lattice in which "holes" are scattered, the process having some of the characteristics of diffusion in a solid.

Agreement between the theory and experiment is poor, but the following theoretical relationship is indicated (Jost, 1952, p. 472):

$$\frac{D_{AB}\mu_B}{T} = f\text{ (molal volume of mixture)} \tag{3.9}$$

The hydrodynamical theory was initiated by Einstein, who applied Stokes' law to describe the drag on large, spherical solute molecules (A) moving through a continuum of small solvent molecules (B). The equation obtained is

$$\frac{D_{AB}\mu_B}{T} = \frac{k_B}{6\pi r_A} \tag{3.10}$$

where k_B is Boltzmann's constant and r_A is the radius of a molecule of A. The expression breaks down, however, for smaller solute molecules.

Progress on the basis of statistical-mechanical theory is reviewed by Reid and Sherwood (1966).

The lack of widespread quantitative success with the theoretical approaches has led to the development of several semiempirical relationships based on equations 3.9 and 3.10. Table 3.2 presents a collection of 10 such expressions, in which A is the solute, B is the solvent, and D^0_{AB} is in cm^2/sec. It is important to note that the equations are for use with the units of each term as prescribed in the table of nomenclature at the end of this chapter. Some of these correlations are more successful in one class of application than another, and this has occasioned the classification of errors in prediction as shown.

The parameter ξ in the Wilke-Chang equation 3.11 is an "association" factor for the solvent. The uncertainty involved in assigning values to ξ for new solvents not covered in the original investigation has resulted in efforts to eliminate this factor from the correlation. To this end, Table 3.2 shows relationships that attempt to allow for intermolecular association forces by introducing functions of the latent heats of vaporization. Allowance for the ratio of solvent size to solute size is made in some cases by introducing the term V_{bB}/V_{bA}.

Specific difficulties have been found in the prediction of diffusivity when water is the solute. For example, discrepancies between measured values and those predicted by the first three equations of Table 3.2 may reach 250 percent. Olander (1961) postulates that polymerization of the water may account for this anomaly. His attempt to rectify this situation is represented by equation 3.15, which is based on a score of data points from nine systems. In this regard, equation 3.20 was found to give acceptable results

Table 3.2. Semiempirical relationships for diffusivity in very dilute binary solutions of non-electrolytes.[a]

Equation number	Restriction	Equation
3.11	Exclude water as solute	$\frac{D^\circ_{AB}\mu_{AB}}{T}=\frac{7.4\times10^{-8}(\xi M_B)^{1/2}}{V_{bA}^{0.6}}$
		For unassociated solvents, $\xi=1.0$; for water, $\xi=2.6$; for methanol, $\xi=1.9$; for ethanol $\xi=1.5$.
3.12	Exclude water as solute	$D^\circ_{AB}=K\left[\frac{T}{\mu_{AB}V_{bA}^{1/3}}\right]=8.2\times10^{-8}\left[1+\left(\frac{3V_{bB}}{V_{bA}}\right)^{2/3}\right]\left[\frac{T}{\mu_{AB}V_{bA}^{1/3}}\right]$
		Special solvent cases: water, $V_{bA}<V_{bB}$, $K=25.2\times10^{-8}$; benzene, $V_{bA}<2V_{bB}$, $K=18.9\times10^{-8}$; other solvents, $V_{bA}<2.5V_{bB}$, $K=17.5\times10^{-8}$.
3.13	Exclude water as solute	$D^\circ_{AB}=\frac{14\times10^{-5}}{V_{bA}^{0.6}\mu'_B\mu_{wT}^{1.1\Delta H_{BT}/\Delta H_{wT}}}$
3.14	Aqueous solutions only	$D^\circ_{AB}=\frac{14.0\times10^{-5}}{\mu_{wT}^{1.1}V_{bA}^{0.6}}$
3.15	Water as solute	$D^\circ_{\text{water as solute}}=(D^\circ_{\text{equation 3.11}})/2.3$
3.16	General	$D^\circ_{AB}=5.4\times10^{-8}\left(\frac{M_B^{1/2}\Delta H_B^{1/3}T}{\mu_B V_{bA}^{0.5}\Delta H_A^{0.3}}\right)^{0.93}$
3.17	General	$\frac{D^\circ_{AB}\mu_B}{T}=4.4\times10^{-8}\left(\frac{V_{bB}}{V_{bA}}\right)^{1/6}\left(\frac{\Delta Hm_B}{\Delta Hm_A}\right)^{1/2}$
3.18	—	$\frac{D^\circ_{AB}\mu_B}{T}=\frac{10\times10^{-8}M_B^{1/2}}{V_{bA}^{1/3}V_{bB}^{1/3}};\quad \frac{V_{bB}}{V_{bA}}\leqslant1.5$
3.19	General	$\frac{D^\circ_{AB}\mu_B}{T}=\frac{8.5\times10^{-8}M_B^{1/2}}{V_{bA}^{1/3}V_{bB}^{1/3}};\quad \frac{V_{bB}}{V_{bA}}>1.5$
3.20	Organic solvents	$\frac{D^\circ_{AB}\mu_B}{T}=8.52\times10^{-8}V_{bB}^{-1/3}\left[1.40\left(\frac{V_{bB}}{V_{bA}}\right)^{1/3}+\frac{V_{bB}}{V_{bA}}\right]$

[a] Units as given in the table of nomenclature must be used.
[b] Numbers in parentheses show how many measurements are compared with prediction.
[c] Comparisons by Reid and Sherwood (1966) with fixed sets of data.
[d] Comparisons by authors with various sets of data.

Table 3.2. (continued)

Average error[b]					
Organic solvents	Water as solvent	Water as solute	Overall	Reference	Equation number
27% (53[c])	11% (46[c])	Up to 200–250%	10% (285[d])	Wilke and Chang (1955)	3.11
25% (53[c])	11% (46[c])	Up to 200–250%	—	Scheibel (1954)	3.12
28% (53[c])	11% (46[c])	Up to 200–250%	20% (120[d])	Othmer and Thakar (1953)	3.13
—	11% (46[c])	—	—	Othmer and Thakar (1953)	3.14
—	—	No anomalies (20[d])	—	Olander (1961)	3.15
26% (42[c])	12% (32[c])	12% (7[d])	13% (115[d])	Sitaraman et al. (1963)	3.16
—	—	No anomalies[e,f]	15.5% (213[d])	King et al. (1965)	3.17
15% (60[d])	9% (16[d])	—	13.5% (76[d])	Reddy and Doraiswamy (1967)	3.18
18% (14[d])	—	26% (6[d])[e,g]	20.5% (20[d])	Reddy and Doraiswamy (1967)	3.19
16% (57[d])	—	See below[h]	—	Lusis and Ratcliff (1968)	3.20

[e] Unsatisfactory for high μ_B.
[f] Error = −72% for water in glycerol.
[g] Error = 100% for water in ethylene glycol.
[h] Acceptable results in most cases, provided that water is assumed to diffuse as a tetramer.

in the majority of cases when applied to water as a solute in organic solvents, provided that water was assumed to diffuse as a tetramer. For organic acids diffusing in organic solvents (excluding alcohols), equation 3.20 showed an average error of 9.9 percent with respect to 18 measurements when the acid was assumed to diffuse as a dimer, in contrast with 44.5 percent when it was assumed to diffuse as a monomer. Acids appeared to diffuse as monomers, however, in methanol, butanol, and ethylene glycol. This is also usual when organic acids diffuse in water. Lusis and Ratcliff (1968) discuss further problems that arise in the prediction of the diffusivity when strong interactions occur between solute and solvent molecules and when long straight-chain hydrocarbon molecules are undergoing diffusion. Special problems arise when complexes are formed, as in the case of iodine-aromatic solutions (Wilke and Chang, 1955).

Six of the ten correlations for D_{AB}^0 in Table 3.2 have appeared in the decade 1960–1970. More time for further extensive testing must elapse before a final selection from among them can be made, although some guidance on the relative effectiveness of these expressions has been indicated. It is to be anticipated that further relationships will appear, pending the development of a more complete theory of the liquid state. In the meantime, tentative recommendations on the basis of the evidence compiled in Table 3.2 are as follows:

For diffusion in organic solvents, use equation 3.12, 3.18 (when applicable), or 3.20. When water is the solvent, use equation 3.14. When water is the solute, use equation 3.15, in conjunction with equation 3.11. The latter result could be checked by applying equation 3.20 in the manner prescribed, and also by the use of equations 3.16 and 3.17.

The relationships in Table 3.2 have generally not been extensively tested outside the temperature range 10 to 30°C.

Illustration 3.2.

Estimate the diffusion coefficient for carbon tetrachloride in very dilute solution in benzene at 25°C. Compare predictions from equations 3.11, 3.12, 3.13, 3.16, 3.17, 3.18, and 3.20 with the experimental value of 1.92×10^{-5} cm²/sec [Horrocks, J. K., and E. McLaughlin, *Trans. Faraday Soc.* **58**, 1357, (1962)].

Convert the experimental value to that corresponding to a temperature of 40°C.

SOLUTION. Carbon tetrachloride is designated component A, and benzene component B. The molal volumes of these two components at their normal boiling points are estimated from the LeBas atomic volumes (Perry, 1963, p. **14**–20) as

$$V_{bA} = 101.2 \text{ cm}^3/\text{gm-mole}, \quad V_{bB} = 96 \text{ cm}^3/\text{gm-mole}.$$

(Calculations are to slide-rule accuracy throughout.)

In equation 3.11

$$\mu_{AB} = 0.6 \text{ cP}, \quad T = 298°\text{K}, \quad M_B = 78.11$$

$$D^0_{AB} = \frac{(7.4 \times 10^{-8})[1(78.11)]^{1/2}}{(101.2)^{0.6}}\left(\frac{298}{0.6}\right) = 2.04 \times 10^{-5} \text{ cm}^2/\text{sec}$$

In equation 3.12,

$$V_{bA} < 2V_{bB}, \text{ so } K = 18.9 \times 10^{-8}$$

$$D^0_{AB} = \frac{(18.9 \times 10^{-8})298}{0.6(101.2)^{1/3}} = 2.015 \times 10^{-5} \text{ cm}^2/\text{sec}$$

In equation 3.13,

$$\mu'_B = 0.65 \text{ cP}, \quad \mu_{wT} = 0.894 \text{ cP}$$

$$\Delta H_{BT} = 8100 \text{ cal/gm-mole}, \quad \Delta H_{wT} = 10{,}500 \text{ cal/gm-mole}$$

$$D^0_{AB} = \frac{14 \times 10^{-5}}{(101.2)^{0.6}(0.65)(0.894)^{1.1(8100)/10{,}500}} = 1.486 \times 10^{-5} \text{ cm}^2/\text{sec}$$

In equation 3.16,

$$\mu_B = 0.6 \text{ cP}, \quad \Delta H_A = 46.42 \text{ cal/gm}, \ \Delta H_B = 94.14 \text{ cal/gm}$$

$$D^0_{AB} = 5.4(10^{-8})\left(\frac{(78.11)^{1/2}(94.14)^{1/3}(298)}{0.6(101.2)^{0.5}(46.42)^{0.3}}\right)^{0.93} = 2.16 \times 10^{-5} \text{ cm}^2/\text{sec}$$

In equation 3.17,

$$\Delta H_{mA} = 7135 \text{ cal/gm-mole}, \quad \Delta H_{mB} = 7355 \text{ cal/gm-mole}$$

$$D^0_{AB} = 4.4(10^{-8})\left(\frac{96}{101.2}\right)^{1/6}\left(\frac{7355}{7135}\right)^{1/2}\left(\frac{298}{0.6}\right) = 2.2 \times 10^{-5} \text{ cm}^2/\text{sec}$$

In equation 3.18 (selected because $V_{bB}/V_{bA} < 1.5$),

$$D^0_{AB} = \frac{10(10^{-8})(78.11)^{1/2}}{(101.2)^{1/3}(96)^{1/3}}\left(\frac{298}{0.6}\right) = 2.065 \times 10^{-5} \text{ cm}^2/\text{sec}$$

In equation 3.20

$$D_{AB}^0 = \frac{8.52(10^{-8})}{(96)^{1/3}} \left[1.40 \left(\frac{96}{101.2} \right)^{1/3} + \frac{96}{101.2} \right] \left(\frac{298}{0.6} \right)$$

$$= 2.15 \times 10^{-5} \text{ cm}^2/\text{sec}$$

The errors in these predictions with respect to the experimental value of 1.92×10^{-5} cm^2/sec are defined as in Illustration 3.1. The resulting errors are +6.25, +4.95, −22.6, +12.5, +14.6, +7.5, and +12 percent for equations 3.11, 3.12, 3.13, 3.16, 3.17, 3.18, and 3.20, respectively.

Equations 3.10 to 3.12 and 3.17 to 3.20 suggest that the quantity $D_{AB}^0 \mu_B / T$ is constant for a given liquid system. This relationship is found to be only approximately true but will be used here to estimate the diffusivity D_{AB}^0 at 40°C (313°K):

$$\mu_B = 0.5 \text{ cP at } 40°\text{C or } 313°\text{K},$$

$$D_{AB}^0 \text{ at } 40°\text{C} = 1.92 \times 10^{-5} \left(\frac{0.6}{0.5} \right) \left(\frac{313}{298} \right) = 2.42 \times 10^{-5} \text{ cm}^2/\text{sec}$$

In binary systems, such as those considered so far, only one diffusivity need be defined. The situation is more complex in multicomponent systems because of interactions between the flows of the various species, and these complications increase with increasing departure of the system from ideality. The special case of diffusion of a dilute species in a mixture of two solvents has been considered by Cullinan and Cusick (1967a) in a treatment confined to completely miscible, nonassociated liquid systems. They note that, although the flux of the dilute component is in this case dependent only on its own gradient, the diffusivity is not directly related to any binary diffusion coefficient. The expression obtained for the diffusivity of the dilute species A is in terms of the limiting binary diffusivities at "infinite" dilution and two thermodynamic factors:

$$\lim_{x_A \to 0} D_A^0 = \left[\frac{x_B}{(D_{AB}^0)^{x_B} (\alpha_{ABC} D_{BC}^0)^{x_C}} + \frac{x_C}{(D_{AC}^0)^{x_C} (\alpha_{ACB} D_{CB}^0)^{x_B}} \right]^{-1} \quad (3.21)$$

The binary diffusivities at "infinite" dilution, D_{AB}^0, D_{BC}^0, D_{CB}^0, and D_{AC}^0, may be estimated by one of the correlations given above, and the thermodynamic factors are calculated by the following relationships:

$$\alpha_{ABC}=\frac{V_C(1-D^0_{AC}/D^0_{BC})}{V_A-V_B},\qquad \left|1-\frac{D^0_{AC}}{D^0_{BC}}\right|>0.25 \tag{3.22}$$

or

$$\alpha_{ABC}=\frac{V_C}{V_A},\qquad \left|1-\frac{D^0_{AC}}{D^0_{BC}}\right|<0.25 \tag{3.23}$$

and

$$\alpha_{ACB}=\frac{V_B(1-D^0_{AB}/D^0_{CB})}{V_A-V_C},\qquad \left|1-\frac{D^0_{AB}}{D^0_{CB}}\right|>0.25 \tag{3.24}$$

or

$$\alpha_{ACB}=\frac{V_B}{V_A},\qquad \left|1-\frac{D^0_{AB}}{D^0_{CB}}\right|<0.25 \tag{3.25}$$

where V_A, V_B, and V_C are the molal volumes of components A, B, and C in the liquid state at the temperature and pressure of the mixture. Agreement was found between equation 3.21 and a limited amount of experimental data.

A simpler and somewhat more effective relationship for dilute solute A in mixed solvents B and C was subsequently offered by Leffler and Cullinan (1970b):

$$\left(\lim_{x_A\to 0} D^0_A\right)\mu_{ABC}=(D^0_{AB}\mu_B)^{x_B}(D^0_{AC}\mu_C)^{x_C} \tag{3.26}$$

where μ_{ABC}, the viscosity of the solution, is essentially that of the solvent mixture B and C.

Concentrated Solutions

Five relationships for the prediction of diffusivity in concentrated binary solutions of nonelectrolytes are given in Table 3.3. Equation 3.27 expresses the concentration dependence of diffusivity in terms of the activity coefficient of the solute, γ_A, and the viscosities of water and the solution, μ_B and μ_{AB}. Gosting and Morris (1949) found that this equation accurately described their measurements of diffusivities in aqueous sucrose solutions at 1 and 25°C at concentrations below 6 gm/100 ml.

Table 3.3. Relationships for diffusivity in concentrated binary solutions of nonelectrolytes.

Equation number	Restrictions	Equation	Reference
3.27	Tested on aqueous solutions	$(D_A)_{\text{conc}} = D^\circ_{AB}\left(1+\frac{d\ln\gamma_A}{d\ln x_A}\right)\frac{\mu_B}{\mu_{AB}}$	Gordon (1937), James et al. (1939)
3.28	Ideal solutions	$\left(\frac{D_A\mu_{AB}}{T}\right)_{\text{conc}} = \left(\frac{D^\circ_{BA}\mu_A}{T} - \frac{D^\circ_{AB}\mu_B}{T}\right)x_A + \frac{D^\circ_{AB}\mu_B}{T}$	Powell et al. (1941), Wilke (1949)
3.29	Nonideal solutions	$\left(\frac{D_A\mu_{AB}}{T}\right)_{\text{conc}} = \left[\left(\frac{D^\circ_{BA}\mu_A}{T} - \frac{D^\circ_{AB}\mu_B}{T}\right)x_A + \frac{D^\circ_{AB}\mu_B}{T}\right]\left(1+\frac{d\ln\gamma_A}{d\ln x_A}\right)$	Powell et al. (1941), Wilke (1949)
3.30	Nonassociated solutions, both ideal and nonideal; also associated solutions if degree of association is constant. Poor for binaries of n-alkanes	$(D_A)_{\text{conc}} = (D^\circ_{AB})^{x_B}(D^\circ_{BA})^{x_A}\left(1+\frac{d\ln\gamma_A}{d\ln x_A}\right)$	Vignes (1966)
3.31	(same as 3.30)	$(D_A\mu_{AB})_{\text{conc}} = (D^\circ_{AB}\mu_B)^{x_B}(D^\circ_{BA}\mu_A)^{x_A}\left(1+\frac{d\ln\gamma_A}{d\ln x_A}\right)$	Leffler and Cullinan (1970a)

Equation 3.28 for ideal solutions expresses a linear variation of the quantity $D_A\mu_{AB}$ with composition at a given temperature. This relationship also provides at least a crude approximation to the dependence of diffusivity on concentration in the measurements by Garner and Marchant (1961) on associated compounds in water. The totally miscible solutes studied were ethanediol, propane 1:2 diol, and glycerol, for which the highly nonlinear variations in diffusivity over the full range of solute concentration were respectively about 5-, 9-, and 100-fold.

The extension of equation 3.28 to nonideal solutions is effected by the introduction of the activity coefficient of the solute, giving equation 3.29. In the case of miscible liquids the term $d\ln\gamma_A/d\ln x_A$ may be evaluated from vapor-liquid equilibrium data in the manner described in thermodynamics texts. Thus for ideal vapors

$$1+\frac{d\ln\gamma_A}{d\ln x_A}=\frac{d\ln p_A}{d\ln x_A}$$

where p_A is the partial vapor pressure of the solute in solution. Limitations on the applicability of equation 3.29 have been indicated by Kincaid, Eyring, and Stearn (1944), and Vignes (1966) shows systems for which the expression is not valid.

The substantial level of agreement between the empirical equation 3.30 (Vignes, 1966) and many experimental data prompted Cullinan (1966, 1968) to attempt a partially theoretical derivation of the expression. This development was expanded by Cullinan and Cusik (1967b) to yield a predictive theory for composition-dependent diffusivities in multicomponent systems that are completely miscible and free from association. The latter contribution has been the subject of further discussion by Vignes (1967) and Cullinan (1967).

Equation 3.30 was modified by Leffler and Cullinan (1970a) to give improved correlation by incorporating the viscosity of the solution and of the pure components, resulting in equation 3.31.

The concentration and temperature dependence of liquid diffusivities has been treated by Gainer (1970) on the basis of absolute rate theory in a manner which does not require thermodynamic data.

Hansen (1967) has provided solutions to the diffusion equation (Fick's second law) for cases in which the diffusivity varies exponentially with concentration. The results were used to correct measurements of diffusivity for solvents in polymer films.

The effects of concentration on diffusivities as indicated by statistical-mechanical theory have been summarized by Reid and Sherwood (1966, p. 546). Further development of this approach is needed, and at present the resulting relationships, while satisfactory for binary ideal solutions, are not reliable for nonideal systems or those in which molecular association is significant.

When considering a new system it is desirable to check whether it is included in the tabulation prepared by Johnson and Babb (1956) of experimentally measured diffusivities in nonelectrolytic solutions. In the absence of experimental values, equations 3.30 and 3.31 appear to be the currently preferred relationships for estimating the effect of concentration on diffusivity.

Illustration 3.3

The following experimental values are available for the diffusivity of very dilute hexane (A) in carbon tetrachloride (B) and of very dilute carbon tetrachloride in hexane at 25°C:

$$D_{AB}^0 = 1.487 \times 10^{-5} \text{cm}^2/\text{sec}$$

$$D_{BA}^0 = 3.858 \times 10^{-5} \text{cm}^2/\text{sec}$$

[D. L. Bidlack and D. K. Anderson, *J. Phys. Chem.*, **68**, 3790, (1964)].

Use equations 3.29, 3.30, and 3.31 to predict the diffusivity at all intermediate compositions, and compare the results with experimentally measured values.

SOLUTION. If experimental measurements of D_{AB}^0 and D_{BA}^0 had not been available it would, of course, have been necessary to predict these quantities from one of the relationships utilized in Illustration 3.2. (For example, equation 3.12 yields the following estimates: $D_{AB}^0 = 1.12 \times 10^{-5}$ cm^2/sec; $D_{BA}^0 = 3.79 \times 10^{-5}$ cm^2/sec.)

Activity coefficients were measured and correlated as a function of composition for this system at 20°C by S. D. Christian, E. Neparko, and H. E. Affsprung [*J. Phys. Chem.*, **64**, 442 (1960)]. The correlation was adjusted to 25°C by Bidlack and Anderson (1964) to give

$$1 + \frac{d \ln \gamma_A}{d \ln x_A} = 1 - 0.354 x_A x_B$$

Solution viscosities are required as a function of composition at 25°C for use in equations 3.29 and 3.31; measurements by Bidlack and Anderson (1964) appear in Figure 3.1, which includes the values

$$\mu_A = 0.2958 \text{ cP}, \qquad \mu_B = 0.8963 \text{ cP}$$

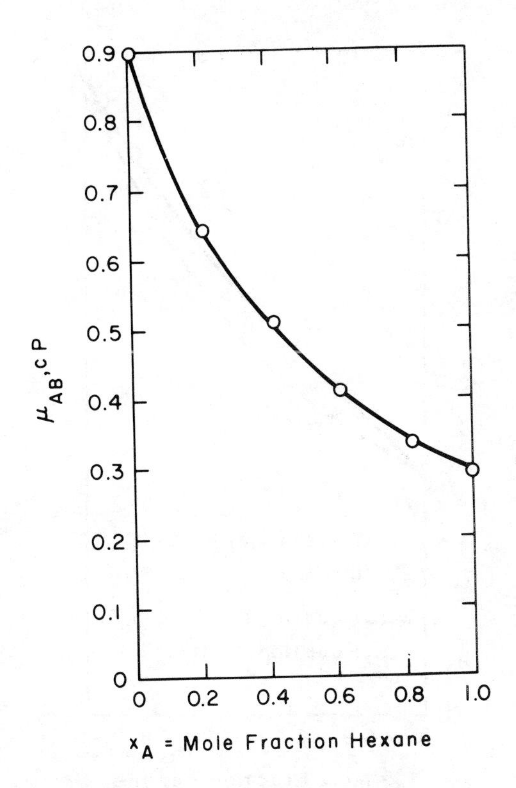

Figure 3.1. Viscosity of hexane–carbon tetrachloride solutions at 25°C (Bidlack and Anderson, 1964).

In equation 3.29, when $x_A = 0.4$,

$$1 + \frac{d \ln \gamma_A}{d \ln x_A} = 1 - 0.354(0.4)(0.6) = 0.915$$

$$\mu_{AB} = 0.511 \text{ cP} \qquad \text{(from Figure 3.1)}$$

$$(D_A)_{\text{conc}} = \frac{298}{0.511}\left\{\left[\frac{(3.858\times10^{-5})0.2958}{298} - \frac{(1.487\times10^{-5})0.8963}{298}\right]0.4 + \frac{(1.487\times10^{-5})0.8963}{298}\right\}(0.915)$$

$$= 2.25\times10^{-5}\ \text{cm}^2/\text{sec}$$

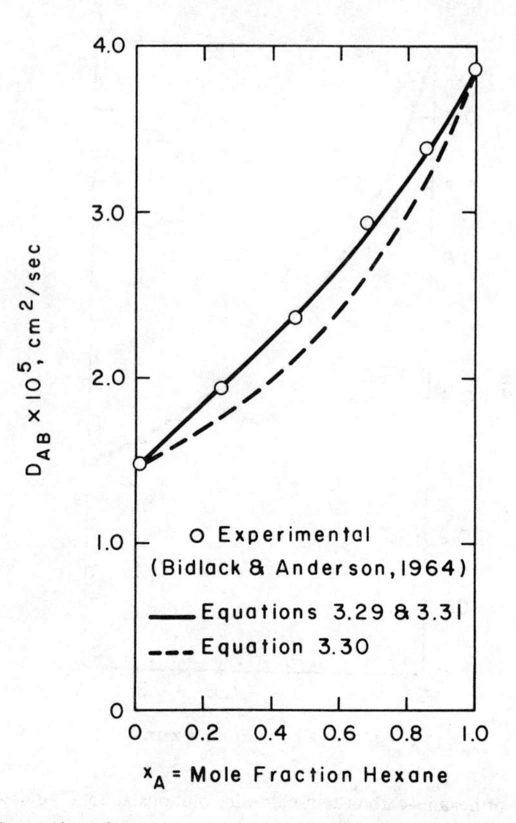

Figure 3.2. Comparison between experimental and predicted diffusivities as a function of composition for the system hexane–carbon tetrachloride at 25°C.

Additional values for other x_A are calculated in the same way and appear as the curve in Figure 3.2.
In Equation 3.30, when $x_A = 0.4$,

$$(D_A)_{\text{conc}} = (1.487 \times 10^{-5})^{0.6}(3.858 \times 10^{-5})^{0.4}(0.915) = 1.99 \times 10^{-5} \text{ cm}^2/\text{sec}$$

Values for other x_A are plotted in Figure 3.2. In equation 3.31, when $x_A = 0.4$,

$$(D_A)_{\text{conc}} = \frac{1}{0.511}[(1.487 \times 10^{-5})0.8963]^{0.6}[(3.858 \times 10^{-5})0.2958]^{0.4}(0.915)$$

$$= 2.24 \times 10^{-5} \text{ cm}^2/\text{sec}$$

Figure 3.2 again shows further values corresponding to other x_A.

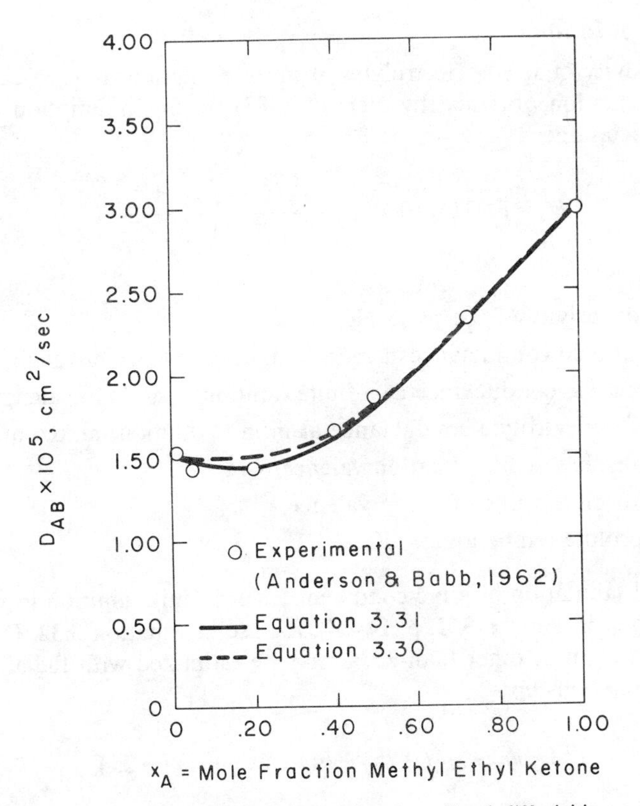

Figure 3.3. Comparison between experimental and predicted diffusivities as a function of composition for the system methyl ethyl ketone–carbon tetrachloride at 25°C.

Many systems have a minimum in the plot of diffusivity versus binary composition. Leffler and Cullinan (1970) examined some such systems, in which both equations 3.30 and 3.31 fitted the data closely and with about the same degree of precision, as exemplified by Figure 3.3. Much poorer agreement was obtained, however, in the system acetone–carbon tetrachloride and with mixtures of *n*-alkanes.

Electrolytes

Molecules of an electrolyte in solution dissociate into cations and anions which, because of their smaller size, diffuse more rapidly than the undissociated molecules. Despite differences between the sizes of the positive and negatively charged ions, however, both types diffuse at the same rate, so that the electrical neutrality of a given solution is preserved.

Dilute Solutions

The diffusivity of *strong* electrolytes at infinite dilution may be calculated from an equation obtained by Nernst (1888) on the assumption of complete dissociation:

$$D_A^0 = 8.931 \times 10^{-10} T \left(\frac{l_+^0 l_-^0}{l_+^0 + l_-^0} \right) \left(\frac{z_+ + z_-}{z_+ z_-} \right) \tag{3.32}$$

where

D_A^0 = diffusivity of the molecule, cm^2/sec,
l_+^0 = cationic conductance at infinite dilution, mho/equivalent,
l_-^0 = anionic conductance at infinite dilution, mho/equivalent,
$l_+^0 + l_-^0$ = electrolyte conductance at infinite dilution, mho/equivalent,
z_+ = absolute value of cation valence,
z_- = absolute value of anion valence,
T = absolute temperature, °K.

A useful tabulation of ionic conductances at infinite dilution in water at 25°C is given in Perry (1963, p. **14**–24), for use in equation 3.32. Diffusivities at temperatures other than 25°C may be estimated with the aid of the following relationship:

$$l_{t^\circ C}^0 = l_{25^\circ}^0 + a(t-25) + b(t-25)^2 + c(t-25)^3 \tag{3.33}$$

Values of a, b, and c for some of the more common ions are tabulated in Perry (1963, p. **14**–24).

Diffusivities of *weak* electrolytes in water were measured by Bidstrup and Geankoplis (1963). The experiments were for concentrations up to 0.1 N in the carboxylic acid series—formic, acetic, propionic, butyric, valeric, and caproic acids. Their resulting correlation, which was shown to be equally applicable to the corresponding α-amino carboxylic acids, is simply the Wilke-Chang equation 3.11 but with the constant 7.4 replaced by 6.6. The average deviation between 25 experimental results and values so calculated was ± 3.7 percent.

Illustration 3.4.

Estimate the diffusivity of potassium chloride in water at infinite dilution and a temperature of 18.5°C.

SOLUTION. The ionic conductances at infinite dilution are adjusted to a temperature of 18.5°C with the aid of equation 3.33 and values tabulated in Perry (1963, p. 14–24):

$$(l_+^0)_{18.5°C} = 73.50 + 1.433(18.5-25) + 0.00406(18.5-25)^2$$
$$-0.0000318(18.5-25)^3$$
$$= 64.36 \text{ mho/equivalent}$$

$$(l_-^0)_{18.5°C} = 76.35 + 1.540(18.5-25) + 0.00465(18.5-25)^2 - 0.0000128 (18.5-25)^3$$
$$= 66.55 \text{ mho/equivalent}$$

and in equation 3.32,

$$D_A^0 = 8.931 \times 10^{-10}(291.5)\left[\frac{(64.36)(66.55)}{64.36+66.55}\right]\left(\frac{1+1}{1}\right) = 1.7 \times 10^{-5} \text{ cm}^2/\text{sec}$$

Experimental and predicted variations in diffusivity with concentration for this system are compared in Illustration 3.5.

Concentrated Solutions

Diffusivities of electrolytes at higher concentrations may be estimated from a semiempirical equation proposed by Gordon (1937):

$$(D_A)_{\text{conc}} = D_A^0\left(1 + \frac{m\,\partial \ln \gamma_\pm}{\partial m}\right)\frac{1}{c'_B \overline{V}_B}\left(\frac{\mu_B}{\mu_{AB}}\right) \tag{3.34}$$

where

D_A^0 is calculated from equation 3.32,
m = molality,
c'_B = number of gm-moles of water per cm^3 of solution,
$\overline{V}_B$ = partial molal volume of water in solution, cm^3/gm-mole,
μ_B = viscosity of water,
μ_{AB} = viscosity of solution,
$\gamma_\pm$ = mean ionic activity coefficient based on molality.

Harned and Owen (1950) and Glasstone (1947) provide tabulations of $\gamma_\pm$ as a function of m for several aqueous solutions, and a method for estimating partial molal volumes $\overline{V}_B$ is described by Lewis and Randall (1923). Equation 3.34 has been found valid up to concentrations of more than 2 N in some systems (Reid and Sherwood, 1966, p. 563).

The effects of ion hydration in nonassociated electrolyte solutions have been examined by J. N. Agar in an extension of the earlier treatment for nonelectrolytic systems by Hartley and Crank (1949). The resulting expression is

$$(D_A)_{\text{conc}} = D'_A\left(1+m\frac{\partial \ln \gamma_\pm}{\partial m}\right)(1-0.018n'm)$$

$$\times\left[1+0.018m\left(\frac{\nu D^*_{H_2O}}{D^0_A}-n'\right)\right]\frac{\mu_B}{\mu_{AB}} \qquad (3.35)$$

The definitions beneath equation 3.34 apply here, and in addition,

n' = "hydration number," namely, the number of moles of water transported with the ions of one mole of solute.

ν = number of ions formed from one molecule of solute.

$D^*_{H_2O}$ = self-diffusion coefficient of water, 2.43×10^{-5} cm^2/sec at 25°C.

$D'_A = D^0_A$ corrected for electrophoretic effects. [According to Hall, Wishaw, and Stokes (1953), D'_A may be replaced by D^0_A with only slight error.]

Equation 3.35 was applied by Hall, Wishaw, and Stokes (1953) and by Wishaw and Stokes (1954) to aqueous solutions of various inorganic electrolytes. It was shown that, with n' values of 2.8 for LiCl, 0.6 for NH_4Cl, and 2.5 for $LiNO_3$, the equation reproduced measurements of $(D_A)_{\text{conc}}$ within 1 to 2 percent up to 4 molal for LiCl, 7 molal for NH_4Cl, and 2 to 3 molal for $LiNO_3$. In the case of NH_4NO_3, however, it was found necessary to invoke the concept of ion-pair formation in order to fit their results by further modification of equation 3.35.

The evaluation of diffusivities in solutions of partially dissociated *weak* electrolytes requires allowance for the contributions due to the ions and the undissociated electrolyte molecules. These effects were averaged by Vitagliano and Lyons (1956) to obtain an expression for simple univalent electrolytes in the form

$$D_A = \frac{(1+\alpha)D_u l_+ l_-}{(1-\alpha)l_+ l_- + (F^2/RT)\alpha D_u (l_+ + l_-)}\left(1+m\frac{\partial \ln \gamma_\pm}{\partial m}\right) \qquad (3.36)$$

where

α = degree of dissociation,
D_u = diffusivity of the undissociated molecule,
F = Faraday, 96,500 coulombs/gm equivalent,
R = gas constant, 8.314 joules/(gm-mole)(°K).

A relationship drawn from the Onsager and Fuoss theory (1932) was utilized by Vitagliano and Lyons (1956) to evaluate l_+ and l_-, the equivalent ionic conductances. At infinite dilution ($\alpha = 1$, $1 + m\partial \ln \gamma_\pm / \partial m = 1$), equation 3.36 reduces to the Nernst equation 3.32, written for univalent electrolytes. The relationship was applied with success to the weak electrolyte system acetic acid–water, for which good activity data and accurate values of α are available at various concentrations.

Experimental measurements of liquid diffusivities, upon which these correlations are based, have been largely confined to the temperature range 10 to 30°C.

The relationships presented above are of course only to be used in the absence of experimentally measured values. The selection of a correlation for use evidently depends on the system in question, the availability of necessary data, and the accuracy required. A comprehensive tabulation of experimental diffusivities for nonelectrolytes has been compiled by Johnson and Babb (1956). Similar extensive data for electrolytes are given by Harned and Owen (1958) and by Robinson and Stokes (1959).

Illustration 3.5.

Estimate the diffusivity of potassium chloride in water as a function of concentration at a temperature of 18.5°C.

SOLUTION. Estimations are made using equation 3.34, and the evaluation of each term on the right-hand side of that expression is considered in turn.

EVALUATION OF D_A^0. The term D_A^0 was calculated to be 1.7×10^{-5} cm^2/sec in Illustration 3.4.

EVALUATION OF $1 + m\partial \ln \gamma_\pm / \partial m$. Values of the mean activity coefficient $\gamma_\pm$ for this system at 18.5°C are interpolated from the table on p. 558 of Harned and Owen (1950) and plotted against molality in Figure 3.4. The slope of the curve is measured at various m and used in the relationship

$$\frac{m\partial \ln \gamma_\pm}{\partial m} = \frac{m}{\gamma_\pm} \frac{\partial \gamma_\pm}{\partial m}$$

The results appear in Table 3.4.

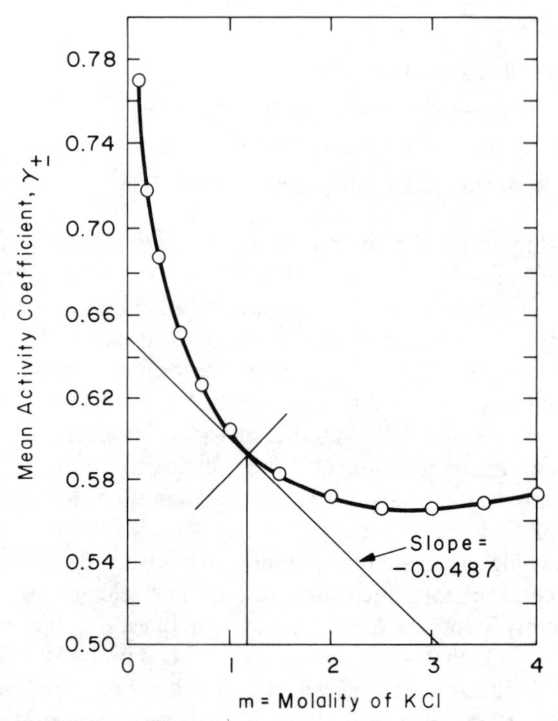

Figure 3.4. Mean activity coefficient $\gamma_\pm$ versus molality of aqueous potassium chloride solutions at 18.5°C.

EVALUATION OF c'_B. For a solution of density ρ it is evident that

$$c'_B = \frac{1000\rho}{(1000 + mM_A)M_B}$$

where A is the solute, B is the solvent, and m is the molality of the solution (i.e., the number of gram-moles of solute per kilogram of solvent). The densities of aqueous solutions of potassium chloride at 18.5°C were graphically interpolated from Perry (1963, p. **3**–76). Reciprocal values—used in the computation of $\overline{V}_B$—are plotted against composition in Figure 3.5. Values of c'_B are listed in Table 3.4.

EVALUATION OF $\overline{V}_B$. The partial specific volumes of water in aqueous KCl solutions of various concentrations were determined by the graphical method of tangent intercepts, as described by Lewis and Randall (1923). For example, at a potassium chloride concentration of 20 mass percent in Figure 3.5, the tangent intercept at zero KCl is 0.993 cm^3 per gram of water. This is the partial specific volume of water in an aqueous solution containing 20 mass percent KCl at 18.5°C. The corresponding partial

Table 3.4. Terms in equation 3.34 for the computation of $(D_A)_{\text{conc}}$ in aqueous solutions of KCl at 18.5°C (Illustration 3.5).

Mass % KCl	1	2	4	8	12	16	20	24
Normality N	0.1349	0.271	0.549	1.125	1.735	2.37	3.04	3.74
Molality m	0.1353	0.2735	0.5585	1.166	1.830	2.555	3.355	4.24
$1 + m \partial \ln \gamma_{\pm} / \partial m$	0.9206	0.8950	0.8911	0.9044	0.9377	0.9951	1.0372	1.0634
c'_B	0.0552	0.0551	0.0546	0.0537	0.0527	0.0515	0.0504	0.0491
$\overline{V}_B$	18.02	18.02	18.02	18.02	18.02	17.94	17.88	17.84
μ_B / μ_{AB}	1.002	1.007	1.011	1.019	1.019	1.010	0.996	0.975
$(D_A)_{\text{conc}}$ (10^5 cm^2/sec)	1.575	1.540	1.552	1.616	1.710	1.851	1.948	2.015

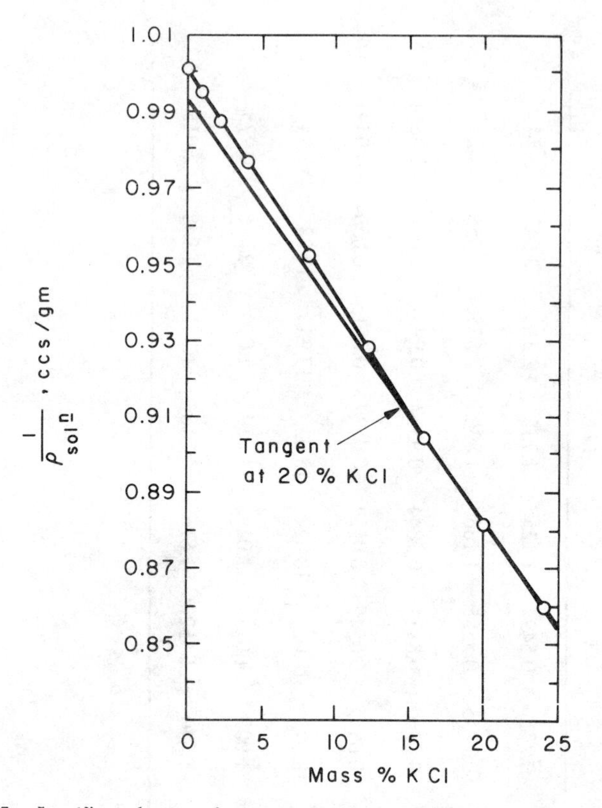

Figure 3.5. Specific volumes of aqueous potassium chloride solutions ($1/\rho_{\text{solution}}$) as a function of composition at 18.5°C.

molal volume is obtained as 18(0.993) or 17.88 cm^3 per gram-mole of water. Table 3.4 contains values of $\overline{V}_B$ corresponding to other solute concentrations.

EVALUATION OF μ_B/μ_{AB}. These ratios at 18.5°C were interpolated from the tabulation in the *International Critical Tables*, Vol. V, (1929), p. 17, and are shown in Table 3.4.

Consider, for example, the calculation of $(D_A)_{\text{conc}}$ for an aqueous solution containing 8 mass percent KCl.

$$w_A = (\text{mass percentage})/100$$

$$m = \frac{1000 w_A}{M_A(1-w_A)} = \frac{80}{74.56(0.92)} = 1.166$$

From Figure 3.4,

$$\gamma_\pm = 0.594, \quad \frac{\partial \gamma_\pm}{\partial m} = -0.0487$$

$$1 + \frac{m \partial \ln \gamma_\pm}{\partial m} = 1 + \frac{m}{\gamma_\pm} \frac{\partial \gamma_\pm}{\partial m} = 1 + \frac{1.166}{0.594}(-0.0487) = 0.9044$$

$$c'_B = \frac{1000(1.0505)}{[1000 + 1.166(74.56)]18} = 0.05375 \text{ gm-mole } H_2O/\text{cm}^3 \text{ of solution}$$

where 1.0505 gm/cm^3 is the density of the 8 mass percent KCl solution at 18.5°C from Figure 3.5.

$$\bar{V}_B = 18 \times (\text{the ordinate intercept of the tangent at 8 mass percent KCl, Figure 3.5})$$
$$= 18(1.001) = 18.02 \text{ cm}^3/\text{gm-mole}$$

$$\left(\frac{\mu_B}{\mu_{AB}}\right)_{18.5°C} = 1.019$$

Substituting in equation 3.34,

$$(D_A)_{\text{conc}} = 1.7 \times 10^{-5}(0.9044) \frac{1}{0.05375(18.02)}(1.019)$$
$$= 1.616 \times 10^{-5} \text{ cm}^2/\text{sec}$$

Estimations for other concentrations are listed in Table 3.4 and plotted in Figure 3.6 for comparison with the experimental measurements of B. W. Clack [*Proc. Phys. Soc.* (Lond.) **36**, 313 (1924)]. The agreement between experimental and predicted values is evidently good over the whole range of concentration up to saturation, including the location of the minimum in the curve at about 2.2 mass percent KCl (0.3 N).

It may be noted that the correction factor $(\mu_B/\mu_{AB})/c'_B \bar{V}_B$ did not depart greatly from unity throughout the calculation, being 1.007 at 1 mass percent of KCl and increasing to 1.111 at 24 mass percent KCl. The corresponding range of the term $1/c'_B \bar{V}_B$ was 1.003 to 1.14.

Figure 3.7, due to Gordon (1937), shows additional comparisons between diffusivities predicted by equation 3.34 (solid lines) and experimental measurements on various systems. The abscissa is in terms of concentration expressed as the square root of normality.

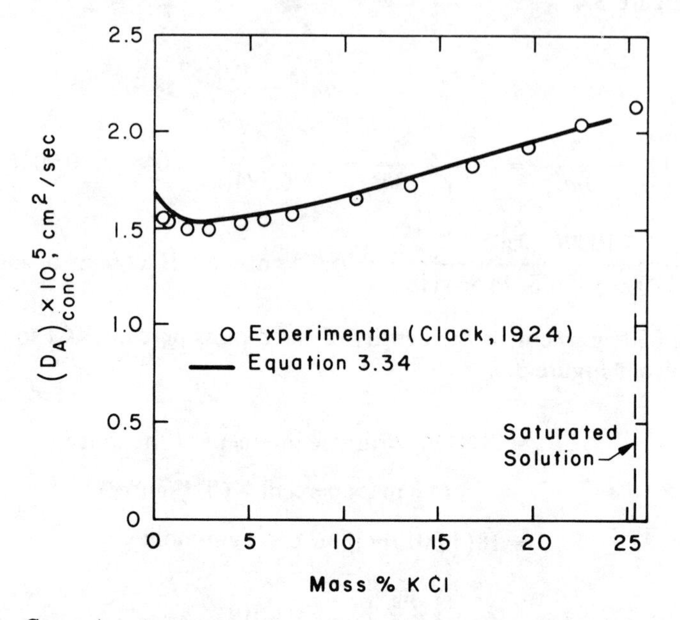

Figure 3.6. Comparison between experimental and predicted diffusivities as a function of composition for the system potassium chloride–water at 18.5°C.

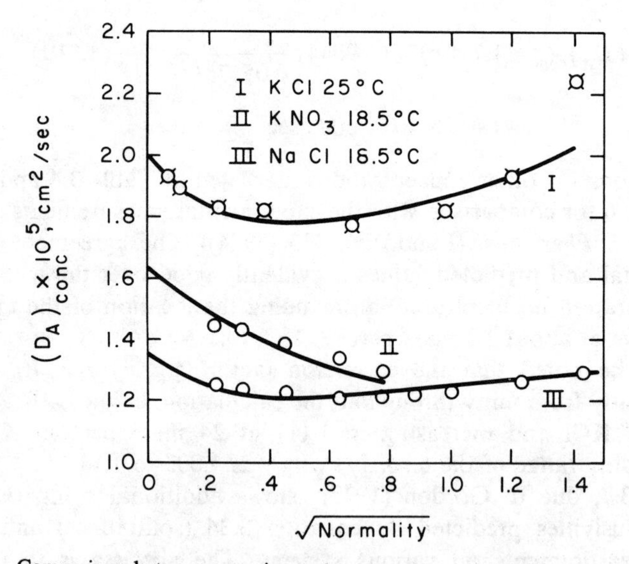

Figure 3.7. Comparison between experimental and predicted diffusivities as a function of composition for aqueous solutions of KCl, KNO_3, and NaCl (Gordon, 1937).

DIFFUSIVITIES IN SOLIDS

Mass transfer of fluids through a solid may be exceptionally complex. In the case of solids which are particulate or which contain large pores, the mass flux may not be proportional to the concentration gradient and may even be against it. This arises when capillary forces are opposed to a liquid concentration gradient. Nevertheless, diffusion relationships are frequently applied to experimental measurements to obtain an empirical effective diffusivity characteristic of that particular fluid and solid structure. The reader is referred to the books by Jost (1960) and Barrer (1941) on this subject.

NOMENCLATURE

A, B	Components.
a, b, c	Constants in equation 3.33.
b	Constant in equation 3.1.
c'_B	Number of gm-moles of water per cm^3 of solution.
D, D^0, D_A, D_{AB}, D_{AC}, D_{BC}, D_{CB}	(Volumetric) molecular diffusivity; in very dilute solution; of species A; of A in B; of A in C; of B in C; of C in B; cm^2/sec. [Note: $3.88 \times D$ in cm^2/sec gives D in ft^2/hr.]
D'_A	D_A^0 corrected for electrophoretic effects—equation 3.35.
$D^*_{H_2O}$	Self-diffusion coefficient of water, 2.43×10^{-5} cm^2/sec at 25°C.
D_{T_1}, D_{T_2}	Molecular diffusivities at temperatures T_1 and T_2, cm^2/sec.
D_u	Molecular diffusivity of the undissociated molecule, cm^2/sec.
d	Distance between centers of unlike molecules on impact, cm.
F	Faraday, 96,500 coulombs/gm equivalent.
ΔH_A, ΔH_B	Latent heats of vaporization of A and B at their normal boiling temperatures, cal/gm.
ΔH_{BT}, ΔH_{wT}	Latent heats of vaporization of solvent (B) and of water at temperature T, cal/gm-mole.
ΔH_{mA}, ΔH_{mB}	As ΔH_A, ΔH_B, but in cal/gm-mole.

k_B	Boltzmann's constant.
l_+^0	Cationic conductance at infinite dilution, mho/equivalent.
l_-^0	Anionic conductance at infinite dilution, mho/equivalent.
$l_+^0 + l_-^0$	Electrolyte conductance at infinite dilution, mho/equivalent.
M_A, M_B	Molecular weights of components A and B.
m	Molality.
n'	"Hydration number," the number of moles of water transported with the ions of one mole of solute.
P	Total pressure, atm.
p_A	Partial vapor pressure of the solute in solution, atm.
R	Gas constant; in equation 3.36 the units are 8.314 Joules/(gm -mole)(°K).
r_A	Radius of a molecule of A.
T	Absolute temperature, °K.
T_c, T_{cA}, T_{cB}	Critical temperature; of A; of B, °K.
t	Temperature, °C.
$\overline{V}_B$	Partial molal volume of water in solution, cm^3/gm-mole.
V_{bA}, V_{bB}	Molal volumes of pure liquid components A and B at their normal boiling temperatures, cm^3/gm-mole.
V_c, V_{cA}, V_{cB}	Critical volume; of components A and B, cm^3/gm-mole.
v	Atomic diffusion volume increment.
x_A, x_B, x_C	Mole fractions of $A, B,$ and C.
z_+	Absolute value of cation valence.
z_-	Absolute value of anion valence.
α	Degree of dissociation.
α_{ABC}, α_{ACB}	Thermodynamic factors; see equations 3.21 to 3.25.
γ_A	Activity coefficient of the solute, A.
$\gamma_\pm$	Mean ionic activity coefficient based on molality.
ϵ_{AB}	Lennard-Jones force constant, equation 3.3.
μ_{AB}, μ_{ABC}	Viscosity of the solution, cP.
μ_{air}	Viscosity of air at temperature of interest, cP.
μ_B, μ_C	Viscosity of solvents B and C, cP.

μ'_B	Viscosity of solvent at 20°C, cP.
μ_{wT}	Viscosity of water at T, cP.
ν	Number of ions formed from one molecule of solute.
ξ	An "association" factor for the solvent, B.
σ_{AB}	Lennard-Jones force constant, equation 3.4, Å.
$\Omega_{D,AB}$	A "collision integral," equation 3.2.

PROBLEMS

3.1 Obtain estimates of the diffusivity of benzene vapor in oxygen at 0°C and a total pressure of 1 atm, using equations 3.2 and 3.5 to 3.7. Compare the results with the experimental value, 0.0797 cm^2/sec, given in Perry (1963, p. **14**-22).

3.2 An experimentally measured value of the diffusivity of toluene vapor in air at 0°C and a total pressure of 1 atm is 0.076 cm^2/sec (Perry, 1963, p. **14**-23). Use the relationships between diffusivity and temperature indicated by equations 3.5 to 3.8 to obtain four estimates of the diffusivity in this system at 30°C and 1 atm. Compare the results with the experimental value given in Perry as 0.088 cm^2/sec.

3.3 Predict the solute diffusivity in very dilute solution in the following binary liquid systems and compare with the indicated experimental values given by Reid and Sherwood (1966):

(a) Ethanol in benzene at 15°C (experimental value $= 2.25 \times 10^{-5}$ cm^2/sec).
(b) Ethanol in water at 25°C (experimental value $= 1.24 \times 10^{-5}$ cm^2/sec).
(c) Water in ethanol at 25°C (experimental value $= 1.132 \times 10^{-5}$ cm^2/sec).

3.4 Estimate the diffusivity of benzoic acid in very dilute solution in benzene at 25°C, using the equations recommended for organic solvents. Assume first that the acid diffuses as a monomer and then as a dimer. Compare the results with the experimental value of 1.38×10^{-5} cm^2/sec given by Reid and Sherwood (1966), to determine which assumption best fits the measurement.

3.5 Repeat Problem 3.4, but for benzoic acid diffusing in water at 25°C, for which the experimental diffusivity is 1.21×10^{-5} cm^2/sec.

3.6 Calculate the diffusivity of very dilute acetone in a binary solvent mixture containing 47.5 mole percent hexane and 52.5 mole percent carbon tetrachloride at 25°C. Viscosity data for mixtures of hexane and carbon tetrachloride are provided by D. L. Bidlack and D. K. Anderson, *J. Phys. Chem.*, **68**, 3790 (1964). Compare the results with an experimentally measured diffusivity of 3.1×10^{-5} cm^2/sec.

3.7 Compute values showing the variation of diffusivity with composition in the binary liquid system methyl ethyl ketone–carbon tetrachloride at 25°C, using equations 3.29, 3.30, and 3.31. Viscosity data for mixtures of methyl ethyl ketone and carbon tetrachloride are provided by D. K. Anderson and A. L. Babb, *J. Phys. Chem.*, **66**, 899, (1962). The following activity-coefficient data for this system as a function of composition are taken from Vignes (1966):

Mole fraction ketone (A):	0	0.2	0.6	1.0
$1 + d\ln\gamma_A / d\ln x_A$:	1.0	0.82	0.88	1.0

Perform computations for these four x_A values and compare the results with Figure 3.3.

3.8 Calculate the diffusivity of potassium nitrate at infinite dilution in water when the temperature is 18.5°C.

3.9 Evaluate the diffusivity as a function of concentration over the range zero to saturation for potassium nitrate in water at 18.5°C. Plot and compare the results with curve II of Figure 3.7.

REFERENCES

Anderson, D. K., and A. L. Babb, *J. Phys. Chem.*, **66**, 899, (1962).

Barrer, R. M., *Diffusion In and Through Solids,* Cambridge University Press, New York, (1941).

Bidlack, D. L., and D. K. Anderson, *J. Phys. Chem.*, **68**, 3790, (1964).

Bidstrup, D. E. and C. J. Geankoplis, *J. Chem. Eng. Data*, **8**, 170–173, (1963).

Bird, R. B., in *Advances in Chemical Engineering*, Vol. 1, T. B. Drew and J. W. Hoopes, Eds., Academic Press, New York (1956), pp. 155–239.

Chapman, S., *Trans. Roy. Soc. (Lond.)* [**A**] **217**, 115–197, (1918).

Chen, N. H., and D. F. Othmer, *J. Chem. Eng. Data*, **7**, 37, (1962).

Clack, B. W., *Proc. Phys. Soc.*, **36**, 313–335, (1924).

Crank, J., *The Mathematics of Diffusion*, Clarendon Press, Oxford, (1956).

Cullinan, H. T., *Ind. Eng. Chem. Fundam.*, **5**, 281–283, (1966).

Cullinan, H. T., *Ind. Eng. Chem. Fundam.*, **6**, 616, (1967).

Cullinan, H. T., *Ind. Eng. Chem. Fundam.*, **7**, 519–520, (1968).

Cullinan, H. T., and M. R. Cusick, *A. I. Ch. E. J.*, **13**, 1171–1174, (1967a).

Cullinan, H. T., and M. R. Cusick, *Ind. Eng. Chem. Fundam.* **6**, 72–77, (1967b).

Fuller, E. N., P. D. Schettler and J. C. Giddings, *Ind. Eng. Chem.*, **58** (5), 19–27, (May, 1966).

Gainer, J. L., *Ind. Eng. Chem. Fundam.* **9**, 381–383, (1970).

Garner, F. H., and P. J. M. Marchant, *Trans. Inst. Chem. Eng. (Lond.)*, **39**, 397–408, (1961).

Glasstone, S., *Thermodynamics for Chemists*, Van Nostrand, Princeton, N. J., (1947), p. 402.

Gordon, A. R., *J. Chem. Phys.*, **5**, 522, (1937).

Gosting, L. J., and M. S. Morris, *J. Am. Chem. Soc.* **71**, 1998–2006, (1949).

Hall, J. R., B. F. Wishaw, and R. H. Stokes, *J. Am. Chem. Soc.*, **75**, 1556–1560, (1953).

Hansen, C. M., *Ind. Eng. Chem. Fundam.*, **6**, 609–614, (1967).

Harned, H. S., and B. B. Owen, *The Physical Chemistry of Electrolytic Solutions*, Am. Chem. Soc. Monograph 95, (1950).

Hartley, G. S., and J. Crank, *Trans. Faraday Soc.*, **45**, 801–818, (1949).

Hirschfelder, J. O., C. F. Curtiss, and R. B. Bird, *Molecular Theory of Gases and Liquids*, Wiley, New York, (1954).

James, J., J. Hollingshead, and A. R. Gordon, *J. Chem. Phys.*, **7**, 89, (1939).

Jeans, J. H., *The Dynamical Theory of Gases*, 3rd. ed., Cambridge University Press, New York, (1921), Chapter 13, p. 307.

Johnson, P. A., and A. L. Babb, *Chem. Rev.*, **56**, 387, (1956).

Jost, W., *Diffusion in Solids, Liquids, and Gases*, Academic Press, New York, (1952).

Jost, W., *Diffusion in Solids, Liquids, and Gases*, Revised Edition, Academic Press, New York, (1960).

Kincaid, J. F., H. Eyring, and A. E. Stearn, *Chem. Rev.*, **28**, 301, (1944).

King, C. J., L. Hsueh, and K. W. Mao, *J. Chem. Eng. Data*, **10**, 348–350, (1965).

Leffler, J., and H. T. Cullinan, *Ind. Eng. Chem. Fundam.*, **9**, 84–88, (1970a).

Leffler, J., and H. T. Cullinan, *Ind. Eng. Chem. Fundam.*, **9**, 88–93, (1970b).

Lewis, G. N., and M. Randall, *Thermodynamics and the Free Energy of Chemical Substances*, McGraw-Hill, New York, (1923), p. 39.

Lusis, M. A., and G. A. Ratcliff, *Can. J. Chem. Eng.*, **46**, 385–387, (1968).

Maxwell, J. C., *Scientific Papers*, Vol. 2, Cambridge University Press, New York (1890), p. 343.

Nernst, W., *Z. Phys. Chem.*, **2**, 613, (1888).

Nienow, A. W., *Brit. Chem. Eng.*, **10**, No. 12, 827–833, (1965).

Olander, D. R., *A. I. Ch. E. J.*, **7**, 175–176, (1961).

Onsager, L., and R. M. Fuoss, *J. Phys. Chem.*, **36**, 2689, (1932).

Othmer, D. F., and H. T. Chen, *Ind. Eng. Chem. Process Des. Dev.*, **1**, 249–254, (1962).

Othmer, D. F., and M. S. Thakar, *Ind. Eng. Chem.*, **45**, 589–593, (1953).

Perry, J. H., Ed., *Chemical Engineers' Handbook*, R. H. Perry, C. H. Chilton, and S. D. Kirkpatrick, Eds., 4th ed., McGraw-Hill, New York, (1963).

Powell, R. E., W. E. Roseveare, and H. Eyring, *Ind. Eng. Chem.*, **33**, 430–435, (1941).

Powell, R. E., W. E. Roseveare, and H. Eyring, *J. Appl. Phys.*, **12**, 669, (1941).

Reddy, K. A., and L. K. Doraiswamy, *Ind. Eng. Chem. Fundam.*, **6**, 77–79, (1967).

Reid, R. C., and T., K. Sherwood, *The Properties of Gases and Liquids,* 2nd ed. McGraw-Hill, New York, (1966), Chapter 11.

Robinson, R. A., and R. H. Stokes, *Electrolyte Solutions*, 2nd. ed., Academic Press, New York, (1959).

Scheibel, E. G., *Ind. Eng. Chem.*, **46**, 2007, (1954).

Sitaraman, R., S. H. Ibrahim, and N. R. Kuloor, *J. Chem. Eng. Data*, **8**, 198–201, (1963).

Sutherland, W., *Phil. Mag.*, **38**, 1, (1894).

Tyrell, H. J. V., *Diffusion and Heat Flow in Liquids*, Butterworths, London, (1961).

Vignes, A., *Ind. Eng. Chem. Fundam.*, **5**, 189–199, (1966).

Vignes, A., *Ind. Eng. Chem. Fundam.*, **6**, 614–616, (1967).

Vitagliano, V., and P. A. Lyons, *J. Am. Chem. Soc.*, **78**, 4538–4542, (1956).

Wilke, C. R., *Chem. Eng. Prog.*, **45**, 218–224, (1949).

Wilke, C. R., and P. Chang, *A. I. Ch. E. J.*, **1**, 264, (1955).

Wishaw, B. F., and R. H. Stokes, *J. Am. Chem. Soc.*, **76**, 2065–2071, (1954).

4

Mass-Transfer Coefficients

Equilibrium.

Consider the distribution of a solute (component A) between two immiscible fluid phases in contact with each other. Under conditions of dynamic equilibrium the rate of transfer of A from the first to the second phase is equal to its rate of transfer in the reverse direction. The equilibrium relationship may be represented by a plot such as Figure 4.1 over a range of compositions of each phase.

The first phase may, for example, be gaseous and the second phase liquid; Y_A will then denote the composition of the gas phase with respect to component A, expressed in any convenient terms such as partial pressure, mole or mass fraction, number of moles or mass per unit volume, or number of moles or mass of A per mole or unit mass of non-A. X_A will then denote the composition of the liquid phase, again in convenient terms.

A point on the equilibrium curve gives the equilibrium compositions of the two phases, a point above the curve indicates that component A is being transferred from the gas to the liquid, and a point below the curve signifies transfer in the opposite direction.

The position of the equilibrium curve is usually determined experimentally for a given system, but its location can sometimes be predicted from thermodynamic considerations. Although a couple of simple examples will be considered here, including one which will be used in a later design illustration, it must be emphasized that phase equilibria constitute a wide

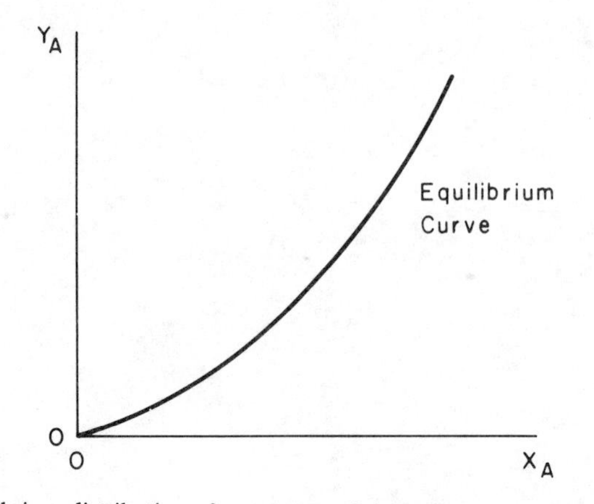

Figure 4.1. Equilibrium distribution of component A between two conjugate phases.

and separate field of study for which the relevant literature should be consulted (e.g., Wilson and Ries, 1956; Smith and Van Ness, 1959; Emmert and Pigford, 1963; Smith, 1963; and Null, 1970).

As an example of the prediction of vapor-liquid equilibria in an ideal binary system, consider the following illustration.

Illustration 4.1.

Calculate the vapor-liquid equilibrium relationships over the whole range of composition at atmospheric pressure for the system cyclopentane-cyclohexane.

Correlate the values by an approximate, empirical equation.

SOLUTION.This system will be considered ideal, which means that the vapor-liquid equilibrium relationships may be predicted from the vapor pressures of the pure substances using Raoult's and Dalton's laws. Raoult's law is

$$p_A = P_A x_A$$

$$p_B = P_B x_B$$

where p_A and p_B are the partial pressures of cyclopentane and cyclohexane, and P_A and P_B are the vapor pressures of these substances at the prevailing temperature. The mole fractions in the liquid phase are x_A and x_B. From Dalton's law,

$$p_A = P y_A$$

$$p_B = P y_B$$

where P is the total pressure, and y_A,y_B are the mole fractions in the vapor phase.

$$P = p_A + p_B, \qquad x_A + x_B = y_A + y_B = 1.0$$

Combining Raoult's and Dalton's laws,

$$y_A = \frac{P_A}{P} x_A = 1 - \frac{p_B}{P} = 1 - \frac{P_B(1 - x_A)}{P}$$

so that

$$x_A = \frac{P - P_B}{P_A - P_B}$$

The relative volatility is defined as

$$\alpha = \alpha_{AB} = \frac{y_A / x_A}{y_B / x_B} = \frac{y_A(1 - x_A)}{x_A(1 - y_A)}$$

Rearranging,

$$y_A = \frac{\alpha x_A}{1 + (\alpha - 1) x_A}$$

For ideal systems α often remains fairly constant over a wide range of composition at constant pressure, and since, ideally, $y_A / x_A = P_A / P$, it follows that $\alpha = P_A / P_B$.

The vapor pressure as a function of temperature for cyclopentane and cyclohexane is given by Maxwell (1955). The atmospheric boiling point of cyclopentane is 120.7°F, and of cyclohexane, 177.3°F. This fixes the temperature range for the calculations. Vapor-pressure readings from the graph in Maxwell (1955) are given for various temperatures in Table 4.1, where the corresponding x_A, y_A, and α values are also tabulated. For example, at 150°F,

$$x_A = \frac{1 - 0.615}{1.63 - 0.615} = 0.379$$

mole fraction of cyclopentane in the boiling or saturated liquid phase, and

$$y_A = \frac{1.63 \times 0.379}{1.00} = 0.618$$

mole fraction of cyclopentane in the saturated vapor.

$$\alpha = \frac{1.63}{0.615} = 2.65$$

Table 4.1. Vapor-pressure and composition data for the system cyclopentane-cyclohexane.

t (°F)	P (atm)	P_A (atm)	P_B (atm)	x_A	y_A	α
120.7	1	1.00	0.345	1.00	1.00	2.90
130.0	1	1.175	0.418	0.768	0.902	2.81
140.0	1	1.390	0.508	0.557	0.775	2.74
150.0	1	1.630	0.615	0.379	0.618	2.65
160.0	1	1.900	0.740	0.224	0.425	2.57
170.0	1	2.210	0.880	0.090	0.200	2.51
177.3	1	2.500	1.00	0.00	0.00	2.50
					Average =	2.67

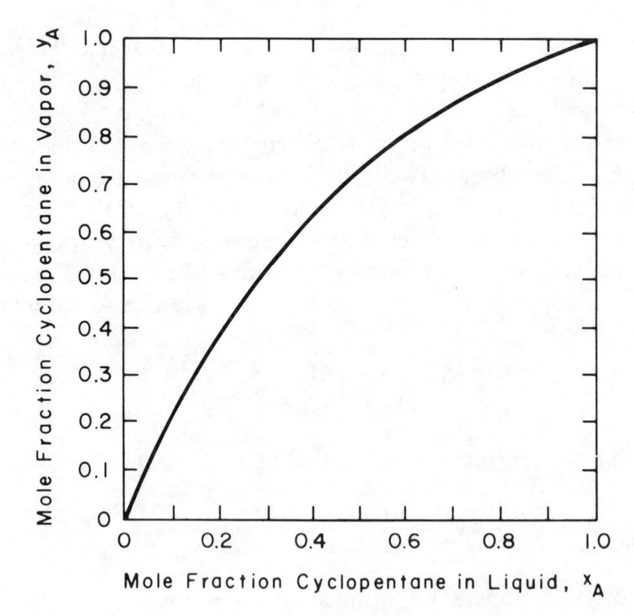

Figure 4.2. Vapor-liquid equilibrium for the system cyclopentane-cyclohexane at 1 atm pressure (Illustration 4.1).

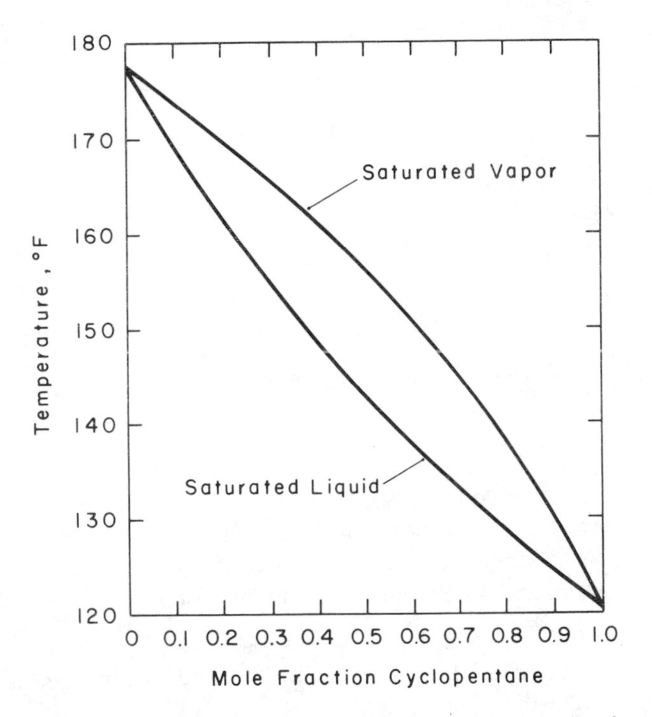

Figure 4.3. The temperature-composition or boiling-point diagram for the system cyclopentane-cyclohexane at 1 atm pressure (Illustration 4.1).

The plot of y_A versus x_A is shown in Figure 4.2, and the temperature-composition diagram is shown in Figure 4.3. The saturated-liquid curve is a plot of t (in °F). versus x_A, and the saturated-vapor curve is a plot of t (in°F) versus y_A. Figure 4.3 is sometimes called the boiling-point diagram.

Table 4.1 shows that the change in α is relatively small over the whole range of concentration, with an average for the computed values of 2.67. Thus an approximate relationship between y_A and x_A for this system at atmospheric pressure is given by

$$y_A = \frac{2.67x_A}{1+(2.67-1)x_A} = \frac{2.67x_A}{1+1.67x_A}$$

Equilibrium data in ternary liquid-liquid systems containing two phases, as used in liquid-liquid extraction, must generally be determined experimentally. The following example shows how such data may be represented on a triangular diagram, how erroneous (or nonexistent) plait-point data may be rectified, and how equilibrium interpolation is facilitated.

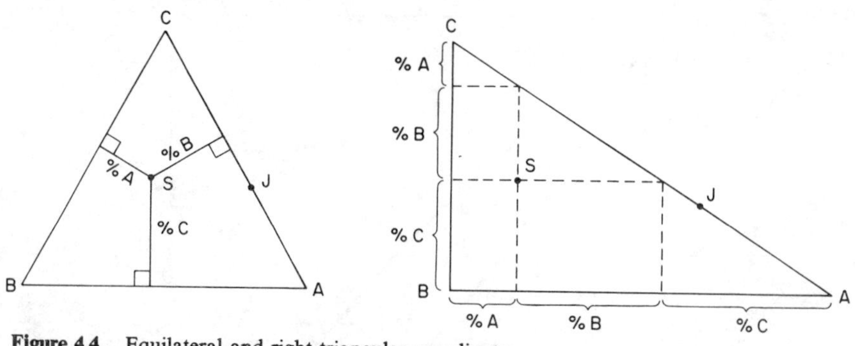

Figure 4.4. Equilateral and right-triangular coordinates.

Illustration 4.2.

Acetone is to be extracted from water using ethyl propionate as solvent at 86°F. The operation will be performed in a continuous, countercurrent, packed extraction column, for which a partial design is developed later in Chapter 7 (Illustration 7.4). Provide the following items for subsequent use in this design procedure:

(a) A representation of the mutual solubility and equilibrium distribution data on a right-triangular diagram.

(b) The corrected location of the plait point (i.e., the point at which the two conjugate equilibrium phases become identical).

(c) The location of a conjugate curve for interpolation of tie lines linking two equilibrium phases.

SOLUTION (a). Ternary systems are often represented on equilateral or right-triangular coordinates, as sketched in Figure 4.4.

Any point on one of the sides of the triangle represents a binary mixture of the components at each end of that side. For example, mixture J contains (line length CJ/line length AC) $\times$ 100% of A, the remainder being C.

A point inside the triangle represents a ternary mixture, such as S. The perpendicular distance of the point S from a given side of the equilateral triangle gives the percentage (or fraction) of the component represented by the apex opposite that side. The vertical height of the equilateral triangle represents 100% (or 1.0). In the right-angled triangle—which need not be isosceles—the composition of a ternary mixture S is conveniently identified as shown, in terms of the scales of axes AB and BC. A point outside the triangle represents an imaginary mixture, of negative composition.

The right-triangular diagram is rather more convenient, since it may be readily constructed to any scale, with enlargement of any particular region of interest. This representation is accordingly adopted here, as shown in Figure 4.5, where the published data of Venkataratnam, Rao, and Rao (1957) for this system are plotted as the binodal curve (mutual solubility data) plus seven tie lines (including the base).

SOLUTION (b). The correct location of the plait point is estimated by the method of Treybal, Weber, and Daley (1946). The notation for this system is as follows:

$$A = \text{acetone}, \qquad B = \text{water}, \qquad C = \text{ethyl propionate}.$$

$$w_{AB} = \text{mass fraction of } A \text{ in the } B\text{-rich phase.}$$

The following quantities are calculated from the coordinates of the terminals of the tie line nearest apex A in Figure 4.5.

$$\frac{w_{AC}}{w_{CC}} = \frac{\text{mass fraction acetone in ethyl propionate-rich layer}}{\text{mass fraction ethyl propionate in ethyl propionate-rich layer}}$$

$$= \frac{0.451}{0.464} = 0.973$$

$$\frac{w_{AB}}{w_{BB}} = \frac{\text{mass fraction acetone in water-rich layer}}{\text{mass fraction water in water-rich layer}} = \frac{0.344}{0.607} = 0.566$$

This procedure is repeated for each of the tie lines in Figure 4.5 and w_{AC}/w_{CC} is then plotted against w_{AB}/w_{BB} on logarithmic coordinates in Figure 4.6. A straight line is drawn through the points.

From points selected at random on the upper portion of the binodal curve, values of w_A/w_C and w_A/w_B are calculated, where w_A, w_B, and w_C are the mass fractions of the three components at any selected point. The curve w_A/w_C versus w_A/w_B is then also plotted on Figure 4.6. The intersection of this curve with the extrapolated straight line corresponds to P, the plait point. Thus, from Figure 4.6 at P,

$$\left(\frac{w_A}{w_C}\right)_P = 2.64, \quad \left(\frac{w_A}{w_B}\right)_P = 1.48$$

and since $w_A + w_B + w_C = 1.0$,

$$w_{AP} = 0.487 = \text{mass fraction acetone}$$

$$w_{BP} = 0.329 = \text{mass fraction water}$$

$$w_{CP} = 0.184 = \text{mass fraction ethyl propionate.}$$

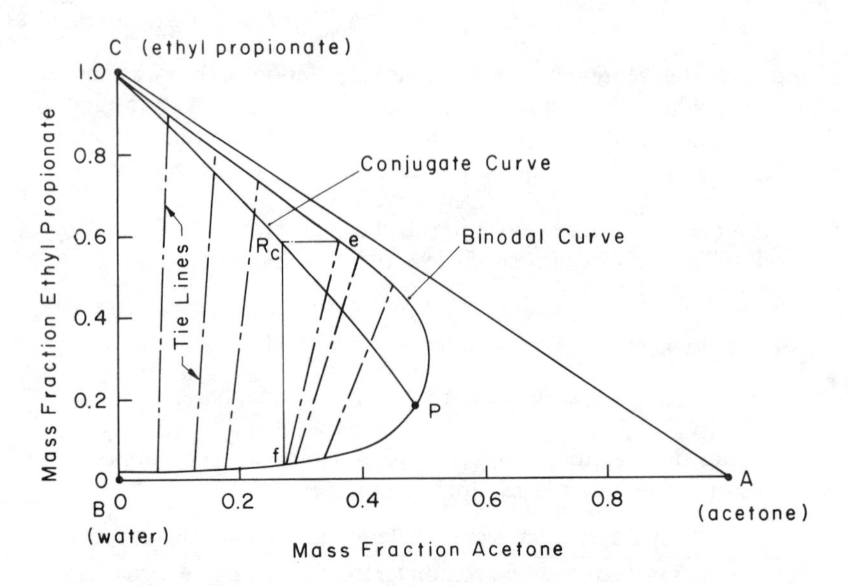

Figure 4.5. Representation of mutual solubility and equilibrium distribution in the system water–acetone–ethyl propionate at 86°F (Illustration 4.2).

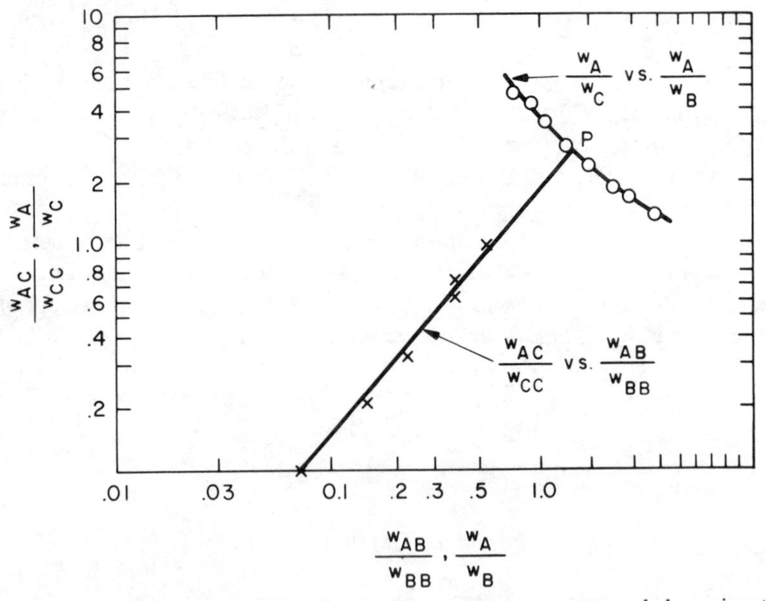

Figure 4.6. Estimation of the plait point in the system water–acetone–ethyl propionate at 86°F (Illustration 4.2).

This is plotted as the estimated plait point P in Figure 4.5. It differs from the published value ($w_{AP}=0.557$, $w_{BP}=0.251$, $w_{CP}=0.192$), which is in error because it does not lie on the binodal curve.

SOLUTION (c). The conjugate curve is constructed in Figure 4.5 from the seven existing tie lines and the estimated plait point. For example, point R_c on the conjugate curve is located at the intersection of line fR_c, parallel to side BC, and line eR_c, parallel to side AB of the triangle. The line ef is an experimentally determined tie line. The conjugate curve passes through the plait point, P, and may be used for tie-line interpolation by reversing the procedure used in its construction.

An alternative representation of ternary equilibrium data is on the distribution diagram, which is a plot of the mass (or mole) fraction of solute A in the C-rich phase against the mass (or mole) fraction of A in the B-rich phase. Such a plot appears in Illustration 7.4.

Individual Mass-Transfer Coefficients

It is assumed that local equilibrium prevails at the interface, where the compositions are Y_A^* and X_A^*. If transfer takes place from the gas to the liquid, the individual coefficients k_Y and k_X for the gas and liquid phases, respectively, are defined as follows:

$$N_A A = k_Y A(Y_A - Y_A^*) = k_X A(X_A^* - X_A) \tag{4.1}$$

A is the area of the interface, where the flux in the direction of decreasing concentration is N_A, and Y_A and X_A are the concentrations of component A in the bulk of the gas and liquid phases. Equation 4.1 shows that

$$\frac{Y_A - Y_A^*}{X_A - X_A^*} = -\frac{k_X}{k_Y} \tag{4.2}$$

This relationship is plotted in Figure 4.7 on the assumption of interfacial equilibrium.

Overall Mass-Transfer Coefficients

Interfacial concentrations (X_A^*, Y_A^*) are often unknown at a given location within a two-phase system, and it is then more convenient to use overall coefficients K_Y and K_X, defined in terms of overall concentration differences or "driving forces," as shown below:

$$N_A A = K_Y A(Y_A - Y_{AL}) = K_X A(X_{AG} - X_A) \tag{4.3}$$

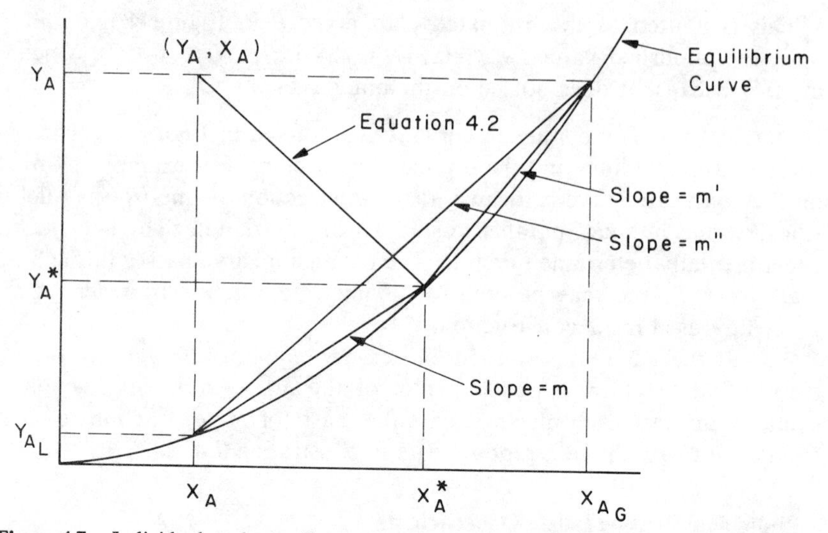

Figure 4.7. Individual and overall concentration "driving forces" for a two-phase system with distributed component A. Bulk concentrations of the two phases are respectively X_A and Y_A, and interfacial equilibrium is assumed.

where Y_{AL} is the gas-phase concentration which would be in equilibrium with the existing liquid-phase concentration, and X_{AG} is the liquid-phase concentration which would be in equilibrium with the existing gas-phase concentration. The location of these quantities is shown in Figure 4.7. The relationships between the individual coefficients of equation 4.1 and the overall coefficients of equation 4.3 are obtained in the following way, noting from Figure 4.7 that

$$m=\frac{Y_A^*-Y_{AL}}{X_A^*-X_A},\qquad m'=\frac{Y_A-Y_A^*}{X_{AG}-X_A^*},\qquad m''=\frac{Y_A-Y_{AL}}{X_{AG}-X_A}$$

$$N_A A=k_Y A(Y_A-Y_A^*)=K_Y A(Y_A-Y_{AL})$$

$$\frac{1}{K_Y}=\frac{1}{k_Y}\left(\frac{Y_A-Y_{AL}}{Y_A-Y_A^*}\right)=\frac{1}{k_Y}\left[\frac{(Y_A-Y_A^*)+(Y_A^*-Y_{AL})}{Y_A-Y_A^*}\right]$$

$$\frac{1}{K_Y}=\frac{1}{k_Y}-\frac{1}{k_Y}\left(\frac{Y_A^*-Y_{AL}}{Y_A^*-Y_A}\right) \tag{4.4}$$

$$Y_A^*-Y_{AL}=m(X_A^*-X_A) \tag{4.5}$$

and from equation 4.2,

$$Y_A^* - Y_A = -\frac{k_X}{k_Y}(X_A^* - X_A) \tag{4.6}$$

Taking the ratio of equation 4.5 to 4.6 and substituting the result in equation 4.4 gives

$$\frac{1}{K_Y} = \frac{1}{k_Y} + \frac{m}{k_X} \tag{4.7}$$

Similarly,

$$\frac{1}{K_X} = \frac{1}{k_X} + \frac{1}{m'k_Y} \tag{4.8}$$

and

$$\frac{1}{K_X} = \frac{1}{m''K_Y} \tag{4.9}$$

By analogy with heat transfer, the term $1/K_Y$ may be regarded as the total resistance to mass transfer based on the driving force $Y_A - Y_{AL}$. Equation 4.7 then shows this to equal the sum of the gas-phase resistance $1/k_Y$ and the liquid-phase resistance m/k_X. A similar interpretation in terms of additive resistances prevails for equation 4.8.

It is evident that even if k_X and k_Y are constants, the overall coefficients will nevertheless vary with concentration unless $m = m' = m'' =$ constant—that is to say, unless the equilibrium curve is linear over the relevant range of concentration. This constitutes a severe limitation on conditions in which constant overall K values may be used with genuine validity. The restriction becomes unimportant, however, in cases which are "gas-phase controlling" ($1/k_Y \gg m/k_X$) or "liquid-phase controlling" ($1/k_X \gg 1/m'k_Y$).

The interfacial area A is often unknown in many types of mass-transfer equipment. In such cases mass-transfer rates are frequently based on unit volume of the equipment, instead of unit interfacial area. The rate equations 4.1 and 4.3 are then modified to the following form:

$$\begin{aligned} N_A A &= k_Y a(Y_A - Y_A^*)_m V_v = k_X a(X_A^* - X_A)_m V_v \\ &= K_Y a(Y_A - Y_{AL})_m V_v = K_X a(X_{AG} - X_A)_m V_v \end{aligned} \tag{4.10}$$

where V_v is the contacting volume of the equipment, a is the interfacial area per unit volume, and the combined quantities $k_X a$, $k_Y a$, $K_X a$, and

$K_Y a$ are called volumetric or capacity coefficients. Equations 4.7, 4.8, and 4.9 may be divided by a to give

$$\frac{1}{K_Y a} = \frac{1}{k_Y a} + \frac{m}{k_X a} \tag{4.11}$$

$$\frac{1}{K_X a} = \frac{1}{k_X a} + \frac{1}{m' k_Y a} \tag{4.12}$$

$$\frac{1}{K_X a} = \frac{1}{m'' K_Y a} \tag{4.13}$$

A distinction is made in subsequent chapters between coefficients under conditions of high and low mass-transfer rates. Coefficients corresponding to low rates are usually denoted with an asterisk, as k^*.

MECHANISMS OF MASS TRANSFER

Many engineering applications involve the transfer of material across the interface between two phases. Several different mechanisms have been proposed to describe conditions in the vicinity of the interface, some of which will now be considered.

The Two-Film Theory

This theory, developed by Lewis (1916) and Whitman (1923), supposes that turbulence in the two phases dies out near the interface, and the entire resistance to transfer is considered as being contained in two fictitious films on either side of the interface, in which transfer occurs by purely molecular diffusion. It is postulated that local equilibrium prevails at the interface and that the concentration gradients in the films are established in a time so short compared to the total time of contact that steady-state diffusion may be assumed. Both mass and molal units will be used in this instance, because the latter will yield expressions which contribute significantly to the development of relationships for the number and height of transfer units in Chapter 7.

Individual or single-phase mass-transfer coefficients k are customarily defined as follows for a mechanism of equimolal counterdiffusion: In the gas phase,

$$\frac{n_A}{M_A} = N_A = \frac{k'_{\rho G} \Delta \rho_{AG}}{M_A} = k'_G \Delta p_A = k'_y \Delta y_A = k'_c \Delta c_A \tag{4.14}$$

and in the liquid phase,

$$\frac{n_A}{M_A} = N_A = \frac{k'_{\rho L}\Delta\rho_{AL}}{M_A} = k'_c\Delta c_A = k'_x\Delta x_A = k'_c\left(\frac{\rho}{M}\right)_{\text{av}}\Delta x_A \tag{4.15}$$

For a mechanism involving bulk molal flow, on the other hand, in the gas phase,

$$N_A = \frac{k_{\rho G}\Delta\rho_{AG}}{M_A} = k_G\Delta p_A = k_y\Delta y_A = k_c\Delta c_A \tag{4.16}$$

and in the liquid phase,

$$N_A = \frac{k_{\rho L}\Delta\rho_{AL}}{M_A} = k_c\Delta c_A = k_x\Delta x_A = k_c\left(\frac{\rho}{M}\right)_{\text{av}}\Delta x_A \tag{4.17}$$

Consider a process of equimolal counterdiffusion between a gas and a liquid phase. The transfer of component A in the gas will be described by equation 2.49 in terms of mole fractions:

$$N_{Az} = \frac{D_G P}{RTz_{fG}}(y_A - y_A^*) = k'_y(y_A - y_A^*) \tag{4.18}$$

and in mass instead of molal units,

$$n_{Az} = \frac{D_G}{z_{fG}}(\rho_{AG} - \rho_{AG}^*) = k'_{\rho G}(\rho_{AG} - \rho_{AG}^*) \tag{4.19}$$

Component A is transferred in the liquid in accordance with equation 2.54:

$$N_{Az} = \frac{D_L c_L}{z_{fL}}(x_A^* - x_A) = k'_x(x_A^* - x_A) \tag{4.20}$$

or in mass units,

$$n_{Az} = \frac{D_L}{z_{fL}}(\rho_{AL}^* - \rho_{AL}) = k'_{\rho L}(\rho_{AL}^* - \rho_{AL}) \tag{4.21}$$

where z_{fG} and z_{fL} are the thicknesses of the fictitious gas and liquid films, and the asterisk denotes local equilibrium values in the gas and liquid phases at the interface.

When a total or bulk molal flow $(N_{Az} + N_{Bz})$ occurs in the direction z in which A is diffusing, equation 2.44 is applicable with nonzero γ and may be written as follows for the gas phase:

$$\frac{n_{Az}}{M_A}=N_{Az}=\left(\frac{N_{Az}}{N_{Az}+N_{Bz}}\right)\frac{D_G P}{RTz_{fG}}\ln\left(\frac{N_{Az}/(N_{Az}+N_{Bz})-y_{A2}}{N_{Az}/(N_{Az}+N_{Bz})-y_{A1}}\right) \quad (4.22)$$

so that from equations 4.16 and 4.22,

$$k_y=\frac{D_G P}{RTz_{fG}}\left(\frac{N_{Az}}{N_{Az}+N_{Bz}}\right)\frac{1}{[N_{Az}/(N_{Az}+N_{Bz})-y_A]_{\text{LM}}} \quad (4.23)$$

where $[N_{Az}/(N_{Az}+N_{Bz})-y_A]_{\text{LM}}$ is the logarithmic mean of the values in the bulk of the phase and at the interface–that is to say, on either side of the fictitious film of thickness z_{fG}. In the special but important case of unimolal unidirectional diffusion, $N_{Bz}=0$, and equation 4.23 becomes

$$k_y=\frac{D_G P}{RTz_{fG}(1-y_A)_{\text{LM}}}=\frac{D_G P}{RTz_{fG}y_{B\text{LM}}} \quad (4.24)$$

$$k_{\rho G}=\frac{k_y RT}{P}=\frac{D_G P}{z_{fG}p_{B\text{LM}}} \quad (4.25)$$

and for the liquid phase, when $N_{Bz}=0$, equation 2.56 corresponds to

$$k_x=\frac{D_L c}{z_{fL}(1-x_A)_{\text{LM}}}=\frac{D_L c}{z_{fL}x_{B\text{LM}}} \quad (4.26)$$

$$k_{\rho L}=\frac{k_x}{c}=\frac{D_L c}{z_{fL}c_{B\text{LM}}}=\frac{D_L}{z_{fL}}\frac{(\rho/M)_{av}}{(\rho_B/M_B)_{\text{LM}}} \quad (4.27)$$

Equations 4.18 to 4.27 show that the two-film theory predicts that the mass transfer coefficient is directly proportional to the molecular diffusivity to the power unity. The complexity of flow normally prevents evaluation of z_f, but it decreases with increasing turbulence.

Relations between the various coefficients in equations 4.14 to 4.27 are given in Problem 4.3 at the end of this chapter.

The Penetration Theory

The work of Higbie (1935) provided the foundation for the penetration theory, which supposes that turbulent eddies travel from the bulk of the phase to the interface, where they remain for a short but constant time before being displaced back into the interior of the phase to be mixed with the bulk fluid. Solute is assumed to "penetrate" into a given eddy during its stay at the interface by a process of unsteady-state molecular diffusion, in accordance with Fick's second law–equation 2.93 in mass-concentration terms:

$$\frac{\partial \rho_A}{\partial t} = D\left(\frac{\partial^2 \rho_A}{\partial z^2}\right) \tag{2.93}$$

where v_0 is negligible and the boundary conditions are

$$t=0, \qquad 0 \leqslant z \leqslant \infty: \qquad \rho_A = \rho_{A\infty}$$

$$t>0, \qquad z=0: \qquad \rho_A = \rho_A^*$$

$$t>0, \qquad z=\infty: \qquad \rho_A = \rho_{A\infty}$$

The interfacial composition of the phase under consideration is constant at ρ_A^*, as is the concentration $\rho_{A\infty}$ at z effectively equal to "infinity." A solution may be obtained by Laplace transforms (Coulson and Richardson, 1964) to give

$$\frac{\rho_A^* - \rho_A}{\rho_A^* - \rho_{A\infty}} = \operatorname{erf} \frac{z}{2\sqrt{Dt}} \tag{4.28}$$

Here erf is the error function,

$$\operatorname{erf} \xi = \frac{2}{\sqrt{\pi}} \int_0^{\xi} \exp(-\zeta^2)\, d\zeta$$

Equation 4.28 gives the concentration within a given eddy at the interface as a function of position z and time t. Under conditions of negligible bulk flow the instantaneous rate of unsteady diffusion from the interface is

$$n_{A0} = -D\left(\frac{\partial \rho_A}{\partial z}\right)_{z=0} \tag{4.29}$$

and evaluating the derivative from equation 4.28,

$$n_{A0} = (\rho_A^* - \rho_{A\infty})\left(\frac{D}{\pi t}\right)^{1/2} \tag{4.30}$$

The total solute penetrating the eddy in an exposure time t_e is

$$\int_0^{t_e} n_{A0}\,dt = (\rho_A^* - \rho_{A\infty})\left(\frac{D}{\pi}\right)^{1/2}\int_0^{t_e} t^{-1/2}\,dt = 2(\rho_A^* - \rho_{A\infty})\left(\frac{Dt_e}{\pi}\right)^{1/2} \tag{4.31}$$

and the average rate of transfer during exposure is obtained by dividing equation 4.31 by t_e,

$$(n_{A0})_{\text{av}} = 2(\rho_A^* - \rho_{A\infty})\left(\frac{D}{\pi t_e}\right)^{1/2} \tag{4.32}$$

which predicts that the mass-transfer coefficient is directly proportional to the square root of the molecular diffusivity—in contrast with the two-film theory. In most applications the exposure time t_e for a given eddy is unknown, but it will decrease with increasing turbulence.

Angelo, Lightfoot, and Howard (1966) have extended the penetration theory to allow for "stretching" surfaces, such as those which occur, for example, in large oscillating droplets. A further modification has been presented by Ruckenstein (1968), who makes allowance for velocity distributions within the eddies during penetration by the solute.

The Theory of Penetration with Random Surface Renewal

The original penetration theory postulates that the period spent by all eddies at the surface is constant. This picture may be modified by proposing an "infinite" range of ages for elements of the surface. The probability of an element of surface being replaced by a fresh eddy is considered to be independent of the age of that element. Danckwerts (1951) introduced this modification by defining a surface age distribution function, $\phi(t)$, such that the fraction of surface with ages between t and $t+dt$ is $\phi(t)\,dt$. If the probability of replacement of a surface element is independent of its age, Danckwerts showed that

$$\phi(t) = se^{-st}$$

where s is the fractional rate of surface renewal. Equation 4.30 gives the instantaneous rate of transfer from the surface for a given element. Thus for all elements with ages between t and $t+dt$ the instantaneous transfer rate is

$$se^{-st}(\rho_A^* - \rho_{A\infty})\left(\frac{D}{\pi t}\right)^{1/2} dt$$

The flux of component A from a surface having elements with ages distributed over the range $0 \ll t \ll \infty$ is therefore

$$(n_{A0})_{av} = s(\rho_A^* - \rho_{A\infty})\left(\frac{D}{\pi}\right)^{1/2} \int_0^\infty t^{-1/2} e^{-st}\, dt$$

or

$$(n_{A0})_{av} = (\rho_A^* - \rho_{A\infty})(sD)^{1/2} \tag{4.33}$$

which again indicates that the mass-transfer coefficient is directly proportional to the square root of the molecular diffusivity. Procedures for predicting s are not available at present, but its value will increase with increasing turbulence. The results of assuming other relationships between s and t are given in Problem 4.5 at the end of this chapter.

The Film-Penetration Theory

The film-penetration model, presented by Toor and Marchello (1958), represents a combination of the three earlier theories reviewed above. The entire transfer resistance is considered to lie in a laminar surface layer of thickness z_L, where ρ_A is uniform at $\rho_{A\infty}$ for all z greater than z_L. Surface renewal occurs by eddies which penetrate the surface from the bulk of the phase. Thus transfer through young elements of surface obeys the penetration theory ($k \propto \sqrt{D}$), transfer through old elements follows the film theory ($k \propto D$), and transfer through elements of intermediate age combines both mechanisms. The transfer equation is again Fick's second law, equation 2.93, but with the boundary conditions

$$t=0, \qquad 0 \leqslant z \leqslant \infty: \qquad \rho_A = \rho_{A\infty}$$

$$t>0, \qquad z=0: \qquad \rho_A = \rho_A^*$$

$$t>0, \qquad z=z_L: \qquad \rho_A = \rho_{A\infty}$$

The key difference from the penetration theory lies in the third boundary condition. A solution for $\rho_A(z,t)$ may again be obtained using Laplace transforms. The result is used to evaluate equation 4.29, giving, for small t,

$$n_{A0} = (\rho_A^* - \rho_{A\infty})\left(\frac{D}{\pi t}\right)^{1/2}\left[1 + 2\sum_{n=1}^{\infty} \exp\left(-\frac{n^2 z_L^2}{Dt}\right)\right] \tag{4.34}$$

and for large t,

$$n_{A0} = (\rho_A^* - \rho_{A\infty})\frac{D}{z_L}\left[1 + 2\sum_{n=1}^{\infty}\exp\left(-n^2\pi^2\frac{Dt}{z_L^2}\right)\right] \tag{4.35}$$

Equations 4.34 and 4.35 are equivalent, but are obtained by different techniques for taking the inverse transform to obtain $\rho_A(z,t)$. The first equation converges rapidly for small t ($t \ll z_L^2/D$) and the second for large t ($t \gg z_L^2/D$). The equations give instantaneous n_{A0} for elements of age t. Average values are obtainable as

$$(n_{A0})_{av} = \int_0^{\infty} n_A \psi(t)\, dt \tag{4.36}$$

where $\psi(t)$ is a general surface age distribution function. When all surface elements have the same exposure time, as in Higbie's penetration theory, the result for short t_e is

$$(n_{A0})_{av} = (\rho_A^* - \rho_{A\infty})2\left(\frac{D}{\pi t_e}\right)^{1/2}\left[1 + 2\sqrt{\pi}\sum_{n=1}^{\infty}\text{ierfc}\frac{nz_L}{\sqrt{Dt_e}}\right] \tag{4.37}$$

and for large t_e,

$$(n_{A0})_{av} = (\rho_A^* - \rho_{A\infty})\frac{D}{z_L}\left\{1 + \frac{2}{\pi^2}\frac{z_L^2}{Dt_e}\left[\frac{\pi^2}{6} - \sum_{n=1}^{\infty}\exp\left(-n^2\pi^2\frac{Dt_e}{z_L^2}\right)\right]\right\} \tag{4.38}$$

The integral of the complementary error function is denoted by ierfc. Equations 4.37 and 4.38 reduce to the penetration theory, equation 4.32, for small t_e, and to the film theory, equation 4.20, for very large t_e.

When the surface age distribution function corresponding to Danckwerts' random-surface-renewal theory is used in equation 4.36, the result for short average element life is

$$(n_{A0})_{av} = (\rho_A^* - \rho_{A\infty})(sD)^{1/2}\left[1 + 2\sum_{n=1}^{\infty}\exp\left(-2nz_L\sqrt{\frac{s}{D}}\right)\right] \tag{4.39}$$

and for long average element life,

$$(n_{A0})_{av} = (\rho_A^* - \rho_{A\infty})\frac{D}{z_L}\left[1 + 2\sum_{n=1}^{\infty}\frac{1}{1 + n^2\pi^2 D/sz_L^2}\right] \qquad (4.40)$$

When the rate of surface renewal is high (sz_L^2/D is large) equation 4.39 reduces to Danckwerts' equation (4.33). Conversely, when sz_L^2/D is low, equation 4.40 reduces to the film theory of equation 4.20. The series terms in equations 4.37 to 4.40 converge rapidly in all cases, but the quantities z_L, t_e, and s are not predictable at present.

The Mass-Flow or Convective-Transfer Theory

In a series of papers, Kishinevskii and co-workers (1949–1954) proposed a surface-renewal mechanism which, in contrast with the theories described above, postulates that transfer into an eddy at the interface occurs predominantly by convective mass flow and not by molecular diffusion. The authors also dispute the suggestion that the probability of replacement of a surface element is independent of its age.

King (1966) has proposed another general model for turbulent liquid-phase mass transfer to and from a free gas-liquid interface. The model requires the evaluation of three parameters and involves concepts of surface renewal in which surface tension exerts a damping effect upon the smaller eddies. Allowance is made for a continuous cddy diffusivity profile near the free interface, thereby avoiding the postulate of a "film" or discontinuity in transport properties as required by the film-penetration theory.

More recently, Wasan and Ahluwalia (1969) have presented a model in which heat or mass transfer between a wall and a fluid stream is postulated to occur by a mechanism of consecutive film and surface renewal. Agreement was obtained between the theory and measurements in both gaseous and liquid fluidized beds.

The form of the experimental dependence of the mass transfer coefficient upon D should indicate which theory is more nearly correct. One difficulty—particularly in the area of gas absorption, to which the theories have been applied with some persistence—is that the diffusivities of the usual solute gases are only inaccurately known and do not vary greatly from each other. Work by Lewis (1954, 1955) on a mechanically agitated cell containing two liquid phases suggests that the coefficient is independent of D. This contrasts with the study of Murdoch and Pratt (1953) on a liquid-liquid wetted-wall column, which indicated that the mass-transfer coefficient is proportional to D raised to the 2/3 power. (The latter dependence occurs quite frequently in a variety of situations.) Dobbins (1956) developed relationships pertaining to the film-penetration theory in a publication which appeared two years earlier than that of Toor and Marchello (1958). He applied the results to experiments simulating the rate of gas absorption by rivers (Dobbins, 1964), and found exponents on D between 0.985 and 0.65, the lower values corresponding to increasing turbulence. Experiments on gas absorption in agitated vessels and packed columns indicate liquid mass-transfer coefficients proportional to D to the 0.5 power (Vivian and King, 1964; Kozinsky and King, 1966; and Tavares da Silva and Danckwerts, 1968). Further evidence favoring the surface-renewal models rather than the two-film theory has been obtained in experiments on gas absorption with chemical reaction in a variety of columns (Danckwerts et al., 1963; Richards et al., 1964; Danckwerts and Gillham, 1966; and Tavares da Silva and Danckwerts, 1968).

There are nevertheless many instances in which the relative simplicity of the two-film theory has contributed to the quantitative representation of complex processes in a way which has been consistently useful in design. Examples of this will be found in Chapter 7.

Relationships between mass-transfer coefficients and diffusivity arising from boundary-layer theory are indicated in Chapters 5 and 6.

NOMENCLATURE

A, B, C	Components $A, B,$ and C.
A	Interfacial area, ft^2.
a	Interfacial area per unit volume, ft^2/ft^3.
c	Total concentration, lb-mole/ft^3.
c_A	Concentration of component A, lb-mole/ft^3.
c_{BLM}	Logarithmic-mean concentration of component B in a film of thickness z_{fL}, lb-mole/ft^3.
D, D_G, D_L	(Volumetric) molecular diffusivity; in the gas and liquid phases, ft^2/hr.
K_X, K_Y	Overall mass transfer coefficients based on ΔX and ΔY,lb-mole/(hr)(ft^2)(unit of ΔX or ΔY).
$k_c, k_G, k_x, k_y,$ $k_{\rho G}, k_{\rho L}$	Individual mass-transfer coefficients defined in equations 4.16 and 4.17.
$k'_c, k'_G, k'_x, k'_y,$ $k'_{\rho G}, k'_{\rho L}$	Individual mass-transfer coefficients for equimolal

	counterdiffusion, defined in equations 4.14 and 4.15.
k_X, k_Y	Individual mass-transfer coefficients based on ΔX and ΔY, lb-mole/(hr)(ft^2)(unit of ΔX or ΔY).
M	Average molecular weight.
M_A, M_B	Molecular weights of components A and B.
m	Slope of equilibrium curve, see also Figure 4.7.
m', m''	Defined in Figure 4.7.
N_A, N_{Az}	Molal flux of component A relative to stationary coordinates, in the z direction, lb-mole/(ft^2)(hr).
N_{Bz}	Molal flux of component B in the z direction relative to stationary coordinates, lb-mole/(ft^2)(hr).
n_A, n_{A0}, n_{Az}	Mass flux of component A relative to stationary coordinates, at the surface $z=0$, in direction z, lb-mass/(ft^2)(hr).
P, P_A, P_B	Total pressure; vapor pressures of components A and B at the prevailing temperature, lb-force/ft^2 *or* atm.
p_A, p_B, p_{BLM}	Partial pressures of components A and B; logarithmic-mean partial pressure of component B in film of thickness z_{fG}, lb-force/ft^2 *or* atm.
R	Gas constant, 1545 ft lb-force/(lb-mole)(°R) *or* 0.73 atm ft^3/(lb-mole)(°R).
s	Fractional rate of surface renewal, hr^{-1}.
T	Absolute temperature, °R.
t, t_e	Time, exposure time, hr.
V_v	Contacting volume, ft^3.
v_0	Mass-average velocity in the z direction at the surface $z=0$, ft/hr.
w_A, w_B, w_C	Mass fractions of components A, B, and C.
$w_{AB}, w_{AC}, w_{BB}, w_{CC}$	Mass fraction of first-subscript component in the phase rich in the second-subscript component.
w_{AP}, w_{BP}, w_{CP}	Mass fractions of A, B, and C in the mixture represented by the plait point.
X_A	Composition of L phase, any convenient units.
X_A^*, x_A^*	Local equilibrium concentration in the L phase at the interface, any convenient units; mole fraction.
X_{AG}	L-phase concentration which would be in equilibrium with the existing G-phase concentration, any convenient units.

x_A, x_B	Mole fractions of components *A* and *B*.
x_{BLM}	Logarithmic-mean mole fraction of component *B* in the film of thickness z_{fL}.
Y_A	Composition of *G* phase, any convenient units.
Y_A^*, y_A^*	Local equilibrium concentration in the *G* phase at the interface, any convenient units; mole fraction.
Y_{AL}	*G*-phase concentration which would be in equilibrium with the existing *L*-phase concentration, any convenient units.
y_A, y_B	Mole fractions of components *A* and *B*.
y_{BLM}	Logarithmic-mean mole fraction of component *B* in the film of thickness z_{fG}.
z, z_{fG}, z_{fL}, z_L	Distance in direction of diffusion; thickness of the fictitious *G*- and *L*-phase films; thickness of the laminar surface layer in the film-penetration theory, ft.
α, α_{AB}	Relative volatility; in the binary system *A*-*B*.
ρ	Total density, lb-mass/ft^3.
$\rho_A, \rho_A^*, \rho_{AG}, \rho_{AG}^*, \rho_{AL}, \rho_{AL}^*, \rho_{A\infty}$	Mass concentration of component *A*; interfacial value in the phase under consideration; in the *G* phase; local equilibrium mass concentration in the *G* phase at the interface; in the *L* phase; local equilibrium mass concentration in the *L* phase at the interface; at $z=\infty$, lb-mass/ft^3.
ρ_B	Mass concentration of component *B*, lb-mass/ft^3.
$(\rho/M)_{av}$, $(\rho_B/M_B)_{LM}$	Mean value for the phase under consideration; see c_{BLM} above, lb-mole/ft^3.
$\phi(t)$	Surface age distribution function.
$\psi(t)$	A general surface age distribution function.
Subscript m	A suitable mean.

PROBLEMS

4.1 The aqueous absorption of sulfur dioxide from air in a countercurrent column packed with 1-in. ceramic rings has been studied by R. P. Whitney and J. E. Vivian [*Chem. Eng. Prog.*, **45**, 323–337 (1949)]. At 70°F and 1 atm, using a column diameter of 8 in. and packed height of 2 ft, the individual capacity coefficients for the liquid and gas phases were correlated as follows:

$$k_L a = 0.044\left(\frac{L'}{S}\right)^{0.82}, \quad \text{lb-mole } SO_2/(\text{hr})(\text{ft}^3)(\text{lb-mole } SO_2/\text{ft}^3 \text{ solution})$$

$$k_G a = 0.028\left(\frac{G'}{S}\right)^{0.7}\left(\frac{L'}{S}\right)^{0.25}, \quad \text{lb-mole } SO_2/(\text{hr})(\text{ft}^3)(\text{partial pressure of } SO_2, \text{ atm})$$

where L' and G' are the mass velocities of the liquid and gas streams in lb/hr, and S is the cross-sectional area of the empty column in ft^2, over the ranges $900 \leq L'/S \leq 12{,}000$ and $65 \leq G'/S \leq 850$. The following measurements are from the authors' run 7 at 70°F:

$L'/S = 4950$ lb/(hr)(ft^2)

$G'/S = 609$ lb/(hr)(ft^2) (average of top and bottom rates)

SO_2 concentration in entering water = 0

SO_2 concentration in leaving water = 0.0111 lb-mole/ft^3 of solution

SO_2 partial pressure in entering air = 0.176 atm

SO_2 partial pressure in leaving air = 0.137 atm

Equilibrium data for this system are available in Perry (1963., pp. 14-6–7), and it will be assumed that the densities of water and of the relevant SO_2 solutions are equal. Estimate the following quantites at the top and bottom of the column under these conditions:

(a) The interfacial concentrations of SO_2 in the gas and liquid phases.

(b) The overall capacity coefficient in the same units as $k_L a$.

(c) The percentage of the total resistance to transfer which resides in the gas and liquid phases, respectively,

(d) The local mass transfer rates.

4.2 Suppose the gas flow rate is increased by 35 percent in Problem 4.1. What liquid rate will restore the overall capacity coefficient to its original value at the top of the column?

4.3 Derive the following relationships between the various mass-transfer coefficients in equations 4.14 to 4.27:

In the gas phase:

$$k_{\rho G}=k_c=RTk_G=\frac{RT}{P}k_y=\frac{P}{p_{BLM}}k'_{\rho G}=\frac{k'_{\rho G}}{y_{BLM}}$$

$$=\frac{P}{p_{BLM}}k'_c=\frac{PRT}{p_{BLM}}k'_G=\frac{RT}{p_{BLM}}k'_y$$

In the liquid phase:

$$k_{\rho L}=k_c=\frac{k_x}{c}=\frac{k_x}{(\rho/M)_{av}}=\frac{c}{c_{BLM}}k'_{\rho L}=\frac{k'_{\rho L}}{x_{BLM}}$$

$$=\frac{c}{c_{BLM}}k'_c=\frac{k'_x}{c_{BLM}}$$

4.4 A long, capsule-shaped bubble of carbon dioxide at 1 atm is introduced into a vertical glass tube which is filled with water at 10°C. The tube has an internal diameter of 3 mm and a length of 65 cm, and the water flows downwards, drawing the bubble along at a velocity of 130 cm/sec. Suppose that the average length of the bubble is 7 cms and that CO_2 diffuses into the water film between the bubble and the wall of the tube in accordance with the penetration theory. Each element of water surface is then exposed to the gas for the time required by the bubble to pass it. If similar bubbles enter the tube at $\frac{1}{4}$-sec intervals, estimate the absorption rate of CO_2.

At 10°C the diffusivity of CO_2 in water is 1.41×10^{-5} cm²/sec, and the solubility, expressed in terms of Henry's law, is given by $p_A=1040x_A$, where p_A is in atmospheres and x_A is the mole fraction of CO_2 in water.

4.5 The theory of penetration with random surface renewal uses a surface age distribution function $\phi(t)=s\,e^{-st}$. The reader should examine Danckwerts' (1951) paper for the development of this expression, which assumes the probability of replacement of a surface element to be independent of its age. Suppose, instead, that the fractional rate of surface renewal s is related to the age of the surface through the constants ξ and n as $s=\xi t^n$, where $\xi, n+1>0$. Prove that the corresponding forms of the surface age distribution function and of equation 4.33 are given by

$$\phi(t)=\frac{[\xi/(n+1)]^{1/(n+1)}}{\Gamma[(n+2)/(n+1)]}\exp\left(-\frac{\xi t^{n+1}}{n+1}\right)$$

$$(n_{A0})_{av}=(\rho_A^*-\rho_{A\infty})\times2\left(\frac{D}{\pi}\right)^{1/2}\left(\frac{\xi}{n+1}\right)^{1/2(n+1)}\frac{\Gamma[(2n+3)/2(n+1)]}{\Gamma[(n+2)/(n+1)]}$$

where the gamma functions are evaluated from standard tables.

4.6 For the conditions of Problem 4.5, obtain expressions for $\phi(t)$ and $(n_{A0})_{av}$ corresponding to n-values of 0, 1, and 2.

4.7 Water droplets form at 2-sec intervals on the tip of a thin-walled vertical nozzle. The nozzle is immersed in ethyl acetate, which is saturated with water at 25°C. It is assumed that the drop grows spherically and that its surface is extended during formation by the addition of fresh elements which remain at the surface until detachment from the nozzle. Determine the amount of ethyl acetate that enters a single drop during growth and before detachment, if the fresh elements continually added to the growing surface are subject to mass transfer in accordance with the penetration theory.

The volume of the drop at detachment is 0.014 cm^3, and the diffusivity and solubility of ethyl acetate in water at 25°C are 0.889×10^{-5} cm^2/sec and 0.01614 mole fraction, respectively.

Note: The reader may wish to compare his solution with the detailed derivation in Chapter 8. The individual mass-transfer coefficient based on the drop surface at detachment under these conditions is obtained there in general form as equation 8.43.

REFERENCES

Angelo, J. B., E. N. Lightfoot, and D. W. Howard, *A. I. Ch. E. J.*, **12**, 751–760, (1966)

Coulson, J. M., and J. F. Richardson, *Chemical Engineering*, 2nd ed., Vol. 1, Pergamon Press, Oxford, (1965), pp. 313–314.

Danckwerts, P. V., *Ind. Eng. Chem.* **43**, 1460, (1951).

Danckwerts, P. V., *Chem. Eng. Sci.* **23**, 1045, (1968).

Danckwerts, P. V., and A. J. Gillham, *Trans. Inst. Chem. Engrs.*, (Lond.), **44**, T 42, (1966).

Danckwerts, P. V., A. M. Kennedy, and D. Roberts, *Chem. Eng. Sci.*, **18**, 63, (1963).

Dobbins, W. E., Part 2-1 in "Biological Treatment of Sewage and Industrial Wastes," M. L. McCabe, and W. W. Eckenfelder, Eds., Reinhold, New York, (1956).

Dobbins, W. E., Int. Conf. on Water Pollut. Res., Lond., Sept., 1962, Pergamon Press, New York, (1964), p. 61.

Emmert, R. E., and R. L. Pigford, in *Chemical Engineers' Handbook*, R. H. Perry, C. H. Chilton, and S. D. Kirkpatrick, Eds., 4th ed. McGraw-Hili, New York, (1963), Section 14.

Higbie, R., *Trans. A. I. Ch. E.*, **31**, 365, (1935).

King, C. J., *Ind. Eng. Chem. Fundam.* **5**, 1–8, (1966).

Kishinevskii, M. Kh., and A. V. Pamfilov, *J. Appl. Chem. U.S.S.R. (Engl. Transl.)*, **22**, 118, (1949).

Kishinevskii, M. Kh., *J. Appl. Chem. U.S.S.R. (Engl. Transl.)*, **24**, 542, (1951).

Kishinevskii, M. Kh., and M. A. Keraivarenko, *J. Appl. Chem. US.S.R. (Engl. Transl.)*, **24**, 413, (1951); *Novik* **26**, 673, (1953).

Kishinevskii, M. Kh., *J. Appl. Chem. U.S.S.R. (Engl. Transl.)*, **27**, 359, (1954).

Kozinsky, A. A., and C. J. King, *A. I. Ch. E. J.*, **12**, 109, (1966).

Lewis, J. B., *Chem. Eng. Sci.*, **3**, 248, (1954); **4**, 290, (1955).

Lewis, W. K., *Ind. Eng. Chem.*, **8**, 825, (1916).

Maxwell, J. B., *Data Book on Hydrocarbons*, Van Nostrand, Princeton, (1955), p. 39.

Murdoch, R., and H. R. C. Pratt, *Trans. Inst. Chem. Eng. (Lond.)*, **31**, 307, (1953).

Null, H. R., *Phase Equilibrium in Process Design*, Wiley-Interscience, New York, (1970).

Richards, G. M., G. A. Ratcliff, and P. V. Danckwerts, *Chem. Eng. Sci.*, **19**, 325–328, (1964).

Ruckenstein, E., *Chem. Eng. Sci.*, **23**, 363–371, (1968).

Smith, B. D., *Design of Equilibrium Stage Processes*, McGraw-Hill, New York, (1963), Chapters 1 and 2.

Smith, J. M., and H. C. Van Ness, *Introduction to Chemical Engineering Thermodynamics*, McGraw-Hill, New York, (1959), Chapter 12.

Tavares da Silva, A., and Danckwerts, P. V., paper presented at Tripartite Chem. Eng. Conf., Montreal, Symp. on Mass Transfer with Chem. React., (1968).

Toor, H. L., and J. M. Marchello, *A I Ch. E. J.*.,**4**, 97–101, (1958).

Treybal, R. E., L. D. Weber, and J. F. Daley, *Ind. Eng. Chem.*, **38**, 817, (1946).

Venkataratnam, A., R. J. Rao, and C. V. Rao, *Chem. Eng. Sci.*, **7**, 102–110, (1957).

Vivian, J. E., and C. J. King, *A. I. Ch. E. J.*, **10**, 221, (1964).

Wasan, D. T., and M. S. Ahluwalia, *Chem. Eng. Sci.*, **24**, 1535–1542, (1969).

Whitman, W. G., *Chem. and Met. Eng.* **29**, 147, (1923).

Wilson, E. D., and H. C., Ries, *Principles of Chemical Engineering Thermodynamics*, McGraw-Hill, New York, (1956), Chapter 13,

5

Mass Transfer in Laminar Flow

Some engineering situations arise in which mass transfer occurs through a fluid that is in laminar flow. These include evaporation, sublimation, or condensation on a plate or on the inner surface of a tube. In other cases the surface of the plate or tube may sublime or dissolve and then diffuse through the fluid. Alternatively, the solute may diffuse or be injected through perforations in the wall.

The final sections of both this chapter and the next review many of the published studies on transfer with a high mass flux. Much—though not all — of the literature on this subject expresses concentration in terms of mass per unit volume or mass fraction. It has therefore been decided to adopt these units here in order to facilitate further literature study of this important topic by the reader. A consistent treatment accordingly requires the use of these units throughout Chapters 5 and 6.

The choice of units is of considerable interest when one is concerned simultaneously with analytical expressions for mass transfer and fluid mechanics; attention is directed to the following somewhat lengthy quotation from Spalding (1962):

> Chemical engineers have long used molal fluxes and concentrations. The two advantages are: simplified arithmetic in stoichiometric calculations; and a simple solution to the Stefan-flow problem when the fluid is an isothermal binary ideal-gas mixture. Aeronautical engineers tend to use mass units. Other unit systems have been employed by particular specialists; for example, the mass of steam per unit mass of dry air is often used in air-conditioning calculations.

If a uniform procedure is to be adopted for all mass-transfer calculations, however, it appears wiser to the present author [Spalding] to adopt both mass fluxes and mass concentrations, for three reasons: when chemical reactions occur within the fluid, the "law of conservation of moles" is not generally obeyed; many substances (e.g. kerosene vapor diffusing into a flame) do not have a recognisable or constant molecular weight; and only *mass* fluxes enter the equations of motion of the fluid, not molal fluxes. It may be that chemical engineers will have to forego the aforesaid advantages of the molal system if they are to make full use of the achievements of boundary-layer theory.

Having elected to express composition in mass rather than molal terms, a choice remains between mass fraction, w_A, and mass concentration, ρ_A. Although the former would allow for density variations with composition (see equations 2.29 to 2.36), the complexity of the resulting differential equations is often prohibitive. Much of this chapter will consequently consider low-mass-flux conditions in constant-property systems and in terms of ρ_A and $\Delta\rho_A$. Procedures for extending the results to high-mass-flux conditions, with and without physical-property variations, are then given in the final section of the chapter.

MASS TRANSFER IN THE LAMINAR BOUNDARY LAYER ON A FLAT PLATE

Consider a fluid—designated component B—which is in laminar flow over a flat plate oriented parallel to the undisturbed stream. Suppose that component A is diffusing from the surface of the plate into the fluid stream. The diffusing component A may originate from sublimation or dissolution of the plate itself, from evaporation or dissolution of liquid held in the pores of the plate surface, or from the injection of a different fluid through perforations in the surface of the plate. This latter procedure is used in transpiration cooling to protect the surface from a hot gas by injection of a different cold gas into the boundary layer. In all these processes a concentration gradient will exist between the plate and the fluid. The thickness of the concentration boundary layer, δ_c, at a given distance x from the leading edge of the plate, is arbitrarily defined as the normal distance from the plate surface at which $\rho_{A0}-\rho_A=0.99(\rho_{A0}-\rho_{A\infty})$, where $\rho_{A\infty}$ is the concentration in the undisturbed stream.

The relative motion in the x direction between the plate and the fluid is zero at the surface, and a velocity gradient normal to the direction of flow extends throughout the fluid because of the drag force resulting from retardation of the fluid at the surface. The thickness of the momentum

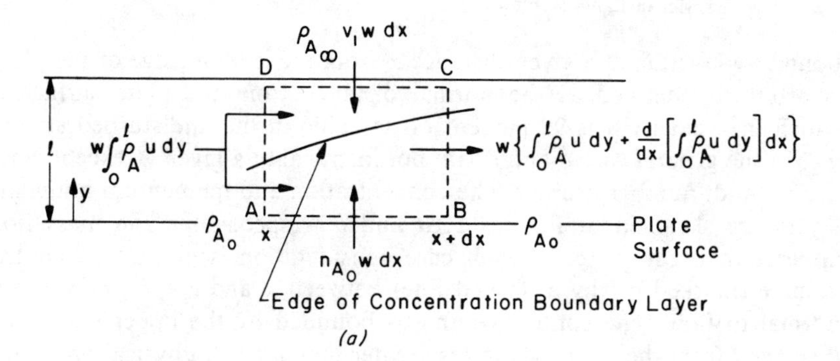

(a)

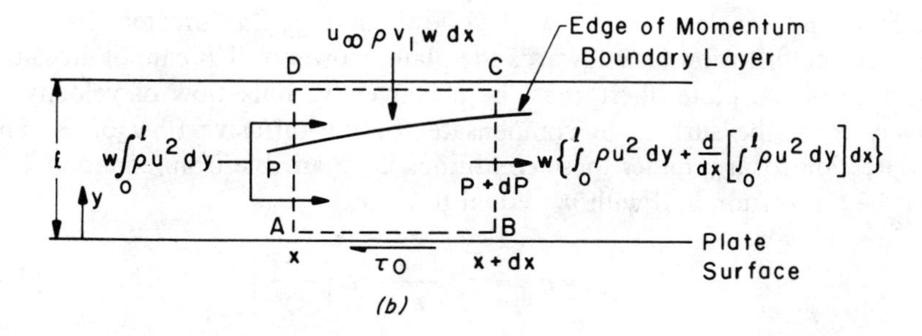

(b)

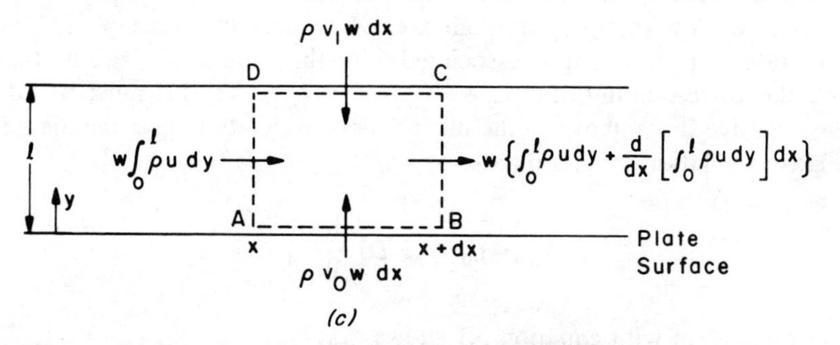

(c)

Figure 5.1. (*a*) Mass flow rates of component A across the boundaries of the control volume $ABCD$; (*b*) Flow rates of x-directed momentum across the boundaries of the control volume $ABCD$; (*c*) Mass flow rates across the boundaries of the control volume $ABCD$.

boundary layer, δ, at a given distance x from the leading edge of the plate, is arbitrarily defined as the normal distance from the plate surface at which the velocity u is 99 percent of its value in the undisturbed stream, u_∞. In the general case δ and δ_c are not identical at a given x, except when the Schmidt number is unity. The concentration and momentum boundary layers are sketched in Figures 5.1*a* and *b*, respectively. The mass flow appears in Figure 5.1*c*; in each case consideration is given to a control volume (marked off by a dashed line) between x and $x+dx$, of width w normal to flow. The control volume is bounded by the upper plane at a distance l from the plate, where l is greater than δ or δ_c. Physical properties are assumed to be constant, unchanged by the mass transfer.

Since $\rho_{A\infty}$ is less than ρ_{A0}, it follows that $\rho_{B\infty}$ is greater than ρ_{B0}, promoting diffusion of B towards the plate. However, if B cannot cross the surface of the plate, there must be a convective bulk flow of velocity v_0 away from the surface to compensate for the diffusive flow of B. The expression for n_{A0} under these conditions is obtainable from equation 2.10 in the y direction and with n_{By} equal to zero:

$$(n_{Ay})_{y=0}=n_{A0}=n_{A0}\frac{\rho_{A0}}{\rho}-D\left(\frac{\partial \rho_A}{\partial y}\right)_{y=0} \tag{5.1}$$

The convective bulk flow moving with mass-average velocity v_0 has a volumetric flow rate Av_0 through area A at the plate surface. The mass flow rate of component A associated with this volumetric rate is $Av_0\rho_{A0}$ and the corresponding flux is $Av_0\rho_{A0}/A=v_0\rho_{A0}$. To this must be added the diffusive flux relative to the mass-average velocity to give the mass flux relative to stationary coordinates as

$$n_{A0}=v_0\rho_{A0}-D\left(\frac{\partial \rho_A}{\partial y}\right)_{y=0} \tag{5.2}$$

Comparison with equation 5.1 shows that

$$v_0=\frac{n_{A0}}{\rho}=\frac{\rho_{A0}v_{A0}}{\rho} \tag{5.3}$$

For component B the equation corresponding to 5.2 is as follows, since $n_B=0$:

$$0=v_0\rho_{B0}-D\left(\frac{\partial \rho_B}{\partial y}\right)_{y=0}$$

but

$$\frac{\partial \rho_B}{\partial y} = -\frac{\partial \rho_A}{\partial y}; \quad \rho = \rho_{A0} + \rho_{B0}$$

so that

$$v_0 = \frac{-D}{\rho - \rho_{A0}}\left(\frac{\partial \rho_A}{\partial y}\right)_{y=0} = \frac{-D}{1 - w_{A0}}\left(\frac{\partial w_A}{\partial y}\right)_{y=0} \tag{5.4}$$

Figure 5.1c shows that, for steady flow with constant physical properties,

$$w\frac{d}{dx}\left[\int_0^l \rho u\,dy\right]dx = \rho v_1 w\,dx + \rho v_0 w\,dx \tag{5.5}$$

Therefore

$$v_1 = \frac{d}{dx}\left[\int_0^l u\,dy\right] - v_0 \tag{5.6}$$

From Figure 5.1a,

$$n_{A0} w\,dx = w\frac{d}{dx}\left[\int_0^l \rho_A u\,dy\right]dx - \rho_{A\infty} v_1 w\,dx \tag{5.7}$$

Inserting equation 5.2 for n_{A0}, equation 5.6 for v_1, and noting that $(\rho_{A\infty} - \rho_A) \doteq 0$ for $\delta_c \leqslant y \leqslant l$,

$$\frac{d}{dx}\int_0^{\delta_c}(\rho_{A\infty} - \rho_A)u\,dy = D\left(\frac{\partial \rho_A}{\partial y}\right)_{y=0} + v_0(\rho_{A\infty} - \rho_{A0}) \tag{5.8}$$

The momentum equation for the control volume $ABCD$ equates the total efflux rate of x-directed momentum to the sum of all external forces in the x direction which act on this volume. Thus from Figure 5.1b,

$$\sum g_c dF_x = -g_c\left(\tau_0 w\,dx + \frac{dP}{dx}dx\,wl\right) = w\frac{d}{dx}\left[\int_0^l \rho u^2\,dy\right]dx - u_\infty \rho v_1 w\,dx \tag{5.9}$$

Substituting equation 5.6 for v_1 and rearranging,

$$\frac{d}{dx}\int_0^l (u_\infty - u)u\,dy = \frac{g_c}{\rho}\left(\tau_0 + \frac{dP}{dx}l\right) + v_0 u_\infty$$

Bernoulli's equation may be applied along a streamline in the frictionless flow just outside the momentum boundary layer to show that dP/dx is zero. But dP/dy is effectively zero inside the boundary layer, so that dP/dx is zero in this region also for this special case—namely, a flat plate parallel to u_∞. Furthermore, the definition of δ means that $(u_\infty - u) \doteq 0$ for $\delta \leqslant y \leqslant l$, so that if the fluid is Newtonian,

$$\frac{d}{dx}\int_0^\delta (u_\infty - u)u\,dy = \frac{\mu}{\rho}\left(\frac{\partial u}{\partial y}\right)_{y=0} + v_0 u_\infty \tag{5.10}$$

Equations 5.8 and 5.10 are the integral equations for the concentration or mass-transfer boundary layer and the momentum boundary layer, respectively.

When the concentration of component A is dilute and mass-transfer rates are low, the corresponding concentration gradients are small. Equation 5.4 shows that v_0 is then slight. Attention will be confined to this special but important case, allowing the terms containing v_0 to be neglected in equations 5.2, 5.8, and 5.10.

The development of mass-transfer relationships will require an expression for δ, the thickness of the momentum boundary layer, as a function of x. This is obtained as follows, assuming a cubic polynomial form for the velocity distribution in the boundary layer:

$$u = a + by + cy^2 + dy^3 \tag{5.11}$$

The boundary conditions are

$$y = 0, \quad u = 0, \quad \left(\frac{\partial^2 u}{\partial y^2}\right)_{y=0} = 0$$

$$y = \delta, \quad u = u_\infty, \quad \left(\frac{\partial u}{\partial y}\right)_{y=\delta} = 0$$

so that equation 5.11 becomes

$$\frac{u}{u_\infty} = \frac{3}{2}\left(\frac{y}{\delta}\right) - \frac{1}{2}\left(\frac{y}{\delta}\right)^3 \tag{5.12}$$

Substituting equation 5.12 for u in equation 5.10 and performing the integration and differentiation as indicated,

$$\delta\frac{d\delta}{dx} = \frac{140}{13}\frac{\mu}{\rho u_\infty}$$

and integrating with $\delta=0$ at $x=0$,

$$\frac{\delta}{x}=4.64\left(\frac{\mu}{\rho u_{\infty}x}\right)^{1/2}=4.64N_{\mathrm{Re},x}^{-1/2} \tag{5.13}$$

Next assume a cubic polynomial form for the concentration distribution in the mass-transfer boundary layer, where ρ_{A0} is constant:

$$\rho_A'=\rho_A-\rho_{A0}=a_1+b_1y+c_1y^2+d_1y^3 \tag{5.14}$$

The boundary conditions are

$$y=0,\qquad \rho_A'=\rho_{A0}'=0,\qquad \left(\frac{\partial^2\rho_A'}{\partial y^2}\right)_{y=0}=0$$

$$y=\delta_c,\qquad \rho_A'=\rho_{A\infty}'=\rho_{A\infty}-\rho_{A0},\qquad \left(\frac{\partial\rho_A'}{\partial y}\right)_{y=\delta_c}=0$$

so that equation 5.14 is written as

$$\frac{\rho_A'}{\rho_{A\infty}'}=\frac{3}{2}\left(\frac{y}{\delta_c}\right)-\frac{1}{2}\left(\frac{y}{\delta_c}\right)^3 \tag{5.15}$$

Equations 5.12 and 5.15 are solved for u and ρ_A, respectively, and the results substituted in equation 5.8 with $v_0=0$ to obtain

$$\rho_{A\infty}'u_{\infty}\frac{d}{dx}\int_0^{\delta_c}\left[1-\frac{3}{2}\left(\frac{y}{\delta_c}\right)+\frac{1}{2}\left(\frac{y}{\delta_c}\right)^3\right]\left[\frac{3}{2}\left(\frac{y}{\delta}\right)-\frac{1}{2}\left(\frac{y}{\delta}\right)^3\right]dy$$

$$=D\rho_{A\infty}'\left\{\frac{\partial}{\partial y}\left[\frac{3}{2}\left(\frac{y}{\delta_c}\right)-\frac{1}{2}\left(\frac{y}{\delta_c}\right)^3\right]\right\}_{y=0} \tag{5.16}$$

The use of equation 5.12 for u in the mass-transfer boundary layer involves the assumption that δ_c/δ is less than unity. Then, denoting δ_c/δ by ϕ, equation 5.16 yields

$$\rho_{A\infty}'u_{\infty}\frac{d}{dx}\left[\delta\left(\frac{3\phi^2}{20}-\frac{3\phi^4}{280}\right)\right]=\frac{3D\rho_{A\infty}'}{2\delta_c}=\frac{3D\rho_{A\infty}'}{2\delta\phi} \tag{5.17}$$

If the term $3\phi^4/280$ can be neglected in comparison with $3\phi^2/20$,

$$\phi^3\delta\frac{d\delta}{dx}+2\phi^2\delta^2\frac{d\phi}{dx}=\frac{10D}{u_\infty} \tag{5.18}$$

Equation 5.13 shows that

$$\delta\frac{d\delta}{dx}=10.76\frac{\mu}{\rho u_\infty}, \qquad \delta^2=21.53\frac{\mu x}{\rho u_\infty}$$

and insertion in equation 5.18 leads to

$$\phi^3+\tfrac{4}{3}x\frac{d\phi^3}{dx}=\frac{0.929}{N_{\text{Sc}}}$$

The solution is

$$\phi^3=\frac{0.929}{N_{\text{Sc}}}+Cx^{-3/4}$$

Suppose that mass transfer begins at distance x_0 from the leading edge of the plate ($\delta_c=0=\phi$ at $x=x_0$). This permits the evaluation of C, giving

$$\frac{\delta_c}{\delta}=\phi=0.976N_{\text{Sc}}^{-1/3}\left[1-\left(\frac{x_0}{x}\right)^{3/4}\right]^{1/3}$$

When $x_0=0$,

$$\delta_c\doteq\delta N_{\text{Sc}}^{-1/3} \tag{5.19}$$

and with equation 5.13,

$$\frac{\delta_c}{x}=4.64N_{\text{Re},x}^{-1/2}N_{\text{Sc}}^{-1/3} \tag{5.20}$$

where the factor 0.976 has been replaced by unity in view of the approximation used to obtain equation 5.18. Equation 5.19 shows that

$$\phi\doteq N_{\text{Sc}}^{-1/3} \tag{5.21}$$

For liquids the Schmidt number N_{Sc} is much greater than unity, while for gases it is usually between 0.6 and 1.1 (Foust et al., 1960). The expressions developed will therefore be valid for liquids, while any errors resulting from application to gases should be slight.

A *local* mass-transfer coefficient at the surface may be defined as

$$-k_\rho^* = \frac{n_{A0}}{\rho'_{A\infty}} = -\frac{D}{\rho'_{A\infty}}\left(\frac{\partial \rho'_A}{\partial y}\right)_{y=0} = \frac{-3D\rho'_{A\infty}}{2\rho'_{A\infty}\delta_c} \tag{5.22}$$

and inserting equation 5.20 for δ_c,

$$(N_{\mathrm{Sh}})_x^* = \frac{k_\rho^* x}{D} = 0.323 N_{\mathrm{Re},x}^{1/2} N_{\mathrm{Sc}}^{1/3} \tag{5.23}$$

where $(N_{\mathrm{Sh}})_x^*$ is the local Sherwood number. The mean mass-transfer coefficient over the range $0 \leqslant x \leqslant L$ is calculated as

$$k_{\rho m}^* = \frac{\int_0^L k_\rho^* \, dx}{\int_0^L dx}$$

Substitution for k_ρ^* from equation 5.23 leads to a mean Sherwood number as follows:

$$(N_{\mathrm{Sh}})_m^* = \frac{k_{\rho m}^* L}{D} = 0.646 N_{\mathrm{Re},L}^{1/2} N_{\mathrm{Sc}}^{1/3} \tag{5.24}$$

The above expressions apply under the assumptions stated, provided that flow in the boundary layer is laminar. Transition to a turbulent boundary layer normally begins in the range $3 \times 10^5 \leqslant x_c u_\infty \rho / \mu \leqslant 3 \times 10^6$.

It should be noted that a Blasius-type of "exact" solution to the differential equations of the mass-transfer boundary layer on a flat plate is available as an alternative to the approximate method given above (see, e.g., Welty, Wicks, and Wilson, 1969). The results from the two approaches are so close, however, as to indicate that the simpler integral method shown here may be extended to situations not amenable to solution by the more exact treatment.

Illustration 5.1

A plate of solid naphthalene is oriented at zero incidence to a pure air stream flowing with a velocity of 15 ft/sec at points remote from the surface of the plate. If the air pressure is 14.7 psia and the system is at a temperature of 113°F, calculate the following for comparison with Illustrations 5.2, 6.1, and 6.4.

(a) The average mass transfer coefficient over the first foot of plate length.

(b) The average rate of mass transfer per unit width over the first foot along the plate.

(c) The local mass-transfer coefficient at a point 1 ft from the leading edge of the plate.

(d) The local naphthalene concentration gradient at the plate surface at a distance of 1 ft from the leading edge.

(e) The local thicknesses of the momentum and concentration boundary layers 1 ft from the leading edge of the plate.

SOLUTION. Correlations of the vapor pressure of naphthalene as a function of temperature are provided by W. J. Christian and S. P. Kezios (*A.I.Ch.E.J.*, **5**, 61–68, 1959) and by S. Uno (Ph.D. thesis, Illinois Institute of Technology, Chicago, 1958):

$$\log P_A = 12.198 - (6881/T) \qquad \text{(Christian and Kezios)}$$

(P_A in lb-force/ft^2, T in °R),

$$\log P_A^+ = 11.84528 - (3857/T) \qquad \text{(Uno)}$$

(P_A^+ in mm Hg, T in °K)

Vapor-pressure measurements were also made on crystal reagent naphthalene at various temperatures by C. H. Bedingfield and T. B. Drew (*Ind. Eng. Chem.*, **42**, 1164–1173, 1950). For a naphthalene vapor pressure of 1.546 lb-force/ft^2, the above three sources indicate corresponding temperatures of 113, 114.1, and 115.34°F, respectively.

These results show that, *for the temperature region under consideration (ca. 113°F)*, uncertainties in the temperature corresponding to a given vapor pressure of naphthalene probably exceed the depression of the surface temperature caused by sublimation. [The experiments of Christian and Kezios (1959) and Bedingfield and Drew (1950) on this system indicate that the temperature depression here is of the order of 1°F.] Additional inaccuracies in the equations for heat- and mass-transfer coefficients prevent reliable evaluation of such a small temperature depression by the methods of Illustration 5.8. The naphthalene surface temperature is accordingly taken as 113°F. Such an assumption may be unacceptable in some cases, however, as shown by Illustration 5.8.

SOLUTION (a). At 113°F and 14.7 psia,

$$\rho_{\text{air}} = \frac{14.7(144)29}{1545(573)} = 0.06935 \text{ lb-mass/ft}^3$$

(Decimal rounding has been avoided, for the purposes of Illustration 5.2.)

$$\mu_{\text{air}}\ (113°\text{F, 1 atm}) = 0.0457 \text{ lb-mass/ft hr.}$$

From Illustration 3.1, since 113°F = 45°C,

$$D = 0.0687(3.88) = 0.2665 \text{ ft}^2/\text{hr}.$$

The Reynolds number at $L = 1$ ft is

$$N_{\text{Re},L} = \frac{1(15)(3600)(0.06935)}{0.0457} = 82{,}000$$

which corresponds to laminar flow in the boundary layer. Substituting in equation 5.24,

$$k^*_{\rho m} = 0.646\left(\frac{0.2665}{1}\right)(82{,}000)^{1/2}\left[\frac{0.0457}{0.06935(0.2665)}\right]^{1/3} = 66.8 \text{ ft/hr}$$

SOLUTION (b). Since the gas stream is pure air, $\rho_{A\infty} = 0$, and ρ_{A0} is obtained from the ideal-gas law using the saturation vapor pressure of naphthalene at 113°F. The correlation by Christian and Kezios gives

$$\log P_A = 12.198 - \frac{6881}{573} = 0.1893$$

$$P_A = 1.546 \text{ lb-force/ft}^2$$

$$\rho_{A0} = \frac{P_A M}{RT} = \frac{1.546(128.16)}{1545(573)} = 0.000224 \text{ lb-mass/ft}^3$$

$$q_{AW} = k^*_{\rho m}(1\times 1)(\rho_{A0} - \rho_{A\infty}) = 66.8(1\times 1)(0.000224 - 0)$$

$$= 0.0149 \text{ lb naphthalene/hr}.$$

SOLUTION (c) Equations 5.23 and 5.24 show that, for a given x,

$$k^*_\rho = k^*_{\rho m}/2 = 66.8/2 = 33.4 \text{ ft/hr}.$$

SOLUTION (d). From equation 5.22,

$$\left(\frac{\partial \rho_A}{\partial y}\right)_{y=0} = \frac{k^*_\rho(\rho_{A\infty} - \rho_{A0})}{D} = \frac{33.4(0 - 0.000224)}{0.2665}$$

$$= -0.028 \text{ lb naphthalene/ft}^4$$

SOLUTION (e). Equation 5.13 shows that, at a distance of 1 ft from the leading edge,

$$\delta = \frac{4.64(1)}{(82{,}000)^{1/2}} = 0.0162 \text{ ft}$$

Combining this with equation 5.19,

$$\delta_c = 0.0162\left[\frac{0.0457}{0.06935(0.2665)}\right]^{-1/3} = 0.012 \text{ ft}$$

MASS TRANSFER IN LAMINAR NATURAL CONVECTION ON A VERTICAL PLATE

Density differences may be associated with variation in concentration from point to point in a fluid. Buoyancy forces are consequently present which lead to free or natural convection flows. The influence of such natural convection on mass transfer may be considerable, particularly in the absence of any forced convection. Attention will here be confined to natural convection arising from gravitation, although it may also occur under centrifugal effects or in an electrically conducting fluid exposed to a magnetic field.

It is assumed that solute concentrations are low and the concentration distribution is such that the changes in density are small in relation to the density itself. This enables the density to be regarded as constant when it is not introduced as a difference.

Consider a flat vertical plate where the coordinate x is measured from the leading edge and the distance y is measured from and normal to the plate surface. The concentration of solute (component A) is constant at ρ_{A0} along the surface of the plate and is $\rho_{A\infty}$ at points remote from the plate. Physically the situation resembles that obtained by rotating the three parts of Figure 5.1 counterclockwise through 90°, setting u_∞ equal to zero, and introducing a term for the gravitational force into equation 5.9. The x-momentum equation for the control volume $ABCD$ then becomes

$$\sum g_c dF_x = -\left(g_c \tau_0 w\, dx + g_c \frac{dP}{dx}\, dx\, wl + w\, dx\, g \int_0^l \rho\, dy\right)$$

$$= w\frac{d}{dx}\left[\int_0^l \rho u^2\, dy\right] dx \qquad (5.25)$$

Now

$$\frac{dP}{dy}=0; \quad g_c\frac{dP}{dx}=-\rho_\infty g \tag{5.26}$$

and at low solute concentrations a volume expansion coefficient β_e may be defined as

$$\beta_e=-\frac{1}{\rho}\frac{(\rho-\rho_\infty)}{(\rho_A-\rho_{A\infty})} \tag{5.27}$$

Also, for Newtonian fluids,

$$g_c(\tau_{yx})_{y=0}=g_c\tau_0=\mu\left(\frac{\partial u}{\partial y}\right)_{y=0} \tag{5.28}$$

The concentration and momentum boundary layers are considered to have equal thickness at a given x in natural convection. This is because density differences (giving rise to convective motion) can exist only where concentration differences are present. Accordingly, for $y>\delta$, $\rho_A=\rho_{A\infty}$, $u=0$. These considerations, together with equations 5.25 to 5.28 and the assumption that $\rho/\rho_\infty \doteq 1$, lead to

$$\frac{d}{dx}\int_0^\delta u^2\,dy=-\frac{\mu}{\rho}\left(\frac{\partial u}{\partial y}\right)_{y=0}+\beta_e g\int_0^\delta(\rho_A-\rho_{A\infty})\,dy \tag{5.29}$$

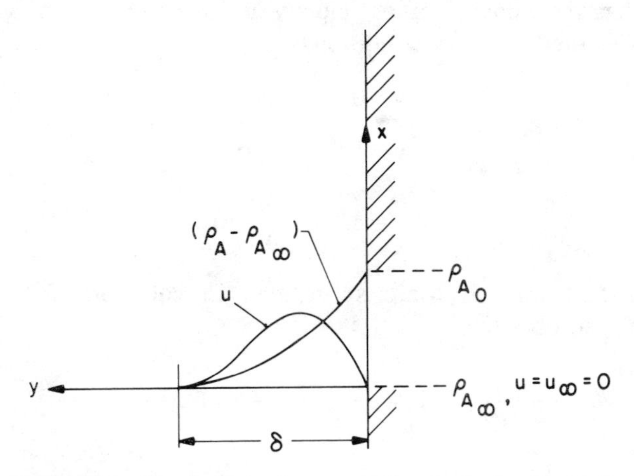

Figure 5.2. Distributions of velocity and concentration in the laminar natural-convection boundary layer on a vertical plate.

The concentration-boundary-layer equation for negligible bulk velocity due to diffusion (v_0) is given by equation 5.8 as

$$\frac{d}{dx}\int_0^{\delta}(\rho_A-\rho_{A\infty})u\,dy=-D\left(\frac{\partial\rho_A}{\partial y}\right)_{y=0} \tag{5.30}$$

The distribution of concentration and velocity in the boundary layer on the plate is as sketched in Figure 5.2. The concentration and velocity profiles will be approximated by the following expressions, which have been used successfully in the heat-transfer analog:

$$\rho_A-\rho_{A\infty}=(\rho_{A0}-\rho_{A\infty})\left(1-\frac{y}{\delta}\right)^2 \tag{5.31}$$

which satisfies the boundary conditions

$$y=0,\qquad \rho_A-\rho_{A\infty}=\rho_{A0}-\rho_{A\infty}$$

$$y=\delta,\qquad \rho_A-\rho_{A\infty}=0$$

$$y=\delta,\qquad \frac{d\rho_A}{dy}=0$$

and

$$u=\Lambda\frac{y}{\delta}\left(1-\frac{y}{\delta}\right)^2 \tag{5.32}$$

where Λ has the dimensions of velocity and is a function of x. Equation 5.32 satisfies the boundary conditions

$$y=0,\qquad u=0$$

$$y=\delta,\qquad u=0$$

$$v=\delta,\qquad \frac{du}{dy}=0$$

Equations 5.31 and 5.32 are substituted in equations 5.29 and 5.30, respectively, to obtain

$$\frac{1}{105}\frac{d}{dx}(\Lambda^2\delta)=-\frac{\mu}{\rho}\frac{\Lambda}{\delta}+\tfrac{1}{3}\beta_e g(\rho_{A0}-\rho_{A\infty})\delta \tag{5.33}$$

$$\frac{1}{30}(\rho_{A0}-\rho_{A\infty})\frac{d}{dx}(\Lambda\delta)=\frac{2D}{\delta}(\rho_{A0}-\rho_{A\infty}) \tag{5.34}$$

The following functional forms are now assumed:

$$\Lambda = B_1 x^m \tag{5.35}$$

$$\delta = B_2 x^n \tag{5.36}$$

Inserting equations 5.35 and 5.36 into equations 5.33 and 5.34 leads to

$$\frac{2m+n}{105} B_1^2 B_2 x^{2m+n-1} = -\frac{B_1}{B_2}\frac{\mu}{\rho} x^{m-n} + \beta_e g(\rho_{A0} - \rho_{A\infty})\frac{B_2}{3} x^n \tag{5.37}$$

$$\frac{m+n}{30} B_1 B_2 x^{m+n-1} = \frac{2D}{B_2} x^{-n} \tag{5.38}$$

For these equations to be valid for any x, the exponents on x must be the same in each equation, or

$$2m+n-1 = m-n = n$$

$$m+n-1 = -n$$

from which

$$m = \tfrac{1}{2}, \qquad n = \tfrac{1}{4}$$

Substitution in equations 5.37 and 5.38 enables solution for B_1 and B_2:

$$B_1 = 5.17\frac{\mu}{\rho}\left(\frac{20}{21} + \frac{\mu}{\rho D}\right)^{-1/2}\left(\frac{\beta_e g(\rho_{A0} - \rho_{A\infty})}{(\mu/\rho)^2}\right)^{1/2} \tag{5.39}$$

$$B_2 = 3.93\left(\frac{20}{21} + \frac{\mu}{\rho D}\right)^{1/4}\left(\frac{\beta_e g(\rho_{A0} - \rho_{A\infty})}{(\mu/\rho)^2}\right)^{-1/4}\left(\frac{\mu}{\rho D}\right)^{-1/2} \tag{5.40}$$

The boundary-layer thickness at x is then

$$\delta = 3.93 N_{Sc}^{-1/2}(0.952 + N_{Sc})^{1/4}\left(\frac{\beta_e g(\rho_{A0} - \rho_{A\infty})}{(\mu/\rho)^2}\right)^{-1/4} x^{1/4} \tag{5.41}$$

Now

$$\beta_e = -\frac{1}{\rho}\left(\frac{\rho - \rho_\infty}{\rho_A - \rho_{A\infty}}\right) = -\frac{1}{\rho_0}\left(\frac{\rho_0 - \rho_\infty}{\rho_{A0} - \rho_{A\infty}}\right) \tag{5.42}$$

and dividing throughout by x,

$$\frac{\delta}{x}=3.93N_{\mathrm{Sc}}^{-1/2}(0.952+N_{\mathrm{Sc}})^{1/4}N_{\mathrm{Gr}D}^{-1/4} \tag{5.43}$$

where $N_{\mathrm{Gr}D}$ is the Grashof number for mass transfer,

$$N_{\mathrm{Gr}D}=\frac{gx^3}{(\mu/\rho)^2}\left(\frac{\rho_\infty}{\rho_0}-1\right) \tag{5.44}$$

The mass flux at the surface of the plate at a given x is

$$i_{A0}\doteq n_{A0}=-D\left(\frac{\partial\rho_A}{\partial y}\right)_{y=0} \tag{5.45}$$

and evaluating the derivative from equation 5.31,

$$n_{A0}=\frac{2D}{\delta}(\rho_{A0}-\rho_{A\infty})=k_\rho^*(\rho_{A0}-\rho_{A\infty}) \tag{5.46}$$

Combination with equation 5.43 gives

$$(N_{\mathrm{Sh}})_x^*=\frac{k_\rho^* x}{D}=\frac{2x}{\delta}=0.508N_{\mathrm{Sc}}^{1/2}(0.952+N_{\mathrm{Sc}})^{-1/4}N_{\mathrm{Gr}D}^{1/4} \tag{5.47}$$

This expression shows that the local coefficient of mass transfer is inversely proportional to the fourth root of x, the distance from the leading edge of the plate. Accordingly, the mean coefficient over the range $x=0$ to L is

$$k_{\rho m}^*=\frac{1}{L}\int_0^L k_\rho^*\,dx=\tfrac{4}{3}(k_\rho^*)_{x=L} \tag{5.48}$$

Equation 5.32 shows that the maximum velocity is located at $y=\delta/3$ and is given by

$$u_{\max}=\tfrac{4}{27}\Lambda=\tfrac{4}{27}B_1x^m \tag{5.49}$$

or

$$u_{\max}=0.766\frac{\mu}{x\rho}(0.952+N_{\mathrm{Sc}})^{-1/2}N_{\mathrm{Gr}D}^{1/2} \tag{5.50}$$

These developments for natural convection require that the boundary layer be laminar. It is found experimentally that, at least in the case of gases and liquid water, the transition to turbulent flow in the heat-transfer analog occurs in the range

$$10^8 \leqslant N_{\mathrm{Gr}} N_{\mathrm{Pr}} \leqslant 10^{10} \tag{5.51}$$

The mass-transfer analog would be

$$10^8 \leqslant N_{\mathrm{Gr}D} N_{\mathrm{Sc}} \leqslant 10^{10} \tag{5.52}$$

Experimental data on flow transition in such mass-transfer systems are scarce.

Illustration 5.2

A plate of solid naphthalene is suspended vertically in pure air which is free from forced convection. If the air pressure is 14.7 psia and the system is at a temperature of 113°F, calculate the following for comparison with Illustrations 5.1, 6.1, and 6.4:

(a) The average mass-transfer coefficient over the first foot down the plate.
(b) The average rate of mass transfer over the first foot from the upper edge.
(c) The local mass-transfer coefficient at a point 1 ft below the upper edge of the plate.
(d) The local naphthalene concentration gradient at the plate surface at a distance of 1 ft from the upper edge.
(e) The local thicknesses of the momentum and concentration boundary layers 1 ft down from the upper edge of the plate.
(f) The maximum velocity in the boundary layer 1 ft below the upper edge.

SOLUTION. Naphthalene-air mixtures are more dense than pure air at a given temperature. The boundary-layer flow due to natural convection will therefore be downwards. Physically the situation resembles that obtained by rotating the three parts of Figure 5.1 clockwise through 90°, setting u_∞ equal to zero, and introducing a term for the gravitational force into equation 5.9. In this case, however, gravity acts in the positive x direction. Summing forces, noting that $g_c\, dP/dx = \rho_\infty g$, and defining β_e as $(\rho - \rho_\infty)/\rho(\rho_A - \rho_{A\infty})$ leads again to equation 5.29. Equation 5.47 then follows, with

$$N_{\mathrm{Gr}D} = g x^3 \frac{(\rho_0 - \rho_\infty)/\rho_0}{(\mu/\rho)^2}$$

The naphthalene surface temperature will be taken to be 113°F, as explained in Illustration 5.1.

SOLUTION (a). From Illustration 5.1, at 113°F and 1 atm,

$$\rho_{air} = \rho_\infty = 0.06935 \text{ lb-mass/ft}^3, \quad \rho_{A0} = 0.000224 \text{ lb-mass/ft}^3$$

$$\mu_{air} = 0.0457 \text{ lb-mass/ft hr}, \quad P_A = 1.546 \text{ lb-force/ft}^2$$

$$D = 0.2665 \text{ ft}^2/\text{hr}$$

so

$$p_{B0} = 14.7(144) - 1.546 = 2115.254 \text{ lb-force/ft}^2$$

$$\rho_{B0} = \frac{2115.254(29)}{1545(573)} = 0.06929 \text{ lb-mass/ft}^3$$

$$\rho_0 = \rho_{A0} + \rho_{B0} = 0.069514 \text{ lb-mass/ft}^3$$

$$N_{GrD} = \frac{4.17(10^8)(1^3)(0.069514 - 0.06935)(0.06935)^2}{0.069514(0.0457)^2} = 2.215 \times 10^6$$

$$N_{Sc} = \frac{0.0457}{0.06935(0.2665)} = 2.475$$

$$N_{GrD} N_{Sc} = 5.49 \times 10^6$$

which corresponds to laminar flow in the boundary layer. Substituting in equation 5.47,

$$(k_\rho^*)_{x=1\text{ ft}} = 0.508\left(\frac{0.2665}{1}\right)(2.475)^{1/2}(0.952 + 2.475)^{-1/4}(2.215 \times 10^6)^{1/4}$$

$$= 6.04 \text{ ft/hr}$$

From equation 5.48,

$$k_{\rho m}^* = \tfrac{4}{3}(6.04) = 8.05 \text{ ft/hr}$$

SOLUTION (b). For a plate width of 1 ft,

$$q_{AW} = k_{\rho m}^*(1 \times 1)(\rho_{A0} - \rho_{A\infty}) = 8.05(1 \times 1)(0.000224 - 0)$$

$$= 0.0018 \text{ lb naphthalene/hr.}$$

SOLUTION (c). From solution (a),

$$(k_\rho^*)_{x=1\text{ ft}} = 6.04 \text{ ft/hr}$$

SOLUTION (d).

$$\left(\frac{\partial \rho_A}{\partial y}\right)_{y=0} = \frac{-k_\rho^*(\rho_{A0} - \rho_{A\infty})}{D} = \frac{-6.04(0.000224 - 0)}{0.2665}$$

$$= -0.00507 \text{ lb naphthalene/ft}^4$$

SOLUTION (e). From equation 5.47,

$$\delta = \delta_c = \frac{2D}{k_\rho^*} = \frac{2(0.2665)}{6.04} = 0.0882 \text{ ft.}$$

SOLUTION (f). Substitution in equation 5.50 gives

$$u_{\max} = 0.766\left[\frac{0.0457}{(1)0.06935}\right](0.952 + 2.475)^{-1/2}(2.215 \times 10^6)^{1/2}$$

$$= 407 \text{ ft/hr} \quad \text{or} \quad 0.113 \text{ ft/sec}$$

This occurs at a normal distance of $y = 0.0882/3 = 0.0294$ ft from the plate.

More refined numerical calculation procedures would reduce the inaccuracies in these figures, occasioned by the small difference between ρ_0 and ρ_∞.

MASS TRANSFER BETWEEN TWO IMMISCIBLE COCURRENT STREAMS IN LAMINAR FLOW

When two immiscible fluids are in cocurrent flow, the shear stresses resulting from their relative motion cause a velocity distribution near the interface which affects the rate of mass transfer between the two phases. This problem was considered by Potter (1957) for the case of laminar boundary layers on either side of a plane, stable interface. The hydrodynamic situation is represented in Figure 5.3.

The upper and lower streams have uniform velocities $u_{\infty 1}$ and $u_{\infty 2}$, respectively, and are brought into contact at O; coordinates are taken as shown, and the interface is a plane at $y = 0$. The laminar momentum

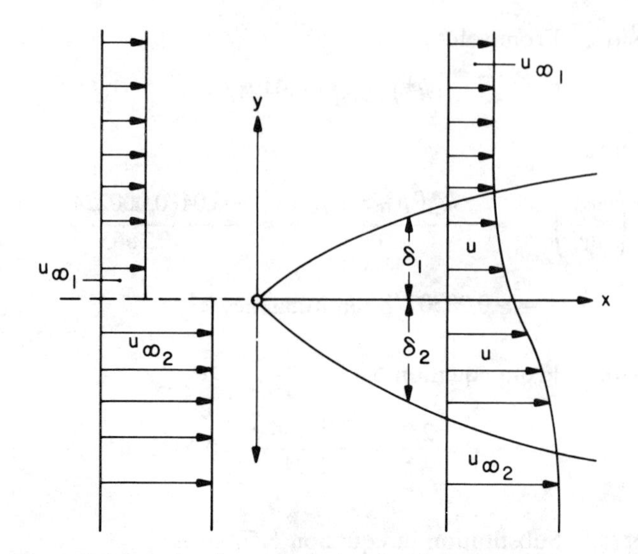

Figure 5.3. Momentum boundary layers at the interface between two cocurrent streams.

boundary layers on either side of the interface have thicknesses δ_1 and δ_2. Equation 5.8 was applied to each mass-transfer boundary layer under conditions of negligible v_0 and assuming quartic polynomial distributions for both velocity and concentration. The approximate solutions for local k_ρ^* values in phases 1 and 2 are as follows for $N_{\mathrm{Sc}} > 1$:

$$(N_{\mathrm{Sh}1})_x^* = \frac{k_{\rho 1}^* x}{D} = \left(\frac{74 + 115(u_0/u_{\infty 1})}{630}\right)^{1/2} \left(\frac{x u_{\infty 1} \rho_1}{\mu_1}\right)^{1/2} N_{\mathrm{Sc}1}^{p} \tag{5.53}$$

$$(N_{\mathrm{Sh}2})_x^* = \frac{k_{\rho 2}^* x}{D} = \left(\frac{74(u_{\infty 2}/u_{\infty 1}) + 115(u_0/u_{\infty 1})}{630}\right)^{1/2} \left(\frac{x u_{\infty 1} \rho_2}{\mu_2}\right)^{1/2} N_{\mathrm{Sc}2}^{p'} \tag{5.54}$$

where x is the distance from the initial point of contact of the streams, and p and p' are functions of $u_{\infty 2}/u_{\infty 1}$ and $\mu_2\rho_2/\mu_1\rho_1$. Evaluation of the interfacial velocity u_0 is described in the original paper. In the limiting case of flow over a flat plate, $u_0 = 0$, $p = \frac{1}{3}$, and equation 5.53 reduces to equation 5.23—with a minor change in the constant because of the use of a quartic rather than a cubic polynomial distribution for u and ρ_A.

When $u_0 = u_{\infty 1}$ the upper fluid is flowing without shear; p then assumes the value of $\frac{1}{2}$, and equation 5.53 reduces to

$$k^*_{\rho 1}=0.548\left(\frac{D_1 u_{\infty 1}}{x}\right)^{1/2}=0.548\left(\frac{D_1}{t}\right)^{1/2} \tag{5.55}$$

where the time that the phases have been in contact at x is $t=x/u_{\infty 1}$. Equation 5.55 is within 3 percent of the exact value given earlier by the "penetration theory" for unsteady-state molecular diffusion (equation 4.30):

$$k^*_{\rho 1}=\left(\frac{D_1}{\pi t}\right)^{1/2} \doteq 0.564\left(\frac{D_1}{t}\right)^{1/2} \tag{5.56}$$

Potter (1957) evaluated p, the exponent on the Schmidt group, for a wide range of $u_0/u_{\infty 1}$ and assuming $N_{Sc1}=100$. The results appear in Figure 5.4 and were considered to be accurate to about 5 percent over the range of Schmidt numbers for liquids.

The treatment is restricted to low rates of mass transfer. Furthermore, when N_{Sc} is high, as in the case of liquids, $\delta_c \ll \delta$. Velocities are then relatively constant in the concentration boundary layers, and comparison between equations 5.55 and 5.56 suggests that the penetration theory will

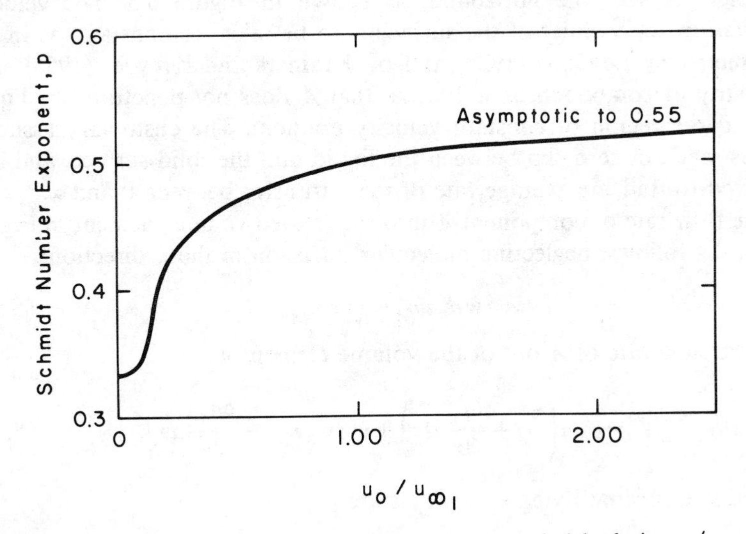

Figure 5.4. Exponent on Schmidt number versus relative interfacial velocity, $u_0/u_{\infty 1}$ (Potter, 1957).

provide a good estimate of the local mass-transfer coefficients, if t is calculated as x/u_0.

The analysis is confined to stable interfaces; unfortunately the criteria for the instability of a horizontal interface between two liquids are not firmly established.

MASS TRANSFER IN A FALLING LIQUID FILM IN LAMINAR FLOW

The situation to be analyzed consists of a constant-property liquid film falling in steady laminar flow down a flat solid surface. The free-falling surface of the film is adjacent to a gas phase. Mass transfer may occur either between the solid surface and the film or between the film and the gas phase. Each possibility will be examined in turn, for conditions of low solute concentrations and mass-transfer rates, such that the bulk velocity due to diffusion may be neglected.

Mass Transfer between an Inclined Plate and a Falling Liquid Film

The following situation is to be studied. A flat plate of width w has an insoluble surface for x less than zero, and for all x greater than zero the plate surface consists of material A with solubility ρ_A^*. A film of liquid is in steady laminar flow in the x direction down the plate, which is inclined at an angle α with the horizontal, as shown in Figure 5.5. The velocity gradient in the vicinity of the surface is to be taken as constant, as in the treatments by Lévêque (1928) and by Kramers and Kreyger (1956). The solubility of component A is low, so that A does not penetrate the liquid beyond the region of constant velocity gradient. The customary assumption is made of zero slip between the liquid and the solid surface, and it is required to find the average rate of mass transfer between 0 and x.

The flow rate of component A into the shaded volume element in Figure 5.5a is as follows, neglecting molecular diffusion in the x direction:

$$w\,dy\,u\rho_A + w\,dx\,n_{Ay} \tag{5.57}$$

and the flow rate of A out of the volume element is

$$w\,dy\,u\left(\rho_A + \frac{\partial \rho_A}{\partial x}dx\right) + w\,dx\left(n_{Ay} + \frac{\partial n_{Ay}}{\partial y}dy\right) \tag{5.58}$$

Equating and simplifying at steady state,

$$u\frac{\partial \rho_A}{\partial x} = -\frac{\partial n_{Ay}}{\partial y} \tag{5.59}$$

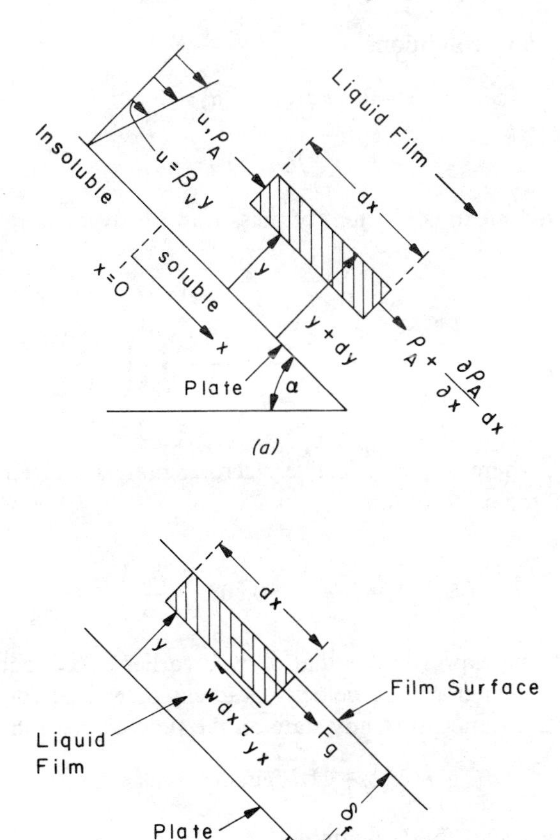

Figure 5.5. (*a*) Enlarged sketch of a falling liquid film near the surface of an inclined plate; (*b*) Forces acting on an element of liquid film falling down an inclined plate.

but n_{Ay}, the flux of component A in the y direction, is given by

$$n_{Ay} = -D\frac{\partial \rho_A}{\partial y} \tag{5.60}$$

so that the differential equation for this process is

$$\beta_v y \frac{\partial \rho_A}{\partial x} = D\frac{\partial^2 \rho_A}{\partial y^2} \tag{5.61}$$

with the boundary conditions

$$x=0, \quad y \geqslant 0, \qquad \rho_A = 0$$

$$x>0, \quad y=0, \qquad \rho_A = \rho_A^*$$

An integrated mean coefficient of mass transfer over the range 0 to x is defined by

$$k_{\rho m}^* = \frac{\int_0^x k_\rho^* \, dx}{\int_0^x dx} = -\frac{D}{(\rho_A^* - \rho_{A\infty})x} \int_0^x \left(\frac{\partial \rho_A}{\partial y} \right)_{y=0} dx \tag{5.62}$$

For this case, where $\rho_{A\infty}$ is effectively zero, Kramers and Kreyger (1956) solved equation 5.61 to obtain

$$(N_{\mathrm{Sh}})_m^* = \frac{k_{\rho m}^* x}{D} = 0.808 x \left(\frac{\beta_v}{Dx} \right)^{1/3} \tag{5.63}$$

which is the same expression as that derived earlier by Lévêque (1928) for the heat-transfer analog. The velocity gradient β_v will be found from the following force balance at steady state on the fluid element in Figure 5.5*b*:

$$dF_g g_c = w(\delta_f - y)\, dx\, \rho g \sin\alpha = w\, dx\, \tau_{yx} g_c \tag{5.64}$$

and for a Newtonian fluid, assuming $\rho, \mu \neq f(\rho_A)$,

$$(\delta_f - y)\rho g \sin\alpha = \mu \frac{du}{dy} \tag{5.65}$$

Integrating,

$$u = \frac{\rho g \sin\alpha}{\mu} \left(\delta_f y - \frac{y^2}{2} \right) \tag{5.66}$$

The volumetric rate of flow down the plate is given by

$$Q = w \int_0^{\delta_f} u\, dy = \frac{1}{3\mu} w \rho g \sin\alpha\, \delta_f^3 \tag{5.67}$$

so that the film thickness is

$$\delta_f = \left(\frac{3\mu Q}{w \rho g \sin\alpha} \right)^{1/3} \tag{5.68}$$

From equations 5.65 and 5.68,

$$\beta_v = \left(\frac{du}{dy}\right)_{y=0} = \left[N_{\text{Ref}} \times \frac{3}{4}\frac{\rho(g\sin\alpha)^2}{\mu}\right]^{1/3} \tag{5.69}$$

where

$$N_{\text{Ref}} = \frac{4Q\rho}{w\mu} \tag{5.70}$$

Combining equations 5.63 and 5.69 gives

$$(N_{\text{Sh}})_m^* = 0.783 N_{\text{Ref}}^{1/9} N_{\text{Sc}}^{1/3} \left(\frac{x^3\rho^2 g\sin\alpha}{\mu^2}\right)^{2/9} \tag{5.71}$$

Equation 5.71 will apply for a film of Newtonian liquid, provided that two restrictions are fulfilled:

1. The film is in laminar flow, so that equations 5.66 and 5.68 are valid.
2. Solute does not penetrate the film beyond the region in which the velocity distribution may be assumed to be linear.

With regard to the first restriction, Dukler and Bergelin (1952) found that, even at N_{Ref} $(=4Q\rho/w\mu)$ of 1000, δ_f exceeds the value given by equation 5.68. In contrast, Jackson (1955) found that for liquids with viscosities up to that of water, equation 5.68 describes the liquid film falling down a vertical wall up to N_{Ref} of at least 4000. Many other investigators have found equation 5.68 to be applicable up to N_{Ref} of about 2000, although even at low N_{Ref} the velocity distribution may deviate from equation 5.66 because of rippling.

Considering both restrictions, Kramers and Kreyger (1956) found "fairly good agreement" between their experimental values and those predicted from equation 5.71 for $N_{\text{Ref}} \equiv 4Q\rho/w\mu < 2000$. Their measurements were on the dissolution of benzoic acid plates in falling water films, for the conditions $5 \leqslant x \leqslant 80$ mm, an entrance length upstream of the benzoic acid surface of 330 mm, and angles α of 6, 45, and 80 degrees.

In many cases the length of soluble surface in the direction of flow, the velocity of fall, or other conditions may be such that solute penetrates the film beyond the region for which $u = \beta_v y$. The appropriate differential equation is obtained by inserting equation 5.66 for $\beta_v y$ into equation 5.61, to give

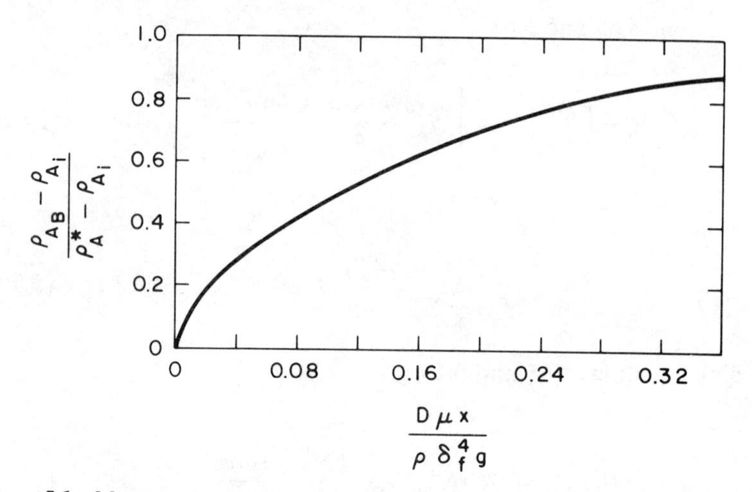

Figure 5.6. Mass transfer from a wall to a falling liquid film in laminar flow (Nusselt, 1923).

$$\frac{\rho g \sin\alpha}{\mu}\left(\delta_f y - \frac{y^2}{2}\right)\frac{\partial \rho_A}{\partial x} = D\frac{\partial^2 \rho_A}{\partial y^2} \tag{5.72}$$

with boundary conditions as before. Nusselt (1923) solved this equation by a difference method for a vertical surface and expressed his results as shown in Figure 5.6.

Experimental results frequently show mass-transfer rates that are substantially higher than predicted by these relationships based on a laminar-flow parabolic velocity distribution. Stirba and Hurt (1955), for example, found "considerably greater" rates of transfer than the theoretical values down to N_{Ref} of 300 using surface lengths of 30 to 75 in. These authors summarize many findings of this kind, which are attributed to eddying and turbulence within the film, associated with rippling of the free-falling surface. It is notable that, according to Stirba and Hurt (1955), this apparently occurs even though the overall thickness of the film may coincide with that predicted from laminar-flow considerations.

Illustration 5.3

A storage vessel formerly used to hold stearic acid has been converted to contain ethanol. Intensive cleaning was undertaken before this new application, in order to avoid contamination with stearic acid. After filling with ethanol, however, it was realized that the inside of a vertical overflow pipe within the vessel had not been cleaned. If the film of solidified stearic acid on the inner wall of the overflow pipe is $\frac{1}{8}$ in. thick, how long would it take to remove this acid with a steady ethanol overflow rate of 4.09 ft^3/hr, and what will be the concentration of stearic acid in the contaminated ethanol?

The system is at a temperature of 77°F, and the overflow pipe has a length of 10 ft and an internal diameter of 3 in. when clean.

Compare the results with those in Illustration 6.2.

SOLUTION. The relevant physical properties at the prevailing temperature are as follows:

$$\rho_{C_2H_5OH} = 49.0 \text{ lb-mass/ft}^3$$

$$\mu_{C_2H_5OH} = 2.66 \text{ lb-mass/(ft)(hr)}$$

$$D = 2.48 \times 10^{-5} \text{ ft}^2/\text{hr} \quad \text{(from Stirba and Hurt, 1955)}$$

$$\rho_A^* = \text{saturation solubility} = 0.814 \text{ lb } C_{17}H_{35}COOH/\text{ft}^3 \text{ of solution in ethanol}$$

(from Stirba and Hurt, 1955)

The density and viscosity will be regarded as constant in view of the low solubility of stearic acid in ethanol.

According to J. A. Tallmadge and C. Gutfinger [*Ind. Eng. Chem.* **59**, No. 11, 19–34, (1967)], thin-film flow down a vertical cylindrical surface is hydrodynamically equivalent to that down a vertical flat plate when the Goucher number N_{Go} is greater than 3. In the present case, using the initial or minimum radius of the coated pipe and cgs units,

$$N_{Go} = r_t \left(\frac{\rho g}{2\sigma} \right)^{1/2} = 3.49 \left[\frac{(0.785)981}{2(21.85)} \right]^{1/2} = 14.65$$

where σ is the surface tension. Evidently the thin-film flow of ethanol down the inner wall of the overflow pipe is equivalent to that on a vertical flat plate.

$$N_{Ref,av} = \frac{4Q\rho}{w_{av}\mu} = \frac{4(4.09)49}{\pi(2.875/12)2.66} = 400$$

where $w_{av} = \pi(d_{t\text{initial}} + d_{t\text{final}})/2$.

Equation 5.71 requires that solute should not penetrate the film beyond the region of linear velocity gradient. It is therefore restricted to short contact times, for which experimental confirmation was obtained by Kramers and Kreyger (1956) and by Oliver and Atherinos (1968). In both investigations the soluble plate length did not exceed about 3 in. in the direction of flow. In the present case, with a soluble length of 10 ft, the use of Figure 5.6, corresponding to equation 5.72, is more appropriate. As noted in the text, however, the application of Figure 5.6 using the molecular diffusivity to evaluate the abscissa results in an underestimate of the rate of transfer because of some degree of turbulence within the film associated with waves and rippling of the free surface. It is therefore proposed to replace D in the abscissa of Figure 5.6 with an "apparent diffusivity," D_A, read from Figure 5.7. This plot represents the correlation obtained by Stirba and Hurt (1955) from the application of Figure 5.6 to measurements of solid dissolution, and equations 5.75 through 5.77 to measurements of gas absorption, in vertically falling liquid films. Thus for $N_{\text{Re}f,\text{av}}$ of 400, Figure 5.7 gives

$$D_A = 13 \times 10^{-5} \text{ ft}^2/\text{hr}$$

which exceeds the molecular diffusivity for this system by a factor of 5.24. From equation 5.68,

$$\delta_{f,\text{av}} = \left[\frac{3(2.66)4.09}{\pi(2.875/12)(49)(4.17\times 10^8)} \right]^{1/3} = 0.001285 \text{ ft}$$

and

$$\frac{D_A \mu x}{\rho \delta_{f,\text{av}}^4 g} = \frac{(13\times 10^{-5})(2.66)10}{49(0.001285)^4(4.17\times 10^8)} = 0.0622$$

The corresponding ordinate of Figure 5.6 shows that

$$\frac{\rho_{AB} - \rho_{Ai}}{\rho_A^* - \rho_{Ai}} = 0.36 = \frac{\rho_{AB} - 0}{0.814 - 0}$$

so

$$\rho_{AB} = 0.293 \text{ lb } C_{17}H_{35}COOH/\text{ft}^3 \text{ of solution in ethanol.}$$

This will be the average concentration of stearic acid dissolved in the ethanol leaving the overflow pipe. The average dissolution rate is estimated as $4.09(0.293) = 1.2$ lb $C_{17}H_{35}COOH$/hr.

Taking the density of stearic acid to be 53 lb-mass/ft^3, the amount initially present is

$$\frac{\pi}{4}\left(\frac{3^2 - 2.75^2}{144}\right)10(53) = 4.16 \text{ lb } C_{17}H_{35}COOH$$

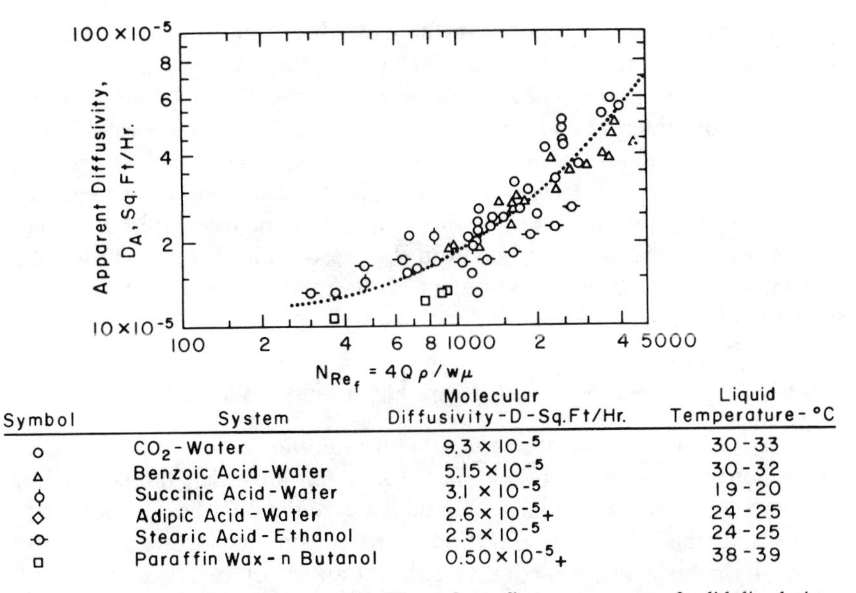

Figure 5.7. Apparent diffusivities causing Figure 5.6 to fit measurements of solid dissolution, and equations 5.75 to 5.77 to fit measurements of gas absorption, in vertically falling liquid films (Stirba and Hurt, 1955).

The time required for the removal of this acid by dissolution in the falling film of ethanol is therefore estimated to be

$$\frac{4.16}{1.2} = 3.47 \text{ hr}$$

Small amounts ($\leqslant 0.05$ percent) of a wetting agent were used in some—though not all—of the runs with the organic-acid–water systems in preparing Figure 5.7. Furthermore, the presence of a free surface means that the characteristics of a failing-film flow may depend on the Reynolds, Weber, and Froude numbers of the flow, rather than on the Reynolds number alone (Fulford, 1964). These considerations suggest the need for further study before Figure 5.7 is accepted as having general validity. In any event, it will be appreciated that the transport properties of a falling liquid film *may* be influenced by variations in velocity of the adjacent gas phase. Hikita et al. [*Chem. Eng. (Tokyo)* **23**, 459, (1959)] found no effect of gas velocity on mass transfer in the film for gas Reynolds numbers $N_{\mathrm{Re},g}$ below 7000. Oliver and Atherinos (1968), however, found some effect of countercurrent gas velocity on liquid-film absorption rates at gas Reynolds numbers of 6000, although the influence up to $N_{\mathrm{Re},g} = 32{,}000$ was described as "modest." According to Stirba (private communication, April 7,

1971) the gas Reynolds number in the CO_2 absorption runs of Figure 5.7 was about 700 (CO_2 velocity $\doteq$ 0.75 ft/sec at 1 atm and 30°C).

The calculation presented here has neglected the effects of the entrance length within which the velocity profile of the falling film is established. Interpolation between the measurements of entrance length made by S. R. Tailby and S. Portalski [*Chem. Eng. Sci.* **17**, 283–290, (1962)] for methanol and isopropanol in vertical film flow suggests that, for a film Reynolds number of 400, the entrance length in the present case is of the order of 3 to 4 in.—that is, less than 3.5 percent of the total length over which mass transfer occurs.

Mass Transfer between a Gas Phase and a Falling Liquid Film

Consider now a situation that could be of importance, for example, in gas absorption in a wetted-wall column for a system in which the controlling resistance is in the liquid phase. For such a case it would be appropriate to assume that the solute concentration at the free-falling surface is constant at ρ_A^*. It is further assumed that concentrations of A are low, so that diffusional velocities normal to the wall are effectively zero, and that diffusion in the x direction is negligible compared to transfer by convective motion. Consequently, for a parabolic velocity distribution in the film, the appropriate differential equation is conveniently written as

$$u_{max}\left[1-\left(\frac{\delta_f-y}{\delta_f}\right)^2\right]\frac{\partial \rho_A}{\partial x}=D\frac{\partial^2 \rho_A}{\partial(\delta_f-y)^2} \tag{5.73}$$

where u_{max} is the velocity of the free-falling surface, shown by equation 5.66 to be

$$u_{max}=\frac{\rho g \sin\alpha\, \delta_f^2}{2\mu} \tag{5.74}$$

The inlet concentration of the liquid is constant at ρ_{Ai}, and Johnstone and Pigford (1942) provide the following solution to equation 5.73 for a vertical film:

$$\frac{\rho_{AB}-\rho_A^*}{\rho_{Ai}-\rho_A^*}=\sum_{j=1}^{\infty}\alpha_j e^{-\beta_j \Upsilon} \tag{5.75}$$

where the constants for the first four terms are given below:

j	α_j	β_j
1	0.7857	5.1213
2	0.1001	39.318
3	0.03599	105.64
4	0.01811	204.75

and

$$\Upsilon = \frac{Dx}{\delta_f^2 u_{\max}} = 2.9351\left(\frac{4Q\rho}{w\mu}\right)^{-4/3}\left(\frac{\mu}{\rho D}\right)^{-1}\left(\frac{x^3\rho^2 g}{\mu^2}\right)^{1/3} \tag{5.76}$$

in terms of the same groups as in equation 5.71. For low values of Υ,

$$\frac{\rho_{AB}-\rho_A^*}{\rho_{Ai}-\rho_A^*} = 1 - \frac{3}{\sqrt{\pi}}\sqrt{\Upsilon} \tag{5.77}$$

An average mass-transfer coefficient over the gas-liquid interface may be evaluated in the following manner. A balance on solute A absorbed by the film over the differential distance of fall, dx, is

$$u_{\text{av}} w \delta_f \, d\rho_{AB} = k_\rho^*(\rho_A^* - \rho_{AB}) w \, dx \tag{5.78}$$

where u_{av} is the average velocity in the film, evaluated with the aid of equation 5.66 as

$$u_{\text{av}} = \frac{1}{\delta_f}\left[\frac{\rho g \sin\alpha}{\mu}\int_0^{\delta_f}\left(\delta_f y - \frac{y^2}{2}\right)dy\right] \tag{5.79}$$

Integrating and dividing by equation 5.74 shows that

$$\frac{u_{\text{av}}}{u_{\max}} = \frac{2}{3} \tag{5.80}$$

Equation 5.78 may then be integrated to give the mean Sherwood number as

$$(N_{\text{Sh}})_m^* = \frac{k_{\rho m}^* x}{D} = \frac{2}{3}\frac{u_{\max}\delta_f}{D}\ln\frac{\rho_A^* - \rho_{Ai}}{\rho_A^* - \rho_{AB}} \tag{5.81}$$

At low rates of flow ($N_{Ref}<100$), only the first term of equation 5.75 is significant. Combination with equation 5.81 gives

$$(N_{Sh})^*_m = \frac{2}{3}\frac{u_{max}\delta_f}{D}\ln\frac{e^{5.1213T}}{0.7857}$$

$$=\frac{2}{3}\frac{u_{max}\delta_f}{D}\left[0.241+5.1213\left(\frac{Dx}{\delta_f^2 u_{max}}\right)\right]\doteq 3.41\frac{x}{\delta_f} \qquad (5.82)$$

where δ_f is given by equation 5.68. The coefficient $k^*_{\rho m}$ is used in the following expression for the mass transferred across the total interface:

$$\tfrac{2}{3}u_{max}w\delta_f(\rho_{AB}-\rho_{Ai}) = k^*_{\rho m}wx(\rho^*_A-\rho_{AB})_{\text{“mean”}} \qquad (5.83)$$

Equation 5.81 is solved for $k^*_{\rho m}$ and the result inserted in equation 5.83 to reveal that

$$(\rho^*_A-\rho_{AB})_{\text{“mean”}} = \frac{(\rho^*_A-\rho_{Ai})-(\rho^*_A-\rho_{AB})}{\ln[(\rho^*_A-\rho_{Ai})/(\rho^*_A-\rho_{AB})]} \qquad (5.84)$$

Although results analogous to equation 5.82 are available for high film Reynolds numbers (Sherwood and Pigford, 1952), they are largely of academic interest only. This is because of the appearance of readily visible waves on the film surface when the Reynolds number N_{Ref} attains the order of 20 (Tailby and Portalski, 1960; Fulford, 1964, pp. 186–189). In consequence, substantial increases in mass transfer occur over that predicted for films in smooth laminar flow (Fulford, 1964, pp. 198–200). Stirba and Hurt (1955) summarize many such findings, which are attributed to the increase in the free surface and the partial turbulence within the film associated with the waves.

Illustration 5.4

Water at 86°F flows steadily in a thin film down the inside wall of a vertical, 2-in.-i.d. tube. Pure carbon dioxide flows countercurrently up the tube at an average velocity of 0.25 ft/sec. Estimate the absorption rate for a tube length of 3 ft when the water rate is 4 ft³/hr, the average CO_2 pressure is 1 atm, and the water is initially pure.

SOLUTION. The relevant physical properties are as follows at 86°F:

$$\rho_{H_2O} = 62.2 \text{ lb-mass/ft}^3$$

$$\mu_{H_2O} = 1.935 \text{ lb-mass/ft hr}$$

$$D = 9.3 \times 10^{-5} \text{ ft}^2/\text{hr (from Stirba and Hurt, 1955)}$$

$$\rho_A^* = \text{saturation solubility} = 0.0817 \text{ lb CO}_2/\text{ft}^3 \text{ of aqueous solution}$$

(calculated from Perry, 1963, p. **14**-4)

As described in Illustration 5.3, film flow on a vertical cylindrical wall is equivalent to that on a flat plate when $N_{Go} > 3$. For the present case, using cgs units,

$$N_{Go} = r_t \left(\frac{\rho g}{2\sigma} \right)^{1/2} = 2.54 \left[\frac{(0.996)981}{2(71.18)} \right]^{1/2} = 6.65$$

where σ is the surface tension.

$$N_{Ref} = \frac{4Q\rho}{w\mu} = \frac{4(4)(62.2)}{\pi(2/12)(1.935)} = 982$$

At this Reynolds number the presence of waves and rippling on the surface of the film will enhance the rate of absorption over that which would be calculated from equations 5.75 and 5.76 using the molecular diffusivity. These expressions will therefore be employed with the "apparent diffusivity" D_A read from Figure 5.7 in Illustration 5.3. Thus for $N_{Ref} = 982$, Figure 5.7 gives $D_A = 18 \times 10^{-5}$ ft^2/hr, which exceeds the molecular diffusivity for this system by a factor of 1.936. From equation 5.68,

$$\delta_f = \left[\frac{3(1.935)(4)}{\pi(2/12)(62.2)(4.17 \times 10^8)} \right]^{1/3} = 0.001196 \text{ ft}$$

Equation 5.74 for this case is

$$u_{max} = \frac{(62.2)(4.17 \times 10^8)(0.001196)^2}{2(1.935)} = 9560 \text{ ft/hr}$$

From equation 5.76,

$$\Upsilon = \frac{(18\times10^{-5})(3)}{(0.001196)^2(9560)} = 0.0395$$

Substituting in equation 5.75,

$$\frac{\rho_{AB}-\rho_A^*}{\rho_{Ai}-\rho_A^*} = 0.7857\exp[-5.1213(0.0395)] + 0.1001\exp[-39.318(0.0395)]$$

$$+0.03599\exp[-105.64(0.0395)] + 0.01811\exp[-204.75(0.0395)]$$

$$= 0.664$$

Alternatively, since Υ is small. equation 5.77 gives

$$\frac{\rho_{AB}-\rho_A^*}{\rho_{Ai}-\rho_A^*} = 1 - \frac{3}{\sqrt{\pi}}\sqrt{0.0395} = 0.664$$

and because $\rho_{Ai}=0$,

$$\rho_{AB} = (1-0.664)\rho_A^* = 0.336(0.0817)$$

$$= 0.02745 \text{ lb } CO_2/\text{ft}^3 \text{ of aqueous solution}$$

The absorption rate is accordingly estimated as 4(0.02745) or 0.1098 lb CO_2/hr.

The gas flow conditions here are comparable to those for which Figure 5.7 was established, and therefore modify the estimated absorption rate to a negligible degree.

MASS TRANSFER IN LAMINAR FLOW THROUGH A TUBE

The analogous case of heat transfer to fluids in either laminar or turbulent flow through tubes is of enormously widespread occurrence in an extensive diversity of processing equipment, including condensers, heat exchangers, and evaporators. The study of mass transfer inside tubes may be similarly justified by a large range of applications, some of which are indicated below. Flow may be either laminar or turbulent in many of these examples, depending upon operating conditions.

1. Transpiration and film cooling is used to protect surfaces of combustion chambers and ducts exposed to very hot gases, as in jet engines and rocket motors. In film cooling, streams of gas (or liquid) are blown through

slots in a direction tangential to the wall. The mass flow of coolant away from the wall partially reverses the direction of heat transfer in the vicinity of the solid surface. When a liquid film is used, mass transfer takes the form of evaporation from the liquid surface. A more even film is obtained in transpiration cooling, where the coolant flows through pores in a porous wall. (See, e.g., Knuth, 1954, 1955; Kinney et al., 1952; Hartnett and Eckert, 1957; Zucrow and Sellers, 1961.) In ablation cooling part of the solid surface evaporates into the adjacent gas phase, thereby cooling the rest of the surface by absorption of the latent heat of vaporization and convective transport.

2. Gas-liquid wetted-wall columns have been widely used, particularly in experimental studies on gas absorption (e.g., Vivian and Peaceman, 1956), distillation (e.g., Kaiser, 1961), and humidification (e.g., Cairns and Roper, 1954).

3. Liquid-liquid wetted-wall columns have been investigated during work on extraction (e.g., Murdoch and Pratt, 1953).

4. Falling-film evaporators are employed to concentrate heat-sensitive substances, which flow in thin films down the inside of heated tubes. Evaporation occurs from the film into the gas core (e.g., Moore and Hesler, 1963; Sinek and Young, 1962).

5. Thin-film reactors are utilized to provide close control of highly exothermic reactions, as in the sulfation of liquid alcohols with an SO_3–inert-gas mixture to produce detergent components (e.g., Hurlbert, Knott, and Cheney, 1967). Related reactions in an *annular* reactor are described in U.S. Patent 1,029,029.

6. Corrosion problems are commonplace in pipelines. Allowance for appreciable corrosion may be incorporated in the design thickness of the pipe walls (e.g., Perry 1963, p. **23**–5).

7. Oxygenation and other mass-transfer processes occur during blood flow, both *in vivo* and in artificial blood oxygenators used to exchange CO_2 for O_2 during open-heart surgery (e.g., Landino et al., 1966).

8. Scaling, salting, and fouling of tubes is a progressive mass-transfer phenomenon during the operation of most evaporators (e.g., Perry, 1963, p. **11**–25).

9. Descaling of evaporator tubes with appropriate chemical solutions is often practiced and relies in part upon diffusional and convective mass transfer (e.g., Badger and Banchero, 1955).

10. The condensation of mixed vapors, which may be either partially or totally condensible, is frequently encountered in many industrial and laboratory processes. As an example, the work of Estrin, Hayes, and Drew (1965) may be noted, in which a study was made of the condensation of mixed acetone and toluene vapors in an *annulus*.

11. Desalination of seawater may be performed by reverse osmosis during flow in tubular membranes (Sourirajan, 1970, Chapter 4).

12. Hemodialysis is carried out in the artificial kidney using tubular dialyzers for which optimum dimensions are presented by Wolf and Zaltzman (1968).

The developments to follow will be restricted to low solute concentrations and transfer rates to permit analogy with the corresponding heat-transfer processes.

The differential equation expressing the overall continuity of matter in flow through a tube may be formulated with reference to the annular differential element of volume $2\pi r\,dr\,dx$ shown in Figure 5.8. The flow rate of mass into the volume element is

$$2\pi r\,dr\,u\rho + 2\pi r\,dx\,v\rho \tag{5.85}$$

The flow rate of mass out of the volume element is

$$2\pi r\,dr\left(u\rho + \frac{\partial(u\rho)}{\partial x}\,dx\right) + 2\pi\,dx\left(rv\rho + \frac{\partial(rv\rho)}{\partial r}\,dr\right) \tag{5.86}$$

The rate of accumulation of mass in the volume element is

$$2\pi r\,dr\,dx\,\frac{\partial\rho}{\partial t} \tag{5.87}$$

The rate of flow in equals the rate of flow out plus the rate of accumulation, from which

$$r\frac{\partial(u\rho)}{\partial x} + \frac{\partial(rv\rho)}{\partial r} + r\frac{\partial\rho}{\partial t} = 0 \tag{5.88}$$

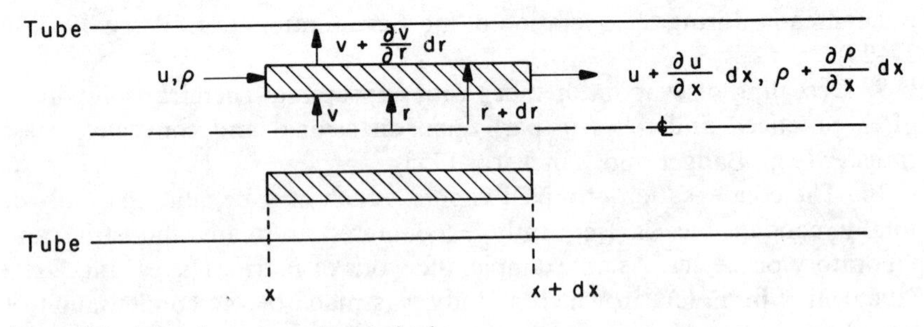

Figure 5.8. Flow through a cylindrical tube.

Expanding and dividing by r,

$$\rho\left(\frac{\partial u}{\partial x}+\frac{v}{r}+\frac{\partial v}{\partial r}\right)+u\frac{\partial \rho}{\partial x}+v\frac{\partial \rho}{\partial r}+\frac{\partial \rho}{\partial t}=0 \tag{5.89}$$

For steady-state conditions, $\partial\rho/\partial t=0$. For constant density, under conditions of either steady or unsteady state, the continuity equation 5.89 becomes

$$\frac{\partial u}{\partial x}+\frac{v}{r}+\frac{\partial v}{\partial r}=0 \tag{5.90}$$

The differential equation describing the axial and radial distribution of concentration of solute component A in a binary system is derived as follows for constant diffusivity and density. Consider a fluid in laminar flow through a cylindrical tube, where component A is being transferred to the fluid from the tube walls which are at concentration ρ_{AW}. The concentration of A is $\rho_A=f(x,r)$ and a balance may be made on A with reference to the annular differential element of volume $2\pi r\,dr\,dx$ shown in Figure 5.9. The flow rate of A into the volume element is

$$2\pi r\,dr\,n_{Ax}+2\pi\left(rn_{Ar}+\frac{\partial(rn_{Ar})}{\partial r}dr\right)dx \tag{5.91}$$

The flow rate of A out of the volume element is

$$2\pi r\,dr\left(n_{Ax}+\frac{\partial n_{Ax}}{\partial x}dx\right)+2\pi rn_{Ar}\,dx \tag{5.92}$$

The rate of accumulation of A in the volume element is

$$2\pi r\,dr\,dx\,\frac{\partial \rho_A}{\partial t} \tag{5.93}$$

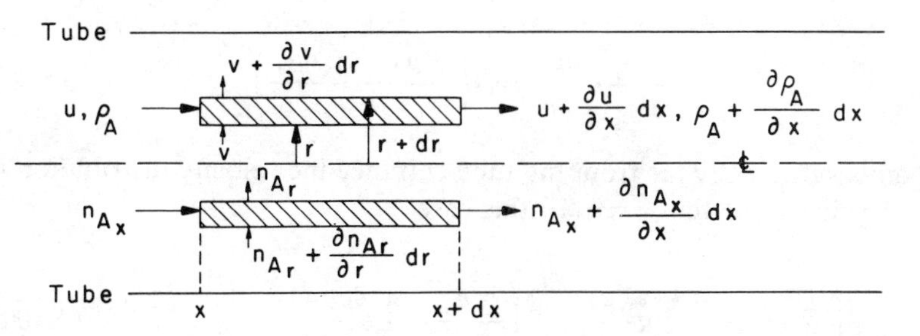

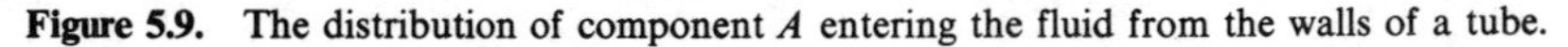

Figure 5.9. The distribution of component A entering the fluid from the walls of a tube.

where

$$n_{Ar} = D\frac{\partial \rho_A}{\partial r} - v\rho_A \tag{5.94}$$

$$n_{Ax} = -D\frac{\partial \rho_A}{\partial x} + u\rho_A \tag{5.95}$$

The rate of flow in equals the rate of flow out plus the rate of accumulation, from which

$$\frac{\partial n_{Ax}}{\partial x} + \frac{\partial \rho_A}{\partial t} = \frac{1}{r}\frac{\partial (rn_{Ar})}{\partial r} \tag{5.96}$$

Equations 5.94, 5.95, and 5.96 may be combined and the indicated differentiations performed to give

$$u\frac{\partial \rho_A}{\partial x} + v\frac{\partial \rho_A}{\partial r} + \frac{\partial \rho_A}{\partial t} + \rho_A\left(\frac{\partial u}{\partial x} + \frac{v}{r} + \frac{\partial v}{\partial r}\right) = D\left[\frac{\partial^2 \rho_A}{\partial r^2} + \frac{1}{r}\frac{\partial \rho_A}{\partial r} + \frac{\partial^2 \rho_A}{\partial x^2}\right] \tag{5.97}$$

When steady-state conditions prevail, $\partial\rho_A/\partial t = 0$. For a constant density, equation 5.90 shows the quantity in parentheses on the left side of equation 5.97 to be zero, so that at steady state,

$$u\frac{\partial \rho_A}{\partial x} + v\frac{\partial \rho_A}{\partial r} = D\left(\frac{\partial^2 \rho_A}{\partial r^2} + \frac{1}{r}\frac{\partial \rho_A}{\partial r} + \frac{\partial^2 \rho_A}{\partial x^2}\right) \tag{5.98}$$

If axial diffusion can be neglected, $\partial^2\rho_A/\partial x^2 = 0$ and

$$u\frac{\partial \rho_A}{\partial x} + v\frac{\partial \rho_A}{\partial r} = D\left(\frac{\partial^2 \rho_A}{\partial r^2} + \frac{1}{r}\frac{\partial \rho_A}{\partial r}\right) \tag{5.99}$$

At points sufficiently far from the tube entrance the velocity distribution is fully developed, so that $v = 0$. In this case,

$$u\frac{\partial \rho_A}{\partial x} = D\left(\frac{\partial^2 \rho_A}{\partial r^2} + \frac{1}{r}\frac{\partial \rho_A}{\partial r}\right) \tag{5.100}$$

Equation 5.99 or equation 5.100 is usually solved with either of two common boundary conditions:

(a) Uniform wall concentration along the tube $[\rho_{AW} \neq \rho_{AW}(x)]$.
(b) Uniform mass flux at the wall $[n_{AW} \neq n_{AW}(x)]$.

Solutions in terms of boundary condition (a) (uniform wall concentration) are often expressed as average Sherwood numbers over the mass-transfer section of the tube. These values are used to obtain the total transfer over a given length and the bulk average concentration at distance x along the mass-transfer section of the tube.

Solutions for boundary condition (b) (uniform mass flux at the wall) are frequently given as local Sherwood numbers at x. The specified uniform mass flux at the wall enables the average or "mixing cup" concentration at any x to be obtained by a simple mass balance on component A. The local value of the Sherwood number then allows estimation of the local concentration at the wall, since

$$(N_{\text{Sh}})_x^* = \frac{n_{AW} d_t}{(\rho_{AW} - \rho_{AB}) D} \tag{5.101}$$

Various degrees of velocity- and concentration-profile development may be identified, depending on the distance downstream from the tube entrance at which mass transfer is being considered. The development of the velocity distribution in the entrance region is shown in Figure 5.10.

At the tube inlet the fluid velocity is equal to V at all radii. The thickness of the momentum boundary layer is zero at the entrance, increasing with distance downstream. Retardation of fluid in the boundary layer is accompanied by acceleration of the axial core, so as to satisfy

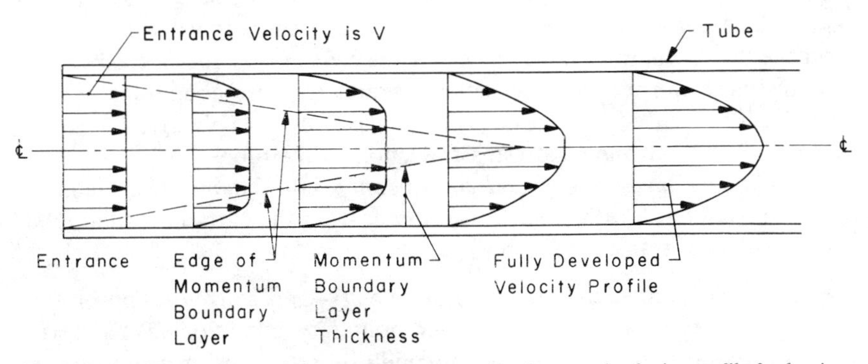

Figure 5.10. Development of the momentum boundary layer and velocity profile for laminar flow in the entrance region of a tube.

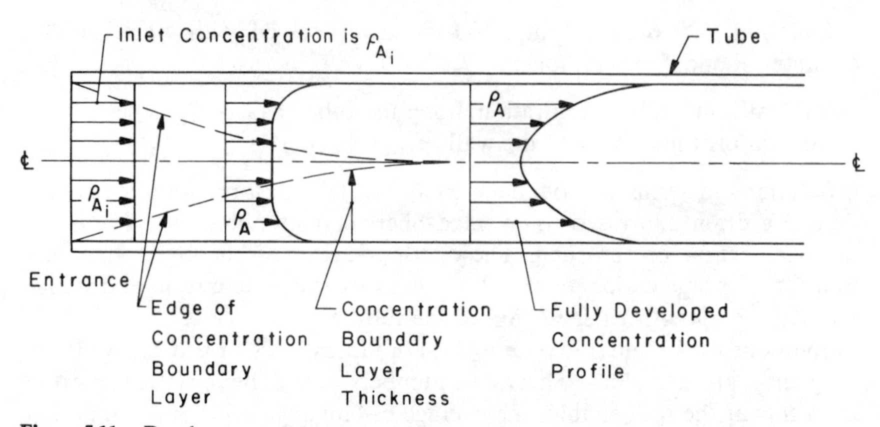

Figure 5.11. Development of the concentration boundary layer and profile for laminar flow in the entrance region of a tube. Transfer is from the tube walls to the fluid.

continuity. The flow is said to be fully developed at all points beyond that at which the boundary layer converges on the centerline.

Similar considerations apply to the development of the concentration profile, shown in Figure 5.11 for the case in which mass transfer is taking place from the tube walls to the fluid, with transfer beginning at the entrance. In general the thickness of the concentration boundary layer is not the same as that of the momentum boundary layer, except when the Schmidt number is unity.

The following are among the combinations of velocity- and concentration-profile development which may be encountered:

1. *Developing velocity and concentration distributions.* These conditions prevail near the inlet to a tube when mass transfer begins at the entrance.
2. *Fully developed velocity distribution and developing concentration distribution.* This set of conditions arises when the mass-transfer section is preceded by a length of tube in which the velocity profile is established without mass transfer. A hypothetical variant of these conditions, which lends itself to analysis, is one in which a *uniform-plug velocity distribution* prevails throughout the mass-transfer section of the tube.
3. *Fully developed velocity and concentration distributions.* These conditions are found at locations far downstream from the entrance to the tube and to the mass-transfer section.

Each of these three combinations of velocity- and concentration-profile development may be solved either for boundary condition (a) (uniform wall concentration) or for boundary condition (b) (uniform mass flux at the wall). These solutions will now be presented in turn. They are largely adapted from the analogous heat-transfer relationships, and accordingly require low concentrations and transfer rates, such that the bulk velocity caused by diffusion is negligible.

Developing Velocity and Concentration Distributions

These conditions represent a combination of Figures 5.10 and 5.11 in the entrance region of the tube, where the momentum and concentration boundary layers are developing simultaneously.

Uniform Wall Concentration

Even in the entrance region the term $v\partial\rho_A/\partial r$ in equation 5.99 becomes negligible within a very short distance beyond the inlet. This term was accordingly omitted by Kays (1955), who used Langhaar's (1942) developing velocity profiles in a numerical integration of equation 5.100 for this boundary condition and a Schmidt number of 0.7. The solution was extended with a digital computer by Goldberg (1958) for Schmidt numbers between 0.5 and 5.0. Figure 5.12 has been prepared from the tabulated results of Goldberg; the ordinate gives the average Sherwood number for use with the logarithmic-mean concentration difference between the wall and the bulk of the fluid over the mass-transfer section in question. The abscissa is $1/x_+$, where x_+ is a dimensionless distance along the tube, defined as

$$x_+ = \frac{x/r_t}{N_{\text{Re}}N_{\text{Sc}}} = \frac{x}{r_t}\bigg/\frac{d_t V\rho}{\mu}\frac{\mu}{\rho D} \tag{5.102}$$

Figure 5.12 includes curves for a uniform-plug velocity distribution and for a fully developed parabolic velocity distribution throughout the mass-transfer section, corresponding to relationships developed in the next section of this chapter. Asymptotic values of the Sherwood number for fully developed velocity and concentration distributions far downstream are also shown. These values are derived in a later section of this chapter.

Figure 5.12 shows that, for given values of N_{Sc}, N_{Re}, and x/r_t, the Sherwood number for a developing velocity profile exceeds—or at least is equal to—that for the case of a fully developed parabolic velocity profile throughout. This is because the velocities in the vicinity of the wall are greater for the developing velocity profile than for the fully developed one.

For decreasing Schmidt numbers, conditions in the entrance region approach those for a uniform-plug velocity distribution, whereas high-Schmidt-number fluids approach conditions corresponding to a fully developed parabolic velocity distribution throughout. The latter conditions are approximated by all fluids at points far enough from the entrance.

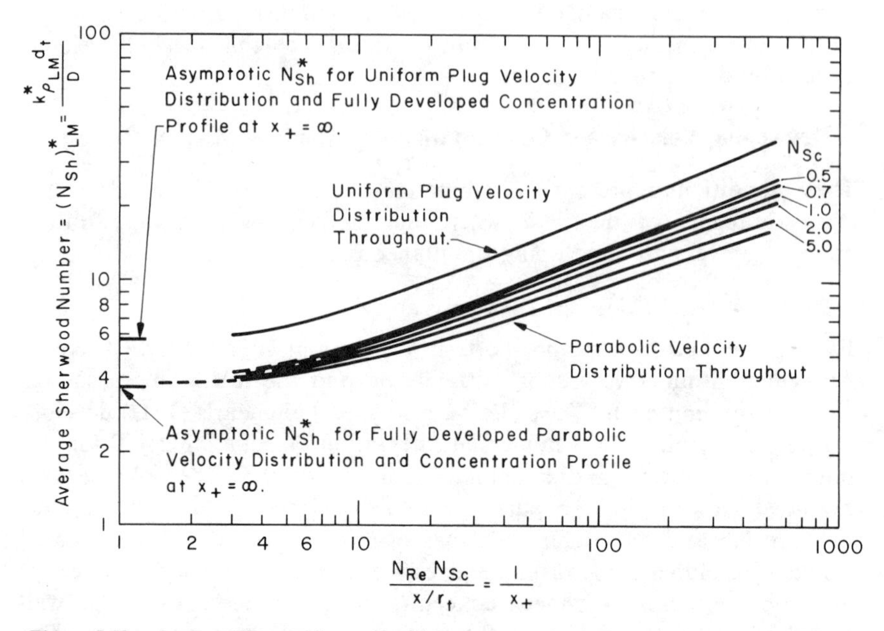

Figure 5.12. Average Sherwood numbers for mass transfer in laminar flow through a tube with uniform wall concentration (Goldberg, 1958).

Uniform Mass Flux at the Wall

Solutions to equation 5.99 for this boundary condition are shown in Figure 5.13, which has been prepared from the tabulated results of Heaton, Reynolds, and Kays (1964). The figure gives local Sherwood numbers (for use in equation 5.101) as a function of $1/x_+$, where x_+ is defined in equation 5.102. Also shown in Figure 5.13 is the asymptotic value of the Sherwood number for fully developed velocity and concentration distributions, as derived in a later section of this chapter.

It may be noted that the Schmidt number of 0.01 is unrealistically low—the results were developed for heat transfer, in which Prandtl numbers of about 0.01 are commonplace for liquid metals. The curve for $N_{Sc} = 0.01$ is retained here to indicate trends and to facilitate interpolation.

Uniform-Plug Velocity Distribution and Developing Concentration Distribution

Uniform Wall Concentration

Solutions to equation 5.100 may be obtained by a procedure analogous to that used by Graetz (1883, 1885) for heat transfer, assuming that all

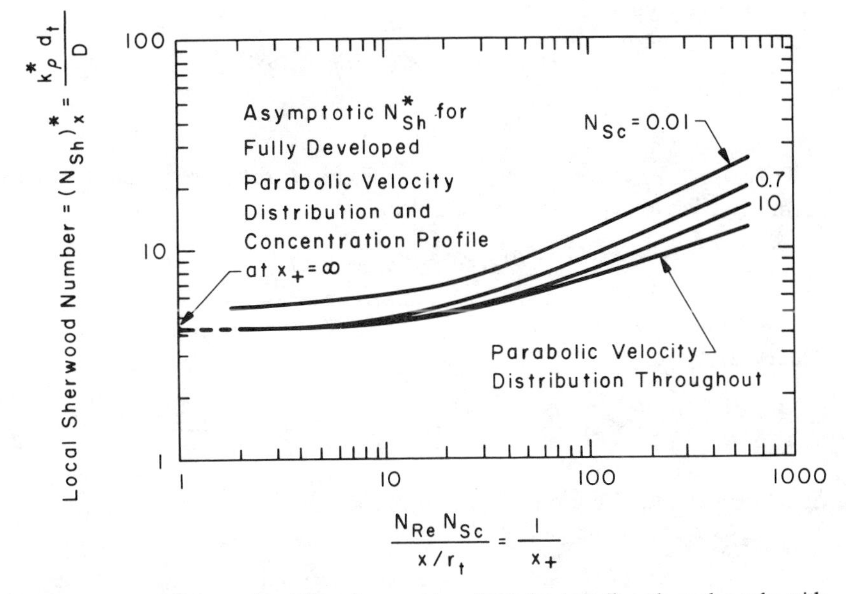

Figure 5.13. Local Sherwood numbers for mass transfer in laminar flow through a tube with uniform mass flux at the wall (Heaton et al., 1964; lowest curve: Seigel et al., 1958).

relevant physical properties remain constant, ρ_{AW} is independent of x, the velocity profile is fully developed when mass transfer begins, and the concentration at the inlet to the mass-transfer section is uniform at ρ_{Ai}. Following Graetz, two cases may be considered: plug flow, and fully developed flow with a parabolic velocity profile. For illustrative purposes the plug-flow case will be considered first in some detail. For plug flow, equation 5.100 becomes

$$\frac{\partial \rho_A}{\partial x} = \frac{D}{V}\left[\frac{\partial^2 \rho_A}{\partial r^2} + \frac{1}{r}\frac{\partial \rho_A}{\partial r}\right]$$

or

$$\frac{\partial \theta}{\partial x} = \frac{D}{V}\left[\frac{\partial^2 \theta}{\partial r^2} + \frac{1}{r}\frac{\partial \theta}{\partial r}\right] \tag{5.103}$$

where

$$\theta = \frac{\rho_{AW} - \rho_A}{\rho_{AW} - \rho_{Ai}}$$

The boundary conditions are

$$\theta=0 \quad \text{at} \quad r=r_t \quad \text{for} \quad x>0$$

$$\theta=1.0 \quad \text{at} \quad x=0$$

Following conventional procedure, assume a solution of the following form in order to separate the variables:

$$\theta(r,x)=R'(r)X'(x) \tag{5.104}$$

where R' is some function of r only, and X' is some function of x only. Then clearly

$$\frac{\partial\theta}{\partial x}=R'\frac{dX'}{dx}, \quad \frac{\partial\theta}{\partial r}=X'\frac{dR'}{dr}, \quad \frac{\partial^2\theta}{\partial r^2}=X'\frac{d^2R'}{dr^2}$$

Substituting into equation 5.103 and rearranging,

$$\frac{V}{D}\frac{1}{X'}\frac{dX'}{dx}=\frac{1}{R'}\left[\frac{d^2R'}{dr^2}+\frac{1}{r}\frac{dR'}{dr}\right]=-b^2 \tag{5.105}$$

The definitions of R' and X' imply that the left side of equation 5.105 is independent of r, while the right side is independent of x. Both sides therefore equal a constant, denoted by $-b^2$. From the first of the two resulting differential equations,

$$X'=\exp\left(-\frac{D}{V}b^2x\right) \tag{5.106}$$

The second differential equation obtainable from equation 5.105 is

$$\frac{d^2R'}{dr^2}+\frac{1}{r}\frac{dR'}{dr}+b^2R'=0 \tag{5.107}$$

This is Bessel's equation of zero order (Li, 1960, p. 328) and, for the boundary conditions

$$\left(\frac{\partial\theta}{\partial r}\right)_{r=0}=0=\left(\frac{dR'}{dr}\right)_{r=0}, \qquad \theta=0=R' \text{ at } r=r_t$$

the particular solution is (Li, 1960, pp. 328–330)

$$R'=AJ_0(br)$$

where $J_0(br)$ is a series—namely the Bessel function of the first kind of order zero. The boundary condition $R'=0$ at $r=r_t$ shows that (*Li*, 1960, pp. 328–330)

$$J_0(br_t)=0 \tag{5.108}$$

so that the constant b is a root of equation 5.108. If these roots are denoted by $b_1, b_2, b_3, \ldots, b_j, \ldots$, the solution of equation 5.107 is

$$R' = \sum_{j=1}^{\infty} A_j J_0(b_j r) \tag{5.109}$$

The constants A_j will next be evaluated from the boundary condition $\theta=1$ at $x=0$. First, one can show that (Churchill, 1941)

$$\int_0^{r_t} rJ_0(b_j r)J_0(b_k r)\,dr=0 \qquad \text{if} \quad j\neq k \tag{5.110}$$

and

$$=\tfrac{1}{2}r_t^2 J_1^2(b_j r_t) \qquad \text{if} \quad j=k \tag{5.111}$$

where J_1 is the Bessel function of the first kind of order 1. Now consider the multiplication of equation 5.109 by $rJ_0(b_j r)$, giving $rR'J_0(b_j r)$. Although the entire series is involved in this multiplication, equation 5.110 shows that the product of $rJ_0(b_j r)$ and all but the jth term is zero upon integrating thus.

$$\int_0^{r_t} rR'J_0(b_j r)\,dr = A_j \int_0^{r_t} rJ_0^2(b_j r)\,dr$$

$$= A_j \times \tfrac{1}{2} r_t^2 J_1^2(b_j r_t)$$

after invoking equation 5.111. Rearranging,

$$A_j = \frac{2}{r_t^2 J_1^2(b_j r_t)} \int_0^{r_t} rR'J_0(b_j r)\,dr$$

Now at $x=0$, $\theta=1=R'$, and from the definitions of J_0 and J_1 (Apostol, 1961) it is evident that

$$\int_0^{r_t} rJ_0(b_j r)\,dr = \frac{r_t}{b_j} J_1(b_j r_t)$$

so that

$$A_j = \frac{2}{b_j r_t} \frac{1}{J_1(b_j r_t)}$$

and inserting this result in equation 5.109,

$$R' = \sum_{j=1}^{\infty} \frac{2J_0(b_j r)}{b_j r_t J_1(b_j r_t)} \tag{5.112}$$

Combining equations 5.104, 5.106, and 5.112,

$$\theta = \frac{\rho_{AW} - \rho_A}{\rho_{AW} - \rho_{Ai}} = \sum_{j=1}^{\infty} \frac{2}{b_j r_t} \frac{J_0(b_j r)}{J_1(b_j r_t)} \exp\left(-\frac{D}{V} b_j^2 x\right) \tag{5.113}$$

Let $a_j = b_j r_t$; then since

$$-\frac{a_j^2 Dx}{V r_t^2} = -\frac{2a_j^2 (x/r_t)}{N_{\text{Re}} N_{\text{Sc}}}$$

it follows that

$$\frac{\rho_{AW} - \rho_A}{\rho_{AW} - \rho_{Ai}} = \sum_{j=1}^{\infty} \frac{2}{a_j J_1(a_j)} J_0\left(a_j \frac{r}{r_t}\right) \exp\left(\frac{-2a_j^2 (x/r_t)}{N_{\text{Re}} N_{\text{Sc}}}\right) \tag{5.114}$$

where a_j is the jth root of equation 5.108—that is, $J_0(a_j) = 0$. Li (1960, p. 330) cites the following values for the first five roots: $a_1 = 2.405$, $a_2 = 5.520$, $a_3 = 8.654$, $a_4 = 11.792$, $a_5 = 14.931$.

Equation 5.114 gives the point value of ρ_A at (x, r). The bulk average or "mixing cup" value of ρ_A at a given x is

$$\rho_{AB} = \frac{\int_0^{r_t} \rho_A u 2\pi r \, dr}{\int_0^{r_t} u 2\pi r \, dr} \tag{5.115}$$

but for plug flow, $u = V$ at all r, so that

$$\rho_{AB} = \frac{2}{r_t^2} \int_0^{r_t} \rho_A r \, dr \tag{5.116}$$

The value of ρ_{AB} is obtained from a combination of equations 5.114 and 5.116, to give

$$\frac{\rho_{AB}-\rho_{Ai}}{\rho_{AW}-\rho_{Ai}}=1-\frac{\rho_{AW}-\rho_{AB}}{\rho_{AW}-\rho_{Ai}}$$

$$=1-4\sum_{j=1}^{j=\infty} a_j^{-2}\exp\left(\frac{-2a_j^2(x/r_t)}{N_{\mathrm{Re}}N_{\mathrm{Sc}}}\right) \tag{5.117}$$

A plot of $(\rho_{AB}-\rho_{Ai})/(\rho_{AW}-\rho_{Ai})$ versus $(\pi/4)(d_t/x)N_{\mathrm{Re}}N_{\mathrm{Sc}}$ obtained in this way for plug flow is shown in Figure 5.14. Data plotted at the top left of this figure were obtained by Gilliland and Sherwood (1934) for the vaporization of eight different liquids in a wetted-wall column at low air rates. Although these data fit the curve for plug flow rather well in this region, Boelter (1943) has explained this surprising result in terms of distortion of the more probable parabolic velocity profile by natural convection effects.

Data at the higher values of $(\pi/4)(d_t/x)N_{\mathrm{Re}}N_{\mathrm{Sc}}$ agree with predictions for a parabolic velocity distribution and will be considered later. It may be noted that

$$\frac{\pi}{4}\frac{d_t}{x}N_{\mathrm{Re}}N_{\mathrm{Sc}}=\frac{W}{\rho Dx}$$

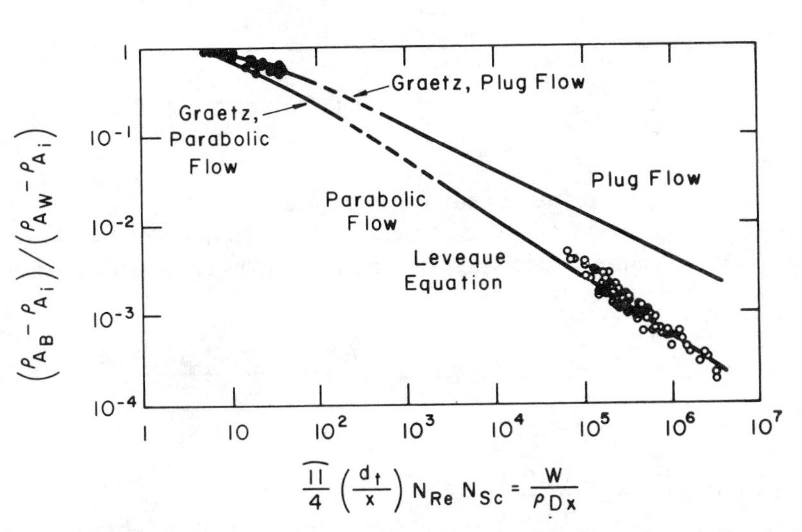

Figure 5.14. Mass transfer in laminar flow through a tube with plug and parabolic velocity distributions (Linton and Sherwood, 1950).

where W is the mass flow rate. The group $W/\rho Dx$ is evidently the mass-transfer equivalent of the Graetz number, Wc_p/kx, used in heat transfer. Figure 5.14 shows that for $W/\rho Dx$ greater than about 100 the plug-flow solution is linear with a slope of $-\frac{1}{2}$ and may evidently be represented as follows:

$$\frac{\rho_{AB}-\rho_{Ai}}{\rho_{AW}-\rho_{Ai}}=4\left(\frac{W}{\rho Dx}\right)^{-1/2} \tag{5.118}$$

The local coefficient of mass transfer at x is defined by the following differential mass balance:

$$\frac{\pi d_t^2}{4}V d\rho_{AB}=k_\rho^*(\pi d_t dx)(\rho_{AW}-\rho_{AB})$$

or

$$(N_{\text{Sh}})_x^*=\frac{k_\rho^* d_t}{D}=\frac{d_t}{4}N_{\text{Re}}N_{\text{Sc}}\frac{d\rho_{AB}}{dx}\frac{1}{(\rho_{AW}-\rho_{AB})} \tag{5.119}$$

Equation 5.119 is used with ρ_{AB} obtained from equations 5.114 and 5.116 to give the local Sherwood number as

$$(N_{\text{Sh}})_x^*=\frac{\sum_{j=1}^{j=\infty}\exp\left(\frac{-2a_j^2(x/r_t)}{N_{\text{Re}}N_{\text{Sc}}}\right)}{\sum_{j=1}^{j=\infty}a_j^{-2}\exp\left(\frac{-2a_j^2(x/r_t)}{N_{\text{Re}}N_{\text{Sc}}}\right)} \tag{5.120}$$

Equation 5.120 is plotted as curve 1 in Figure 5.15.

An average mass-transfer coefficient $k_{\rho a}^*$ between the inlet and the outlet of the mass-transfer section may be defined in terms of an arithmetic-mean driving force as follows:

$$\frac{\pi d_t^2}{4}V(\rho_{ABO}-\rho_{Ai})=k_{\rho a}^*(\pi d_t L)\frac{(\rho_{AW}-\rho_{Ai})+(\rho_{AW}-\rho_{ABO})}{2} \tag{5.121}$$

or

$$(N_{\text{Sh}})_a^*=\frac{k_{\rho a}^* d_t}{D}=\frac{1}{4}\left(\frac{d_t}{L}\right)N_{\text{Re}}N_{\text{Sc}}\frac{2(\rho_{ABO}-\rho_{Ai})/(\rho_{AW}-\rho_{Ai})}{2-(\rho_{ABO}-\rho_{Ai})/(\rho_{AW}-\rho_{Ai})} \tag{5.122}$$

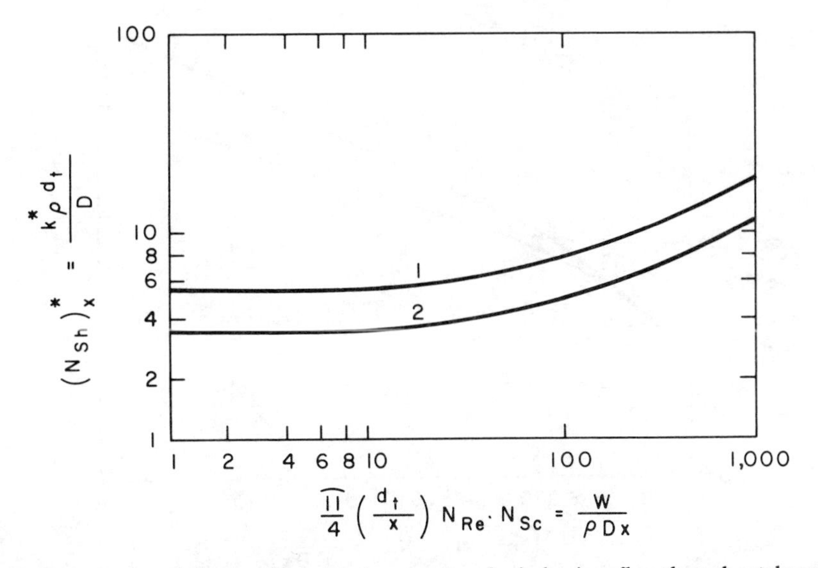

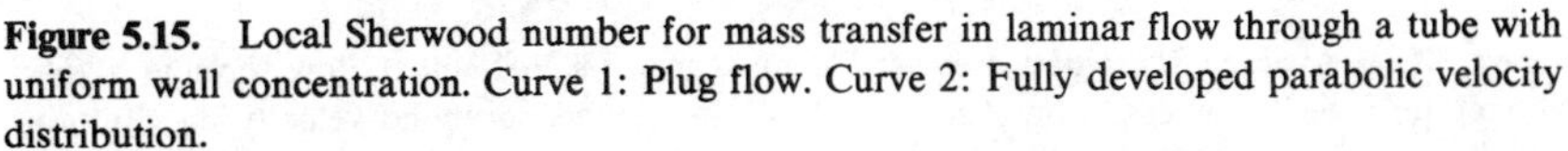

Figure 5.15. Local Sherwood number for mass transfer in laminar flow through a tube with uniform wall concentration. Curve 1: Plug flow. Curve 2: Fully developed parabolic velocity distribution.

so that from equation 5.117,

$$(N_{\mathrm{Sh}})^*_a = \frac{1}{2}\left(\frac{d_t}{L}\right) N_{\mathrm{Re}} N_{\mathrm{Sc}} \left[\frac{1 - 4 \sum_{j=1}^{j=\infty} a_j^{-2} \exp\left(\frac{-2a_j^2 (x/r_t)}{N_{\mathrm{Re}} N_{\mathrm{Sc}}} \right)}{1 + 4 \sum_{j=1}^{j=\infty} a_j^{-2} \exp\left(\frac{-2a_j^2 (x/r_t)}{N_{\mathrm{Re}} N_{\mathrm{Sc}}} \right)} \right] \tag{5.123}$$

The first five values of a_j are given below equation 5.114. Equation 5.123 is shown graphically as curve 3 in Figure 5.16.

An alternative form of average mass transfer coefficient $k^*_{\rho \mathrm{LM}}$ may be written for use with the logarithmic-mean concentration difference over the mass-transfer section; thus

$$\frac{\pi d_t^2}{4} V(\rho_{ABO} - \rho_{Ai}) = k^*_{\rho \mathrm{LM}} (\pi d_t L) \frac{(\rho_{AW} - \rho_{Ai}) - (\rho_{AW} - \rho_{ABO})}{\ln\left(\frac{\rho_{AW} - \rho_{Ai}}{\rho_{AW} - \rho_{ABO}} \right)} \tag{5.124}$$

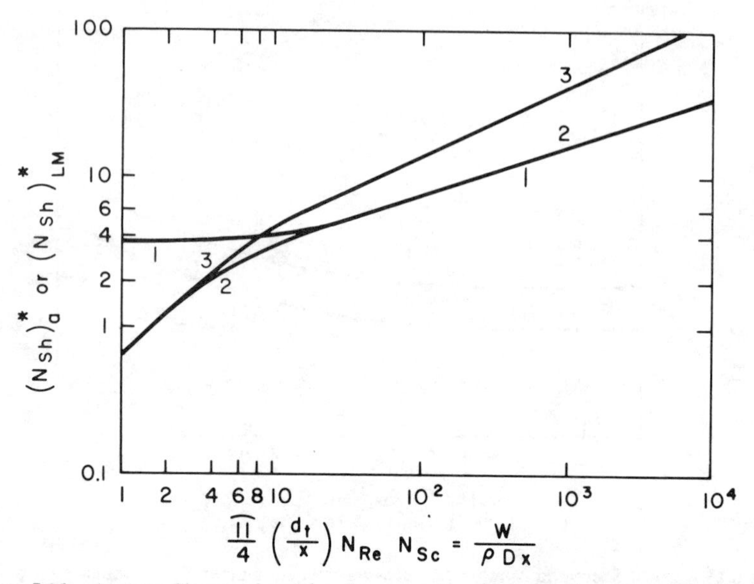

Figure 5.16. Average Sherwood number for mass transfer in laminar flow through a tube with uniform wall concentration. Curve 1: Fully developed parabolic velocity distribution, $(N_{Sh})_{LM}^*$. Curve 2: Fully developed parabolic velocity distribution, $(N_{Sh})_a^*$. Curve 3: Plug flow, $(N_{Sh})_a^*$.

or

$$(N_{Sh})_{LM}^* = \frac{k_{\rho LM}^* d_t}{D} = \frac{1}{4}\left(\frac{d_t}{L}\right) N_{Re} N_{Sc} \ln\left(\frac{\rho_{AW} - \rho_{Ai}}{\rho_{AW} - \rho_{AB0}}\right)$$

and after combination with equation 5.117,

$$(N_{Sh})_{LM}^* = \frac{k_{\rho LM}^* d_t}{D} = \frac{1}{4}\left(\frac{d_t}{L}\right) N_{Re} N_{Sc} \ln\left[4 \sum_{j=1}^{j=\infty} a_j^{-2} \exp\left(\frac{-2a_j^2 (x/r_t)}{N_{Re} N_{Sc}}\right)\right]^{-1}$$

(5.125)

This relationship is plotted as the top curve in Figure 5.12 for comparison with solutions for the case of simultaneously developing velocity and concentration profiles.

Fully Developed Parabolic Velocity Distribution and Developing Concentration Distribution

Two different boundary conditions will be considered in turn for the situation in which the parabolic velocity profile is fully developed at the entrance to the mass-transfer section. The first consists of a uniform solute concentration along the wall of the tube $[\rho_{AW} \neq \rho_{AW}(x)]$. The alternative boundary condition, that of uniform mass flux at the wall $[n_{AW} \neq n_{AW}(x)]$, is encountered more rarely, and in consequence only the final results of the analysis will be presented for this case.

Uniform Wall Concentration

The complexity of analysis is increased in this situation because of the replacement of plug flow by a fully developed and constant parabolic velocity profile. Other assumptions are the same as in the plug flow case. The original analysis for heat transfer by Graetz (1885) has been recalculated and extended by many investigators, including Drew (1931), Jakob (1949, p. 451), and Sellars, Tribus, and Klein (1956). Perhaps the most detailed restatement of the development is that given by Boelter, Cherry, and Johnson (1937), a highly condensed version of which is given here in order to clarify the analytical procedure.

Inserting the parabolic velocity distribution for u into equation 5.100 gives

$$\frac{\partial \rho_A}{\partial x} = \frac{D}{2V\left[1-(r/r_t)^2\right]}\left[\frac{\partial^2 \rho_A}{\partial r^2} + \frac{1}{r}\frac{\partial \rho_A}{\partial r}\right] \tag{5.126}$$

It is convenient to express this relationship in dimensionless form as follows:

$$(1-r_+^2)\frac{\partial \theta}{\partial x_+} = \frac{\partial^2 \theta}{\partial r_+^2} + \frac{1}{r_+}\frac{\partial \theta}{\partial r_+} \tag{5.127}$$

where

$$\theta = \frac{\rho_{AW}-\rho_A}{\rho_{AW}-\rho_{Ai}} \tag{5.128}$$

$$r_+ = \frac{r}{r_t} \tag{5.129}$$

$$x_+ = \frac{x/r_t}{N_{\mathrm{Re}}N_{\mathrm{Sc}}} \tag{5.130}$$

The boundary conditions are

$$\theta(r_+,0)=1,$$

$$\theta(1,x_+)=0,$$

$$\frac{\partial\theta}{\partial r_+}(0,x_+)=0,$$

$$\theta(r_+,\infty)=0.$$

As in the case of plug flow, an application of the technique of separation of variables,

$$\theta(r_+,x_+)=\phi(r_+)\Gamma(x_+)$$

in equation 5.127 leads to a simple first-order ordinary differential equation in Γ and to a boundary value problem of the Sturm-Liouville type for $\phi(r_+)$, namely

$$\frac{d^2\phi}{dr_+^2}+\frac{1}{r_+}\frac{d\phi}{dr_+}+\beta^2(1-r_+^2)\phi=0$$

$$\phi(1)=0$$

$$\frac{d\phi(0)}{dr_+}=0$$

If the eigenvalues of this system are labeled $\beta_1,\beta_2,\beta_3,\ldots,\beta_j,\ldots$ in increasing order and $\phi_j(r_+)$ denotes the eigenfunction corresponding to β_j, then the solution for the concentration distribution must accordingly be of the form [Drew, (1931, p. 65), Jakob, (1949, p. 453), Boelter, Cherry, and Johnson, (1937, p. X-21, Eq. Xd.-20)]

$$\theta(r_+,x_+)=\frac{\rho_{AW}-\rho_A}{\rho_{AW}-\rho_{Ai}}=\sum_{j=1}^{j=\infty}B_j\phi_j(r_+)\exp(-\beta_j^2x_+) \qquad (5.131)$$

The coefficients B_j in this expansion may be calculated from the equation

$$1=\sum_{j=1}^{j=\infty}B_j\phi_j(r_+) \qquad (5.132)$$

which results from requiring the series in equation 5.131 to satisfy the boundary condition $\phi(r_+,0)=1$. Both sides of 5.132 are multiplied by

$\phi_k r_+(1-r_+^2)$ and then integrated with respect to r_+ from 0 to 1 to obtain

$$B_j = \frac{\int_0^1 \phi_j r_+ (1-r_+^2)\, dr_+}{\int_0^1 \phi_j^2 r_+ (1-r_+^2)\, dr_+} = \frac{-2}{\beta_j \left(\dfrac{\partial \phi}{\partial \beta}\right)_{j, r_+ = 1}} \tag{5.133}$$

The calculations utilize the following integral relations used by Graetz:

$$\int_0^1 \phi_j \phi_k r_+ (1-r_+^2)\, dr_+ = \begin{cases} 0 & \text{for } j \neq k, \\ \dfrac{1}{2\beta_j}\left[\dfrac{d\phi_j}{dr_+}\left(\dfrac{\partial \phi}{\partial \beta}\right)_j\right]_{r_+=1} & \text{for } j = k \end{cases} \tag{5.134}$$

$$\int_0^1 \phi_j r_+ (1-r_+^2)\, dr_+ = -\frac{1}{\beta_j^2}\left(\frac{d\phi_j}{dr_+}\right)_{r_+=1} \tag{5.135}$$

The function ϕ_j may be represented by the infinite series

$$\phi_j(r_+) = \sum_{i=0}^{i=\infty} \alpha_{ji} r_+^i \tag{5.136}$$

where

$$\alpha_{ji} = 0 \qquad \text{for} \quad i < 0$$

$$\alpha_{ji} = 1 \qquad \text{for} \quad i = 0$$

$$\alpha_{ji} = -\beta_j^2(\alpha_{i-2} - \alpha_{i-4})/i^2$$

The eigenvalues can be calculated from equation 5.136 using a trial-and-error procedure. The method has been described by Brown (1960), who outlines a digital-computer program for the evaluation of the eigenvalues β_j, the eigenfunctions ϕ_j for $0 \leqslant r_+ \leqslant 1$, and the derivatives $(\partial\phi/\partial\beta)_{j,r_+=1}$ and $(d\phi_j/dr_+)_{r_+=1}$. Eigenvalues and derivatives for the first 11 terms in equation 5.131 as calculated by Brown are given in Table 4 of the Appendix. The corresponding coefficients B_j of course follow from this table and equation 5.133. Brown's values for the first six eigenfunctions appear in Table 5 of the Appendix. Higher eigenvalues, eigenfunctions, and related constants may be estimated from relationships presented by

Sellars, Tribus, and Klein (1956). Their expressions for β_j and B_j in the present notation are

$$\beta_j = 4(j-1) + \tfrac{8}{3}; \qquad j = 1,2,3,\ldots \tag{5.137}$$

$$B_j = (-1)^{j-1} \times 2.84606 \beta_j^{-2/3} \tag{5.138}$$

Also

$$\frac{-B_j}{2}\left(\frac{d\phi_j}{dr_+}\right)_{r_+=1} = 1.01276 \beta_j^{-1/3} \tag{5.139}$$

The bulk-average or mixing-cup concentration at x is given by equation 5.115, which may be written as

$$\frac{\rho_{AW}-\rho_{AB}}{\rho_{AW}-\rho_{Ai}} = \theta_b = \frac{1}{\pi V}\int_0^1 \theta\left[2V(1-r_+^2)\right]2\pi r_+\, dr_+ \tag{5.140}$$

Combining equations 5.131 and 5.140 gives

$$\theta_b = \sum_{j=1}^{j=\infty} 4B_j\left[\exp(-\beta_j^2 x_+)\right]\int_0^1 \phi_j r_+(1-r_+^2)\, dr_+ \tag{5.141}$$

and with equations 5.130 and 5.135,

$$\frac{\rho_{AW}-\rho_{AB}}{\rho_{AW}-\rho_{Ai}} = \sum_{j=1}^{j=\infty} \frac{-4B_j}{\beta_j^2}\left(\frac{d\phi_j}{dr_+}\right)_{r_+=1} \exp\left(\frac{-\beta_j^2(x/r_t)}{N_{\mathrm{Re}}N_{\mathrm{Sc}}}\right) \tag{5.142}$$

The quantity $1-(\rho_{AW}-\rho_{AB})/(\rho_{AW}-\rho_{Ai})$ equals $(\rho_{AB}-\rho_{Ai})/(\rho_{AW}-\rho_{Ai})$, and this is plotted versus $(\pi/4)(d_t/x)N_{\mathrm{Re}}N_{\mathrm{Sc}}$ for a parabolic velocity profile in Figure 5.14. As noted earlier, values in the lower range of $W/\rho Dx$ show fortuitous agreement with predictions for plug flow because of natural convection effects. These data were obtained for evaporation from a wetted-wall column into an airstream, and in this region Boelter (1943) has replaced the curve corresponding to a parabolic velocity distribution in Figure 5.14 with a family of curves in which the following quantity is a parameter:

$$\frac{d_t^3 \rho g_c}{\mu^2}\left(\frac{M_i T_W}{M_W T_i} - 1\right)\frac{\mu}{\rho D}\frac{d_t}{L}$$

These curves provide adequate correction of the data for distortion of the parabolic velocity profile because of density gradients. The quantities M_i and T_i are the molecular weight and absolute temperature of the inlet gas, and M_W and T_W are the average molecular weight and temperature of the gas saturated with the diffusing solute at the liquid surface temperature.

Measurements shown at the higher values of $W/\rho Dx$ are in agreement with the predictions for a parabolic velocity distribution. These data were obtained by Linton and Sherwood (1950) for the dissolution of soluble tube walls, cast from benzoic acid, cinnamic acid, and β-naphthol; water flowed through the tubes.

For values of $W/\rho Dx$ above about 400, the plot for the parabolic velocity profiie becomes linear in Figure 5.14 and coincides with the approximate solution given by Lévêque as

$$\frac{\rho_{AB}-\rho_{Ai}}{\rho_{AW}-\rho_{Ai}}=5.5\left(\frac{W}{\rho Dx}\right)^{-2/3} \tag{5.143}$$

Equation 5.143 is valid for those cases where the tube is short enough that the concentration distribution does not become fully developed—that is to say, the outer edges of the concentration boundary layers do not reach the tube axis. That this is a common situation will be demonstrated later via equations 5.149 to 5.151.

The local value of the Sherwood number, based on $\rho_{AW}-\rho_{AB}$, has the following form in terms of the Graetz solution:

$$(N_{\text{Sh}})_x^* = \frac{\displaystyle\sum_{j=1}^{j=\infty} \frac{B_j}{2}\left(\frac{d\phi_j}{dr_+}\right)_{r_+=1} \exp\left(\frac{-\beta_j^2(x/r_t)}{N_{\text{Re}}N_{\text{Sc}}}\right)}{\displaystyle 2\sum_{j=1}^{j=\infty} \frac{B_j}{2\beta_j^2}\left(\frac{d\phi_j}{dr_+}\right)_{r_+=1} \exp\left(\frac{-\beta_j^2(x/r_t)}{N_{\text{Re}}N_{\text{Sc}}}\right)} \tag{5.144}$$

Local values of Sherwood number are plotted against the group $(\pi/4)(d_t/x)N_{\text{Re}}N_{\text{Sc}}$ as curve 2 in Figure 5.15.

Procedures analogous to those leading to equations 5.123 and 5.125 may be used with equation 5.142 to obtain the following two expressions for average Sherwood numbers:

$$(N_{\mathrm{Sh}})_a^* = \frac{1}{2}\left(\frac{d_t}{L}\right) N_{\mathrm{Re}} N_{\mathrm{Sc}} \left[\frac{1 - \sum_{j=1}^{j=\infty} \frac{-4B_j}{\beta_j^2}\left(\frac{d\phi_j}{dr_+}\right)_{r_+=1} \exp\left(\frac{-\beta_j^2(x/r_t)}{N_{\mathrm{Re}} N_{\mathrm{Sc}}}\right)}{1 + \sum_{j=1}^{j=\infty} \frac{-4B_j}{\beta_j^2}\left(\frac{d\phi_j}{dr_+}\right)_{r_+=1} \exp\left(\frac{-\beta_j^2(x/r_t)}{N_{\mathrm{Re}} N_{\mathrm{Sc}}}\right)} \right] \tag{5.145}$$

The coefficient k_{pa}^* in $(N_{\mathrm{Sh}})_a^*$ of equation 5.145 is for use with the arithmetic-mean concentration difference of equation 5.121. A convenient graphical representation of equation 5.145 appears as curve 2 in Figure 5.16.

$$(N_{\mathrm{Sh}})_{\mathrm{LM}}^* = \frac{1}{4}\left(\frac{d_t}{L}\right) N_{\mathrm{Re}} N_{\mathrm{Sc}} \ln \left[\sum_{j=1}^{j=\infty} \frac{-4B_j}{\beta_j^2}\left(\frac{d\phi_j}{dr_+}\right)_{r_+=1} \exp\left(\frac{-\beta_j^2(x/r_t)}{N_{\mathrm{Re}} N_{\mathrm{Sc}}}\right) \right]^{-1} \tag{5.146}$$

The Sherwood number in the above relationship is for use with the logarithmic-mean concentration difference of equation 5.124. Equation 5.146 is plotted as the bottom curve in Figure 5.12 for comparison with solutions for the case of simultaneously developing velocity and concentration profiles. The expression is also shown graphically as curve 1 in Figure 5.16, which compares the behavior of $(N_{\mathrm{Sh}})_{\mathrm{a}}^*$ and $(N_{\mathrm{Sh}})_{\mathrm{LM}}^*$ along the tube. Alternatively one may use Hausen's (1943) representation of equation 5.146:

$$(N_{\mathrm{Sh}})_{\mathrm{LM}}^* = 3.66 + \frac{0.0668(d_t/x) N_{\mathrm{Re}} N_{\mathrm{Sc}}}{1 + 0.04[(d_t/x) N_{\mathrm{Re}} N_{\mathrm{Sc}}]^{2/3}} \tag{5.147}$$

The series in equation 5.144 for the local Sherwood number converges rapidly for large values of $(x/r_t)/N_{\mathrm{Re}} N_{\mathrm{Sc}}$, so that only the first term is significant when $(x/r_t)/N_{\mathrm{Re}}N_{\mathrm{Sc}}$ exceeds 0.1. Under these conditions equation 5.144 reduces to

$$(N_{\mathrm{Sh}})_x^* = \frac{\dfrac{B_1}{2}\left(\dfrac{d\phi_j}{dr_+}\right)_{r_+=1} \exp\left(\dfrac{-\beta_1^2(x/r_t)}{N_{\mathrm{Re}}N_{\mathrm{Sc}}}\right)}{\dfrac{B_1}{\beta_1^2}\left(\dfrac{d\phi_j}{dr_+}\right)_{r_+=1} \exp\left(\dfrac{-\beta_1^2(x/r_t)}{N_{\mathrm{Re}}N_{\mathrm{Sc}}}\right)}$$

$$= \frac{\beta_1^2}{2} = \frac{7.312}{2} = 3.656 \tag{5.148}$$

This value coincides with that for a fully developed concentration profile (Rohsenow and Choi, 1961, p. 141, 402), indicating that the entry length in which the concentration profile becomes fully developed is approximately given by

$$\left(\frac{x}{d_t}\right)_{\mathrm{dev}} \doteq 0.05 N_{\mathrm{Re}} N_{\mathrm{Sc}} \tag{5.149}$$

The corresponding thermal entry length is of course obtained by replacing N_{Sc} by N_{Pr} in equation 5.149:

$$\left(\frac{x}{d_t}\right)_{\mathrm{dev}} \doteq 0.05 N_{\mathrm{Re}} N_{\mathrm{Pr}} \tag{5.150}$$

The entry length in which the velocity profile becomes fully developed is given by Prandtl and Tietjens (1931) as

$$\left(\frac{x}{d_t}\right)_{\mathrm{dev}} \doteq 0.05 N_{\mathrm{Re}} \tag{5.151}$$

Consider, for example, the case of water flowing through a tube at a Reynolds number of 100 and a temperature of 68°F. At this temperature the Prandtl number is 7. The Schmidt number for various materials in water is substantially dependent upon the solute (Perry, 1950, p. 540), but a typical value of 1200 is suggested by Foust et al. (1960, p. 111). Equations 5.149, 5.150, and 5.151 show that the concentration, thermal, and hydrodynamic entry lengths are then, respectively, as follows:

$$\left(\frac{x}{d_t}\right)_{\text{dev, concn distribution}} = 6000$$

$$\left(\frac{x}{d_t}\right)_{\text{dev, temp. distribution}} = 35$$

$$\left(\frac{x}{d_t}\right)_{\text{dev, velocity distribution}} \doteq 5$$

These calculations serve to show why a fully developed concentration distribution is frequently not attained for fluids with high Schmidt number. Lévêque's approximate solution for $(\rho_{AB}-\rho_{Ai})/(\rho_{AW}-\rho_{Ai})$ in such cases was given as equation 5.143. The local and mean Sherwood numbers corresponding to Lévêque's solution will now be presented, since they coincide with the respective Graetz solutions at high $W/\rho Dx$ ($>$c.a. 400), but are simpler in form.

Lévêque's (1928) approximate solution to equation 5.100 for a constant-property fluid is obtained, in mass-transfer terms, by assuming that the concentration boundary layer is confined to a thin zone near the wall of the tube in cases of high mass velocities through relatively short tubes in laminar flow. Linear velocity distributions are assumed within such thin concentration boundary layers, so that $u=\beta_v(r_t-r)$, where β_v is the velocity gradient at the wall. Lévêque's expression for local Sherwood number with uniform concentration along the wall is then

$$(N_{\text{Sh}})_x^* = \frac{k_\rho^* d_t}{D} = \frac{d_t}{0.893}\left(\frac{\beta_v}{9Dx}\right)^{1/3} \tag{5.152}$$

The velocity distribution in fully developed laminar flow of Newtonian fluids in a tube is $u=2V[1-(r/r_t)^2]$, and assuming this to hold in the laminar boundary layer in the entry region,

$$\left(\frac{du}{dy}\right)_{y=0} = \left(-\frac{du}{dr}\right)_{r=r_t} = \frac{4V}{r_t} = \frac{8V}{d_t} \tag{5.153}$$

Inserting equation 5.153 for β_v in equation 5.152,

$$(N_{\text{Sh}})_x^* = 1.077\left(\frac{d_t}{x}\right)^{1/3}(N_{\text{Re}}N_{\text{Sc}})^{1/3} \tag{5.154}$$

where k_ρ^* is for use with $\rho_{AW}-\rho_{AB}$.

Lévêque's expression for the average Sherwood number is

$$(N_{\mathrm{Sh}})_a^* = \frac{k_{\rho a}^* d_t}{D} = 1.615 d_t \left(\frac{\beta_v}{8DL} \right)^{1/3} \tag{5.155}$$

and substituting equation 5.153 for β_v,

$$(N_{\mathrm{Sh}})_a^* = 1.615 \left(\frac{d_t}{L} \right)^{1/3} (N_{\mathrm{Re}} N_{\mathrm{Sc}})^{1/3} \tag{5.156}$$

where $k_{\rho a}^*$ is for use as in equation 5.121. The assumptions regarding the velocity distribution within the concentration boundary layer restrict the use of equations 5.154 and 5.156 to values of $W/\rho Dx$ above about 400.

Uniform Mass Flux at the Wall

The more esoteric boundary condition of uniform mass flux at all points along the tube wall may occasionally be encountered. The heat-transfer analog is of course more commonly found—for example, in tubes which are uniformly heated electrically, in radiant or nuclear heating, and in those countercurrent heat exchangers in which the product of the mass flow rate and specific heat is the same for each fluid. In mass transfer the condition would arise, for instance, if the walls of the tube were porous and the solute were forced through these walls at a fixed rate per unit surface all the way along the tube. The appropriate differential equation is again equation 5.126, and the boundary conditions are

$$x = 0, \quad 0 \leqslant r \leqslant r_t, \qquad \rho_A = \rho_{Ai}$$

$$x > 0, \qquad n_{AW} = \text{constant}$$

$$\frac{\partial \rho_{AB}}{\partial x} = \text{constant}$$

The solution for the local Sherwood number was provided by Sellars, Tribus, and Klein (1956) as

$$(N_{\mathrm{Sh}})_x^* = \frac{k_\rho^* d_t}{D} = \left(\frac{11}{48} - \frac{1}{2} \sum_{j=1}^{j=\infty} \frac{\exp\left[-\lambda_j^2 (x/r_t)/N_{\mathrm{Re}} N_{\mathrm{Sc}} \right]}{C_j \lambda_j^4} \right)^{-1} \tag{5.157}$$

The quantity $\frac{48}{11}$ is the asymptotic value of N_{Sh}^* corresponding to fully developed concentration and velocity distributions. Values of the first three eigenvalues and constants were given by Sellars et al. (1956), and

Table 5.1. Eigenvalues and constants in equation 5.157.

j	λ_j^2	C_j
1	25.68	7.630×10^{-3}
2	83.86	2.058×10^{-3}
3	174.2	0.901×10^{-3}
4	296.5	0.487×10^{-3}
5	450.9	0.297×10^{-3}

these were subsequently revised and extended by Seigel, Sparrow, and Hallman (1958), as shown in Table 5.1. Eigenvalues and constants C_j for j greater than five may be estimated from

$$\lambda_j = 4j + \tfrac{4}{3} \tag{5.158}$$

$$C_j = 0.358\lambda_j^{-2.32} \tag{5.159}$$

The local Sherwood number from equation 5.157 is plotted as the bottom curve in Figure 5.13. A material balance using the specified uniform n_{AW} gives ρ_{AB} at any x. The local value of ρ_{AW} may then be calculated from the corresponding $(N_{\text{Sh}})_x^*$ using equation 5.101.

Fully Developed Velocity and Concentration Distributions

Attention is directed under this heading to the asymptotic or limiting values to which the Sherwood number reduces in the region where laminar flow is fully developed, both in terms of mass transfer and hydrodynamically. These conditions are located downstream from the points at which the outer edges of the hydrodynamic and concentration boundary layers, respectively, meet at the centerline of the flow channel. The boundary conditions of uniform wall concentration and of uniform mass flux at the wall are considered in turn.

Uniform Wall Concentration

The asymptotic or limiting value reached by the Sherwood number for this boundary condition was given earlier by equation 5.148 as

$$N_{\text{Sh}}^* = \frac{k_\rho^* d_t}{D} = 3.656 \tag{5.160}$$

This result is plotted in Figure 5.12.

Uniform Mass Flux at the Wall

The uniform flux at the wall is given by $n_{AW} = k_\rho^*(\rho_{AW} - \rho_{AB})$; with constant k_ρ^* the quantity $\rho_{AW} - \rho_{AB}$ is also independent of x and

$$\frac{\partial \rho_A}{\partial x} = \frac{\partial \rho_{AW}}{\partial x} = \frac{\partial \rho_{AB}}{\partial x} = \text{constant} \tag{5.161}$$

Equation 5.126 may be written in this case as

$$\frac{2V}{D}\left[1 - \left(\frac{r}{r_t}\right)^2\right]\left(\frac{\partial \rho_{AB}}{\partial x}\right)_{\text{const}} = \frac{1}{r}\frac{\partial}{\partial r}\left(r\frac{\partial \rho_A}{\partial r}\right) \tag{5.162}$$

Also,

$$\text{at } r = 0, \quad \frac{\partial \rho_A}{\partial r} = 0$$

$$\text{at } r = r_t, \quad \rho_A = \rho_{AW}$$

Integrating equation 5.162 for these conditions at a given x gives

$$\rho_{AW} - \rho_A = \frac{2r_t^2 V}{D}\left(\frac{\partial \rho_{AB}}{\partial x}\right)_{\text{const}}\left[\frac{3}{16} - \frac{1}{4}\left(\frac{r}{r_t}\right)^2 + \frac{1}{16}\left(\frac{r}{r_t}\right)^4\right] \tag{5.163}$$

This expression gives ρ_A at any r for a cross section where the local concentration at the wall is ρ_{AW}, the latter varying with x. The mass-transfer coefficient is defined as

$$k_\rho^* = \frac{n_{AW}}{\rho_{AW} - \rho_{AB}} = \frac{D\left(\dfrac{\partial \rho_A}{\partial r}\right)_{r=r_t}}{\rho_{AW} - \rho_{AB}} \tag{5.164}$$

where

$$\rho_{AW} - \rho_{AB} = \frac{\int_0^{r_t} (\rho_{AW} - \rho_A) ur\,dr}{\int_0^{r_t} ur\,dr} \tag{5.165}$$

Combination of equations 5.163 and 5.165 with the parabolic velocity distribution $2V[1-(r/r_t)^2]$ for u gives

$$\rho_{AW}-\rho_{AB}=\frac{11}{48}\frac{r_t^2V}{D}\left(\frac{\partial\rho_A}{\partial x}\right)_{\text{const}} \tag{5.166}$$

From equation 5.163,

$$\left(\frac{\partial\rho_A}{\partial r}\right)_{r=r_t}=\frac{r_tV}{2D}\left(\frac{\partial\rho_A}{\partial x}\right)_{\text{const}} \tag{5.167}$$

Equations 5.166 and 5.167 are substituted in equation 5.164 to obtain

$$N_{\text{Sh}}^*=\frac{k_\rho^* d_t}{D}=\frac{48}{11}=4.36 \tag{5.168}$$

This value is shown as the asymptote in Figure 5.13.

Assuming a Cubic Polynomial Concentration Distribution

The radial distribution of component A is assumed to follow a cubic polynomial form:

$$\rho_A=\rho_{AW}-a_1y-a_2y^2-a_3y^3, \qquad y=r_t-r \tag{5.169}$$

To evaluate a_1, a_2, and a_3, consider a differential length of tube dx, where the mass transfer through a layer immediately adjacent to the wall occurs by molecular diffusion, so that

$$dq_{AW}=-D2\pi(r_t-y)dx\frac{\partial\rho_A}{\partial y} \tag{5.170}$$

Then

$$\frac{\partial^2\rho_A}{\partial y^2}=-\frac{dq_{AW}}{D2\pi(r_t-y)^2dx}=\frac{1}{r_t-y}\frac{\partial\rho_A}{\partial y}$$

but y equals zero at the wall, so that

$$\left(\frac{\partial^2\rho_A}{\partial y^2}\right)_{y=0}=\frac{1}{r_t}\left(\frac{\partial\rho_A}{\partial y}\right)_{y=0} \tag{5.171}$$

The derivatives in this expression are evaluated from equation 5.169 to obtain $a_2=a_1/2r_t$. At the axis of the tube,

$$\left(\frac{\partial\rho_A}{\partial y}\right)_{y=r_t}=0=-2a_1-3a_3r_t^2, \qquad a_3=\frac{-2a_1}{3r_t^2}$$

Substitution for a_2 and a_3 in equation 5.169 gives

$$\rho_A = \rho_{AW} - a_1 y\left[1 + \frac{1}{2}\left(\frac{y}{r_t}\right) - \frac{2}{3}\left(\frac{y}{r_t}\right)^2\right] \tag{5.172}$$

The ratio of $\rho_A - \rho_{AW}$ at y to $\rho_A - \rho_{AW}$ at the centerline is therefore

$$\frac{\rho_A - \rho_{AW}}{\rho_{As} - \rho_{AW}} = \frac{6y}{5r_t} + \frac{3}{5}\left(\frac{y}{r_t}\right)^2 - \frac{4}{5}\left(\frac{y}{r_t}\right)^3 \tag{5.173}$$

For Newtonian fluids a combination of the momentum and rheological equations results in

$$\tau_{rx} = \frac{r\Delta P}{2L} = \frac{\mu}{g_c}\left(-\frac{du}{dr}\right) \tag{5.174}$$

Integrating with $u=0$ at $r=r_t$,

$$u = \frac{\Delta P g_c}{4\mu L}(r_t^2 - r^2) = \frac{\Delta P g_c}{4\mu L}(2r_t y - y^2)$$

The ratio of u at y to u at the centerline is therefore

$$\frac{u}{u_s} = \frac{2y}{r_t} - \left(\frac{y}{r_t}\right)^2 \tag{5.175}$$

Equations 5.173 and 5.175 are solved for $\rho_A - \rho_{AW}$ and u, respectively, and the results inserted in equation 5.165. After expanding, integrating, and simplifying, one obtains

$$\rho_{AB} - \rho_{AW} = 0.583(\rho_{As} - \rho_{AW}) \tag{5.176}$$

The flux of component A from the wall is as follows at low concentrations and transfer rates:

$$n_{AW} = -k_\rho^*(\rho_{AB} - \rho_{AW}) = -D\left(\frac{\partial \rho_A}{\partial y}\right)_{y=0} = -\frac{6}{5}\frac{D}{r_t}(\rho_{As} - \rho_{AW})$$

and introducing equation 5.176,

$$N_{\text{Sh}}^* = \frac{k_\rho^* d_t}{D} = \frac{k_\rho^*(2r_t)}{D} = 4.12 \tag{5.177}$$

Equation 5.177 gives a value for the Sherwood number for mass transfer under fully developed conditions. It is evident that the Sherwood number calculated from the assumed cubic-polynomial concentration distribution —with unspecified wall conditions—falls between the values for uniform wall concentration and uniform mass flux at the wall. Extensions to provide asymptotic Nusselt and Sherwood numbers for heat and mass transfer in non-Newtonian fluids have been provided by Skelland (1967).

Illustration 5.5.

A tube with an internal diameter of 1 in. is cast from solid naphthalene. Pure air enters the tube at a velocity of 3 ft/sec. If the average air pressure is 14.7 psia and the system is at 113°F, estimate the tube length required for the average concentration of naphthalene vapor in the air to reach a value of

(a) 2.24×10^{-5} lb-mass/ft^3 (i.e., 10 percent of saturation),

(b) 5.6×10^{-5} lb-mass/ft^3 (i.e., 25 percent of saturation).

Compare with Illustration 6.3 for turbulent conditions.

SOLUTION (a). The relevant physical properties for this system at these conditions of temperature and pressure are available from Illustrations 5.1 and 5.2. The naphthalene surface temperature will be considered to be 113°F (see Illustration 5.1).

$$N_{\mathrm{Re}}=\frac{(1/12)(3)(3600)(0.0694)}{0.0457}=1365$$

Flow is therefore laminar.

$$\frac{N_{\mathrm{Re}}N_{\mathrm{Sc}}}{x/r_t}=\frac{1365(2.475)(0.5)}{12x}=\frac{140.8}{x}$$

$$\frac{q_{AW}}{A}=\frac{\pi d_t^2 V\Delta\rho_A}{4\pi d_t x}=\frac{3(3600)(2.24\times10^{-5})}{12(4)x}=\frac{0.00504}{x}$$

$$\Delta\rho_{A\mathrm{LM}}=\frac{(0.000224-0)-(0.000224-0.0000224)}{2.303\log[0.000224/(0.000224-0.0000224)]}$$

$$=0.000211 \text{ lb naphthalene/ft}^3$$

$$(N_{\mathrm{Sh}})^*_{\mathrm{LM}}=\frac{k^*_{\rho\mathrm{LM}}d_t}{D}=\frac{q_{AW}}{A\Delta\rho_{A\mathrm{LM}}}\frac{d_t}{D}=\frac{0.00504(1/12)}{0.000211(0.2665)x}=\frac{7.47}{x}$$

Therefore

$$x=\frac{7.47}{(N_{\rm Sh})^*_{\rm LM}}, \qquad \frac{N_{\rm Re}N_{\rm Sc}}{x/r_t}=\frac{140.8}{x}$$

Various values are assumed for x, and the corresponding $(N_{\rm Sh})^*_{\rm LM}$ is read from the interpolated curve for $N_{\rm Sc}=2.475$ in Figure 5.12. The correct x is found when the assumed value equals $7.47/(N_{\rm Sh})^*_{\rm LM}$. The result is

$$x=0.45 \text{ ft}, \qquad \frac{N_{\rm Re}N_{\rm Sc}}{x/r_t}=313, \quad (N_{\rm Sh})^*_{\rm LM}=16.6$$

The value of $(N_{\rm Sh})^*_{\rm LM}$ is 4.54 times greater than the limiting or asymptotic value of 3.656 (equation 5.160), demonstrating the very substantial contribution of the entrance effects to the transfer process in this case. (Equations 5.149 and 5.151 show that the entrance lengths required to achieve fully developed concentration and velocity distributions are about 14.08 and 5.69 ft, respectively, along the tube.)

SOLUTION (b). Proceeding as in Solution (a),

$$x=\frac{20.2}{(N_{\rm Sh})^*_{\rm LM}}, \qquad \frac{N_{\rm Re}N_{\rm Sc}}{x/r_t}=\frac{140.8}{x}$$

and by trial-and-error use of Figure 5.12,

$$x=2.04 \text{ ft}, \qquad \frac{N_{\rm Re}N_{\rm Sc}}{x/r_t}=69, \quad (N_{\rm Sh})^*_{\rm LM}=9.9$$

The average flux at the wall over the first 2.04 ft along the tube is

$$\frac{q_{AW}}{A}=\frac{\pi d_t^2 V\Delta\rho_A}{4\pi d_t x}=\frac{3(3600)(5.6\times10^{-5})}{12(4)2.04}$$

$$=0.00617 \text{ lb naphthalene}/(\text{ft}^2)(\text{hr}).$$

Illustration 5.6

Pure water flows at an average velocity of 0.05 ft/sec through a $\frac{1}{4}$-in.-diameter metal tube. At a distance of 1 ft from the inlet the metal tube is replaced by a tube of the same diameter, cast from benzoic acid, and having a length of 4 ft. If the system is at a temperature of 77°F, estimate the average concentration of benzoic acid in the water at the outlet from the cast tube.

SOLUTION. The relevant physical properties are as follows at the prevailing temperature:

$$\rho_{H_2O} = 62.24 \text{ lb-mass/ft}^3$$

$$\mu_{H_2O} = 2.16 \text{ lb-mass/(ft)(hr)}$$

$$D = 4.695 \times 10^{-5} \text{ ft}^2/\text{hr}\dagger$$

$$\rho_A^* = \text{saturation solubility} = 0.213 \text{ lb } C_6H_5COOH/\text{ft}^3 \text{ of aqueous solution}\ddagger$$

The density and viscosity will be regarded as constant in view of the low solubility of benzoic acid in water.

$$N_{Re} = \frac{(0.25/12)(0.05)(3600)(62.24)}{2.16} = 108$$

The flow is therefore laminar, and from equation 5.151,

$$x_{dev} \doteq 0.05(108)\frac{0.25}{12} = 0.1125 \text{ ft}$$

Thus the velocity profile is fully developed to parabolic form at the entrance to the benzoic acid section of the tube.

$$N_{Sc} = \frac{2.16}{62.24(0.00004695)} = 740$$

$$\frac{N_{Re}N_{Sc}}{x/r_t} = \frac{108(740)(0.125)}{4(12)} = 208$$

Figure 5.12 yields $(N_{Sh})^*_{LM} = 12$.

Alternatively, from equation 5.147,

$$(N_{Sh})^*_{LM} = 3.66 + \frac{0.0668[2(208)]}{1 + 0.04[2(208)]^{2/3}} = 12.26$$

†From R. C. Reid and T. K. Sherwood, *The Properties of Gases and Liquids*, 2nd ed., McGraw Hill, New York, (1966), p. 555.

‡From A. Seidell, *Solubilities of Organic Compounds*, 3rd ed., Vol. II, Van Nostrand, New York, (1941), pp. 500–501.

This exceeds the asymptotic value of 3.656 (equation 5.160), corresponding to complete development of both velocity and concentration distributions, by a factor of 3.35.

$$k^*_{\rho LM} = \frac{12.26(0.00004695)}{0.25/12} = 0.0276 \text{ ft/hr}$$

$$A = \pi d_t x = \pi\left(\frac{0.25}{12}\right)4 = 0.262 \text{ ft}^2$$

$$q_{AW} = 0.0276(0.262)\frac{(0.213-0)-(0.213-\rho_{ABO})}{2.303\log[0.213/(0.213-\rho_{ABO})]} = \pi\frac{d_t^2}{4}V(\rho_{ABO}-\rho_{Ai})$$

$$= \frac{\pi}{4}\left(\frac{0.25}{12}\right)^2 0.05(3600)\rho_{ABO}$$

$$\log\left(\frac{0.213}{0.213-\rho_{ABO}}\right) = 0.0512$$

$$\rho_{ABO} = 0.024 \text{ lb } C_6H_5COOH/\text{ft}^3 \text{ of solution.}$$

As an alternative procedure the mass-transfer equivalent of the Graetz number may be calculated as follows:

$$\frac{W}{\rho D x} = \frac{\pi}{4}\frac{d_t}{x}N_{Re}N_{Sc} = \frac{\pi}{4}(2)(208) = 327$$

The corresponding ordinate of the parabolic-flow curve in Figure 5.14 is

$$\frac{\rho_{ABO}-\rho_{Ai}}{\rho_{AW}-\rho_{Ai}} \doteq 0.116$$

in agreement with the value obtained from equation 5.143. Since $\rho_{Ai}=0$ and $\rho_{AW}=0.213$, this corresponds to $\rho_{ABO}=0.0247$ lb C_6H_5COOH/ft^3 of solution, which, of course, agrees satisfactorily with the previous estimate.

Dispersion in Tube Flow

When a soluble material is injected locally into a fluid flowing through a tube, it is dispersed by a combination of two mechanisms: convection associated with the velocity distribution across the tube, and either molecular or eddy diffusion, depending upon the flow regime. In laminar flow, when the injected material has spread to occupy a length of tube much greater than the diameter, Taylor (1953) finds that axial convection and radial diffusion combine to disperse the solute axially (relative to a plane moving at the mean velocity of flow) by a mechanism which follows the same law as one-dimensional molecular diffusion in a stationary fluid. The governing expression parallels equation 2.93, with z replaced by $x-Vt$ and with D replaced by $r_t^2V^2/48D$. A solution is presented for the case of momentary injection, giving the solute concentration after a time t at a distance x downstream from the point of injection. A similar result is provided for the case in which a solution containing a given concentration of solute starts to flow at $t=0, x=0$ into a tube initially filled with pure solvent. The reader is referred to the original papers for details, for application to diffusivity measurement, and for extension to turbulent conditions (Taylor, 1953, 1954a, b, c).

MASS TRANSFER IN LAMINAR FLOW BETWEEN FLAT PARALLEL PLATES

Applications of this geometry in mass transfer under either laminar or turbulent conditions include the desalination of sea water by reverse osmosis during flow between flat, parallel membranes (Sourirajan, 1970, Chapter 4), hemodialysis for the replacement of kidney function during renal failure (Wolf and Zaltzman, 1968; Colton et al., 1971), transpiration and film cooling, gas absorption (e.g., Sherwood and Woertz, 1939), and the condensation of mixtures (e.g., Rohsenow and Choi, 1961, Chapter 10). Corrosion, scaling, and descaling problems may arise in heat exchangers of the plate and plate-fin varieties and of the flat-plate type sometimes used in evaporators (e.g., Perry, 1963, pp. **11**-13, **11**-28).

Consideration will be confined to low solute concentrations and transfer rates. The development is analogous to that given earlier for tubes, and the results are summarized in Table 5.2 and Figures 5.17 and 5.18. The coefficients $k^*_{\rho a}$, $k^*_{\rho \mathrm{LM}}$, and k^*_{ρ} in equations 5.182, 5.183, and 5.184 are for use in the following expressions:

$$H_p wV(\rho_{ABO}-\rho_{Ai})=k^*_{\rho a}(2wL)\times\tfrac{1}{2}[(\rho_{AW}-\rho_{Ai})+(\rho_{AW}-\rho_{ABO})]$$

$$H_p wV(\rho_{ABO}-\rho_{Ai})=k^*_{\rho \mathrm{LM}}(2wL)\frac{(\rho_{AW}-\rho_{Ai})-(\rho_{AW}-\rho_{ABO})}{\ln[(\rho_{AW}-\rho_{Ai})/(\rho_{AW}-\rho_{ABO})]}$$

$$\rho_{AW,\text{local at }x}=\rho_{AB}+\frac{n_{AW}}{k^*_{\rho}}$$

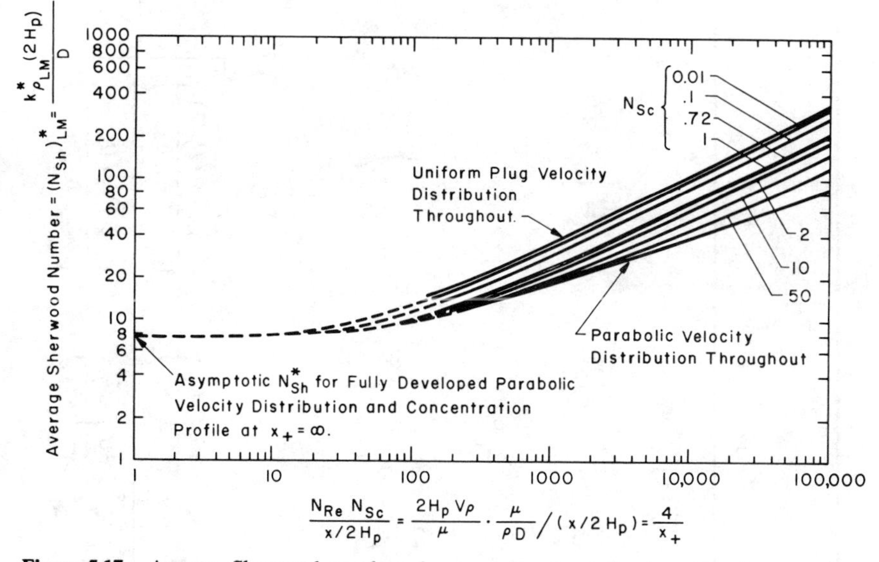

Figure 5.17. Average Sherwood numbers for mass transfer in laminar flow between flat parallel plates with uniform wall concentration (Sparrow, 1955).

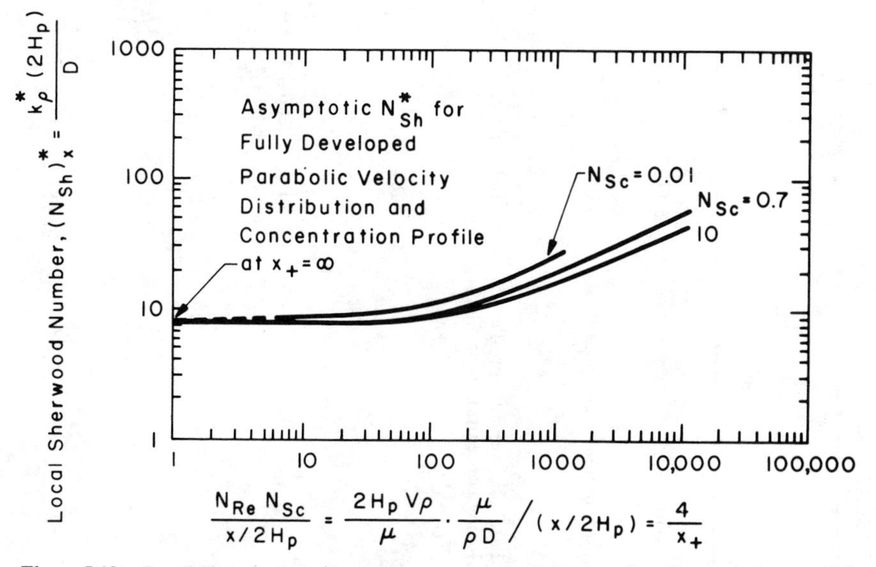

Figure 5.18. Local Sherwood numbers for mass transfer in laminar flow between flat parallel plates with uniform mass flux at the wall (Heaton, Reynolds, and Kays, 1964).

Table 5.2. Mass transfer in laminar flow between flat parallel plates.

Conditions	Eq. no.	Equation[a]	Analog from tube analysis
Continuity	5.178	$\rho\left(\frac{\partial u}{\partial x}+\frac{\partial v}{\partial h}\right)+u\frac{\partial \rho}{\partial x}+v\frac{\partial \rho}{\partial h}+\frac{\partial \rho}{\partial t}=0$	Eq. 5.89
Distribution of component A	5.179	$u\frac{\partial \rho_A}{\partial x}+v\frac{\partial \rho_A}{\partial h}+\frac{\partial \rho_A}{\partial t}+\rho_A\left(\frac{\partial u}{\partial x}+\frac{\partial v}{\partial h}\right)=D\left[\frac{\partial^2 \rho_A}{\partial h^2}+\frac{\partial^2 \rho_A}{\partial x^2}\right]$	Eq. 5.97
Parabolic velocity profile; developing concentration profile; $\rho_{AW} \neq \rho_{AW}(x)$	5.180	$\frac{\rho_{AW}-\rho_A}{\rho_{AW}-\rho_{Ai}}(h_+,x_+)=\sum_{j=1}^{j=\infty} G_j\psi_j(h_+)\exp\left(-\frac{8}{3}\gamma_j^2 x_+\right)$	Eq. 5.131
" " "	5.181	$\frac{\rho_{AW}-\rho_{AB}}{\rho_{AW}-\rho_{Ai}}=\sum_{j=1}^{j=\infty}\frac{-3G_j}{2\gamma_j^2}\left(\frac{d\psi_j}{dh_+}\right)_{h_+=1}\exp\left(-\frac{8}{3}\frac{\gamma_j^2 x/(H_p/2)}{N_{Re}N_{Sc}}\right)$	Eq. 5.142
" " "	5.182	$(N_{Sh})^*_a=\frac{k^*_{\rho a}(2H_p)}{D}$ $=\left(\frac{H_p}{L}\right)N_{Re}N_{Sc}\left[\frac{1-\sum_{j=1}^{j=\infty}\frac{-3G_j}{2\gamma_j^2}\left(\frac{d\psi_j}{dh_+}\right)_{h_+=1}\exp\left(-\frac{8}{3}\frac{\gamma_j^2 x/(H_p/2)}{N_{Re}N_{Sc}}\right)}{1+\sum_{j=1}^{j=\infty}\frac{-3G_j}{2\gamma_j^2}\left(\frac{d\psi_j}{dh_+}\right)_{h_+=1}\exp\left(-\frac{8}{3}\frac{\gamma_j^2 x/(H_p/2)}{N_{Re}N_{Sc}}\right)}\right]$	Eq. 5.145

Parabolic velocity profile; developing concentration profile; $\rho_{AW} \neq \rho_{AW}(x)$	5.183	$(N_{Sh})^*_{LM} = \dfrac{k^*_{\rho LM}(2H_p)}{D} = \dfrac{1}{2}\left(\dfrac{H_p}{L}\right) N_{Re} N_{Sc} \ln\left[\sum_{j=1}^{j=\infty} \dfrac{-3G_j}{2\gamma_j^2}\left(\dfrac{d\psi_j}{dh_+}\right)_{h_+=1} \exp\left(-\dfrac{8}{3}\dfrac{\gamma_j^2 x/(H_p/2)}{N_{Re}N_{Sc}}\right)\right]^{-1}$	Eq. 5.146
Parabolic velocity profile; developing concentration profile; $n_{AW} \neq n_{AW}(x)$. (Cess and Shaffer, 1959)	5.184	$(N_{Sh})^*_x = \dfrac{k^*_\rho(2H_p)}{D} = \left[\dfrac{17}{140} + \dfrac{1}{4}\sum_{j=1}^{j=\infty} F_j X_j(1) \exp\left(-\dfrac{8}{3}\dfrac{\eta_j^2 x/(H_p/2)}{2VH_p/D}\right)\right]^{-1}$	Eq. 5.157
Fully developed velocity and concentration profiles	5.185	$N^*_{Sh} = \dfrac{k^*_\rho(2H_p)}{D} = 7.6, \quad \rho_{AW} \neq \rho_{AW}(x)$ (Tien, 1962)	Eq. 5.160
" " "	5.186	$N^*_{Sh} = \dfrac{k^*_\rho(2H_p)}{D} = 8.23, \quad n_{AW} \neq n_{AW}(x)$	Eq. 5.168
" " "	5.187	$N^*_{Sh} = \dfrac{k^*_\rho(2H_p)}{D} = 7.86, \quad \rho_A = \rho_{AW} - a_1 y - a_2 y^2 - a_3 y^3$	Eq. 5.177

[a] $x_+ = \dfrac{x/(H_p/2)}{N_{Re}N_{Sc}}, \; h_+ = \dfrac{h}{H_p/2}, \; N_{Re} = \dfrac{2H_p V\rho}{\mu}, \; G_j = \dfrac{-2}{\gamma_j(\partial\psi/\partial\gamma)_{j,h_+=1}}.$

Values of γ, η, ψ, $(\partial\psi/\partial\gamma)_{j,h_+=1}$, $(d\psi_j/dh_+)_{h_+=1}$, F_j, and X_j are given in Tables 6, 7, and 8 of the Appendix.

To emphasize the analogy with the tube development, the last column of Table 5.2 indicates the equation in the tube analysis which is analogous to the expression for parallel plates. Thus equation 5.145 for a tube corresponds to equation 5.182 for parallel plates.

Equation 5.183 is plotted as the bottom curve in Figure 5.17, where it may be compared with solutions for simultaneously developing velocity and concentration distributions.

MASS TRANSFER IN LAMINAR FLOW THROUGH CONCENTRIC ANNULI

The annulus represents a geometry of frequent occurrence in many heat-transfer applications, ranging from sophisticated nuclear reactors to simple heat exchangers. Mass-transfer problems of corrosion, scaling, and descaling may arise under laminar or turbulent conditions in some of these devices, while further mass-transfer phenomena are involved in annular reactors (e.g., U.S. Patent 1,029,029), in annular condensers for mixed vapors (e.g., Estrin, Hayes, and Drew, 1965), and in transpiration and film cooling of annular ducts.

Mass transfer during flow through an annulus may occur under a variety of conditions of concentration and mass flux at each of the two cylindrical surfaces. Thus a specified distribution of concentration or mass flux along one surface may be combined with a possibly different distribution of either quantity along the other surface, to yield a proliferation of possibilities which will not be considered in detail here. Solutions of the relevant equations have been obtained for systematic combinations of these alternatives by Lundberg, Reynolds, and Kays (1963), for the analogous case of heat transfer in hydrodynamically fully developed laminar flow of incompressible, constant-property fluids. Superposition of the solutions provided by these authors enables any general axially symmetric boundary conditions to be satisfied. The heat-transfer solutions can of course be converted to the corresponding mass-transfer ones, under conditions for which the bulk velocity due to diffusion is low, by the substitution of molecular diffusivity for thermal diffusivity, Sherwood number for Nusselt number, and Schmidt number for Prandtl number. The reader is referred to the original 194-page report for details of the various solutions.

Relationships in the entrance region of an annulus, where velocity and concentration distributions are both developing, have been presented by Heaton, Reynolds, and Kays (1964).

Asymptotic solutions for fully developed velocity and concentration distributions far downsteam from the entrance are given by Lundberg, Reynolds, and Kays (1963).

Further annulus studies are reported in a series of papers by Reynolds, Lundberg, and McCuen (1963); Lundberg, McCuen, and Reynolds (1963); and Kays and Leung (1963)—the last paper dealing with turbulent flow.

TRANSFER WITH A HIGH MASS FLUX

The mass-transfer relationships developed so far in this chapter have been for processes in which the solute concentrations and mass flux are low. Under these conditions the velocity components associated with the mass transfer are small enough to exert a negligible influence on the velocity field. A simple analogy then prevails between the processes of momentum, heat, and mass transfer. Physical properties in such systems are independent of location, since concentration variations are small.

Consideration is now given to high-mass-flux conditions, in which the associated components of velocity introduce significant differences between mass transfer on the one hand and heat- or momentum-transfer processes without mass transfer on the other. The influence of mass transfer when superimposed on momentum and heat transfer processes also receives mention. A useful review of this field is given by Nienow (1967), who provides a bibliography of nearly fifty references.

Much of the published work on high-mass-flux problems expresses concentrations in terms of mass per unit volume or mass fraction. These units will accordingly be used here to facilitate further literature study of the subject.

Consider the transfer of a single component A from the surface of a flat plate to a laminar stream of fluid B flowing parallel to the plate. The momentum- and concentration-boundary-layer equations have already been developed as equations 5.10 and 5.8, respectively, and may be written as follows with constant physical properties:

$$\frac{g_c}{\rho}\tau_0 = \frac{\mu}{\rho}\left(\frac{\partial u}{\partial y}\right)_{y=0} = \frac{d}{dx}\int_0^{\delta}(u_\infty - u)u\,dy - v_0 u_\infty \tag{5.10}$$

$$-D\left(\frac{\partial \rho_A}{\partial y}\right)_{y=0} = \frac{d}{dx}\int_0^{\delta_c}(\rho_A - \rho_{A\infty})u\,dy - v_0(\rho_{A0} - \rho_{A\infty}) \tag{5.8}$$

The first term on the right-hand side of equation 5.10 is always positive, whereas the second term is negative for mass flux *from* the wall. The gradient $(\partial u/\partial y)_{y=0}$ is therefore reduced, and since u changes by the fixed amount u_∞ to zero, this corresponds to an increase in the momentum-boundary-layer thickness for a finite v_0. The local shear stress at the wall, τ_0, is also reduced under these conditions.

Similar reasoning applied to equation 5.8 indicates that a finite mass flux from the wall increases the thickness of the concentration boundary layer and reduces the mass-transfer coefficient, defined as

$$k_\rho^0 = -\frac{D}{\rho_{A0}-\rho_{A\infty}}\left(\frac{\partial \rho_A}{\partial y}\right)_{y=0} \tag{5.188}$$

The reverse conclusions hold for a significant mass flux *to* the wall, resulting in reduced thickness of the boundary layers, accompanied by increases in the mass-transfer coefficient and surface shear stress. Corresponding qualitative conclusions may be obtained for the thermal boundary layer in the presence of high mass-transfer rates, as outlined by Nienow (1967).

These qualitative indications of the influence of a high mass flux are in accordance with the results of more rigorous analyses by Schuh (1950), Mickley et al. (1954), Hartnett and Eckert (1957), and Stewart (1962). The results of Hartnett and Eckert are perhaps the most frequently cited, but the phenomena are also well illustrated by the findings of Schuh (1950), and these will be outlined here.

Schuh's development is for constant-property fluids in steady flow over a flat plate oriented parallel to the undisturbed stream, and with component A being transferred between the plate and the fluid. The treatment begins with the familiar differential equations for the boundary layer, the derivation of which may be found, for example, in Schlichting (1967) and Knudsen and Katz (1958). For the two-dimensional steady-state process considered here, neglecting frictional heat generation and assuming that transport in the x direction by diffusion and by conduction are both negligible, the starting equations are: the continuity equation

$$\frac{\partial u}{\partial x}+\frac{\partial v}{\partial y}=0 \tag{5.189}$$

the momentum conservation equation

$$u\frac{\partial u}{\partial x}+v\frac{\partial u}{\partial y}=\frac{\mu}{\rho}\frac{\partial^2 u}{\partial y^2} \tag{5.190}$$

the equation for the conservation of component A

$$u\frac{\partial \rho_A}{\partial x}+v\frac{\partial \rho_A}{\partial y}=D\frac{\partial^2 \rho_A}{\partial y^2} \tag{5.191}$$

and the energy conservation equation

$$u\frac{\partial T}{\partial x}+v\frac{\partial T}{\partial y}=\alpha_t\frac{\partial^2 T}{\partial y^2} \tag{5.192}$$

The mass-transfer process is assumed to introduce negligible fluxes of energy and momentum, and the boundary conditions are

$$y=0, \qquad u=0, \qquad v=v_0, \qquad \rho_A=\rho_{A0},$$

$$T=T_0, \qquad \rho_{A0}, T_0 \text{ independent of } x$$

$$y=\infty, \qquad u=u_\infty, \qquad \rho_A=\rho_{A\infty}, \qquad T=T_\infty$$

Schuh converted the velocity, concentration, temperature, and y coordinate in these equations into the following dimensionless forms:

$$\omega(\xi)=\frac{u}{u_\infty}, \quad C(\xi)=\frac{\rho_A-\rho_{A0}}{\rho_{A\infty}-\rho_{A0}}$$

$$\theta(\xi)=\frac{T-T_0}{T_\infty-T_0}, \quad \xi=\frac{y}{2}\sqrt{\frac{\rho u_\infty}{\mu x}}=\frac{y}{2x}\sqrt{N_{\mathrm{Re},x}}$$

The boundary conditions for the velocity, concentration, and temperature fields then read

$$y=0, \qquad \xi=0, \omega=0, C=0, \theta=0$$

$$y=\infty, \qquad \xi=\infty, \omega=1, C=1, \theta=1$$

The bulk velocity in the y direction at the wall, v_0, was expressed, after Nusselt, as

$$v_0=\frac{-D}{(P/p_{A0}-1)\rho_{A0}}\left(\frac{\partial\rho_A}{\partial y}\right)_{y=0} \tag{5.193}$$

where P is the total pressure and p_{A0} is the partial pressure of component A at the plate surface. The appearance of the mole fraction (p_{A0}/P) in this expression for the mass-average velocity may perhaps be surprising at first sight. The development, however, is for fluids with constant properties. In the case of gases (to which Schuh's results are confined), constant density at fixed P and T requires constancy of the average molecular weight, $\overline{M}$, as in $\rho_{\text{ideal}} = P\overline{M}/RT$. This necessitates equal molecular weights for all components of the gas mixture, so that mass and mole fractions of a given component are the same. The mass- and molal-average velocities of the mixture are then equal. The bulk velocity in dimensionless form becomes

$$\frac{v_0}{u_\infty} = -\frac{1}{2}\frac{D}{\rho_{A0}}\frac{(\rho_{A\infty}-\rho_{A0})}{(P/p_{A0}-1)u_\infty}\left(\frac{dC}{d\xi}\right)_{y=0}\sqrt{\frac{\rho u_\infty}{\mu x}} \tag{5.194}$$

Solutions obtained by Schuh (1950) for the velocity and concentration fields are presented in dimensionless form in Figures 5.19 and 5.20 for a Schmidt number of 0.6, with M as parameter. The value 0.6 for the Schmidt number is applicable, for example, to the diffusion of water or ammonia in air, although the assumption of constant physical properties (independent of composition) would not be completely satisfied for such systems. The quantity M is defined by

$$M = -\frac{1}{N_{\text{Sc}}}\frac{(\rho_{A\infty}-\rho_{A0})}{\rho_{A0}(P/p_{A0}-1)}\left(\frac{dC}{d\xi}\right)_{y=0} = \frac{2v_0}{u_\infty}\sqrt{N_{\text{Re},x}} \tag{5.195}$$

Negative values of M correspond to mass transfer towards the plate; positive values, to transfer from the plate to the fluid. Figures 5.19 and 5.20 show that both velocity and concentration profiles are strongly influenced by M. The profile slopes increase with increasing mass transfer towards the surface, corresponding to reduced boundary-layer thicknesses and increased surface shear stress and mass-transfer coefficient. The reverse is true when mass transfer occurs from the surface. These conclusions also follow from Figure 5.21, which shows the dimensionless concentration gradient at the wall and the quantity M as functions of N, a dimensionless quantity proportional to the driving force $\rho_{A0}-\rho_{A\infty}$ and therefore to the rate of mass transfer to or from the plate. The local mass-transfer coefficient at a given value of N follows from the definitions of C and ξ as

$$k_\rho^0 = \frac{D}{2x}\sqrt{N_{\text{Re},x}}\left(\frac{dC}{d\xi}\right)_{y=0} \tag{5.196}$$

where $(dC/d\xi)_{y=0}$ is obtained from Figure 5.21.

It appears that the velocity and concentration boundary layers are lifted off the plate when M reaches a value of 1.238 (Emmons and Leigh, 1953; Hartnett and Eckert, 1957). Boundary-layer equations will therefore not apply as v_0 approaches $0.619u_\infty/\sqrt{N_{\text{Re},x}}$.

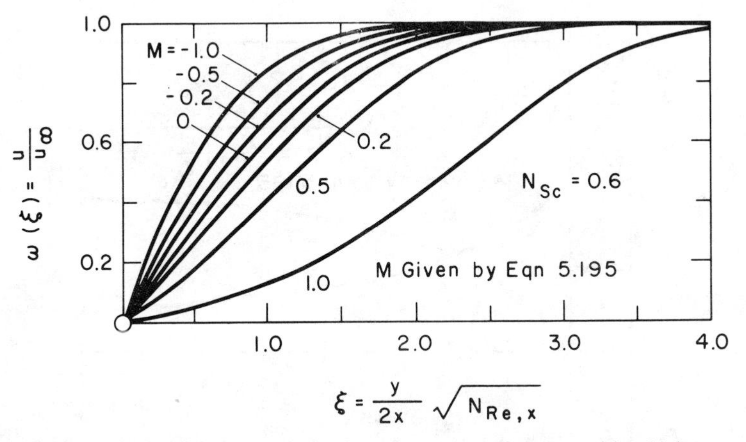

Figure 5.19. Velocity field in steady laminar flow over a flat plate with a high rate of mass transfer between the plate and the fluid (Schuh, 1950).

Illustration 5.7

Pure air at a temperature of 150°F and a pressure of 14.7 psia flows with a velocity of 20 ft/sec along the surface of a porous flat plate. The plate is thoroughly wetted with water and is maintained at a temperature of 150°F by a technique similar to that used in the experiments of Mendelson and Yerazunis (1965). If the plate is continually supplied from beneath with water at 150°F, estimate the rate of evaporation at a point 9 in. from the leading edge.

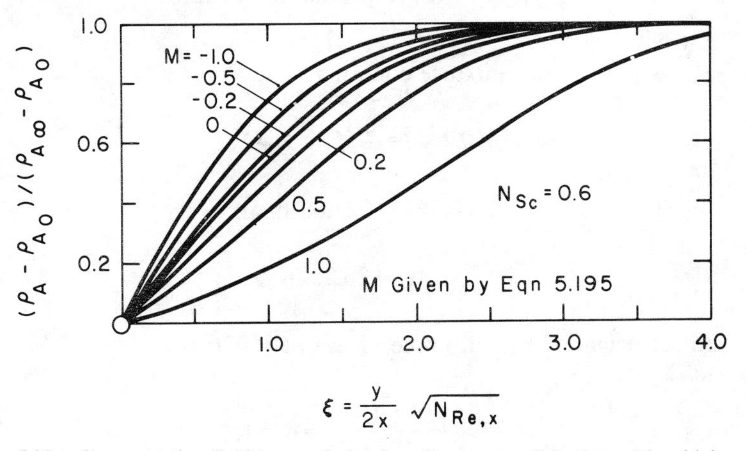

Figure 5.20. Concentration field in steady laminar flow over a flat plate with a high rate of mass transfer between the plate and the fluid (Schuh, 1950).

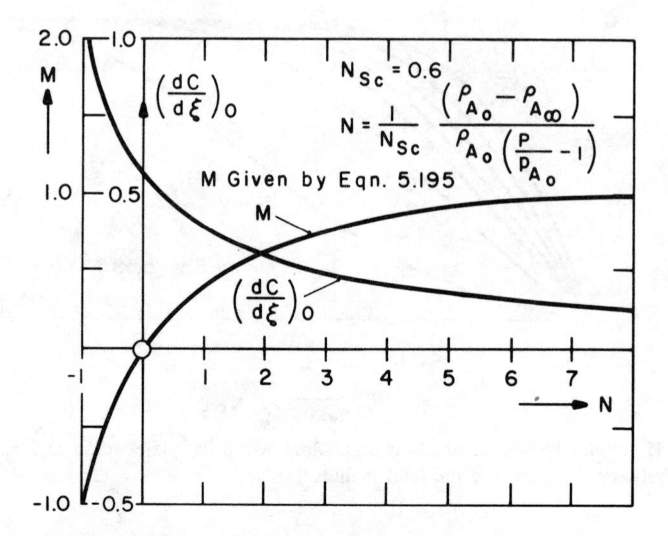

Figure 5.21. Dimensionless concentration gradient at the wall and the quantity M as functions of N (Schuh, 1950).

SOLUTION. At a temperature of 150°F the vapor pressure of water is found from steam tables to be 3.718 lb-force/in.^{2}. Thus, at the surface,

$$y_A = \frac{3.718}{14.7} = 0.253 = \text{mole fraction } H_2O$$

$$y_B = 1 - y_A = 0.747 = \text{mole fraction air}$$

One pound-mole of this mixture contains

$$18(0.253) = 4.55 \text{ lb } H_2O$$

$$29(0.747) = \underline{21.68} \text{ lb air}$$

$$M_0 = 26.23 \text{ lb}$$

The volume occupied by 1 lb-mole of gas at 150°F is

$$359\left(\frac{460+150}{492}\right) = 445 \text{ ft}^3$$

so

$$\rho_{A0} = \frac{4.55}{445} = 0.01022 \text{ lb-mass/ft}^3$$

$$\rho_{A\infty} = 0$$

$$\rho_0 = \frac{26.23}{445} = 0.059 \text{ lb-mass/ft}^3$$

$$\rho_\infty = \frac{29}{445} = 0.0652 \text{ lb-mass/ft}^3$$

$$\rho_{av} = \tfrac{1}{2}(0.059 + 0.0652) = 0.0621 \text{ lb-mass/ft}^3$$

The viscosity of dry air at 150°F and 1 atm is 0.0195 cP (Perry, 1963, p. 3-197), and for moist air containing a mole fraction 0.253 of water vapor it is 0.0188 cP (*ASHRAE Handbook of Fundamentals*, 1967, p. 109). Therefore

$$\mu_{av} = \tfrac{1}{2}(0.0195 + 0.0188) = 0.0192 \text{ cP}$$

The diffusivity for the air–water-vapor system at 150°F and atmospheric pressure is 0.32 cm^2/sec (*ASHRAE Handbook of Fundamentals*, 1967, p. 109). Thus

$$N_{Sc} = \frac{\mu}{\rho D} = \frac{0.0192(2.42)}{0.0621(0.32)(3.88)} = 0.602$$

which coincides with the N_{Sc} value for which Schuh's solution was developed; evidently his requirement of constant physical properties is also approximately satisfied. The term N in Figure 5.21 is evaluated as

$$N = \frac{1}{0.602} \frac{0.01022 - 0}{0.01022\left(\dfrac{14.7}{3.718} - 1\right)} = 0.563$$

From Figure 5.21,

$$\left(\frac{dC}{d\xi}\right)_{y=0} = 0.444$$

$$M = N\left(\frac{dC}{d\xi}\right)_{y=0} = 0.563(0.444) = 0.25$$

$$N_{Re,x} = \frac{0.75(20)(3600)(0.0621)}{0.0192(2.42)} = 72{,}100$$

which corresponds to laminar flow in the boundary layer. Equation 5.196 gives

$$k_\rho^0 = \frac{0.32(3.88)}{2(0.75)}(72{,}100)^{1/2}(0.444) = 98.6 \text{ ft/hr}$$

Equation 5.188 shows that

$$-\rho_0 D\left(\frac{\partial w_A}{\partial y}\right)_{y=0} = -D\left(\frac{\partial \rho_A}{\partial y}\right)_{y=0}$$

$$= k_\rho^0(\rho_{A0} - \rho_{A\infty}) = 98.6(0.01022 - 0)$$

$$= 1.01 \text{ lb-mass/(ft}^2\text{)(hr)}$$

The quantity v_0 is obtained from equation 5.195:

$$v_0 = \frac{Mu_\infty}{2\sqrt{N_{\text{Re},x}}} = \frac{0.25(20)(3600)}{2(72{,}100)^{1/2}} = 33.5 \text{ ft/hr}$$

and

$$w_{A0} = \frac{4.55}{26.23} = 0.1734$$

so that substitution in equation 5.197 yields

$$n_{A0} = 33.5(0.059)(0.1734) + 1.01 = 1.353 \text{ lb } H_2O \text{ vapor/(ft}^2\text{)(hr)}$$

This is the local mass flux at the specified position on the plate.

Some correction procedures are now indicated which attempt to relate mass-transfer coefficients under high-mass-flux conditions to the corresponding—and somewhat more easily predicted—coefficients at low mass flux. Theoretical approaches have been mainly in terms of the stagnant film and the laminar boundary layer concepts. Experimental work has been very limited and indicates only partial success for the theoretical expressions to be noted below.

If all components in a mixture have the same molecular weight, then mass fractions and mole fractions are of course equal. This is the case, for instance, in a constant-property ideal-gas mixture, for which the diffusivity is essentially constant, viscosity variations with composition are slight, and the density $\rho = P\overline{M}/RT$. Constancy of ρ at a given temperature and pressure requires that $\overline{M}$, the average molecular weight of the mixture, is independent of composition (i.e., $M_A = M_B = \cdots$). Several of the theoretical results to follow are for constant-property fluids.

A commonly used definition of the driving force under high-mass-flux conditions will first be given.

The Driving Force B

During transfer of the single component A into the nontransferring stream B, the mass transfer from the surface ($y=0$) is the sum of two mechanisms as outlined above equation 5.1, namely the diffusive flux caused by the concentration gradient and the convective flux associated with the bulk flow. Thus

$$N_{A0}M_A = n_{A0} = v_0\rho_0 w_{A0} - \rho_0 D\left(\frac{\partial w_A}{\partial y}\right)_{y=0} \tag{5.197}$$

$$= n_{A0}w_{A0} - \rho_0 D\left(\frac{\partial w_A}{\partial y}\right)_{y=0} \tag{5.198}$$

A mass-transfer coefficient is commonly defined in terms of the diffusive flux as

$$k_w^0 = \frac{-\rho_0 D(\partial w_A/\partial y)_{y=0}}{w_{A0} - w_{A\infty}} \tag{5.199}$$

and

$$w_{A0} = \rho_{A0}/\rho_0; \quad w_{A\infty} = \rho_{A\infty}/\rho_\infty$$

Combination with equation 5.198 gives

$$n_{A0} = k_w^0\left(\frac{w_{A0} - w_{A\infty}}{1 - w_{A0}}\right) \tag{5.200}$$

The allowance for convective flux in equation 5.200 should render k_w^0 less dependent on mass flux than the more commonly defined coefficient in which the denominator $1 - w_{A0}$ is replaced by unity. Equation 5.200 also defines a driving force B for the transfer of a single component A as

$$B = \frac{w_{A0} - w_{A\infty}}{1 - w_{A0}} \tag{5.201}$$

Spalding (1960, 1963) has generalized the driving force B in equation 5.201 for a variety of situations. For the case in which more than one component is being transferred without reaction, he shows that

$$B = \frac{w_{A0} - w_{A\infty}}{w_{AT} - w_{A0}} \tag{5.202}$$

where w_{AT} is the mass fraction of component A in the transferred material. Clearly w_{AT} equals unity for the transfer of a single component. In the aqueous dissolution of crystals of sodium thiosulfate pentahydrate ($Na_2S_2O_3 \cdot 5H_2O$), if component A represents the anhydrous salt, then $w_{AT} = 0.637$.

At low concentrations and transfer rates k_w^0 approaches k_w^*, the coefficient when bulk flow due to diffusion is negligible, and $\rho_0 = \rho_\infty = \rho$. In this case

$$i_{A0} \doteq n_{A0} = k_w^* (w_{A0} - w_{A\infty}) \tag{5.203}$$

$$= \frac{k_w^*}{\rho} (\rho_{A0} - \rho_{A\infty}) \tag{5.204}$$

$$= k_\rho^* (\rho_{A0} - \rho_{A\infty}) \tag{5.205}$$

so

$$k_w^* = \rho k_\rho^* \tag{5.206}$$

The coefficients k_w^0 and k_w^* may be compared with k_w, defined in general as

$$n_{A0} = k_w (w_{A0} - w_{A\infty}) \tag{5.207}$$

Equation 5.207 is often used without regard for the degree of convective flux or bulk flow which may be occurring. Equations 5.200 and 5.207 show that

$$k_w^0 = (1 - w_{A0}) k_w \tag{5.208}$$

and for low concentrations and transfer rates,

$$k_w = k_w^0 = k_w^* \qquad (w_{A0} \to 0) \tag{5.209}$$

The question of interest is the relationship between k_w or k_w^0 for high mass flux and the corresponding k_w^* for low mass flux. This problem is considered in outline below.

Corrections from Film Theory

Transfer of the single component A through nontransferring B is postulated as occurring by molecular diffusion and associated convective flux through a stagnant film between the surface and the bulk of the fluid. Thus at some point in the film the mass flux of A with respect to stationary coordinates is

$$n_{Ay} = n_{Ay} w_A - \rho D \frac{dw_A}{dy}$$

and integrating over the film thickness for constant properties at steady state,

$$n_{A0} = \frac{\rho D (w_{A0} - w_{A\infty})}{\Delta_L (1 - w_A)_{\text{LM}}} \tag{5.210}$$

where

$$(1 - w_A)_{\text{LM}} = \frac{w_{A0} - w_{A\infty}}{\ln[(1 - w_{A\infty})/(1 - w_{A0})]} \tag{5.211}$$

Then

$$k_w = \frac{n_{A0}}{w_{A0} - w_{A\infty}} = \frac{\rho D}{\Delta_L (1 - w_A)_{\text{LM}}} \tag{5.212}$$

so

$$k_w^* = \lim_{w_A \to 0} k_w = \frac{\rho D}{\Delta_L} \tag{5.213}$$

Thus

$$\frac{k_w}{k_w^*} = \frac{1}{(1 - w_A)_{\text{LM}}} \tag{5.214}$$

and with equations 5.201 and 5.208,

$$\frac{k_w^0}{k_w^*} = \frac{1 - w_{A0}}{(1 - w_A)_{\text{LM}}} = \frac{\ln(1 + B)}{B} \tag{5.215}$$

Equation 5.215 is independent of Schmidt number and has received experimental confirmation in the case of molecular diffusion without forced convection. It has also been widely used with considerable success in correlating results from complex flow systems under both laminar and turbulent conditions where rigorous analysis is not feasible (see Chapter 6). Although k_w^0/k_w^* values from equation 5.215 do not coincide with those indicated by laminar-boundary-layer theory, they are nevertheless comparable in form and magnitude, as shown in Figure 5.22 (after Nienow, 1967). (B of course assumes positive or negative values according to whether the transfer is from or to the surface at $y = 0$.)

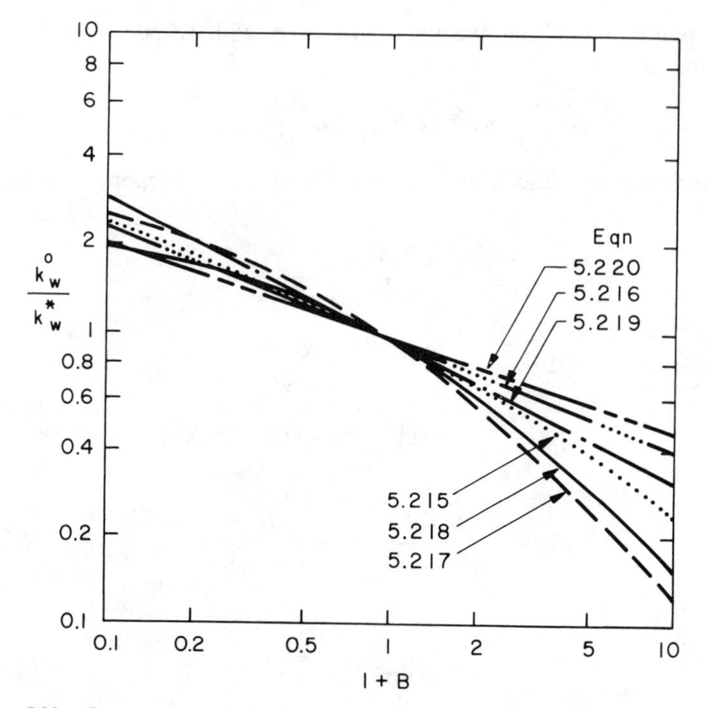

Figure 5.22. Comparison between various corrections for effects of a high mass flux (Nienow, 1967).

Corrections from Laminar-Boundary-Layer Theory

The complexity of boundary-layer problems with a high mass flux of course stems from the need for simultaneous solution of the coupled momentum and mass conservation equations. In a series of analytical papers Spalding et al. (1960; 1961a, b, c; 1963) presented laminar-boundary-layer developments for plane and axisymmetrical flow with a high mass flux for a large range of N_{Sc} and constant physical properties. The following expression was offered as giving good approximate correction for the effect of a mass flux with a given N_{Sc} and surface geometry (Spalding and Evans, 1961c):

$$\frac{k_w^0}{k_w^*} = (1+B)^{-0.4} \tag{5.216}$$

Equation 5.216 is plotted in Figure 5.22. For mass transfer *into* the fluid it gives higher correction factors than those obtained from the film theory (equation 5.215). This is opposite to the finding in Spalding's earliest paper (1954), in which independent polynomial expressions had been assumed for the momentum and concentration distributions in the boundary layers. This work had indicated corrections lower than equation 5.215 for positive B.

Spalding's earlier studies were followed by Merk (1959), who considered flow with constant physical properties over a flat plate. Two approximate expressions were proposed, with respective errors of up to 3 and 4 percent from the exact solution:

$$\frac{k_w^0}{k_w^*}=(1+0.724B)^{-1} \tag{5.217}$$

for $-0.5\leqslant B\leqslant 1.5$, $0.6\leqslant N_{\mathrm{Sc}}\leqslant 2$, and

$$\frac{k_w^0}{k_w^*}=(1+0.566B)^{-1} \tag{5.218}$$

for $-0.5\leqslant B\leqslant 1.0$, $N_{\mathrm{Sc}}\to\infty$ (effectively $N_{\mathrm{Sc}}>40$). These relationships appear graphically in Figure 5.22.

Ranz and Dickson (1965) present a variety of correction factors relating low- to high-mass-flux conditions in the vicinity of a stagnation point with either bilateral or axial symmetry. For single-component transfer they give the following simple expressions corresponding to each end of the N_{Sc} range:

$$\frac{k_w^0}{k_w^*}=(1+B)^{-1/2}, \qquad N_{\mathrm{Sc}}\to 0 \tag{5.219}$$

$$\frac{k_w^0}{k_w^*}=(1+B)^{-1/3}, \qquad N_{\mathrm{Sc}}\to\infty \tag{5.220}$$

Equations 5.219 and 5.220 are plotted in Figure 5.22.

Additional theoretical studies on high-flux mass transfer in laminar boundary layers are presented by Acrivos (1962a, b); Hanna (1963); Spalding, Pun, and Chi (1962); Erickson, Fan, and Fox (1966); Evans (1962); Stewart and Prober (1962, 1963); Stewart (1963); Cho (1963); and Anfimov and Altov (1966). High-flux mass transfer from stationary and slowly moving spheres has been treated by Fuchs (1947) and Faeth (1965), respectively.

Corrections from Penetration Theory

In addition to corrections based on the film and laminar-boundary-layer theories, further corrections for the effect of high mass flux on transfer rates have been provided by Bird et al. (1960) on the basis of the penetration theory. As in the case of the film theory, the result is independent of N_{Sc} because, in contrast with boundary-layer solutions, no allowance is made for velocity distributions. The corrective function arising from the penetration theory is approximately enclosed by the band of curves in Figure 5.22, which also shows that the effects of N_{Sc} are in any case fairly small.

Experimental work at high mass flux has been very limited, with principal orientation towards the laminar boundary layer. The studies by Mickley et al. (1954) were later found by the authors to be unacceptable because of defects in the apparatus. Ranz and Dickson (1965) provide some experimental verification of equations 5.219 and 5.220, and the theoretical conclusions of Stewart and Prober (1962) were confirmed experimentally by Christie (1962), Nicholson (1961), and Mendelson and Yerazunis (1965, evaporation of CCl_4). In contrast, studies disagreeing with Stewart and Prober's (1962) predictions were conducted by Spalding and Christie (1965) and Mendelson and Yerazunis (1965, evaporation of water). Related effects of high mass flux on heat transfer from a cylinder and a sphere were measured by Johnson and Hartnett (1963) and Short and Dana (1963), respectively. Nienow et al. (1969) measured high-flux mass transfer from single spheres and concluded that equation 6.132 (Chapter 6) is preserved only when the coefficient in N_{Sh} is evaluated from the data using driving force B of equation 5.201.

Corrections for Variation in Physical Properties

The large variations in concentration associated with transfer under high-mass-flux conditions may give rise to important variations in the relevant physical properties. Before attempting to make allowance for these effects, however, it must be recognized that the magnitude of these variations depends upon whether the phase under consideration is a liquid or a gas. Thus in liquids the viscosity and diffusivity may vary substantially with concentration, whereas the density changes only slightly. In gases, on the other hand, strong variations in concentration may be accompanied by significant changes in density, with only minor effects on viscosity and diffusivity. Correctional emphasis will therefore be determined by the nature of the phase under consideration.

The case of isothermal binary diffusion in a perfect-gas mixture in which the density varies with composition, whereas the viscosity and diffusivity are constant, has been considered by Hanna (1962). He presented the following approximate correction:

$$\frac{(k_w^0)_{\text{var }\rho}}{(k_w^0)_{\text{const }\rho}}=\frac{\ln(M_0/M_\infty)}{M_0/M_\infty-1} \tag{5.221}$$

where M_0 and M_∞ are the molecular weights of the mixture at the wall and in the bulk or free stream respectively. Equation 5.221 is valid for both laminar and turbulent flow, for any geometry, and for all relevant Schmidt numbers. The approximate nature of this equation stems from the assumption that the mass flux at the wall is relatively insensitive to variations in the velocity distribution caused by the changes in density with concentration.

In a subsequent study Hanna (1965) considered in greater detail the problem of isothermal binary diffusion in a laminar boundary layer consisting of a perfect-gas mixture with variable density. The viscosity and diffusivity were considered constant. An approximate formula was presented as follows to correct for high rates of mass transfer under these conditions for all Schmidt numbers near unity:

$$\frac{k_x^0}{k_x^*}=\frac{\{[\ln(1+B_x)]/B_x\}^{4/3}}{[1+\frac{4}{21}\ln(1+B_x)]^{4/9}}\left[\frac{M_0/M_\infty-1}{\exp(M_0/M_\infty-1)-1}\right]^{1/3} \tag{5.222}$$

This expression was considered to apply reasonably well to all geometries. However, it is for use with concentrations expressed in mole fractions. Thus for the transfer of component A through a nontransferring component,

$$B_x=\frac{x_{A0}-x_{A\infty}}{1-x_{A0}} \tag{5.223}$$

also

$$k_x^*=\frac{N_{A0}}{x_{A0}-x_{A\infty}}, \quad k_x^0=\frac{-c_0D(\partial x_A/\partial y)_{y=0}}{x_{A0}-x_{A\infty}} \tag{5.224}$$

Knuth (1963) devised reference compositions for gaseous laminar boundary layers with variable fluid properties. Physical properties are then evaluated at the reference composition, which is a function of the concentrations at the wall and in the bulk of the fluid. It was shown that the arithmetic average of these two values is suitable only for low transfer rates or small differences in molecular weight.

Large variations in concentration in liquid systems could cause significant changes in viscosity, depending upon the components involved. Little direct investigation of such effects has been reported, although for a constant-density laminar boundary layer on a flat plate, Hanna (1962) suggests correction by analogy with his analysis for heat transfer in a liquid of variable viscosity (Hanna and Myers, 1961). The work of Schuh (1950) outlined earlier also included solutions for variable physical properties.

The influence on the mass-transfer coefficient of variation in diffusivity with concentration has been studied by Nienow et al. (1966, 1968, 1969) for cases of aqueous dissolution of solid electrolytes. Mass-transfer rates

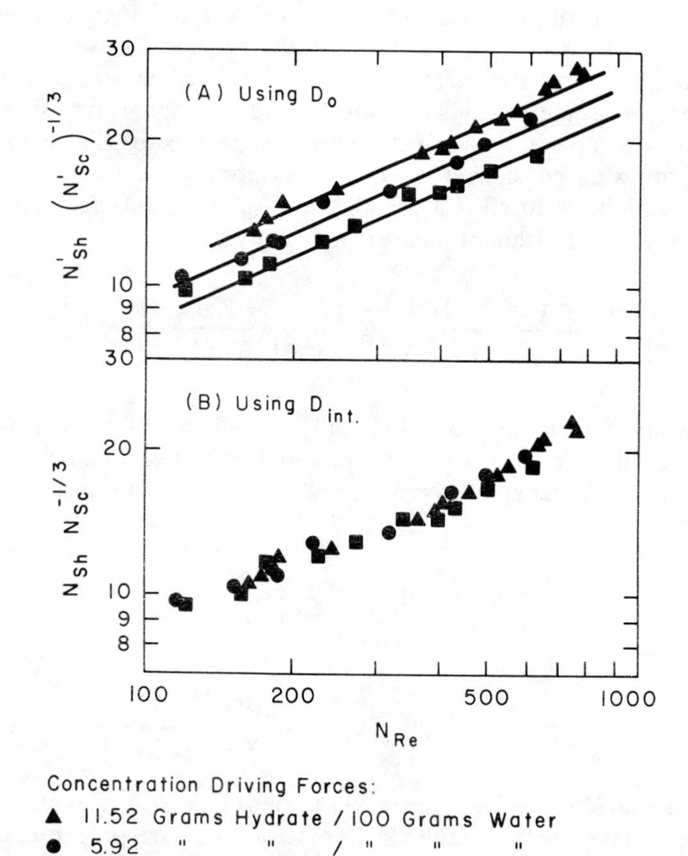

Figure 5.23. Dissolution of single ammonium alum spheres in aqueous solution at 20°C (Nienow et al., 1966).

were measured from single spheres of ammonium alum [$NH_4 \cdot Al(SO_4)_2 \cdot 12H_2O$] suspended in vertically upward-flowing streams of solution at 20°C. For this temperature the concentration range from saturation to infinite dilution corresponds to changes in density, viscosity, and diffusivity of −4, −30, and +110 percent, respectively. Figure 5.23*A* correlates their results for three different concentration driving forces, using bulk-solution physical properties except for the diffusivity, which was evaluated at the surface concentration in accordance with the suggestion of Ranz and Dickson (1965) (i.e., $N'_{\text{Sh}} = k_w^0 d_s / \rho_\infty D_0$, $N'_{\text{Sc}} = \mu_\infty / \rho_\infty D_0$, $N_{\text{Re}} = d_s u_\infty \rho_\infty / \mu_\infty$). It is important to note in Figure 5.23*A* that an increase in mass flux from the surface is accompanied by an increase in the Sherwood number, instead of the decrease predicted by boundary-layer theory when physical properties are constant. Improved correlation of the same data is achieved in Figure 5.23*B* by evaluating the density and viscosity at the arithmetic mean of the surface and bulk concentrations, and particularly by using the integrated mean diffusivity over the concentration range ρ_{A0} to $\rho_{A\infty}$, so that in Figure 5.23*B*, $N_{\text{Sh}} = k_w^0 d_s / \rho_a D_{\text{int}}$, $N_{\text{Sc}} = \mu_a / \rho_a D_{\text{int}}$, and $N_{\text{Re}} = d_s u_\infty \rho_\infty / \mu_\infty$. Entirely similar results were obtained for potassium alum.

Allowance for simultaneous variation in density, viscosity, and diffusivity with concentration in the presence of a high mass flux was attempted by Olander (1962) for various laminar-boundary-layer-type flows of binary liquid solutions. The analysis was primarily for solid-liquid systems with N_{Sc} values greater than 100, so that the velocity distribution approximates a linear form within the concentration boundary layer. A perturbation method was used to solve the diffusion equation, assuming that density and diffusivity vary exponentially with mass fraction over the relevant concentration range, as follows:

$$e_\rho = -(w_{A0} - w_{A\infty}) \frac{d \ln \rho}{dw_A} = \ln\left(\frac{\rho_\infty}{\rho_0}\right) \tag{5.225}$$

$$e_D = -(w_{A0} - w_{A\infty}) \frac{d \ln D}{dw_A} = \ln\left(\frac{D_\infty}{D_0}\right) \tag{5.226}$$

The solutions for three flow geometries—the rotating disk, the flat plate, and the falling film—were shown to be the same if the geometry was characterized appropriately in each case. Olander therefore considered that his result might be applicable to any external, laminar boundary-layer flow, regardless of geometry. The general relationship obtained was

$$\frac{N_{\text{Sh}0}}{N^*_{\text{Sh}}} = [1+0.5e_\rho+0.262e_D-0.566B$$

$$+0.166e_\rho^2+0.046e_D^2+0.407B^2+0.215e_\rho e_D$$

$$-0.566e_\rho B-0.232e_D B]\left(\frac{\rho_\infty \mu_\infty}{\rho_0 \mu_0}\right)^{m/3} \tag{5.227}$$

For the flat plate and the rotating disk, $m=\frac{1}{2}$; for the falling film, $m=\frac{1}{3}$. The left-hand side of equation 5.227 defines the ratio of Sherwood numbers based on properties at the surface. The numerator corresponds to varying physical properties and to finite bulk velocity normal to the surface, whereas the denominator corresponds to physical properties which are constant at their surface values and to zero bulk velocity due to diffusion.

The theories for constant physical properties reviewed earlier all indicate that k_w^0/k_w^* is less than unity for B greater than zero. In contrast, equation 5.227 shows that certain variations in physical properties may exist for which $N_{\text{Sh}0}/N^*_{\text{Sh}}$ is greater than unity for positive B, that is, for mass transfer from the surface. Evidently this could occur, for example, if D decreased sufficiently with increasing concentration. These indications are in accordance with the experimental findings of Nienow et al. (1966) in Figure 5.23*A*.

Equation 5.227 is simple to use, in conjunction with equations 5.225 and 5.226, but it holds only for the range $-0.4 \leqslant B \leqslant 0.4$. Furthermore, equation 5.226 is an oversimplification in many cases (e.g., it will not represent minima in D). These considerations led Emanuel and Olander (1964) to apply a modified integral technique to solve a revised form of the diffusion equation for high-Schmidt-number systems in boundary-layer-type flow. Equation 5.225 was retained to characterize the density-composition relationship, but the variation of diffusivity with composition was written as

$$\frac{D}{D_0} = 1-\beta\left[1-\frac{w_A-w_{A\infty}}{w_{A0}-w_{A\infty}}\right]-\gamma\left[1-\left(\frac{w_A-w_{A\infty}}{w_{A0}-w_{A\infty}}\right)^2\right] \tag{5.228}$$

The result of the analysis was

$$\frac{N_{\mathrm{Sh}0}}{N^*_{\mathrm{Sh}}}=[1-0.222\gamma-0.167\beta]^2[1-0.027\gamma+B(0.544+0.012\gamma)$$

$$-e_\rho(0.656-0.545\gamma-0.388\beta)]^{-1}\left(\frac{\rho_\infty\mu_\infty}{\rho_0\mu_0}\right)^{m/3} \qquad (5.229)$$

for the ranges

$$-0.7\leqslant B\leqslant 3.0,\quad -0.4\leqslant e_\rho\leqslant 0.3,\quad 0.33\leqslant D_\infty/D_0\leqslant 3.0$$

Experimental results for mass transfer from rotating disks of potassium bromide and copper sulfate pentahydrate into water showed good agreement with equation 5.229 ($m=\frac{1}{2}$). Marked deviations were found, however, in the case of sucrose at higher levels of driving force ($w_{A0}-w_{A\infty}>0.1$). This was attributed to the 200-fold change in viscosity of this solution from saturation to infinite dilution. Satisfactory agreement was found even in this system, though, for $\mu_0/\mu_\infty\leqslant 10$.

Conclusions Concerning Transfer with a High Mass Flux

A number of conclusions may be drawn regarding transfer under high-mass-flux conditions; many of them have been summarized by Nienow (1967). Most studies to date have been restricted to constant-property flow in laminar boundary layers. Extensions to allow for variation in physical properties with concentration have also been presented in a very limited number of papers. Some experimental confirmation of the theoretical studies has been noted, but substantial deviations between theory and experiment are also common.

The driving force B most frequently assumes a value between -0.25 and 1.0 in applications commonly encountered by the chemical engineer. Effects resulting from variations in physical properties may outweigh those associated with high mass flux in this range, particularly for liquid systems. Clearly, more experimental effort is needed in this area, in order to discriminate between the various theoretical relationships available. When constant-property assumptions are applicable, this might be done for a given geometry and N_{Sc} by comparison between experimental plots of k^0_w/k^*_w versus $1+B$ and theoretical curves such as those in Figure 5.22. The similarity between the various curves in this figure, irrespective of theory, geometry, and Schmidt number, implies that a generalized correction procedure may eventually be found which will be adequate for many engineering needs.

Illustration 5.8

A very porous solid plate is oriented at zero incidence to a pure air stream flowing with a velocity of 30 ft/sec at points remote from the plate surface. The underside of the plate is immersed in a reservoir of carbon tetrachloride, the temperature of which is equal to that at the air-plate interface. The latter is thoroughly wetted with CCl_4 as a consequence of capillary action through the plate. If the air pressure is 14.7 psia and the temperature in the bulk of the air stream is 262°F, estimate the surface temperature of the CCl_4, the local mass-transfer coefficient, and the rate of evaporation at a point 6 in. from the leading edge of the plate.

SOLUTION. The local sensible heat flux q_S in general contains the following contributions [A. P. Colburn and T. B. Drew, *Trans. A. I. Ch. E.*, **33**, 197–215, (1937)]:

1. sensible heat carried by convection in the normal manner without mass transfer,
2. sensible heat carried bodily by the transferring substances.

Thus

$$q_S = h_G \frac{dt}{d\eta} + (n_A c_{pA} + n_B c_{pB})(t - t_0)$$

where t_0 is the surface temperature, h_G is the gas phase heat transfer coefficient, and $\eta = y/y_{fG}$ (y being distance measured into the gas stream from—and normal to—the surface, and y_{fG} the thickness of a fictitious gas-phase film; see Chapter 4). Then

$$\int_{t_0}^{t_\infty} \frac{dt}{q_S - (n_A c_{pA} + n_B c_{pB})(t - t_0)} = \frac{1}{h_G} \int_0^{1.0} d\eta$$

Integrating,

$$q_S = \left(\frac{a}{1 - e^{-a}} \right) h_G (t_\infty - t_0)$$

where

$$a = \frac{n_A c_{pA} + n_B c_{pB}}{h_G}$$

and $a/(1 - e^{-a})$ is the *Ackermann factor*, allowing for the effect of mass transfer on heat transfer.

It may be noted that n_A, n_B will be negative if A, B diffuse in a direction opposite to the sensible heat transfer. The total heat flux to the surface is

$$q_t = q_S + \lambda_{A0} n_A + \lambda_{B0} n_B$$

where λ_{A0} and λ_{B0} are the latent heats of vaporization of A and B at t_0. In the present case $n_B = 0$, so that

$$q_t = \left(\frac{a}{1 - e^{-a}} \right) h_G (t_\infty - t_0) + \lambda_{A0} n_A \tag{i}$$

$$a = \frac{n_A c_{pA}}{h_G} \tag{ii}$$

and from equation 5.200,

$$n_A = n_{A0} = (k_w^0)_{\text{var } \rho} \left(\frac{w_{A0} - w_{A\infty}}{1 - w_{A0}} \right) \tag{iii}$$

The procedure for evaluating the coefficient in this expression will be as follows:

1. Calculate k_ρ^* from equation 5.23, using suitably evaluated physical properties.
2. Compute k_w^* from k_ρ^* via equation 5.206.
3. Estimate $(k_w^0)_{\text{const } \rho}$ from k_w^* with the appropriate relationship selected from equations 5.215 to 5.220.
4. Evaluate $(k_w^0)_{\text{var } \rho}$ from $(k_w^0)_{\text{const } \rho}$ by applying equation 5.221, extended to nonisothermal conditions as described later.

The quantity h_G is obtained from the heat-transfer analog of equation 5.23. The steady-state temperature of the plate surface will be such that all sensible heat arriving at the surface is carried away by the evaporating carbon tetrachloride. The quantity q_t is therefore zero. The temperature t_0 is selected by trial and error so as to satisfy equations i to iii above with $q_t = 0$.

Assume, for the final trial, that $t_0 = 80°F$.

Graphical interpolation of vapor-pressure data for CCl_4 as a function of temperature (from Perry, 1963, p. 3-49) shows $P_{CCl_4(80°F)} = P_A = 119$ mm Hg. Thus at the surface,

$$y_A = \frac{119}{760} = 0.1565 = \text{mole fraction } CCl_4$$

$$y_B = 1 - y_A = 0.8435 = \text{mole fraction air}$$

One pound-mole of this mixture contains

$$153.84(0.1565) = 24.06 \text{ lb } CCl_4$$

$$29(0.8435) = \underline{24.45} \text{ lb air}$$

$$M_0 = 48.51 \text{ lb}$$

$$w_{A0} = \frac{24.06}{48.51} = 0.496$$

$$B = \frac{w_{A0} - w_{A\infty}}{1 - w_{A0}} = \frac{0.496 - 0}{1 - 0.496} = 0.986$$

In comparing high experimental rates of mass transfer near a stagnation point with theoretical developments for constant-property mixtures, the physical properties were evaluated as:

1. *Interfacial* values (except for mainstream ρ in N_{Re}) by Ranz and Dickson (1965).
2. *Averages* of interfacial and mainstream values (except for interfacial D in N_{Sh}) by Mendelson and Yerazunis (1965).
3. *Mainstream* values by Spalding and Christie (1965).

Various other approaches have also been used (e.g., see text; also Schuh, 1950; Nienow et al., 1966, 1967, 1968, 1969; Hanna, 1965; Knuth, 1963; Olander, 1962; Emanuel and Olander, 1964; and Treybal, 1968).

With regard to the approach by Mendelson and Yerazunis (1965), the use of the interfacial or wall value of D in N_{Sh} is certainly consistent with equations 5.22 and 5.23. Several ways of evaluating the physical properties in N_{Re} and N_{Sc} were investigated by these authors for both isothermal and nonisothermal conditions. The use of averages of the interfacial and mainstream values, followed by application of Hanna's equation (equation 5.221), was found to give the closest agreement between theory and experiment for the evaporation of CCl_4 into an air stream (Mendelson, 1964, 1971). The extension of Hanna's isothermal equation (5.221) to nonisothermal conditions in this way has been described by Hanna (private communication, September 24, 1971) as "a reasonable expedient." This procedure will accordingly be followed here, with corresponding physical-property evaluations for the heat-transfer analog of equation 5.23.

Average Viscosity

The viscosity of pure CCl_4 vapor at 80°F (26.65°C) will be estimated by the method of corresponding states (R. C. Reid and T. K. Sherwood, *The*

Properties of Gases and Liquids, 2nd ed., McGraw-Hill, New York, 1966, pp. 404–405):

$$\mu = \frac{34.0\times 10^{-5} T_r^{0.94} M_A^{1/2} P_c^{2/3}}{T_c^{1/6}} \text{ cP} \qquad \text{for } T_r \leqslant 1.5$$

where the critical constants are $T_c = 556.4°K$, $P_c = 45$ atm (Reid and Sherwood, p. 582).

$$\mu_{CCl_4}, (80°F) = \frac{34.0\times 10^{-5}[(273.2+26.65)/556.4]^{0.94}(153.84)^{1/2}(45)^{2/3}}{(556.4)^{1/6}}$$

$$= 0.01035 \text{ cP}$$

For pure air at 80°F (Perry, 1963, p. 3-197)

$$\mu_{air}, (80°F) = 0.018 \text{ cP}$$

The viscosity of the air-CCl_4 mixture at the interfacial or surface conditions will be estimated by means of Wilke's equation, written by Reid and Sherwood (p. 421) as

$$\mu_m = \frac{\mu_A}{1+(y_B/y_A)\phi_{AB}} + \frac{\mu_B}{1+(y_A/y_B)\phi_{BA}}$$

where

$$\phi_{AB} = \frac{\left[1+(\mu_A/\mu_B)^{1/2}(M_B/M_A)^{1/4}\right]^2}{\sqrt{8}\,(1+M_A/M_B)^{1/2}}$$

$$\phi_{BA} = \phi_{AB}(\mu_B/\mu_A)(M_A/M_B)$$

Substituting,

$$\phi_{AB} = \frac{\left[1+(0.01035/0.018)^{1/2}(29/153.84)^{1/4}\right]^2}{\sqrt{8}\,(1+153.84/29)^{1/2}} = 0.3156$$

$$\phi_{BA} = 0.3156\left(\frac{0.018}{0.01035}\right)\left(\frac{153.84}{29}\right) = 2.91$$

$$\mu_{m(\text{surface})} = \frac{0.01035}{1+(0.8435/0.1565)0.3156} + \frac{0.018}{1+(0.1565/0.8435)2.91}$$

$$= 0.0155 \text{ cP}$$

The viscosity of the bulk gas (pure air) at 262°F is 0.0221 cP (Perry, 1963, p. 3-197), so that the average viscosity is

$$\tfrac{1}{2}(0.0155+0.0221)=0.0188 \text{ cP}$$

Average Density

As calculated earlier, $M_0=48.51$ lb, which at 80°F occupies 359(460+80)/492=394 ft³. The gas density at the surface is then 48.51/394=0.1231 lb-mass/ft³. In the bulk of the gas stream (pure air) 29 lb occupies 359(460+262)/492=526 ft³. The gas density in the bulk is therefore 29/526=0.0551 lb-mass/ft³, and the average density is

$$\tfrac{1}{2}(0.1231+0.0551)=0.0891 \text{ lb-mass/ft}^3$$

Average Diffusivity

The diffusivity will be estimated by the expression of Fuller, Schettler, and Giddings (equation 3.5), where, from Table 3 of the Appendix, $(\Sigma v)_A=94.5$ and $(\Sigma v)_B=20.1$. Thus at the surface temperature of 80°F (299.81°K)

$$D_{AB}\,(80°\text{F})=\frac{0.001(299.81)^{1.75}}{1\left[(94.5)^{1/3}+(20.1)^{1/3}\right]^2}\sqrt{\frac{1}{153.84}+\frac{1}{29}}=0.082 \text{ cm}^2/\text{sec.}$$

A similar calculation at the bulk temperature of 262°F (400.86°K) gives a diffusivity of 0.1366 cm²/sec. The average diffusivity is then

$$\tfrac{1}{2}(0.082+0.1366)=0.1093 \text{ cm}^2/\text{sec.}$$

Average Thermal Conductivity

The thermal conductivities of pure CCl_4 vapor and pure air at 80°F were graphically interpolated as 0.00348 and 0.01517 Btu/(ft²)(hr)(°F/ft) respectively. The respective sources were *Int. Crit. Tables* **V**, 215 (1929) and R. E. Bolz and G. L. Tuve, Eds., *Handbook of Tables for Appl. Eng. Science*, C.R.C., (1970), pp. 38–39.

The thermal conductivity of the air-CCl_4 mixture at the interfacial or surface conditions is estimated by means of Mason and Saxena's equation, written by Reid and Sherwood (p. 483) as

$$k_m=\frac{k_A}{1+1.065(y_B/y_A)\phi_{AB}}+\frac{k_B}{1+1.065(y_A/y_B)\phi_{BA}}$$

where ϕ_{AB} and ϕ_{BA} are the same as those used in estimating the viscosity of the surface gas mixture. Substituting:

$$k_{m(\text{surface})} = \frac{0.00348}{1+1.065(0.8435/0.1565)0.3156} + \frac{0.01517}{1+1.065(0.1565/0.8435)2.91}$$

$$= 0.01087 \text{ Btu}/(\text{ft}^2)(\text{hr})(°\text{F}/\text{ft})$$

The thermal conductivity of the bulk gas (pure air) at 262°F is found from the above source to be 0.0195 Btu/(ft^2)(hr)(°F/ft). The average thermal conductivity is therefore

$$\tfrac{1}{2}(0.01087+0.0195) = 0.0152 \text{ Btu}/(\text{ft}^2)(\text{hr})(°\text{F}/\text{ft})$$

Average Specific Heat

The specific heats of pure CCl_4 vapor and pure air at 80°F were interpolated graphically to be 0.133 and 0.249 Btu/(lb)(°F), respectively. The respective sources were *Int. Crit. Tables*, V, 81 (1929) and Perry (1963), p. 3-127. The mass composition of the gas mixture was calculated earlier as $w_{A0} = 0.496$, $w_{B0} = 0.504$, so that the specific heat of the gas mixture at the surface is

$$0.496(0.133) + 0.504(0.249) = 0.1915 \text{ Btu}/(\text{lb})(°\text{F})$$

The specific heat of pure air at 262°F is 0.25 Btu/(lb)(°F) (Perry, 1963, p. 3-127), so that the average specific heat is

$$\tfrac{1}{2}(0.1915+0.25) = 0.221 \text{ Btu}/(\text{lb})(°\text{F}).$$

Equation 5.23 will be evaluated with interfacial D in N_{Sh} and with average physical properties in N_{Re} and N_{Sc}, as discussed earlier:

$$N_{\text{Re},x} = \frac{0.5(30)(3600)(0.0891)}{0.0188(2.42)} = 105{,}900$$

corresponding to laminar flow in the boundary layer.

$$N_{\text{Sc}} = \frac{0.0188(2.42)}{0.0891(0.1093)3.88} = 1.2$$

$$N_{\text{Pr}} = \frac{0.221(0.0188)2.42}{0.0152} = 0.661$$

From equation 5.23,

$$k_\rho^* = 0.323\frac{(0.082)(3.88)}{0.5}(105{,}900)^{1/2}(1.2)^{1/3} = 71.1 \text{ ft/hr}$$

and the heat-transfer analog is

$$h_G = 0.323\frac{0.01087}{0.5}(105{,}900)^{1/2}(0.661)^{1/3} = 1.99 \text{ Btu/(ft}^2\text{)(hr)(°F)}.$$

Equation 5.206 gives

$$k_w^* = \rho k_\rho^* = 0.0891(71.1) = 6.345 \text{ lb-mass/(ft}^2\text{)(hr)}$$

One of the most extensive studies of high mass flux through a laminar boundary layer on a flat plate is that by Spalding et al., leading to equation 5.216. This result will be used here:

$$(k_w^0)_{\text{const }\rho} = k_w^*(1+B)^{-0.4} = 6.345(1+0.986)^{-0.4} = 4.82 \text{ lb-mass/(ft}^2\text{)(hr)}$$

Then using equation 5.221,

$$(k_w^0)_{\text{var }\rho} = \frac{4.82(2.303)\log(48.51/29)}{48.51/29-1} = 3.69 \text{ lb-mass/(ft}^2\text{)(hr)}$$

The local value of the mass flux of CCl_4 is obtainable from equation iii above as

$$n_{A0} = 3.69(0.986) = 3.64 \text{ lb-mass/(ft}^2\text{)(hr)}$$

The influence of mass transfer on the heat-transfer process is allowed for by combining h_G with the Ackermann factor, as in equation i above:

$$h_G\left(\frac{a}{1-e^{-a}}\right) = h_G + \left(\frac{1}{1-e^{-a}} - \frac{1}{a}\right)n_A c_{pA}$$

In this case the flux n_A takes place in a direction opposite to the sensible heat transfer, so that

$$a = -\frac{3.64(0.115)}{1.99} = -0.2104$$

where 0.115 is the specific heat of CCl_4 at the average temperature of 160°F, interpolated as before from *Int. Crit. Tables*, Vol. V, (1929), p. 81.

Then

$$h_G\left(\frac{a}{1-e^{-a}}\right)=1.99+\left(\frac{1}{1-e^{0.2104}}-\frac{1}{-0.2104}\right)(-3.64)(0.115)$$

$$=1.789\ \text{Btu}/(\text{ft}^2)(\text{hr})(°\text{F})$$

Evidently mass transfer reduces the heat transfer in this case by $100 \times(1.99-1.789)/1.99=10.1$ percent. The latent heat of vaporization of CCl_4 at 80°F is 89.4 Btu/lb (Perry, 1963, p. **3**-114). Substitution in equation (i) above gives

$$q_t=1.789(262-80)+89.4(-3.64)$$

$$=325.5-325.5=0$$

This balance demonstrates that the value selected for the surface temperature is correct. The answers to the problem at the specified location on the plate are therefore

$$\text{Surface temperature } t_0=80°\text{F}$$

$$(k_w^0)_{\text{var}\,\rho}=3.69\ \text{lb-mass}/(\text{ft}^2)(\text{hr})$$

$$n_{A0}=3.64\ \text{lb-mass}/(\text{ft}^2)(\text{hr})$$

Illustration 5.9

Rework Illustration 5.7 in terms of equations 5.23, 5.200, 5.201, 5.206, 5.216, 5.217, and 5.221. Compare the results with those obtained using Schuh's analysis.

SOLUTION. The bulk, average, and surface values of the compositions and physical properties will be as in Illustration 5.7, from which $N_{\text{Re},x}=72{,}100$, $N_{\text{Sc}}=0.602$. Then using equation 5.23,

$$k_\rho^*=0.323\frac{(0.32)(3.88)}{0.75}(72{,}100)^{1/2}(0.602)^{1/3}=121\ \text{ft/hr}$$

Equation 5.206 gives

$$k_w^*=\rho k_\rho^*=0.0621(121)=7.51\ \text{lb-mass}/(\text{ft}^2)(\text{hr})$$

$$w_{A0}=0.1734$$

so from equation 5.201,

$$B=\frac{w_{A0}-w_{A\infty}}{1-w_{A0}}=\frac{0.1734-0}{1-0.1734}=0.21$$

and from equation 5.216,

$$(k_w^0)_{\text{const }\rho}=k_w^*(1+B)^{-0.4}=7.51(1+0.21)^{-0.4}=6.96 \text{ lb-mass}/(\text{ft}^2)(\text{hr})$$

Applying the isothermal equation 5.221,

$$(k_w^0)_{\text{var }\rho}=\frac{6.96(2.303)\log(26.23/29)}{26.23/29-1}=7.36 \text{ lb-mass}/(\text{ft}^2)(\text{hr})$$

The local mass flux is obtainable from equation 5.200 as

$$n_{A0}=7.36(0.21)=1.545 \text{ lb } H_2O \text{ vapor}/(\text{ft}^2)(\text{hr})$$

This is 14.2 percent higher than predicted by Schuh's solution, which required constant physical properties. If the correction for varying density provided by equation 5.221 is omitted, then $n_{A0}=6.96(0.21)=1.46$ lb H_2O vapor/(ft^2)(hr), which is 7.9 percent higher than the value found in Illustration 5.7. Alternatively, if equation 5.217 is selected instead of equation 5.216 to evaluate $(k_w^0)_{\text{const }\rho}$, the corresponding estimates of local n_{A0} with and without application of equation 5.221 are, respectively, 6.9 and 1.1 percent higher than predicted by Schuh's solution.

NOMENCLATURE

A, B	Components A and B.
A	Area, ft^2.
A_j	Coefficient in the jth term of an infinite series.
a, b, c, d	Constants in equation 5.11.
a_1, b_1, c_1, d_1	Constants in equation 5.14.
a_j	The jth root of the equation $J_0(a_j)=0$.
B	A mass-transfer driving force defined by equations 5.201 and 5.202.
B_j	Coefficient in the jth term of an infinite series.
B_x	A mass-transfer driving force defined by equation 5.223.

B_1, B_2	Coefficients in equations 5.35 and 5.36.
b_j	The jth root of equation 5.108.
$C = C(\xi)$	$(\rho_A - \rho_{A0})/(\rho_{A\infty} - \rho_{A0})$
C_j	Coefficient in the jth term of an infinite series.
$c,\ c_0,\ c_\infty$	Total concentration, at $y=0$, in free stream, lb-mole/ft^3.
c_A	Concentration of component A, lb-mole/ft^3.
$c_{A0},\ c_{A\infty}$	Concentration of component A at the surface ($y=0$), in the bulk or free stream, lb-mole/ft^3.
c_p	Specific heat, Btu/(lb-mass)(°F).
D	(Volumetric) molecular diffusivity, ft^2/hr.
D_{int}	Integrated mean molecular diffusivity over a given concentration range, ft^2/hr.
D_0	Molecular diffusivity at the surface concentration, ft^2/hr.
D_∞	Molecular diffusivity in the bulk of the fluid, remote from the surface, or outside the concentration boundary layer, ft^2/hr.
d_s	Diameter of a sphere, ft.
d_t	Tube diameter, ft.
e_D, e_ρ	Defined by equations 5.225 and 5.226.
F_g	Gravitational force, lb-force.
F_x	External force in the x direction, lb-force.
G_j	Coefficient in the jth term of an infinite series.
g	Acceleration due to gravity, ft/hr^2.
g_c	Conversion factor, 32.174 lb-mass ft/(lb-force)(sec^2) *or* 4.17×10^8 lb-mass ft/(lb-force)(hr^2).
H_p	Normal distance separating flat, parallel plates, ft.
h	Normal distance from midplane, ft.
h_+	$h/(H_p/2)$
J_0, J_1	Bessel functions of first kind of order zero and one.
k	Thermal conductivity, Btu/(ft^2)(hr)(°F/ft).
k_w	Individual mass-transfer coefficient for any concentration range, $n_A/\Delta w_A$, lb-mass/(ft^2)(hr).
$k_w^0,\ k_x^0,\ k_\rho^0$	Individual mass-transfer coefficients for any concentration range: $i_A/\Delta w_A$, lb-mass/(ft^2)(hr); $J_A/\Delta x_A$, lb-mole/(ft^2)(hr); $i_A/\Delta \rho_A$, ft/hr. (For i_A and J_A see Chapter 2.)

k_w^*, k_x^*, k_ρ^*, $k_{\rho a}^*$, $k_{\rho \mathrm{LM}}^*$ — Individual mass transfer coefficients for low concentrations and transfer rates: $n_A/\Delta w_A$, lb-mass/(ft^2)(hr); $N_A/\Delta x_A$, lb-mole/(ft^2)(hr); $n_A/\Delta\rho_A$, ft/hr; $n_A/(\Delta\rho_A)_{\text{arith. mean}}$, ft/hr; $(n_A)_{\text{av}}/(\Delta\rho_A)_{\text{log. mean}}$, ft/hr.

$k_{\rho m}^*$ — Integrated mean value of k_ρ^*, defined above equation 5.24, ft/hr.

L — Length; plate length in direction x, ft.

Δ_L — Film thickness, ft.

L_c — Characteristic length dimension, ft.

l — Height of control volume, ft.

$\overline{M}$ — Average molecular weight.

M — Parameter defined by equation 5.195.

M_A, M_B — Molecular weights of components A and B.

M_i — Molecular weight of inlet gas.

M_0 — Molecular weight of mixture at the wall.

M_w — Average molecular weight of gas saturated with solute at the liquid surface temperature.

M_∞ — Molecular weight of mixture in bulk of free stream.

m — Constant.

N — $(\rho_{A0}-\rho_{A\infty})/N_{\text{Sc}}\rho_{A0}(P/p_{A0}-1)$

N_A, N_{A0} — Molal flux of component A relative to stationary coordinates; at the surface $y=0$, lb-mole/(ft^2)(hr).

$N_{\text{Gr}D}$ — Grashof number for mass transfer, equation 5.44.

N_{Pr} — Prandtl number.

N_{Re} — Reynolds number, $d_t V\rho/\mu$, $2H_p V\rho/\mu$.

$N_{\text{Re}f}$ — Film Reynolds number, equation 5.70.

$N_{\text{Re},L}$, $N_{\text{Re},x}$ — $Lu_\infty\rho/\mu$; $xu_\infty\rho/\mu$.

N_{Sc} — Schmidt number, $\mu/\rho D$.

N'_{Sc} — $\mu_\infty/\rho_\infty D_0$.

N_{Sh}, N_{Sh}^* — Sherwood number: kL_c/D, k^*L_c/D.

N'_{Sh} — $k_w^0 d_s/\rho_\infty D_0$.

$N_{\text{Sh}0}$ — Sherwood number in terms of k_w^0 and in the presence of property variations and interfacial velocity (Olander, 1962; Emmanuel and Olander, 1964).

$(N_{Sh})_a^*$, $(N_{Sh})_{LM}^*$, $(N_{Sh})_m^*$, $(N_{Sh})_x^*$	Sherwood numbers in which the coefficient k is, respectively, $k_{\rho a}^*$, $k_{\rho LM}^*$, $k_{\rho m}^*$, and k_ρ^*, the latter being a local value.
n	Constant.
n_A, n_{Ah}, $n_{A0}=(n_{Ay})_{y=0}$, n_{Ar}, n_{Aw}, n_{Ax}, n_{Ay}	Mass flux of component A relative to stationary coordinates; in direction h; at the surface ($y=0$); in direction r; at the wall; in directions x and y, lb-mass/(ft^2)(hr).
P	Total pressure, lb-force/ft^2.
p_{A0}	Partial pressure of component A at the plate surface, lb-force/ft^2.
Q	Volumetric flow rate, ft^3/hr.
q_{AW}	Rate of transfer of A at the wall, lb-mass/hr.
R	Gas constant, 1545 ft lb-force/(lb-mole)(°R).
R'	Some function of r only, equation 5.104.
r, r_t	Radius, radius of a tube, ft.
r_+	r/r_t.
T, T_i, T_0, T_w, T_∞	Absolute temperature; of inlet gas; at the surface ($y=0$); of the liquid surface; outside the boundary layer, °R.
t	Time, hr, *or* temperature, °F.
u, u_0, u_s, u_∞	Mass-average velocity in the x direction; of the liquid-liquid interface; at the centerline; outside the momentum boundary layer, ft/hr.
V	Mean velocity in the x direction. W/ρ(cross-sectional area), ft/hr.
v	Mass-average velocity in direction h, r, or y, ft/hr.
v_{A0}	Statistical mean velocity of component A in the y direction relative to stationary coordinates at $y=0$, ft/hr.
v_0, v_1	Mass-average velocity in the y or $-y$ direction at the surface ($y=0$) and at $y=l$, respectively, ft/hr.
W	Mass flow rate, lb-mass/hr.
w	Width normal to flow, ft.
w	Velocity component in the z direction, ft/hr.

w_A, w_{A0}, w_{AT}, $w_{A\infty}$	Mass fraction of component A, ρ_A/ρ; at the surface ($y=0$); in the transferred material; in the bulk of the stream or outside the concentration boundary layer.
X'	Some function of x only, equation 5.104.
x	Distance in direction of flow, ft.
x_+	Dimensionless distance in the direction of flow, equation 5.102 and Table 5.2.
x_A, x_{A0}, $x_{A\infty}$	Mole fraction of component A, c_A/c; at the surface, c_{A0}/c_0; in the bulk or free stream, $c_{A\infty}/c_\infty$.
x_c	Critical value of x at which laminar flow ends in the boundary layer, ft.
x_0	Distance along plate at which mass transfer begins, ft.
y	Distance from and normal to the surface, ft.
z	Coordinate direction.
α	Angle between an inclined plane and the horizontal.
α_j	Coefficient in the jth term of an infinite series.
α_t	Thermal diffusivity, ft^2/hr.
β	Constant in equation 5.228.
β_e	Volume expansion coefficient, equation 5.27.
β_j	Eigenvalues.
β_v	Velocity gradient at the wall, hr^{-1}.
γ	Constant in equation 5.228.
γ_j	Eigenvalues.
δ, δ_c	Thickness of the momentum and concentration boundary layers, ft.
δ_f	Film thickness, ft.
θ, θ_b	$(\rho_{AW}-\rho_A)/(\rho_{AW}-\rho_{Ai})$; $(\rho_{AW}-\rho_{AB})/(\rho_{AW}-\rho_{Ai})$.
$\theta=\theta(\xi)$	$(T-T_0)/(T_\infty-T_0)$.
Λ	Factor in equation 5.32, ft/hr.
λ_j	Eigenvalues.
μ	Viscosity, lb-mass/(ft)(hr).
μ_a	Viscosity at the arithmetic mean of the surface and bulk concentrations, lb-mass/(ft)(hr).
μ_∞	Viscosity at bulk concentration, lb-mass/(ft)(hr).
ξ	$y\sqrt{N_{\mathrm{Re},x}}/2x$.
ρ, ρ_a, ρ_0, ρ_∞	Total density; at the arithmetic mean of the surface and bulk concentrations; at the surface ($y=0$); at the bulk concentration, $lb\text{-}mass/ft^3$.

ρ_A, ρ_A^*	Mass concentration of component A; saturation or equilibrium concentration, lb-mass/ft^3.
$\rho'_A, \rho'_{A0}, \rho'_{A\infty}$	$\rho_A - \rho_{A0}, \rho_{A0} - \rho_{A0} = 0, \rho_{A\infty} - \rho_{A0}$, lb-mass/ft^3.
$\rho_{AB}, \rho_{AB0}, \rho_{Ai}, \rho_{A0}, \rho_{As}, \rho_{AW}, \rho_{A\infty}$	Bulk average or "mixing cup" value of ρ_A; ρ_{AB} at the outlet; mass concentration of A at the inlet; at the surface ($y=0$); at the centerline; at the wall; in the bulk or outside the concentration boundary layer, lb-mass/ft^3.
$\rho_B, \rho_{B0}, \rho_{B\infty}$	Mass concentration of component B; at the surface ($y=0$); in the bulk or outside the concentration boundary layer, lb-mass/ft^3.
$\tau_0 = (\tau_{yx})_{y=0}$	Shear stress at the surface ($y=0$), lb-force/ft^2.
τ_{yx}	Shear stress in the x direction on a surface normal to y, lb-force/ft^2.
Υ	Quantity defined by equation 5.76.
ϕ	δ_c/δ
ϕ_j	Eigenfunctions.
ψ_j	Eigenfunctions.
$\omega = \omega(\xi)$	u/u_∞.

Subscripts: 1 and 2 refer to streams 1 and 2 in equations 5.53 to 5.56.

PROBLEMS

5.1 A flat plate measuring 1 ft×2 ft is immersed at zero incidence in a stream of ethanol, with the shorter sides of the plate parallel to the direction of flow. The surface of the plate contains a semicircular disc of solidified stearic acid with a diameter of 2 ft, the latter constituting the rear edge of the plate. The remainder of the plate surface is insoluble in ethanol. If the velocity of the ethanol at points remote from the surface is 1.2 ft/sec and the temperature of the system is 77°F, determine the total rate of dissolution of stearic acid from one surface of the plate. Relevant physical properties may be taken from Illustration 5.3.

5.2 Repeat Problem 5.1 with the direction of flow reversed.

5.3 Water at a temperature of 77°F flows parallel to a flat plate of benzoic acid which is 1 ft long in the direction of flow. If the velocity of the water at points remote from the surface is 0.75 ft/sec, plot the local mass-transfer coefficient, the thicknesses of the momentum and concentration boundary layers, and the local Sherwood number as functions of distance from the leading edge of the plate. Relevant physical properties appear in Illustration 5.6.

5.4 Determine the average rate of mass transfer per unit width of plate in Problem 5.3.

5.5 A flat plate with low thermal conductivity is coated on one side with an 0.1-in.-thick layer of solid naphthalene and on the other side with a layer of benzoic acid of the same thickness. The coated plate forms one vertical side of a tank which is completely filled with pure water at 77°F. The naphthalene-coated surface is in contact with pure air at 1 atm and 110°F. The water and the air are both free from forced convection, and for present purposes the solid surface temperatures will be assumed equal to the bulk temperatures of the fluids with which they are in contact. Confining attention to the uppermost 2 in. of the plate, on which side will the average coating thickness first be reduced to 90 percent of its original value? What will be the average thickness of coating on the other side at this instant?

Relevant physical properties may be found in Illustrations 5.1, 5.2, 5.6, and 6.5, although adjustment will be needed for the new air temperature.

5.6 A vertical tube with an i.d. of 2.5 in. and a length of 4 ft is cast from benzoic acid. Pure water enters the tube at a rate of 5 ft^3/hr and flows in a thin film down the inner wall. Dry air at 1 atm and 85°F flows into the tube and moves countercurrent to the water at an average velocity of 0.75 ft/sec. The benzoic acid and the water remain at a constant temperature of 50°F throughout. Estimate the following quantities:

(a) The water film thickness.
(b) The average and maximum velocities of the water film.
(c) The concentration of benzoic acid in the outlet water.
(d) The concentration of dissolved air in the outlet water.

For the solubility of benzoic acid in water at the appropriate temperature see Seidell (reference in Illustration 5.6). The solubility of air in water appears in Perry (1963), p. 14-3.

5.7 Estimate the average concentration of water vapor in the air stream leaving the wetted-wall column in Problem 5.6.

5.8 Calculate and plot the radial concentration distribution of benzoic acid at the outlet from the mass-transfer section of the tube in Illustration 5.6.

5.9 The differential equations expressing continuity and solute distribution during mass transfer in laminar flow between flat parallel plates are given by equations 5.178 and 5.179 in Table 5.2. Present full derivations of these expressions with labeled sketches of the relevant control volumes. To what forms do these equations reduce for the cumulative conditions of (a) steady state, (b) negligible axial diffusion, and (c) fully developed velocity distribution?

5.10 Equations 5.180 and 5.181 in Table 5.2 express local and bulk-average values of the solute concentration in mass transfer during laminar flow between flat parallel plates under the conditions indicated. Give full derivations of these two relationships.

5.11 Beginning with equation 5.181, develop equations 5.182 and 5.183 of Table 5.2 for mass transfer in laminar flow between flat parallel plates under the specified conditions.

5.12 Equation 5.186 in Table 5.2 expresses the asymptotic value of N^*_{Sh} for mass transfer in laminar flow between flat parallel plates when n_{AW} is independent of x. Derive this relationship, beginning with the appropriate differential equation.

5.13 Equation 5.187 in Table 5.2 gives the asymptotic value of N^*_{Sh} for mass transfer in laminar flow between flat parallel plates, assuming a cubic-polynomial concentration distribution. Present a complete derivation of this expression.

5.14 The two larger sides of a rectangular duct are made from solid naphthalene, separated by a normal distance of $\frac{1}{2}$ in. Pure air at an average pressure of 14.7 psia flows through the duct with a mean velocity of 2 ft/sec. Estimate the average concentration of naphthalene vapor in the air at a distance of 2 ft downstream from the entrance, if the system is at 113°F and the width-to-height ratio of the duct is great enough for the flow to be regarded as occurring between flat parallel plates.

The relevant physical properties may be found in Illustration 5.1.

5.15 A flat rectangular duct has two large sides separated by $\frac{1}{2}$ in. and a width-to-height ratio sufficiently great that the two small sides exert negligible influence on the transfer processes. Pure ethanol flows through the duct at a mean velocity of 0.1 ft/sec. At a point downstream from that at which the velocity profile is fully developed, the two insoluble large sides are replaced by plates of solidified stearic acid. Estimate the distance downstream of the start of the soluble section at which the mean concentration of stearic acid in the ethanol reaches 30 percent of saturation. The temperature of the system is 77°F, and the physical properties are as in Illustration 5.3.

5.16 For the conditions of Problem 5.15, calculate and plot the concentration distribution of stearic acid over the duct cross section located 5 ft downstream from the start of the soluble section.

5.17 Potassium chloride in the form of a solid flat plate is submerged in a stream of pure water flowing parallel to the plate with an upstream velocity of 0.5 ft/sec. If the plate is 1 ft square and the system is at a temperature of 18.5°C, estimate the total dissolution rate of KCl from a single surface. Compare results with and without allowance for the high mass flux. Some relevant physical properties are obtainable from Illustration 3.5.

5.18 A tube with an internal diameter of $\frac{1}{4}$ in. is made from solid potassium chloride. Water flows through the tube with an average velocity of 0.25 ft/sec. If the system is at 18.5°C, what is the average concentration of KCl in solution at a point 2 ft downstream from the tube inlet? (See Illustration 3.5 for physical properties.)

5.19 A short tube has a mean diameter of 1 ft and a length of 8 in., and its thin, porous walls are kept saturated with liquid water. Dry air at 14.7 psia and 200°F flows with a velocity of 15 ft/sec over and through the tube in the axial direction. Estimate the combined rate of evaporation from the outer and inner cylindrical surfaces when the tube and water supply are maintained at 200°F by a separate heat source.

5.20 Repeat Problem 5.19 for the case in which the separate heat source is removed and after the temperature of the wet surface has assumed an appropriate equilibrium value.

REFERENCES

Acrivos, A., *A. I. Ch. E. J.*, **6**, 410, (1962a).

Acrivos, A., *J. Fluid Mech.*, **12**, 337, (1962b).

Anfimov, N. A., and V. V. Altov, *Int. Chem. Eng.*, **6**, 137, (1966).

Apostol, T. M., *Calculus*, Vol. 1, Blaisdell Publishing Co., (Random House), New York, (1961), p. 458.

Badger, W. L., and J. T. Banchero, *Introduction to Chemical Engineering*, McGraw-Hill, New York, (1955), p. 219.

Bird, R. B., W. E. Stewart, and E. N. Lightfoot, *Transport Phenomena*, Wiley, New York, (1960), pp. 668–672, 675.

Boelter, L. M. K., *Trans. A. I. Ch. E.*,, **39**, 557, (1943).

Boelter, L. M. K., V. H. Cherry, and H. A. Johnson, *Heat Transfer* (Supp. Notes for M.E. 267A–B), University of California Press, Berkeley, (1937), **X**-18–**X**-28.

Brown, G. M., *A. I. Ch. E. J.*, **6**, 179–183, (1960).

Cairns, R. C., and G. H. Roper, *Chem. Eng. Sci.*, **3**, 97–109, (1954).

Cess, R. D., and E. C. Shaffer, *Appl. Sci. Res.*, *A* **8**, 339–344, (1959).

Cho, D. H., Tech. Rep. No. 3, 00R Proj. No. 2340, (1963).

Christie, F. A., D. I. C. Dissertation, Imp. College, University of London, (1962).

Churchill, R. V., *Fourier Series and Boundary Value Problems*, McGraw-Hill, New York, (1941), pp. 158, 162.

Colton, C. K., K. A. Smith, P. Stroeve, and E. W. Merrill, *A. I. Ch. E. J.*, **17**, 773–780, (1971).

Drew, T. B., *Trans. A. I. Ch. E.*, **26**, 26–80, (1931).

Dukler, A. E., and O. P. Bergelin, *Chem. Eng. Prog.*, **48**, 557, (1952).

Eckert, E. R. G., and J. F. Gross, *Introduction to Heat and Mass Transfer*, McGraw-Hill, New York, (1963), pp. 129–30.

Emanuel, A. S., and D. R. Olander, *Int. J. Heat and Mass Transfer*, **7**, 539–548, (1964).

Emmons, H. W., and D. C. Leigh, Fluid Motion Sub-Comm., Aero. Res. Counc., Rep. No. F. M. 1915, June, 1953.

Erickson, L. E., L. T. Fan, and V. G. Fox, *Ind. Eng. Chem. Fundam.*, **5**, 19, (1966).

Estrin, J. E., T. W. Hayes, and T. B. Drew, *A. I. Ch. E. J.*, **11**, 800–803, (1965).

Evans, H. L., *Int. J. Heat Mass Transfer*, **5**, 373, (1962).

Faeth, G. M., *A. I. Ch. E. J.*, **11**, 1133–1135, (1965).

Foust, A. S., L. A. Wenzel, C. W. Clump, L. Maus, and L. B. Andersen, *Principles of Unit Operations*, Wiley, New York, (1960), p. 111.

Fuchs, N., N.A.C.A., Tech. Memo. 1160, (1947).

Fulford, G. D., in *Advances in Chemical Engineering*, T. B. Drew, J. W. Hoopes, T. Vermeulen, and G. R. Cokelet, Eds., Vol. 5, Academic Press, (1964), pp. 151–236.

Gilliland, E. R., and T. K. Sherwood, *Ind. Eng. Chem.*, **26**, 516, (1934).

Goldberg, P., M. S. thesis, Department of Mechanical Engineering, Massachusetts Institute of Technology, January, 1958.

Graetz, L., *Ann. Physik*, **18**, 79, (1883).

Graetz, L., *Ann. Physik*, **25**, 337, (1885).

Hanna, O. T., *A. I. Ch. E. J.*, **8**, 278–279, (1962).

Hanna, O. T., *A. I. Ch. E. J.*, **9**, 704–706, (1963).

Hanna, O. T., *A. I. Ch. E. J.*, **11**, 706–712, (1965).

Hanna, O. T., and J. E. Myers, *A. I. Ch. E. J.*, **7**, 437, (1961).

Hartnett, J. P., and E. R. G. Eckert, *Trans. Am. Soc. Mech. Eng.*, **79**, 247–254, (1957).

Hausen, H., *Verfahrenstech. Beih. Z. Ver. dtsch. Ing.*, **4**, 91, (1943).

Heaton, H. S., W. C. Reynolds, and W. M. Kays, *Int. J. Heat Mass Transfe* 763–781, (1964).

Hurlbert, R. C., R. F. Knott, and H. A. Cheney, *Soap Chem. Specialties,* 122, (May 1967).

Hurlbert, R. C., R. F. Knott, and H. A. Cheney, *Soap Chem. Specialties*, 88, (June 1967).

Jackson, M. L., *A. I. Ch. E. J.*, **1**, 231, (1955).

Jakob, M., *Heat Transfer*, Vol. 1, Wiley, New York, (1949).

Johnson, B. V., and J. P. Hartnett, *Trans. Am. Soc. Mech. Eng.*, Ser. C, **85**, 173, (1963).

Johnstone, H. F., and R. L. Pigford, *Trans. A. I. Ch. E.*, **38**, 25–51, (1942).

Kaiser, L., *Int. J. Heat Mass Transfer*, **3**, 175, (1961).

Kays, W. M., *Trans. Am. Soc. Mech. Eng.*, **58**, 1265–1274, (1955).

Kays, W. M., and E. Y. Leung, *Int. J. Heat Mass Transfer*, **6**, 537–557, (1963).

Kinney, G. R., A. E. Abramson, and J. L. Sloop, N.A.C.A. Rep. 1087, (1952).

Knudsen, J. G., and D. L. Katz, *Fluid Dynamics and Heat Transfer*, McGraw-Hill, New York, (1958).

Knuth, E., *Jet Propulsion*, **24**, 359, (1954).

Knuth, E., *Jet Propulsion*, **25**, 16, (1955).

Knuth, E., *Int. J. Heat Mass Transfer*, **6**, 1–22, (1963).

Kramers, H., and P. J. Kreyger, *Chem. Eng. Sci.*, **6**, 42–48, (1956).

Landino, E., J. G. McCreary, W. A. Thompson, and J. E. Powers, *A. I. Ch. E. J.*, **12**, 117–124, (1966).

Langhaar, H. L., *J. App. Mech.*, **64**, A-55, (1942).

Lévêque, J., *Ann. Mines* [12], **13**, 201, 305, 381, (1928).

Li, Wen Hsiung, *Engineering Analysis*, Prentice-Hall, Englewood Cliffs, N. J., (1960).

Linton, W. H., and T. K. Sherwood, *Chem. Eng. Prog.*, **46**, 258, (1950).

Lundberg, R. E., P. A. McCuen, and W. C. Reynolds, *Int. J. Heat Mass Transfer*, **6**, 495–529, (1963).

Lundberg, R. E., W. C. Reynolds, and W. M. Kays, N.A.S.A. Tech. Note D-1972, Washington, August, 1963.

Mendelson, H. D., Ph.D. thesis, Rensselaer Polytechnic Institute, Troy, N. Y., 1964.

Mendelson, H. D., private communication, August 5, 1971.

Mendelson, H. D., and S. Yerazunis, *A. I. Ch. E. J.*, **11**, 834, (1965).

Merk, H. J., *App. Sci. Res.*, **8A**, 237, 261, (1959).

Mickley, H. S., R. C. Ross, A. L. Squyers, and W. E. Stewart, N.A.C.A. Tech. Note 3208, (1954).

Moore, J. G., and W. E. Hesler, *Chem. Eng. Prog.*, **59** (2), 87, (1963).

Murdoch, R., and H. R. C. Pratt, *Trans. Inst. Chem. Eng. (Lond.)*, **31**, 307, (1953).

Nicholson, V., D.I.C. Dissertation, Imperial College, University of London, (1961).

Nienow, A. W., *Brit. Chem. Eng.*, **12**, 1737–1743, (1967).

Nienow, A. W., R. Unahabhokha, and J. W. Mullin, *Ind. Eng. Chem. Fundam.*, **5**, 578–579, (1966).

Nienow, A. W., R. Unahabhokha, and J. W. Mullin, *J. Appl. Chem.*, **18**, 154–156, (1968).

Nienow, A. W., R. Unahabhokha, and J. W. Mullin, *Chem. Eng. Sci.*, **24**, 1655–1660, (1969).

Nusselt, W. Z., *Z. Ver. dtsch. Ing.*, **67**, 206–210, (1923).

Olander, D. R., *Int. J. Heat Mass Transfer*, **5**, 765–780, (1962).

Oliver, D. R., and T. E. Atherinos, *Chem. Eng. Sci.*, **23**, 525–536, (1968).

Perry, J. H., Ed., *Chemical Engineers' Handbook,* 3rd ed., McGraw-Hill, New York, (1950); 4th ed., (1963).

Potter, O. E., *Chem. Eng. Sci.*, **6**, 170–182, (1957).

Prandtl, L., and O. Tietjens, *Hydro und Aeromechanik*, Vol. 2, Springer-Verlag, Berlin, (1931), p. 28.

Ranz, W. E., and P. F. Dickson, *Ind. Eng. Chem. Fundam.*, **4**, 345, (1965).

Reynolds, W. C., R. E. Lundberg, and P. A. McCuen, *Int. J. Heat Mass Transfer*, **6**, 483–493, (1963).

Rohsenow, W. M., and H. Y. Choi, *Heat, Mass, and Momentum Transfer*, Prentice-Hall, Inc., Englewood Cliffs, N. J., (1961).

Schlichting, H., *Boundary Layer Theory*, Mc Graw-Hill, New York, (1967).

Schuh, H., N.A.C.A. Tech. Memo. 1275, (1950).

Seigel, R., E. M. Sparrow, and T. M. Hallman, *App. Sci. Res.*, Ser. A, **7**, 386–392, (1958).

Sellars, J. R., M. Tribus, and J. S. Klein, *Trans. Am. Soc. Mech. Eng.*, **78**, 441, (1956).

Sherwood, T. K., and R. L. Pigford, *Absorption and Extraction*, McGraw-Hill, New York, (1952).

Sherwood, T. K., and B. B. Woertz, *Ind. Eng. Chem.*, **31**, 1034, (1939).

Short, W. W., and T. A. Dana, *A. I. Ch. E. J.*, **9**, 509–513, (1963).

Sinek, J. R., and E. H. Young, *Chem. Eng. Prog.*, **58** (12), 74, (1962).

Skelland, A. H. P., *Ind. Eng. Chem. Fundam.*, **6**, 148–151, (1967).

Sourirajan, S., *Reverse Osmosis*, Logos Press, London, (1970).

Spalding, D. B., *Proc. Roy. Soc.*, **A221**, 78, 104, (1954).

Spalding, D. B., *Int. J. Heat Mass Transfer*, **1**, 192, (1960).

Spalding, D. B., *Int. J. Heat Mass Transfer*, **2**, 15, (1961a).

Spalding, D. B., *Appl. Mech. Rev.*, **15** (7), 505–508, (1962).

Spalding, D. B., *Convective Mass Transfer*, Arnold, London, (1963).

Spalding, D. B., and S. W. Chi, *Int. J. Heat Mass Transfer*, **6**, 363, (1963).

Spalding, D. B., and F. A. Christie, *Int. J. Heat Mass Transfer*, **8**, 511, (1965).

Spalding, D. B., and H. L. Evans, *Int. J. Heat Mass Transfer*, **2**, 199, (1961b).

Spalding, D. B., and H. L. Evans, *Int. J. Heat Mass Transfer*, **2**, 314, (1961c).

Spalding, D. B., W. M. Pun, and S. W. Chi, *Int. J. Heat Mass Transfer*, **5**, 79, (1962).

Sparrow, E. M., N.A.C.A. Tech. Note 3331, (1955).

Stewart, W. E., *A. I. Ch. E. J.*, **8**, 421, (1962).

Stewart, W. E., *A. I. Ch. E. J.*, **9**, 528, (1963).

Stewart, W. E., and R. Prober, *Int. J. Heat Mass Transfer*, **5**, 1149, (1962).

Stewart, W. E., and R. Prober, *Int. J. Heat Mass Transfer*, **6**, 221, (1963).

Stirba, C., and D. M. Hurt, *A. I. Ch. E. J.*, **1**, 178–184, (1955).

Tailby, S. R., and S. Portalski, *Trans. Inst. Chem. Eng. (Lond.)*, **38**, 324–330, (1960).

Taylor, G. I., *Proc. Roy. Soc.*, A **219**, 186–203, (1953).

Taylor, G. I., *Proc. Roy. Soc.*, A **223**, 446–468, (1954a).

Taylor, G. I., *Proc. Roy. Soc.*, A **225**, 473–477, (1954b).

Taylor, G. I., *Proc. Phys. Soc.*,B **LXVII**, 857–869, (1954c).

Tien, C., *Can. J. Chem. Eng.*, **40**, 130–134, (1962).

Treybal, R. E., *Mass Transfer Operations*, 2nd ed., McGraw-Hill, New York, (1968), pp. 59–60.

Vivian, J. E., and D. W. Peaceman, *A. I. Ch. E. J.*, **2**, 437, (1956).

Welty, J. R., C. E. Wicks, and R. E. Wilson, *Fundamentals of Momentum, Heat, and Mass Transfer*, Wiley, New York, (1969), pp. 541–546.

Wolf, L., and S. Zaltzman, *Chem. Eng. Prog. Symp. Ser.*, No. 84, **64**, 104–111, (1968).

Zucrow, M. J., and J. P. Sellers, *Am. Rocket Soc. J.*, **31**, 668, (1961).

6

Mass Transfer in Turbulent Flow

This chapter deals with mass transfer in the turbulent boundary layer on a flat plate under conditions of either forced or natural convection, with mass transfer into a falling liquid film, with analogies between mass and momentum transfer, with the phenomena of interfacial turbulence, and with transfer when the mass flux is high.

Much of the chapter considers low-mass-flux conditions in constant-property systems and in terms of compositions expressed as ρ_A and $\Delta\rho_A$, as explained at the beginning of Chapter 5. Procedures for extending the results to high-mass-flux conditions, with and without physical-property variations, are then given in the final section of the chapter.

MASS TRANSFER IN THE TURBULENT BOUNDARY LAYER ON A FLAT PLATE

Measurements on the time-averaged velocity distribution in the turbulent boundary layer on a smooth flat plate at zero incidence to flow are usually correlated as follows for Reynolds numbers $(xu_\infty\rho/\mu)$ up to about 10^7:

$$\frac{u}{u_\infty} = \left(\frac{y}{\delta}\right)^{1/7} \tag{6.1}$$

Equation 6.1 is known as *Prandtl's one-seventh-power law.* Corresponding expressions for concentration distribution have not been established, but results consistent with observations are obtainable by assuming a relationship analogous to equation 6.1:

$$\frac{\rho_A - \rho_{A0}}{\rho_{A\infty} - \rho_{A0}} = \frac{\rho'_A}{\rho'_{A\infty}} = \left(\frac{y}{\delta_c}\right)^{1/7} \tag{6.2}$$

Neither equation 6.1 nor 6.2 can be valid at the surface—as shown, for example, by differentiating 6.2 with respect to y, which erroneously indicates an infinite concentration gradient at $y=0$. The mass-transfer boundary-layer equation 5.8 may be written as follows for negligible v_0:

$$\frac{d}{dx}\int_0^{\delta_c} (\rho'_{A\infty} - \rho'_A) u\, dy = D\left(\frac{\partial \rho'_A}{\partial y}\right)_{y=0} = k^*_\rho \rho'_{A\infty} \tag{6.3}$$

where the inability to evaluate the concentration gradient at $y=0$ from equation 6.2 has necessitated the introduction of $k^*_\rho \rho'_{A\infty}$. The development of mass-transfer relationships will again require an expression for δ, the thickness of the momentum boundary layer. In the boundary-layer equation 5.10, the velocity gradient at $y=0$ cannot be evaluated from equation 6.1, so the expression developed experimentally by Schultz-Grunow (1940; see also Eckert and Drake, 1959, p. 143) for flat plates will be used for τ_0:

$$\tau_0 g_c = 0.0228 \rho u_\infty^2 \left(\frac{\mu}{\delta u_\infty \rho}\right)^{1/4} \tag{6.4}$$

This expression holds up to $xu_\infty\rho/\mu$ of about 10^7, and for negligible v_0 equation 5.10 becomes

$$\frac{d}{dx}\int_0^{\delta} (u_\infty - u) u\, dy = 0.0228 u_\infty^2 \left(\frac{\mu}{\delta u_\infty \rho}\right)^{1/4} \tag{6.5}$$

Combining equations 6.1 and 6.5 leads to

$$\delta = 0.376 x N_{\mathrm{Re},x}^{-0.2} \tag{6.6}$$

The important assumption is now made that $\delta = \delta_c$ at a given x, requiring restriction to Schmidt numbers of unity. The development is therefore confined to gaseous fluids. Combining equations 6.1, 6.2, and 6.3 under these conditions gives

$$k^*_\rho = u_\infty \frac{d}{dx}\left\{\delta \int_0^{1.0} \left[1 - \left(\frac{y}{\delta}\right)^{1/7}\right]\left(\frac{y}{\delta}\right)^{1/7} d\left(\frac{y}{\delta}\right)\right\} \tag{6.7}$$

or

$$k_\rho^* = \frac{7u_\infty}{72}\frac{d\delta}{dx} \tag{6.8}$$

where $d\delta/dx$ is evaluated from equation 6.6 to obtain the following result, after multiplying 6.8 throughout by $N_{\mathrm{Re},x}N_{\mathrm{Sc}}$:

$$(N_{\mathrm{Sh}})_x^* = \frac{k_\rho^* x}{D} = 0.0292 N_{\mathrm{Re},x}^{0.8} \tag{6.9}$$

since $N_{\mathrm{Sc}} = 1.0$. The integrated mean $k_{\rho m}^*$ over the range $0 \leqslant x \leqslant L$ is obtained in a manner similar to that used for the laminar boundary layer, leading to

$$(N_{\mathrm{Sh}})_m^* = \frac{k_{\rho m}^* L}{D} = 0.0365 N_{\mathrm{Re},L}^{0.8} \tag{6.10}$$

Approaches for systems in which N_{Sc} is not unity are noted later beneath equations 6.57 and 6.123, and in Problem 6.6 at the end of the chapter.

Momentum, heat, and mass transfer in turbulent non-Newtonian boundary layers have been treated by Skelland (1966, 1967).

Illustration 6.1

A smooth plate of solid naphthalene is oriented at zero incidence to a pure air stream flowing with a velocity of 115 ft/sec at points remote from the surface of the plate. If the air pressure is 14.7 psia and the system is at a temperature of 113°F, calculate the following for comparison with Illustrations 5.1, 5.2, and 6.4, assuming that the boundary layer is turbulent from the leading edge of the plate and that the influence of the non-unity Schmidt number can be neglected (Illustration 6.4 shows the result of allowing for the deviation of N_{Sc} from unity):

(a) The average mass-transfer coefficient over the first foot of plate length.

(b) The average rate of mass-transfer per unit width over the first foot of plate length.

(c) The local mass-transfer coefficient at a point 1 ft from the leading edge of the plate.

(d) The local naphthalene concentration gradient at the plate surface at a distance of 1 ft from the leading edge.

(e) The local thicknesses of the momentum and concentration boundary layers 1 ft from the leading edge of the plate.

If the plate is smooth enough to sustain a laminar boundary layer over the leading portion of the surface, derive the following quantities, assuming that transition to a turbulent boundary layer takes place at the commonly occurring value of $x_c u_\infty \rho/\mu = 3.2 \times 10^5$, and that the influence of the non-unity Schmidt number can be neglected (check the latter assumption against Illustration 6.4):

(f) The relationship for the mean Sherwood number when $L > x_c$.
(g) The average mass-transfer coefficient over the first foot of plate length under these new conditions.

Take the naphthalene surface temperature to be 113°F (see Illustration 5.1).

SOLUTION (a). The physical properties at the temperature and pressure of this problem are available from Illustrations 5.1 and 5.2. Thus

$$N_{\mathrm{Re},L} = \frac{1(115)(3600)(0.06935)}{0.0457} = 6.28 \times 10^5$$

The N_{Sc} for this system is 2.475 (Illustration 5.2); if this departure from $N_{\mathrm{Sc}} = 1.0$ is neglected, substitution in equation 6.10 gives

$$k^*_{\rho m} = 0.0365\left(\frac{0.2665}{1}\right)(6.28 \times 10^5)^{0.8} = 425 \text{ ft/hr}$$

SOLUTION (b).

$$q_{AW} = k^*_{\rho m}(1 \times 1)(\rho_{A0} - \rho_{A\infty}) = 425(1 \times 1)(0.000224 - 0)$$
$$= 0.0951 \text{ lb naphthalene/hr}$$

SOLUTION (c). From equations 6.9 and 6.10,

$$k^*_\rho = \frac{0.0292}{0.0365}(425) = 340 \text{ ft/hr}$$

SOLUTION (d). Assuming that naphthalene leaves the solid surface by molecular diffusion, then for the differential element of surface $w\,dx$,

$$dq_{AW} = k^*_\rho w\,dx\,(\rho_{A0} - \rho_{A\infty}) = -Dw\,dx\left(\frac{\partial \rho_A}{\partial y}\right)_{y=0}$$

At 1 ft from the leading edge,

$$\left(\frac{\partial \rho_A}{\partial y}\right)_{y=0} = \frac{340(0-0.000224)}{0.2665} = -0.285 \text{ lb naphthalene/ft}^4$$

SOLUTION (e). From equation 6.6,

$$\delta = 0.376(1)(6.28\times10^5)^{-0.2} = 0.026 \text{ ft}$$

Neglect of the deviation of N_{Sc} from unity implies that $\delta=\delta_c$. In fact, however, since $N_{Sc}=2.475$, δ_c will be less than the value calculated for δ.

SOLUTION (f). If the turbulent boundary layer extrapolates to zero thickness at the leading edge of the plate, then

$$k^*_{\rho m} = \frac{1}{L}\left[\int_0^{x_c} k^*_\rho(\text{equation 5.23})\,dx + \int_{x_c}^{L} k^*_\rho(\text{equation 6.9})\,dx\right]$$

For a Schmidt number of unity this gives

$$\frac{k^*_{\rho m}L}{D} = 0.646 N_{\text{Re},x_c}^{1/2} - 0.0365 N_{\text{Re},x_c}^{0.8} + 0.0365 N_{\text{Re},L}^{0.8}$$

If $N_{\text{Re},x_c} = 3.2\times10^5$,

$$\frac{k^*_{\rho m}L}{D} = 0.0365(N_{\text{Re},L}^{0.8} - 15{,}350)$$

SOLUTION (g).

$$k^*_{\rho m} = 0.0365\left(\frac{0.2665}{1}\right)\left[(6.28\times10^5)^{0.8} - 15{,}350\right] = 276 \text{ ft/hr}$$

The reader is again cautioned to note the differences resulting from allowance for the non-unity Schmidt number, as shown in Illustration 6.4.

MASS TRANSFER IN TURBULENT NATURAL CONVECTION ON A VERTICAL PLATE

The process of mass transfer in turbulent natural convection on a vertical plate may be analyzed by a procedure essentially analogous to that described earlier for laminar natural convection in Chapter 5. The approach given here is for low solute concentrations and transfer rates and is based on that presented by Eckert and Jackson (1950) for the heat-transfer analog. The concentration of solute (component A) is constant at ρ_{A0} along the surface of the plate and is $\rho_{A\infty}$ at points remote from the plate.

The momentum equation for the turbulent boundary layer is given by equation 5.29, but with the term $\mu(\partial u/\partial y)_{y=0}$ replaced by $g_c\tau_0$:

$$\frac{d}{dx}\int_0^\delta u^2\,dy = -\frac{g_c\tau_0}{\rho} + \beta_e g\int_0^\delta(\rho_A - \rho_{A\infty})\,dy \tag{6.11}$$

The equation for the concentration boundary layer ($\delta = \delta_c$) with negligible bulk velocity due to diffusion is given by equation 5.30, but with the term $-D(\partial\rho_A/\partial y)_{y=0}$ replaced by n_{A0}:

$$\frac{d}{dx}\int_0^\delta(\rho_A - \rho_{A\infty})u\,dy = n_{A0} \tag{6.12}$$

These replacements are made because the momentum and mass fluxes at the wall (τ_0 and n_{A0}) cannot be estimated from the gradients at $y=0$ of the expressions that are used to approximate the distributions of velocity and concentration in the boundary layer. Instead, experimental values will be used for τ_0 and n_{A0}.

The time-averaged concentration and velocity profiles in the turbulent boundary layer will be approximated by the following equations, which have been found effective in the heat-transfer analog:

$$\rho_A - \rho_{A\infty} = (\rho_{A0} - \rho_{A\infty})\left[1 - \left(\frac{y}{\delta}\right)^{1/7}\right] \tag{6.13}$$

$$u = \Lambda\left(\frac{y}{\delta}\right)^{1/7}\left(1 - \frac{y}{\delta}\right)^4 \tag{6.14}$$

An assumption will now be made corresponding to that of Eckert and Jackson (1950) in the heat transfer case—namely, that the momentum and mass fluxes at the wall are governed by the same respective relationships in both natural and forced convection boundary layers. In consequence, τ_0 is obtained from equation 6.4. The mass flux n_{A0} is taken to be that given by Colburn's analogy between momentum and mass transfer for a flat plate,

$$\frac{n_{A0}}{(\rho_{A0} - \rho_{A\infty})u_\infty}N_{\mathrm{Sc}}^{2/3} = \frac{k_\rho^*}{u_\infty}N_{\mathrm{Sc}}^{2/3} = j_{Dx} = \frac{C_{dx}}{2} \tag{6.15}$$

where ($C_{dx}=2g_c\tau_0/\rho u_\infty^2$) is the local drag coefficient. The flux n_{A0} occurs at $y/\delta \to 0$, where u_∞ and Λ are similar, as shown by equations 6.1 and 6.14. Thus, combining with equation 6.4 and substituting Λ for u_∞,

$$n_{A0}=0.0228(\rho_{A0}-\rho_{A\infty})\Lambda\left(\frac{\mu}{\delta\Lambda\rho}\right)^{1/4}N_{\mathrm{Sc}}^{-2/3} \tag{6.16}$$

Equations 6.4, 6.13, 6.14, and 6.16 are inserted into equations 6.11 and 6.12 and the integrations performed to give

$$0.0523\frac{d}{dx}(\Lambda^2\delta)=-0.0228\Lambda^2\left(\frac{\mu}{\delta\Lambda\rho}\right)^{1/4}+0.125\beta_e g(\rho_{A0}-\rho_{A\infty})\delta \tag{6.17}$$

$$0.0366\frac{d}{dx}(\Lambda\delta)=0.0228\Lambda\left(\frac{\mu}{\delta\Lambda\rho}\right)^{1/4}N_{\mathrm{Sc}}^{-2/3} \tag{6.18}$$

The following functional forms are now assumed:

$$\Lambda=B_3x^p \tag{6.19}$$

$$\delta=B_4x^q \tag{6.20}$$

Equations 6.19 and 6.20 are incorporated into equations 6.17 and 6.18 and the resulting expressions solved for B_3, B_4, p, and q in a manner similar to that shown in Chapter 5 for the laminar-flow case. Eckert and Jackson (1950) give the results as

$$p=\tfrac{1}{2}, \qquad q=\tfrac{7}{10}$$

$$B_3=0.0689\frac{\mu}{\rho}B_4^{-5}N_{\mathrm{Sc}}^{-8/3}$$

$$B_4^{10}=0.00338\frac{\mu^2}{\beta_e g\rho^2(\rho_{A0}-\rho_{A\infty})}\left[1+0.494N_{\mathrm{Sc}}^{2/3}\right]N_{\mathrm{Sc}}^{-16/3}$$

If the Grashof number is introduced from equation 5.44, together with the definition of β_e from equation 5.27, the expressions for Λ and δ at x are, respectively,

$$\Lambda=1.185\frac{\mu}{\rho x}N_{\mathrm{Gr}D}^{1/2}\left(1+0.494N_{\mathrm{Sc}}^{2/3}\right)^{-1/2} \tag{6.21}$$

$$\frac{\delta}{x}=0.565N_{\mathrm{Gr}D}^{-0.1}N_{\mathrm{Sc}}^{-8/15}\left(1+0.494N_{\mathrm{Sc}}^{2/3}\right)^{0.1} \tag{6.22}$$

The local Sherwood number, defined as

$$(N_{\text{Sh}})_x^* = \frac{n_{A0}x}{(\rho_{A0}-\rho_{A\infty})D} = \frac{k_\rho^* x}{D}$$

is therefore obtained from equations 6.16, 6.21, and 6.22:

$$(N_{\text{Sh}})_x^* = 0.0299 N_{\text{Gr}D}^{2/5} N_{\text{Sc}}^{7/15} \left(1+0.494 N_{\text{Sc}}^{2/3}\right)^{-2/5} \tag{6.23}$$

This expression shows that the local coefficient is proportional to the 0.2 power of x, the distance from the leading edge of the plate. Accordingly, if the boundary layer is assumed to be turbulent from the leading edge, the mean coefficient over the range $x=0$ to L is

$$k_{\rho m}^* = \frac{1}{L}\int_0^L k_\rho^*\,dx = \frac{1}{1.2}(k_\rho^*)_{x=L} \tag{6.24}$$

and the corresponding mean Sherwood number is

$$(N_{\text{Sh}})_m^* = 0.0249 N_{\text{Gr}D}^{2/5} N_{\text{Sc}}^{7/15} \left(1+0.494 N_{\text{Sc}}^{2/3}\right)^{-2/5} \tag{6.25}$$

Equation 6.14 shows that the maximum velocity in the boundary layer is given by

$$u_{\max} = 0.537\Lambda = 0.537 B_3 x^p$$

or

$$u_{\max} = 0.636 \frac{\mu}{x\rho}\left(1+0.494 N_{\text{Sc}}^{2/3}\right)^{-1/2} N_{\text{Gr}D}^{1/2} \tag{6.26}$$

In reality the turbulent boundary layer is preceded by a section in which the flow is laminar, in accordance with "equations" 5.51 and 5.52. Equations 6.24 and 6.25 therefore hold only at Grashof numbers large enough for the laminar boundary layer to occupy just a small fraction of the total length L. This appears to be the case for Grashof numbers higher than about 10^{10}.

MASS TRANSFER IN A FALLING LIQUID FILM IN TURBULENT FLOW

The situation to be considered is analogous to that described under this heading in Chapter 5, where the film was in laminar flow. Attention is confined to the following case.

Mass Transfer between an Inclined Plate and a Falling Liquid Film

The physical situation is that detailed in Chapter 5 with Figure 5.5. Equation 5.61 still applies, provided that solute penetration is confined to the laminar sublayer adjacent to the solid surface. Solute concentrations are low, so that bulk velocities due to diffusion are negligible. Equation 5.65 is also valid with these restrictions, so that

$$\beta_v = \left(\frac{du}{dy}\right)_{y=0} = \frac{\rho g \sin\alpha\, \delta_f}{\mu} \tag{6.27}$$

In the case of turbulent films falling down a vertical wall, Brötz (1954) found experimentally that the film thickness is given by

$$\delta_f = 0.172\left(\frac{Q^2}{w^2 g}\right)^{1/3}, \qquad \frac{4Q\rho}{w\mu} > 2360 \tag{6.28}$$

which is similar to the finding of Kamei and Oishi (1955). From equations 5.63, 6.27, and 6.28,

$$(N_{\mathrm{Sh}})_m^* = \frac{k_{\rho m}^* x}{D} = 0.327 N_{\mathrm{Re}f}^{2/9} N_{\mathrm{Sc}}^{1/3}\left(\frac{x^3\rho^2 g}{\mu^2}\right)^{2/9} \tag{6.29}$$

where $N_{\mathrm{Re}f} = 4Q\rho/w\mu$. Equation 6.29 requires conditions such that solute does not penetrate beyond the laminar sublayer in the region where mass is being transferred from the plate surface. It will be shown later that the thickness of the laminar sublayer is believed to be given by

$$y^+ = \frac{y u^* \rho}{\mu} = 5$$

where the friction velocity u^* is

$$u^* = \sqrt{\frac{\tau_0 g_c}{\rho}}, \qquad \tau_0 g_c = \mu\left(\frac{du}{dy}\right)_{y=0} = \mu\beta_v$$

so flow will be laminar in the region

$$y \leqslant 5\sqrt{\frac{\mu}{\rho\beta_v}}$$

in which β_v is obtained from equations 6.27 and 6.28. Kramers and Kreyger (1956) reported "fairly good agreement" between values predicted from equation 6.29 and measurements made on the rate of dissolution of benzoic acid plates in falling water films. Their data were in the range

$$3200 \leqslant N_{\mathrm{Re}f} = \frac{4Q\rho}{w\mu} \leqslant 7000, \qquad 5\ \mathrm{mm} \leqslant x \leqslant 80\ \mathrm{mm}.$$

The values were obtained with an entrance length upstream from the benzoic acid surface of 330 mm.

Illustration 6.2

Repeat Illustration 5.3, but with the ethanol overflow rate increased to 40.9 ft^3/hr.

SOLUTION.

$$N_{\mathrm{Re}f,\mathrm{av}} = \frac{4Q\rho}{w_{\mathrm{av}}\mu} = \frac{4(40.9)(49)}{\pi(2.875/12)(2.66)} = 4000$$

As in the case of Illustration 5.3, the relatively long contact time precludes the use of equation 6.29. However, C. Stirba and D. M. Hurt [*A. I. Ch. E. J.*, **1**, 178–184, (1955)] used Figure 5.6 with Figure 5.7 of Illustration 5.3 successfully up to film Reynolds numbers of about 4000, and this procedure is followed here. Figure 5.6 involves use of the film thickness estimated from equation 5.68; it may be noted that M. L. Jackson [*A. I. Ch. E. J.*, **1**, 231, (1955)] found experimentally that, for liquids with viscosities around that of water or less, equation 5.68 describes the thickness of a liquid film falling down a vertical wall up to $N_{\mathrm{Re}f}$ of at least 4000. Some controversy on this point was noted below equation 5.71, but equation 5.68 is used for the present purpose. For an $N_{\mathrm{Re}f,\mathrm{av}}$ of 4000, Figure 5.7 of Illustration 5.3 gives $D_A = 56\times 10^{-5}$ ft^2/hr, which exceeds the molecular diffusivity for this system by a factor of 22.6. From equation 5.68,

$$\delta_{f,\mathrm{av}} = \left[\frac{3(2.66)(40.9)}{\pi(2.875/12)(49)(4.17\times 10^8)}\right]^{1/3} = 0.002767\ \mathrm{ft}$$

and

$$\frac{D_A\mu x}{\rho\delta_{f,\mathrm{av}}^4 g} = \frac{(56\times 10^{-5})(2.66)(10)}{49(0.002767)^4(4.17\times 10^8)} = 0.0124$$

The corresponding ordinate of Figure 5.6 shows that

$$\frac{\rho_{AB}-\rho_{Ai}}{\rho_A^*-\rho_{Ai}}=0.12=\frac{\rho_{AB}-0}{0.814-0}$$

so

$$\rho_{AB}=0.0977 \text{ lb } C_{17}H_{35}COOH/\text{ft}^3 \text{ of solution in ethanol}$$

This is the average concentration of stearic acid dissolved in the ethanol leaving the overflow pipe. The average dissolution rate is estimated as

$$40.9(0.0977)=3.99 \text{ lb } C_{17}H_{35}COOH/\text{hr}$$

From Illustration 5.3, the amount of stearic acid initially present is 4.16 lb, so that the time required for its removal by dissolution in the falling film of ethanol is estimated as

$$\frac{4.16}{3.99}=1.04 \text{ hr}$$

The amount of ethanol contaminated under the slower overflow conditions of Illustration 5.3 is only 14.2 ft^3, compared to 42.6 ft^3 for the present case.

Questions concerning the generality of Figure 5.7 were posed in Illustration 5.3; as in that calculation, entrance effects have again been neglected here. Interpolation as before between the measurements of entrance length by Tailby and Portalski [*Chem. Eng. Sci.*, **17**, 283–290, (1962)] for vertically falling films of methanol and isopropanol, respectively, suggests that, for a film Reynolds number of 4000, the entrance length in the present case is somewhat less than a foot—that is, less than 10 percent of the total length over which mass transfer occurs.

ANALOGIES BETWEEN MOMENTUM AND MASS TRANSFER

Rigorous solution of the equations governing the transport of matter in a turbulent stream is complicated because of the unknown fluctuations in components of the velocity. Substantial progress has been made in this field, however, by the continuing development and refinement of analogies between heat, mass, and momentum transfer. Although attention has centered to a greater extent on the analogy between heat and momentum transfer, the resulting expressions are in general readily converted into terms of mass and momentum transfer, and it is the latter analogy that will be primarily considered here.

Some of the principal contributors to this area include the following, in chronological order:

Reynolds (1874); Prandtl (1910, 1928); Taylor (1916); Murphree (1932); Colburn (1933); Chilton and Colburn (1934); von Kármán (1939); Sherwood (1940); Hoffman (1940); Reichardt (1940); Boelter, Martinelli, and Jonassen (1941); Martinelli (1947); Jenkins (1951); Lyon (1951); Seban and Shimazaki (1951); Deissler (1952, 1954, 1955); Lin, Moulton, and Putnam (1953); Metzner and W. L. Friend (1958, 1958a); Metzner and P. S. Friend (1959); Clapp (1961); Gowariker and Garner (1962); Skelland (1970); and Hanna and Sandall (1972). Some excellent reviews are given by Knudsen and Katz (1958, Chapter 15) and by Sherwood (1959).

The analogies to be described here will follow a historical pattern so as to provide a clearly evolving picture of the concepts which have been developed for describing turbulent mass transfer. Of key importance in the formulation of nearly all of these analogies is the structure of the velocity distribution and the corresponding distribution of eddy diffusivities between a solid boundary and the bulk or core of the turbulent stream. The second, third, and fourth analogies to be described make use of part or all of the so-called "universal velocity distribution" for turbulent fluids in smooth tubes, and this is accordingly developed below. Constant-property fluids, with low solute concentrations and mass-transfer rates, are considered throughout.

A dozen examples showing important applications of mass transfer in internal flows through tubes and ducts of other geometries under laminar or turbulent conditions are given in Chapter 5 under the heading "Mass Transfer in Laminar Flow Through a Tube." The subject is clearly of sufficient importance to merit extensive study.

The Universal Velocity Distribution in Smooth Tubes

The relationship described in this section was derived by Prandtl (1933) from his mixing-length theory. It is first necessary to develop the expression for stress distribution in the flow of any fluid through a cylindrical tube, as shown in Figure 6.1.

The fluid is flowing upward in the inclined tube of radius r_t, and for steady, uniform flow the sum of all the forces acting on the fluid between sections 1 and 2 will be zero. The relevant forces are those due to static pressure, gravity, and shear, so that

$$\pi r_t^2 P_1 - \pi r_t^2 P_2 - \pi r_t^2 L\rho \frac{g}{g_c}\cos\theta - 2\pi r_t L\tau_W = 0$$

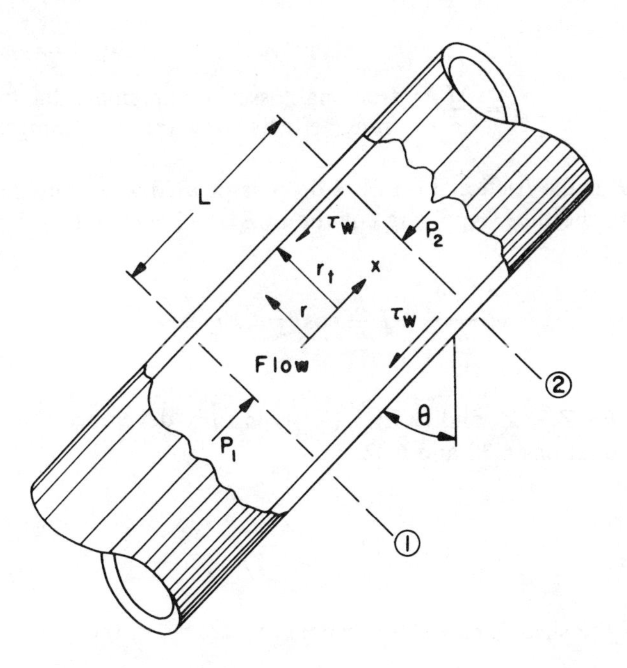

Figure 6.1. Stresses acting on a cylindrical element of fluid of radius r_t in steady flow through an inclined tube.

or

$$\frac{P_1 - P_2}{\rho} - L\frac{g}{g_c}\cos\theta = \frac{2L\tau_W}{r_t\rho} \tag{6.30}$$

Now the total mechanical-energy balance for the steady incompressible flow of a unit mass between points 1 and 2 may be written in general as shown by Perry (1950, p. 377):

$$Z_1\frac{g}{g_c} + \frac{P_1}{\rho} + (\text{KE})_1 - W_s' = Z_2\frac{g}{g_c} + \frac{P_2}{\rho} + (\text{KE})_2 + \Sigma F \tag{6.31}$$

where Z is vertical height above an arbitrary datum plane, P is static pressure, KE is the average kinetic energy per unit mass, and W_s' is the shaft work done by a unit mass of fluid. W_s' is negative when work is done on the fluid by an outside source such as a pump. The term ΣF represents mechanical energy converted into thermal energy as a result of fluid friction, and is given by

$$\sum F = \frac{\Delta P}{\rho} + \text{frictional losses per unit mass due to entrance effects, flow through fittings, etc.}$$

in which ΔP is the frictional pressure drop associated with fully developed flow through the tube. Applying equation 6.31 to Figure 6.1,

$$\frac{P_1 - P_2}{\rho} - L\frac{g}{g_c}\cos\theta = \sum F = \frac{\Delta P}{\rho} \tag{6.32}$$

where $L\cos\theta = Z_2 - Z_1$ and $(\mathrm{KE})_1 = (\mathrm{KE})_2$ in fully developed flow. Therefore, from equations 6.30 and 6.32,

$$(\tau_{rx})_{r=r_t} = \tau_W = \frac{r_t \Delta P}{2L} \tag{6.33}$$

Similarly, if the shear stress at any r is τ_{rx}, where $r \leqslant r_t$, then

$$\tau_{rx} = \frac{r\Delta P}{2L} \tag{6.34}$$

and from equations 6.33 and 6.34,

$$\tau_{rx} = \tau_W \frac{r}{r_t} \tag{6.35}$$

which indicates a linear distribution of shear stress in the fluid, regardless of the flow regime.

At any point y in the fluid in Figure 6.1 the shear stress is as follows, where $y = r_t - r$:

$$\tau_{rx} = \tau_{rx,\,\mathrm{laminar}} + \tau_{rx,\,\mathrm{turbulent}}$$

and for a Newtonian fluid,

$$\tau_{rx} = \frac{\mu}{g_c}\frac{du}{dy} + \frac{\rho}{g_c}\epsilon_M\frac{du}{dy}$$

where ϵ_M is the eddy momentum diffusivity or eddy kinematic viscosity. In isotropic turbulence the main-stream velocity in the x direction will have superimposed upon it a mean eddy velocity u_e, acting randomly in the x, y, and z directions.

Prandtl defined his "mixing length" l as the mean distance that an eddy travels in a direction perpendicular to the main flow before losing its identity by mingling with adjacent fluid. It is possible to show that the eddy momentum diffusivity ϵ_M is proportional to u_e multiplied by l. A further assumption by Prandtl is that $u_e \propto l\,du/dy$, from which $\epsilon_M \propto l^2 du/dy$. The mixing length is not readily evaluated, and a proportionality constant of unity is therefore convenient, implying a modified significance for l. The result is $\epsilon_M = l^2\, du/dy$, and the relationship for τ_{rx} becomes

$$\tau_{rx} = \frac{\mu}{g_c}\frac{du}{dy} + \frac{\rho}{g_c} l^2 \left(\frac{du}{dy}\right)^2 \tag{6.36}$$

Prandtl considered the flow in the tube to be made up of three regions: a very thin region of laminar flow immediately adjacent to the wall, called the laminar sublayer; a turbulent core in the bulk of the fluid; and a thin buffer zone between the laminar sublayer and the turbulent core. The effects of turbulence are regarded as negligible in the laminar sublayer, turbulence and viscous shear exert comparable effects in the buffer zone, and the effects of viscosity are assumed to be negligible in the turbulent core. The second term on the right-hand side of equation 6.36 is therefore negligible compared with the first in the laminar sublayer, where l is effectively zero. The velocity gradient is approximately linear in the laminar sublayer, because the latter is so thin. In consequence τ_{rx} is constant, and

$$\tau_{rx} = \tau_W = \frac{\mu}{g_c}\frac{u}{y}$$

$$u^+ = \frac{u}{u^*} = \frac{\tau_W g_c}{\mu u^*} y = \frac{y u^* \rho}{\mu} = y^+ \tag{6.37}$$

Equation 6.37 is found to hold for values of y^+ up to 5.

The first term on the right-hand side of equation 6.36 is negligible compared with the second in the turbulent core. Prandtl assumed that the mixing length is directly proportional to the distance from the wall: $l = ky$, where k is a constant. The shear stress at y may therefore be expressed in terms of τ_W from equation 6.35 to give

$$\tau_W\left(1 - \frac{y}{r_t}\right) = \frac{\rho}{g_c}\left(ky\frac{du}{dy}\right)^2 \tag{6.38}$$

For regions not far removed from the wall, the term $1-y/r_t$ is near unity. In these regions equation 6.38 may be restated as

$$\sqrt{\frac{\tau_W g_c}{\rho}} = u^* = ky\frac{du}{dy} \tag{6.39}$$

where u^* is called the friction velocity. Integrating,

$$u = \frac{u^*}{k}\ln y + \text{constant} = u^*\left(\frac{\ln y}{k} + \text{constant}'\right) \tag{6.40}$$

Let constant′ equal

$$k' - \frac{1}{k}\ln\frac{\mu}{u^*\rho} \tag{6.41}$$

in which k' is another constant, so that

$$\frac{u}{u^*} = \frac{1}{k}\ln\frac{yu^*\rho}{\mu} + k' \tag{6.42}$$

Equation 6.42 was obtained by assuming that $1-y/r_t$ is close to unity, and it might therefore be expected that the expression would be confined to regions of the turbulent core near the wall. It is found, however, that equation 6.42 correlates experimental measurements over nearly all the turbulent core, except at the centerline, as shown below. Equation 6.42 may be rewritten with the definitions $u^+ = u/u^*$ and $y^+ = yu^*\rho/\mu$, together with experimentally evaluated constants, to give

$$u^+ = 2.5\ln y^+ + 5.5 \tag{6.43}$$

Equations 6.42 and 6.43 indicate a value of $-\infty$ for u at the wall ($y=0$). This erroneous result arises from neglecting the viscous forces represented by the first term on the right-hand side of equation 6.36. Equation 6.43 may be differentiated and the definitions of u^+, u^*, and y^+ inserted to obtain

$$\frac{du}{dy} = \frac{2.5}{y}\sqrt{\frac{\tau_W g_c}{\rho}}$$

which is in error in showing a nonzero value for du/dy at the tube axis ($y = r_t$). In spite of these inadequacies at the wall and at the centerline, equation 6.43 has been very successful in correlating turbulent velocity distributions for y^+ greater than 30.

An equation widely used in the buffer zone for y^+ values between 5 and 30 is

$$u^+ = 5.0 \ln y^+ - 3.05 \tag{6.44}$$

The universal velocity distribution for the turbulent flow of Newtonian fluids in smooth tubes is therefore as follows:

Laminar sublayer, $y^+ \leqslant 5$: $\quad u^+ = y^+ \quad$ (6.37)

Buffer zone, $5 \leqslant y^+ \leqslant 30$: $\quad u^+ = 5.0 \ln y^+ - 3.05 \quad$ (6.44)

Turbulent core, $30 < y^+$: $\quad u^+ = 2.5 \ln y^+ + 5.5 \quad$ (6.43)

At the edge of the laminar sublayer ($y^+ = 5$), equations 6.37 and 6.44 intersect and are continuous in slope, since du^+/dy^+ is the same for both equations at this point. At the junction between the buffer zone and turbulent core ($y^+ = 30$), however, equations 6.43 and 6.44 intersect, but with a discontinuity in slope. Equations 6.37, 6.44, and 6.43 are compared with experimental data in Figure 6.2.

Although developed for tubes, this time-averaged velocity distribution is found to give good representation of the flow between parallel plates and

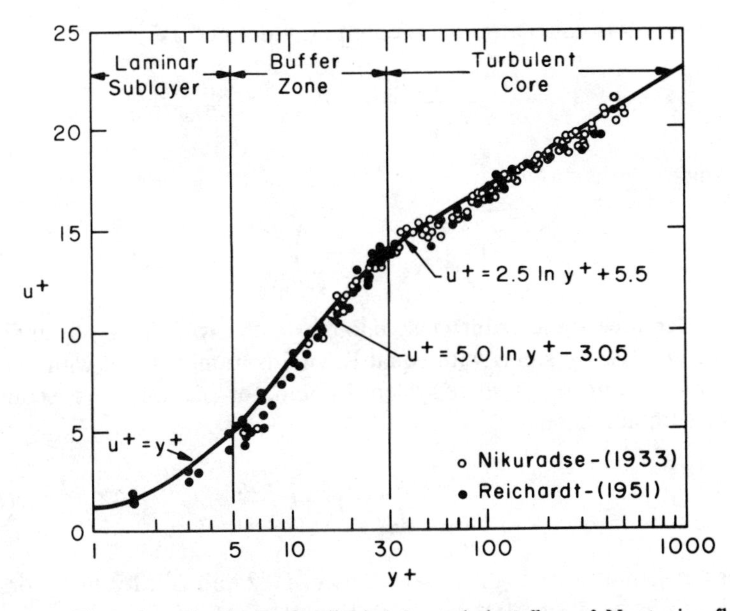

Figure 6.2. The universal velocity distribution for turbulent flow of Newtonian fluids in smooth tubes.

in boundary layers on flat plates, but with some deviation towards the outer edge of the boundary layer (Schlichting, 1955, p. 440). Corrections to equation 6.43 for small but systematic deviations from experimental measurements in tubes have been made by several investigators, notably Millikan, Reichardt, and Hinze. These are reviewed by Bogue and Metzner (1963), who also evolved an effective correction procedure from a comprehensive examination of available data.

The Distribution of Resistance to Mass Transfer in Turbulent Flow

For a Newtonian fluid in turbulent flow past a rigid boundary, the flux of component A at y in the direction y normal to the wall may be written as

$$n_{Ay} = -(D+\epsilon_D)\frac{d\rho_A}{dy} \tag{6.45}$$

where ϵ_D is an eddy mass diffusivity. For regions of small y near the wall, $n_{Ay} \doteq n_{AW}$, and so

$$1 = -\left(\frac{1}{N_{\text{Sc}}} + \frac{\epsilon_D}{\mu/\rho}\right)\frac{\mu}{n_{AW}\rho}\frac{d\rho_A}{dy} \tag{6.46}$$

A dimensionless concentration is commonly defined as

$$\rho_A^+ = \frac{(\rho_{AW}-\rho_A)u^*}{n_{AW}}; \quad d\rho_A = -\frac{n_{AW}d\rho_A^+}{u^*}$$

and, since $y^+ = yu^*\rho/\mu$,

$$1 = \left(\frac{1}{N_{\text{Sc}}} + \frac{\epsilon_D}{\mu/\rho}\right)\frac{d\rho_A^+}{dy^+} \tag{6.47}$$

Consider now some point close enough to the wall for ϵ_D to approach zero for two fluid systems with equal Reynolds numbers and with Schmidt numbers $N_{\text{Sc}1}$ and $N_{\text{Sc}2}$, where system 1 could, for example, be gaseous and system 2 liquid. Then

$$\frac{(d\rho_A^+/dy^+)_{W,2}}{(d\rho_A^+/dy^+)_{W,1}} = \frac{N_{\text{Sc}2}}{N_{\text{Sc}1}} \tag{6.48}$$

For inorganic gases N_{Sc} is usually between 0.2 and 2.2; for many organic liquids N_{Sc} is between 300 and 2000. Equation 6.48 shows that increasing the Schmidt group leads to a corresponding increase in the dimensionless concentration gradient in the wall region (where $\epsilon_D \to 0$, $n_{Ay} \to n_{AW}$). This of course corresponds to locating more and more of the major resistance to mass transfer within the sublayers near the wall as N_{Sc} increases. For this reason assumptions concerning ϵ_D near the wall, the nature and thickness of the laminar sublayer, and other quantities in the wall region become of crucial importance to the success of mass-transfer relationships based on analogies when the Schmidt number is high, as in liquid systems.

Conversely, assumptions concerning wall conditions are less critical when N_{Sc} is small, because concentration gradients then assume only moderate values near the wall and remain significantly above zero to much greater y^+. In the analogous process of heat transfer, N_{Pr} can attain much lower values than N_{Sc}—as, for example, in the case of liquid metals. Thus N_{Pr} for liquid sodium is 0.0072 at 400°F. The heat-transfer analog of equation 6.47 is

$$1 = \left(\frac{1}{N_{Pr}} + \frac{\epsilon_H}{\mu/\rho} \right) \frac{dT^+}{dy^+} \tag{6.49}$$

where

$$T^+ = \frac{(T_W - T)\rho c_p u^*}{q_W}$$

In such cases of very low N_{Pr}, the quantity $1/N_{Pr}$ may exceed $\epsilon_H/(\mu/\rho)$ at all locations, which means that molecular conduction may exert a substantial and possibly dominant influence on the temperature profile and transfer mechanism even in the turbulent core.

Relationships based upon several different assumptions about conditions in the vicinity of the wall are now considered. A more comprehensive study would of course require examination of all the references cited on page 232.

Analogy Assuming Only a Turbulent Core

Osborne Reynolds (1874) was the first to recognize an analogy between the convective transfer of heat and momentum, and his expressions are readily extended to mass transfer. In Reynolds' model the turbulent core extends all the way up to the solid wall; the buffer zone and laminar sublayer are absent, although the fluid velocity is zero immediately adjacent to the wall, in accordance with the requirement of zero slip. Consider the turbulent flow of a constant-property fluid in the vicinity of a solid surface from which mass transfer is occurring. Velocity and concentration changes are significant only in the y direction, as shown in Figure 6.3.

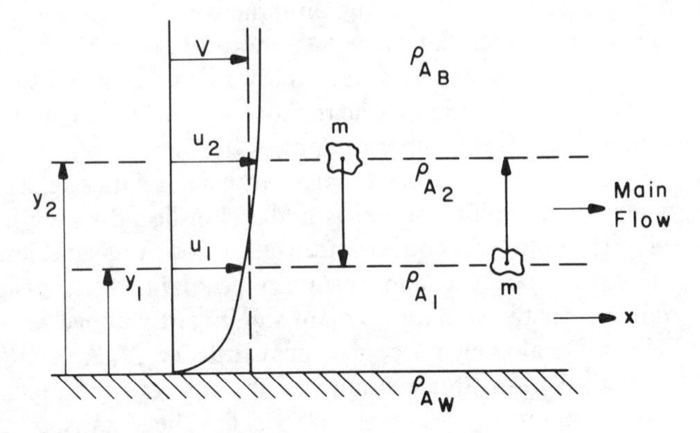

Figure 6.3. Momentum and mass transfer in turbulent flow near a solid surface, assuming only a turbulent core.

It is convenient to visualize the randomly turbulent, fluctuating motion as caused by small aggregates of fluid moving back and forth across the stream in the y direction. This movement involves transverse transfer of momentum because the time-averaged velocity u is a function of y, with the result that turbulent shearing stresses exist in the fluid. Similarly, if the concentration of component A is a function of y, then a net transfer of component A must occur by the *same* mechanism. In particular, suppose a fluid aggregate of mass m travels from a level y_1 to another level y_2, carrying with it the momentum and concentration corresponding to level y_1. Continuity relationships require that another fluid aggregate, also of mass m, travels from level y_2 to level y_1, carrying the momentum and concentration corresponding to level y_2. The net transfer between levels y_1 and y_2 is then

$$\text{Net transfer of } x\text{-directed momentum towards the surface} = m(u_2 - u_1)$$

$$\text{Net transfer of component } A \text{ away from the surface} = -\frac{m}{\rho}(\rho_{A2} - \rho_{A1})$$

Suppose that, on the average, the fluid aggregates travel between the solid surface ($u=0$, $\rho_A = \rho_{AW}$) and locations where velocity and concentration have bulk average values ($u = V$, $\rho_A = \rho_{AB}$); then

$$\frac{\text{Rate of mass transfer}}{\text{Rate of } x\text{-momentum transfer}} = \frac{n_{AW}}{\tau_W g_c} = -\frac{(\rho_{AB} - \rho_{AW})}{\rho V}$$

or

$$\frac{-n_{AW}}{V(\rho_{AB}-\rho_{AW})}=\frac{\tau_W g_c}{\rho V^2}=\frac{f}{2}$$

where f is Fanning's friction factor. The mass-transfer coefficient k_ρ^* is defined as $n_{AW}/(\rho_{AW}-\rho_{AB})$, so that

$$N_{\mathrm{Sh}}^*=\frac{k_\rho^* d_t}{D}=\frac{-n_{AW}N_{\mathrm{Re}}N_{\mathrm{Sc}}}{V(\rho_{AB}-\rho_{AW})} \tag{6.50}$$

and

$$N_{\mathrm{Sh}}^*=\frac{f}{2}N_{\mathrm{Re}}N_{\mathrm{Sc}} \tag{6.51}$$

Equation 6.51 is found to be a good approximation only when the Schmidt number is close to unity. This is a consequence of the simplifications involved in the derivation, and particularly the neglect of the so-called laminar sublayer. Thus it is incorrect to assume that momentum and mass transfer occur by fully turbulent motion right up to the wall. Transfer through the laminar sublayer will be mainly by momentum and molecular diffusion processes. The relationship therefore applies best when momentum and molecular diffusivities are equal, that is to say, when the Schmidt number is unity. This, in fact, is approximately the case in most gas-phase transfer processes.

Analogy Assuming a Laminar Sublayer and a Turbulent Core

The limitations which have been noted in the Reynolds analogy and the resulting equation 6.51 were pointed out by Prandtl (1910, 1928) and by Taylor (1916), who introduced an extension to allow for the effects of the laminar sublayer. This, coupled with the turbulent core, constitutes two regions of flow in the vicinity of a solid surface, as shown in Figure 6.4.

In the laminar sublayer ($0 \leqslant y \leqslant \delta_L$), the shearing stresses are purely viscous, and mass transfer from the solid surface is by molecular diffusion. In the turbulent core ($y > \delta_L$) the transfer of momentum and mass is as postulated in the Reynolds analogy.

Consider first the transfer in the laminar sublayer, which is supposed to be thin enough to permit assuming linear distributions of velocity and concentration.

$$\tau_W g_c=\mu\frac{du}{dy}=\mu\frac{u_L}{\delta_L}$$

$$n_{AW}=-D\frac{d\rho_A}{dy}=-D\frac{(\rho_{AL}-\rho_{AW})}{\delta_L}$$

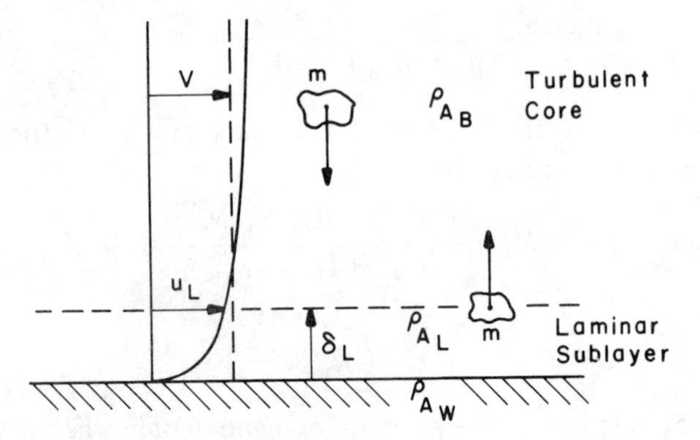

Figure 6.4. Momentum and mass transfer in turbulent flow near a solid surface, assuming a laminar sublayer and a turbulent core.

from which

$$\frac{n_{AW}}{\tau_W g_c} = -\frac{D(\rho_{AL}-\rho_{AW})}{\mu u_L} = -\frac{(\rho_{AL}-\rho_{AW})}{N_{\text{Sc}}\rho u_L} \tag{6.52}$$

Transfer in the turbulent core is described by the Reynolds analogy, in which aggregates of fluid travel, on the average, between the edge of the laminar sublayer ($u=u_L$, $\rho_A=\rho_{AL}$) and locations where velocity and concentration have bulk average values ($u=V$, $\rho_A=\rho_{AB}$), so that

$$\frac{n_{AW}}{\tau_W g_c} = -\frac{(\rho_{AB}-\rho_{AL})}{\rho(V-u_L)} \tag{6.53}$$

Equations 6.52 and 6.53 both apply at $y=\delta_L$. Solving 6.52 for $\rho_{AW}-\rho_{AL}$ and 6.53 for $\rho_{AL}-\rho_{AB}$ and adding the results,

$$\frac{n_{AW}}{V(\rho_{AW}-\rho_{AB})} = \frac{\tau_W g_c}{\rho V^2}\left[\frac{1}{1+(u_L/V)(N_{\text{Sc}}-1)}\right]$$

or, in terms of equation 6.50,

$$N_{\text{Sh}}^{*} = \frac{(f/2)N_{\text{Re}}N_{\text{Sc}}}{1+(u_L/V)(N_{\text{Sc}}-1)} \tag{6.54}$$

This expression reduces to that resulting from the Reynolds analogy (equation 6.51) for a Schmidt number of unity. The problem remaining is to evaluate u_L/V, and this depends on the geometry of the flow system. In the case of turbulent flow through a smooth tube, the laminar sublayer is considered to be determined by equation 6.37:

$$y^+ = u^+ = \frac{u}{u^*} = \frac{u/V}{\sqrt{f/2}}, \qquad y^+ \leqslant 5$$

and at the edge of the laminar sublayer, $u = u_L$, $y^+ = u^+ = 5$, so

$$\frac{u_L}{V} = 5\sqrt{\frac{f}{2}}$$

Substituting in equation 6.54,

$$N_{\mathrm{Sh}}^* = \frac{k_\rho^* d_t}{D} = \frac{(f/2) N_{\mathrm{Re}} N_{\mathrm{Sc}}}{1 + 5\sqrt{f/2}\,(N_{\mathrm{Sc}} - 1)} \tag{6.55}$$

A material balance may be made on component A for the flow through an element of tube having length dx in the direction of flow:

$$\frac{\pi d_t^2}{4} V\, d\rho_{AB} = k_\rho^* \pi d_t (\rho_{AW} - \rho_{AB})\, dx \tag{6.56}$$

Integrating over length L between sections 1 and 2 for constant ρ_{AW} and substituting for k_ρ^* from equation 6.55,

$$\ln \frac{\rho_{AW} - \rho_{AB1}}{\rho_{AW} - \rho_{AB2}} = \frac{2fL/d_t}{1 + 5\sqrt{f/2}\,(N_{\mathrm{Sc}} - 1)} \tag{6.57}$$

Equation 6.55 is found to be a distinct improvement over equation 6.51, although deviations from experimental measurements increase with increasing Schmidt number.

An expression analogous to 6.55 is readily developed for a flat plate from equation 6.54 (Knudsen and Katz, 1958, p. 486–487), and this could replace equation 6.9 when N_{Sc} does not equal unity (see Problem 6.6 at the end of the chapter).

Analogy Assuming a Laminar Sublayer, a Buffer Zone, and a Turbulent Core

A further substantial refinement was obtained by von Kármán (1939), who allowed for three regions of flow—the laminar sublayer, the buffer zone, and the turbulent core. The development in mass-transfer terms will now be given in some detail. Transfer will occur from the walls to a constant-property fluid in turbulent flow through a tube, under conditions such that the mass-average velocity v_0 in the y direction due to diffusion is negligible. Changes in velocity and concentration are assumed to be substantial in the y direction only (Eckert and Gross, 1963, p. 134).

In the laminar sublayer,

$$\tau_W = \frac{\mu}{g_c}\frac{du}{dy}$$

$$(n_{Ay})_{y=0} = n_{AW} = -D\frac{d\rho_A}{dy} = -\frac{Du^*\rho}{\mu}\frac{d\rho_A}{dy^+}$$

Integrating over the limits $\rho_{AW} \leqslant \rho_A \leqslant \rho_{AL}$, $0 \leqslant y^+ \leqslant 5$,

$$\rho_{AL} - \rho_{AW} = -\frac{5N_{\text{Sc}}}{u^*}n_{AW} \tag{6.58}$$

Next, consider the buffer zone. The laminar sublayer and buffer zone are so thin that τ_{yx} and n_{Ay} are assumed constant at τ_W and n_{AW} in these regions; thus

$$\tau_W = \frac{\rho}{g_c}\left(\frac{\mu}{\rho} + \epsilon_M\right)\frac{du}{dy} \tag{6.59}$$

$$n_{AW} = -(D + \epsilon_D)\frac{d\rho_A}{dy} \tag{6.60}$$

where ϵ_M is the eddy momentum diffusivity and ϵ_D is the eddy mass diffusivity. It is assumed that $\epsilon_M = \epsilon_D = \epsilon$. Equation 6.44 shows that

$$\frac{du^+}{dy^+} = \frac{5}{y^+}, \qquad \frac{du}{dy} = \frac{5\tau_W g_c}{\mu y^+}$$

and substituting in equation 6.59,

$$\epsilon = \epsilon_M = \frac{\mu}{\rho}\left(\frac{y^+}{5} - 1\right) \tag{6.61}$$

Inserting this expression in equation 6.60,

$$\frac{d\rho_A}{dy^+} = \frac{d\rho_A}{dy}\frac{\mu}{\rho u^*} = \frac{-n_{AW}N_{\text{Sc}}}{u^*[N_{\text{Sc}}y^+/5 - (N_{\text{Sc}} - 1)]}$$

The integration limits are $\rho_{AL} \leqslant \rho_A \leqslant \rho_{Ab}$, $5 \leqslant y^+ \leqslant 30$, giving

$$\rho_{Ab} - \rho_{AL} = -\frac{5n_{AW}}{u^*} \ln(5N_{\text{Sc}} + 1) \tag{6.62}$$

In the turbulent core, for y^+ somewhat greater than 30, μ/ρ and D may be neglected in comparison with ϵ, and dividing equation 6.60 by equation 6.59 under these conditions,

$$\frac{\rho n_{AW}}{\tau_W g_c} = \frac{-d\rho_A/dy}{du/dy}$$

If the velocity and concentration distributions are similar in form (Rohsenow and Choi, 1961, p. 185) this relationship can be written

$$\frac{n_{AW}}{(u^*)^2} = \frac{-(\rho_{AB} - \rho_{Ab})}{V - u_b} \tag{6.63}$$

where ρ_{AB} and V are the concentration and velocity in a region having bulk average values of these quantities, and u_b is the velocity at the outer edge of the buffer zone ($y^+ = 30$). From equation 6.63,

$$\rho_{AB} - \rho_{Ab} = \frac{-n_{AW}(V^+ - u_b^+)}{u^*} \tag{6.64}$$

where $V^+ = V/u^*$. Equation 6.44 shows that

$$u_b^+ - u_L^+ = 5\ln\frac{30}{5} = 5\ln 6 \tag{6.65}$$

and from equation 6.37, $u_L^+ = y_L^+ = 5$. Therefore,

$$u_b^+ = 5(1 + \ln 6) \tag{6.66}$$

Substituting in equation 6.64,

$$\rho_{AB} - \rho_{Ab} = \frac{-n_{AW}}{u^*}\left[\frac{V}{u^*} - 5(1 + \ln 6)\right] \tag{6.67}$$

Equations 6.58, 6.62, and 6.67 may be added to obtain the overall concentration difference as follows:

$$\rho_{AB} - \rho_{AW} = -n_{AW}\left\{\frac{V}{(u^*)^2} + \frac{5}{u^*}\left[(N_{\text{Sc}} - 1) + \ln\left(\frac{5N_{\text{Sc}} + 1}{6}\right)\right]\right\} \tag{6.68}$$

The Sherwood number for mass transfer is defined by equation 6.50. Inserting equation 6.68 for $\rho_{AB}-\rho_{AW}$, $\sqrt{\tau_W g_c/\rho}$ for u^*, and then replacing $\tau_W g_c/\rho V^2$ by $f/2$,

$$N_{\text{Sh}}^* = \frac{(f/2)N_{\text{Re}}N_{\text{Sc}}}{1+5\sqrt{f/2}\,\{N_{\text{Sc}}-1+\ln[1+\frac{5}{6}(N_{\text{Sc}}-1)]\}} \tag{6.69}$$

which again reduces to Reynolds' equation 6.51 when N_{Sc} equals one. Equation 6.69 enables mass-transfer rates to be predicted from the use of the conventional friction-factor–Reynolds-number chart for smooth tubes.

The development of equation 6.69 was performed assuming that $\epsilon_M=\epsilon_D$. Sherwood (1940) refined the analysis to the extent of eliminating this assumption; his result is expressed as

$$N_{\text{Sh}}^* = \frac{(f/2)N_{\text{Re}}(\epsilon_D/\epsilon_M)N_{\text{Sc}}}{1+5\sqrt{f/2}\left\{\frac{\epsilon_D}{\epsilon_M}N_{\text{Sc}}-1+\ln\left[1+\frac{5}{6}\left(\frac{\epsilon_D}{\epsilon_M}N_{\text{Sc}}-1\right)\right]\right\}} \tag{6.70}$$

Alternative expressions to equations 6.69 and 6.70, resulting from the approaches of Reichardt (1940) and of Boelter, Martinelli, and Jonassen (1941), are well summarized in Perry (1950, pp. 542–543).

The ratio ϵ_D/ϵ_M was evaluated by Sherwood and Woertz (1939) in measurements of the rate of transfer of water vapor across a vertical rectangular duct. Transfer occurred between a water film flowing down one wall of the duct and a calcium chloride solution flowing down the opposite wall. Pitot-tube sampling traverses across the duct established velocity and concentration distributions, enabling du/dy and $d\rho_A/dy$ to be evaluated. Measurements of the frictional pressure drop in the duct then permitted the calculation of ϵ_M and ϵ_D from equations 6.59 and 6.60. For these experiments ϵ_D/ϵ_M was found to be roughly constant at about 1.6. Using this value for ϵ_D/ϵ_M, Sherwood (1940) showed equation 6.70 to give good agreement with Gilliland's (1934) data on vaporization of nine different liquids into a turbulent air stream in a tubular wetted-wall column for Reynolds numbers between 1840 and 30,000.

The von Kármán analogy, equation 6.69, becomes a much poorer approximation at higher Schmidt numbers (greater than 40), and particularly for N_{Sc} values occurring in liquid-phase mass transfer—as found, for example, by Lin, Denton, Gaskill, and Putnam (1951). This is also the case for the modification given by equation 6.70 (Linton and Sherwood, 1950), and is, of course, consistent with the conclusions reached earlier (below equation 6.48) regarding the crucial importance of assumptions on conditions in the wall region for systems with high Schmidt number.

Analogy Assuming an Eddying Sublayer, a Buffer Zone, and a Turbulent Core

Although the concept of three regions of flow—the laminar sublayer, the buffer zone, and the turbulent core—has been widely adopted, most investigators have conceded that this division into three layers represents a simplification of reality. In particular, the existence of the laminar sublayer has been challenged from time to time. Fage and Townend (1932) made ultramicroscopic observations of the colloidal particles in tap water and found that sinuous motion involving velocity components normal to the solid surface occurred as close as 0.000025 in. from the wall. When molecular diffusivities are low, as in the liquid phase, the influence on mass transfer of slight eddies in the wall layer would be significant. These considerations led Lin, Moulton, and Putnam (1953) to abandon the concept of the laminar sublayer, necessitating the development of a new velocity distribution.

In the immediate vicinity of the wall ($0 \leqslant y^+ \leqslant 5$) they introduced an eddy momentum diffusivity formulated as

$$\epsilon_M = \frac{\mu}{\rho}\left(\frac{y^+}{14.5}\right)^3 \tag{6.71}$$

This relationship is arbitrary and was chosen on the basis of simplicity and the best agreement between experimental measurements and their equation for the mass transfer coefficient. Equation 6.59 may then be written for the eddying sublayer or wall layer and integrated, assuming the shearing stress is constant and equal to τ_W near the wall:

$$\frac{u}{\tau_W g_c/\rho} = \frac{u}{(u^*)^2} = \int \frac{dy}{(\mu/\rho)\left[1+(y^+/14.5)^3\right]} + C \tag{6.72}$$

The integration is performed with dy^+ in place of dy and with C evaluated in accordance with the boundary condition $u^+ = 0$ for $y^+ = 0$, to obtain

$$u^+ = \frac{14.5}{3}\left[\frac{1}{2}\ln\frac{(1+y^+/14.5)^2}{1-y^+/14.5+(y^+/14.5)^2} + \sqrt{3}\tan^{-1}\frac{2y^+/14.5-1}{\sqrt{3}} + \frac{\pi\sqrt{3}}{6}\right] \tag{6.73}$$

This replaces equation 6.37 for the eddying sublayer or wall layer ($y^+ \leqslant 5$). Equation 6.43 is retained to describe the turbulent core in this approach, but equation 6.44 for the buffer zone is replaced by the following expression in order to satisfy the velocity and eddy conditions at $y^+ = 5$:

$$u^+ = -3.27 + 5\ln(y^+ + 0.205) \tag{6.74}$$

or

$$u^+ = 4.77 + 5\ln\left(\frac{y^+}{5} + 0.041\right) \tag{6.75}$$

Equation 6.74 for the buffer zone and equation 6.43 for the turbulent core intersect at a y^+ of about 33, and the relationships for the three flow regions are compared with experimental data in Figure 6.5. The curve corresponding to equation 6.73 differs only slightly from that corresponding to equation 6.37 ($u^+ = y^+$). The point, however, is that the concept of a purely laminar sublayer has been discarded.

Equation 6.75 may be differentiated to obtain

$$\frac{du}{dy} = \frac{\rho(u^*)^2}{\mu(y^+/5 + 0.041)} \tag{6.76}$$

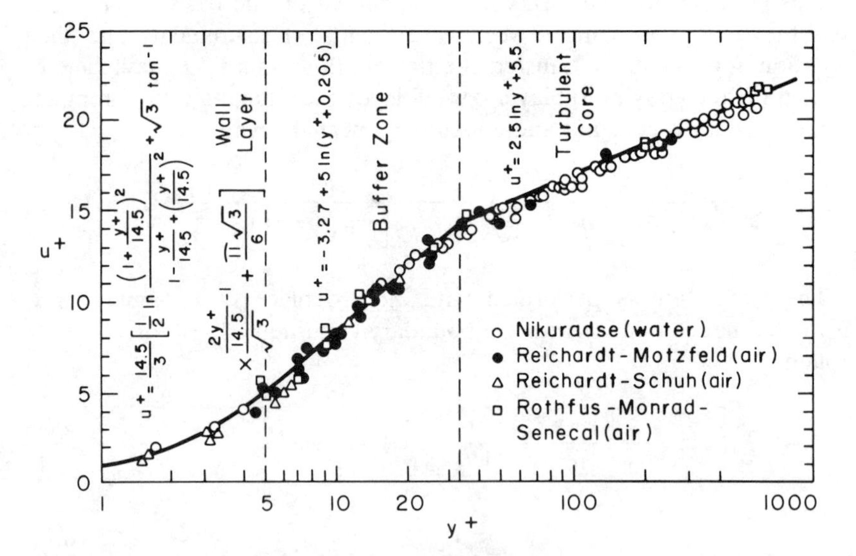

Figure 6.5. Generalized velocity distribution for turbulent flow of Newtonian fluids in smooth tubes, assuming an eddying sublayer, a buffer zone, and a turbulent core (Lin et al., 1953).

The factor $(u^*)^2$ is replaced by $\tau_W g_c/\rho$, and equations 6.59 and 6.76 are combined, assuming the shear stress in the buffer zone is also constant and equal to τ_W, to give

$$\epsilon_M = \frac{\mu}{\rho}\left(\frac{y^+}{5} - 0.959\right) \tag{6.77}$$

The distribution of eddy momentum diffusivity from the outer edge of the turbulent core to the tube wall is now provided by equations 6.71 (for $y^+ \leqslant 5$) and 6.77 (for $5 < y^+ \leqslant 33$).

The flux of component A at any point in the fluid is given by

$$n_{Ay} = -(D + \epsilon_D)\frac{d\rho_A}{dy} \tag{6.78}$$

Equation 6.78 has been integrated in the wall layer and buffer zone assuming τ is constant at τ_W, n_{Ay} is constant at n_{AW}, and $\epsilon_M = \epsilon_D$, as given by equations 6.71 and 6.77, respectively. In the turbulent core it is assumed that τ/n_{Ay} is constant and that eddy diffusivities are dominant, with $\epsilon_M = \epsilon_D$. If the Sherwood number is defined by equation 6.50 Lin, Moulton, and Putnam (1953) present the following result:

$$N^*_{\mathrm{Sh}} = \frac{k^*_\rho d_t}{D} = \frac{1}{\phi_D}\left(\frac{f}{2}\right) N_{\mathrm{Re}} N_{\mathrm{Sc}} \tag{6.79}$$

where

$$\phi_D = 1 + \sqrt{\frac{f}{2}}\left[\frac{14.5}{3}\left(\frac{\mu}{\rho D}\right)^{2/3} F\left(\frac{\mu}{\rho D}\right) + 5\ln\frac{1 + 5.64\mu/\rho D}{6.64(1 + 0.041\mu/\rho D)} - 4.77\right] \tag{6.80}$$

in which

$$F\left(\frac{\mu}{\rho D}\right) = \frac{1}{2}\ln\frac{\left[1 + \frac{5}{14.5}\left(\frac{\mu}{\rho D}\right)^{1/3}\right]^2}{1 - \frac{5}{14.5}\left(\frac{\mu}{\rho D}\right)^{1/3} + \left(\frac{5}{14.5}\right)^2\left(\frac{\mu}{\rho D}\right)^{2/3}} + \sqrt{3}\tan^{-1}\frac{\frac{10}{14.5}\left(\frac{\mu}{\rho D}\right)^{1/3} - 1}{\sqrt{3}} + \frac{\pi\sqrt{3}}{6} \tag{6.81}$$

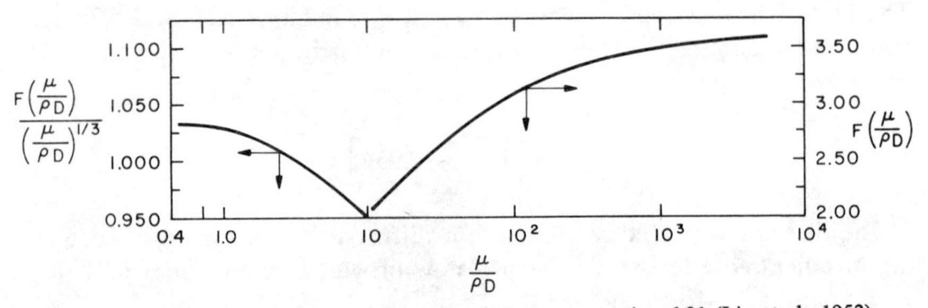

Figure 6.6. Relation between $F(\mu/\rho D)$ and $\mu/\rho D$ from equation 6.81 (Lin et al., 1953).

Equation 6.81 is shown graphically as a plot of $F(\mu/\rho D)$ versus $\mu/\rho D$ in Figure 6.6.

Equations 6.79 to 6.81 are of course readily converted into terms of heat transfer if equality of the eddy momentum diffusivity ϵ_M and eddy thermal diffusivity ϵ_H is assumed, by the substitution of Nusselt number for Sherwood number and Prandtl number for Schmidt number.

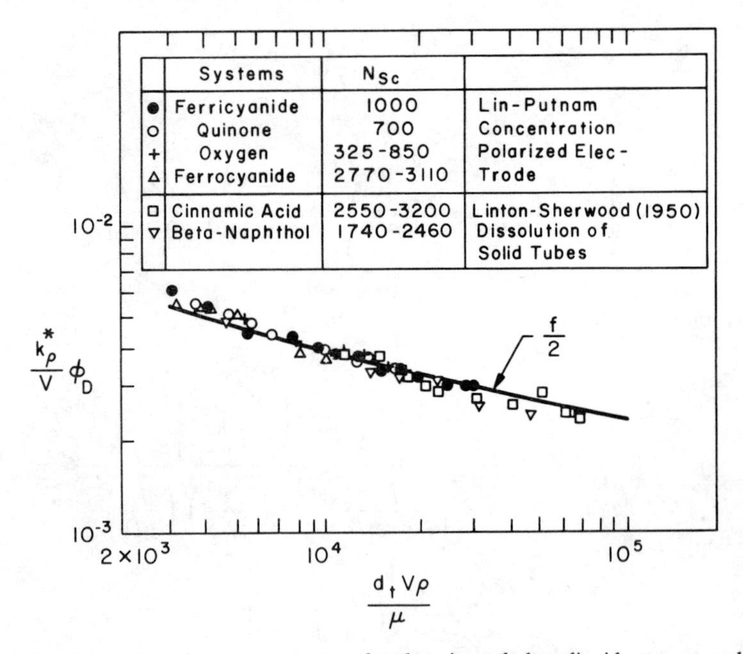

Figure 6.7. Comparison between mass-transfer data in turbulent liquid streams and the relationship $k^*_\rho \phi_D / V = f/2$ indicated by equation 6.79 (Lin et al., 1953).

The agreement between equation 6.79 and a large variety of experimental data is remarkably good, as shown in Figures 6.7 and 6.8 for turbulent liquid and turbulent gas streams, respectively. The two figures cover Schmidt numbers from 0.54 to 3200 and Reynolds numbers from 3000 to 70,000. The analysis approaches that of von Kármán for Schmidt numbers in the vicinity of unity, because of the increase in the rate of molecular diffusion compared to eddy diffusion in the wall layer.

The concentration distributions given by Lin, Moulton, and Putnam (1953) were found by integrating equation 6.78 with the appropriate expression for ϵ_D, namely, equations 6.71 and 6.77 for the wall layer and buffer zone, respectively. The results are as follows: In the wall layer,

$$\frac{\rho_A-\rho_{AW}}{\rho_{AB}-\rho_{AW}}=\frac{\dfrac{14.5}{3}\left(\dfrac{\mu}{\rho D}\right)^{2/3}F\left(\dfrac{\mu}{\rho D},y^+\right)}{\sqrt{\dfrac{2}{f}}+\left\{\dfrac{14.5}{3}\left(\dfrac{\mu}{\rho D}\right)^{2/3}F\left(\dfrac{\mu}{\rho D}\right)+5\ln\dfrac{1+5.64\mu/\rho D}{6.64(1+0.041\mu/\rho D)}-4.77\right\}} \tag{6.82}$$

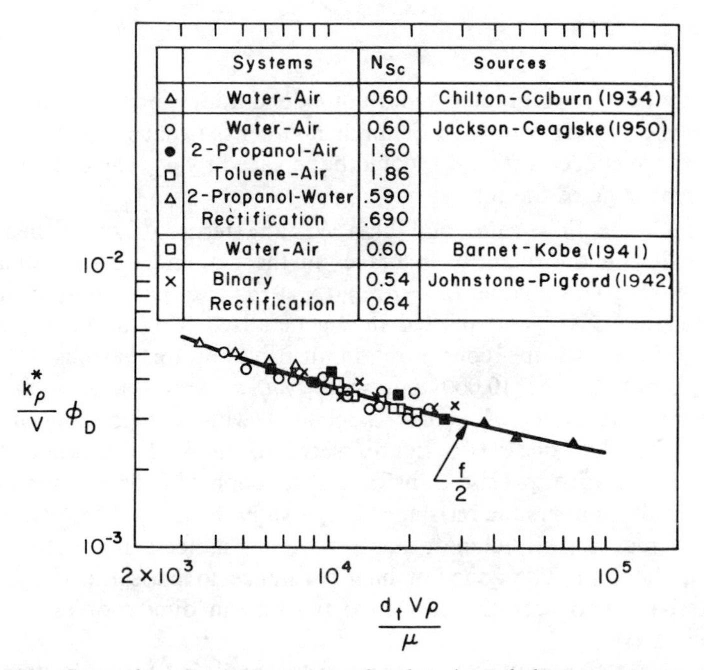

Figure 6.8. Comparison between mass-transfer data in turbulent gas streams and the relationship $k_\rho^*\phi_D/V=f/2$ indicated by equation 6.79 (Lin et al., 1953).

where

$$F\left(\frac{\mu}{\rho D},y^+\right)=\frac{1}{2}\ln\frac{\left[1+\left(\frac{\mu}{\rho D}\right)^{1/3}\left(\frac{y^+}{14.5}\right)\right]^2}{1-\left(\frac{\mu}{\rho D}\right)^{1/3}\left(\frac{y^+}{14.5}\right)+\left(\frac{\mu}{\rho D}\right)^{2/3}\left(\frac{y^+}{14.5}\right)^2}$$

$$+\sqrt{3}\,\tan^{-1}\frac{(\mu/\rho D)^{1/3}(2y^+/14.5)-1}{\sqrt{3}}+\frac{\pi\sqrt{3}}{6} \tag{6.83}$$

In the buffer region,

$$\frac{\rho_A-\rho_{AW}}{\rho_{AB}-\rho_{AW}}=\frac{\frac{14.5}{3}\left(\frac{\mu}{\rho D}\right)^{2/3}F\left(\frac{\mu}{\rho D}\right)+5\ln\frac{1+(\mu/\rho D)(y^+/5-0.959)}{1+0.041\mu/\rho D}}{\sqrt{\frac{2}{f}}+\left\{\frac{14.5}{3}\left(\frac{\mu}{\rho D}\right)^{2/3}F\left(\frac{\mu}{\rho D}\right)+5\ln\frac{1+5.64\mu/\rho D}{6.64(1+0.041\mu/\rho D)}-4.77\right\}} \tag{6.84}$$

The concentration gradient does not extend appreciably into the turbulent core for fluids with high Schmidt number, so that in such cases the bulk average concentration ρ_{AB} could be replaced by ρ_{As}, the concentration at the centerline of the tube.

At moderate flow rates and high N_{Sc} the term $\sqrt{2/f}$ is negligible compared with the quantity in braces in the denominators of equations 6.82 and 6.84. This enables $(\rho_A-\rho_{AW})/(\rho_{AB}-\rho_{AW})$ to be expressed directly as a function of y^+ and plotted in a generalized form as in Figure 6.9. Figure 6.10 shows the concentration distribution for various N_{Sc} at a Reynolds number of 10,000 and with y/d_t as abscissa. A high mass-transfer resistance is of course associated with a high concentration gradient, so that these two figures serve to show the location of the principal resistance to mass transfer as a function of Schmidt number. At low Schmidt numbers the resistance is evidently distributed throughout the fluid, whereas at high Schmidt numbers the resistance is mainly in regions close to the wall. The zones of high resistance to mass transfer become progressively narrower and nearer to the wall in dimensionless terms as N_{Sc} increases.

Optical interferometric techniques were used by Lin, Moulton, and Putnam (1953) to measure concentration distributions close to the surfaces of concentration-polarized electrodes in turbulent liquid streams at a

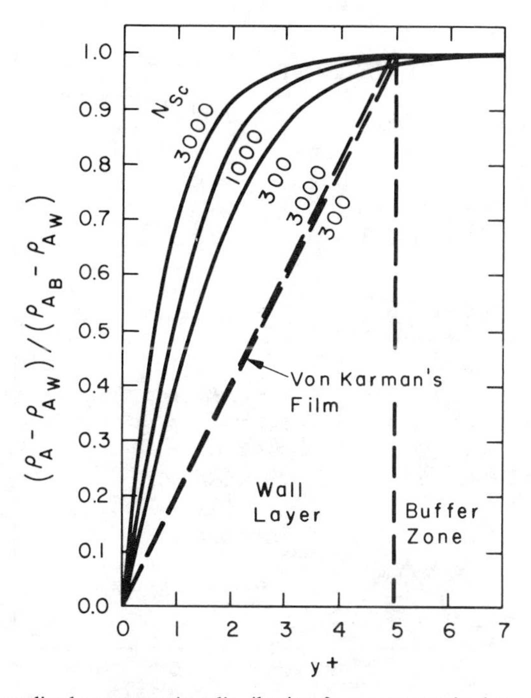

Figure 6.9. Generalized concentration distribution for mass transfer in turbulent streams at moderate flow rates along smooth tubes, assuming an eddying sublayer, a buffer zone, and a turbulent core (Lin et al., 1953).

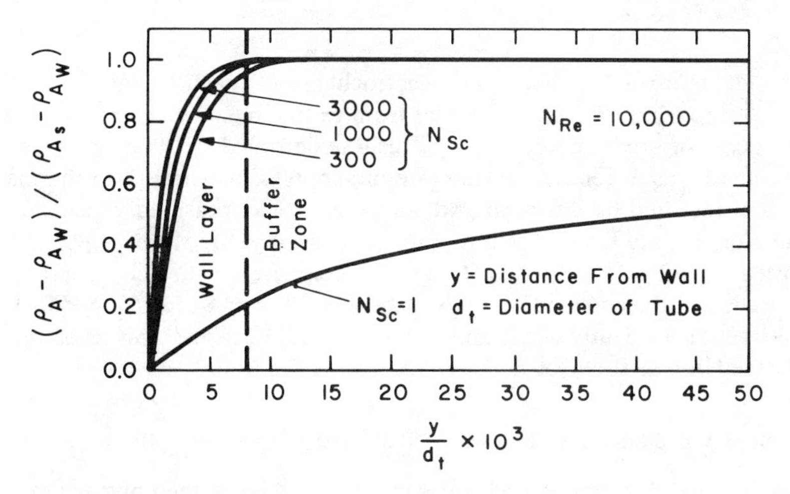

Figure 6.10. Concentration distribution for mass transfer in turbulent streams at a Reynolds number of 10,000 along smooth tubes, assuming an eddying sublayer, a buffer zone, and a turbulent core (Lin et al., 1953).

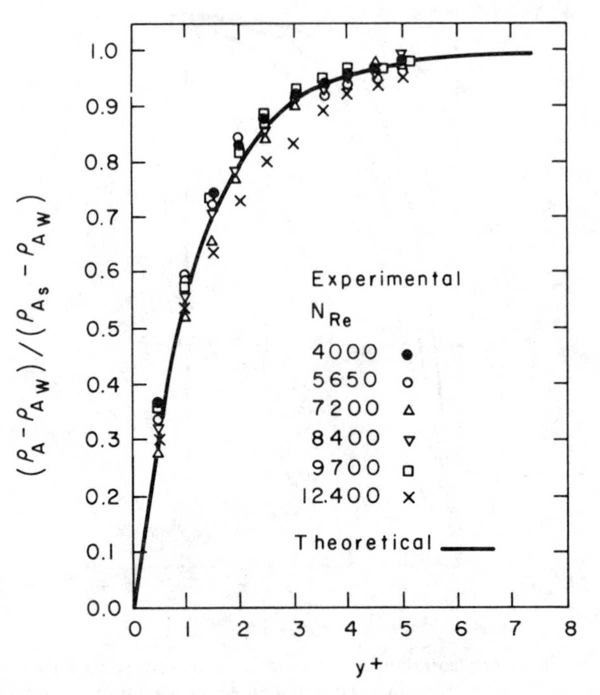

Figure 6.11. Comparison between experimental data and the generalized concentration distribution at moderate flow rates along smooth tubes, assuming an eddying sublayer, a buffer zone, and a turbulent core (Lin et al., 1953).

Schmidt number of 900. The electrochemical reaction involved—the electrodeposition of cadmium metal from cadmium sulfate solution onto a thin layer of fresh mercury—is diffusion-controlled, so that the rate of deposition was dependent on the rate of supply of material from the main body of the fluid by diffusion and convection. The measured values of the cadmium sulfate concentration, expressed as $(\rho_A - \rho_{AW})/(\rho_{As} - \rho_{AW})$, are plotted versus y^+ in Figure 6.11, and compared with the generalized theoretical curve corresponding to equations 6.82 to 6.84 for $N_{Sc} = 900$. The agreement is evidently good, and Lin et al. (1953) consider this to be proof that the laminar sublayer does not exist near the wall.

Analogy Assuming No Discrete Fluid Layers Near the Wall

The artificiality of the three-layer concept involved in the universal velocity distribution of equations 6.37, 6.43, and 6.44 has already been noted, together with discontinuities in slope of the profile at $y^+ = 30$. The exact character of the so-called laminar sublayer for $y^+ < 5$ has also been the subject of much controversy, and the likelihood of some eddying in this layer was indicated, for example, in the previous analogy, due to Lin, Moulton, and Putnam. These considerations prompted Gowariker and Garner (1962) to abandon the concept of three fluid layers. Instead, they developed continuous expressions for the distribution of velocity and eddy diffusivity in smooth tubes as follows.

The starting point is the familiar relationship

$$\tau_{yx} = \frac{\rho}{g_c}\left(\frac{\mu}{\rho} + \epsilon_M\right)\frac{du}{dy} \tag{6.85}$$

where u is the time-average velocity in the x direction. In the vicinity of the wall, where most of the resistance to transfer is assumed to be, $\tau_{yx} g_c/\rho = \tau_W g_c/\rho = (u^*)^2$, so that

$$\frac{du^+}{dy^+} = \frac{du}{dy}\frac{\mu}{(u^*)^2\rho} = \frac{1}{\dfrac{\epsilon_M}{\mu/\rho} + 1} \tag{6.86}$$

Corcoran and Sage's (1956) measurements on velocity distribution show that ϵ_M is dependent on the Reynolds number. This, together with the form of equation 6.86, suggests that an appropriate form of expression for $\epsilon_M/(\mu/\rho)$ might be

$$\frac{\epsilon_M}{\mu/\rho} = F_1(N_{\text{Re}})\left[\frac{y^+}{u^+} - F_2(N_{\text{Re}})\right] \tag{6.87}$$

where $F_1(N_{\text{Re}})$ and $F_2(N_{\text{Re}})$ are to be determined. Inserting equation 6.87 into 6.86 gives

$$\frac{dy^+}{du^+} - \frac{F_1(N_{\text{Re}})}{u^+}y^+ = 1 - F_1(N_{\text{Re}})F_2(N_{\text{Re}}) \tag{6.88}$$

This equation has the solution

$$y^+ = \left(\frac{F_1(N_{\text{Re}})F_2(N_{\text{Re}}) - 1}{F_1(N_{\text{Re}}) - 1}\right)u^+ + c(u^+)^{F_1(N_{\text{Re}})} \tag{6.89}$$

where c is the constant of integration. The functions and the constant c were evaluated from velocity-distribution data, to obtain

$$F_1(N_{\text{Re}}) = 7 - \exp\left[-2.3\times10^{-8}(N_{\text{Re}})^2\right] \tag{6.90}$$

$$F_2(N_{\mathrm{Re}}) = 1 + \exp\left[-3.83(N_{\mathrm{Re}})^{0.05}\right] \tag{6.91}$$

$$c = 3.0 \times 10^{-7} \tag{6.92}$$

Equation 6.89 has considerable merit in accommodating the significant variations in the u^+ vs y^+ plot which occur up to N_{Re} of 14,000; in eliminating the concept of sharply defined fluid layers, thereby avoiding the previous discontinuities in the relation between $\epsilon_M/(\mu/\rho)$ and y^+; and in describing both upper laminar and turbulent flow, as demonstrated in Figures 6.12 and 6.13. This follows because the data used in evaluating equations 6.90 to 6.92 covered the range $2\times 10^3 \leqslant N_{\mathrm{Re}} \leqslant 40\times 10^3$. The division of equation 6.86 by equation 6.47 with the assumption that $\epsilon_M = \epsilon_D = \epsilon$, followed by integration and the substitution of equation 6.87 for $\epsilon/(\mu/\rho)$, leads to

$$\rho_A^+ = u^+ + \frac{N_{\mathrm{Sc}} - 1}{N_{\mathrm{Sc}}} \int_0^{u^+} \frac{du^+}{F_1(N_{\mathrm{Re}})\left[y^+/u^+ - F_2(N_{\mathrm{Re}})\right] + 1/N_{\mathrm{Sc}}} \tag{6.93}$$

where $F_1(N_{\mathrm{Re}})$ and $F_2(N_{\mathrm{Re}})$ are given by equations 6.90 and 6.91, and u^+ and y^+ are related by equation 6.89. Equation 6.93 defines the concentration distribution in the tube as a continuous function of y^+ over at least the range $0 \leqslant y^+ \leqslant 700$, enabling evaluation of ρ_{AB} and hence the Sherwood number from equation 6.50. A closed solution is not obtainable from equation 6.93; Gowariker and Garner (1962) accordingly expressed the relationship in the following generalized form:

$$\rho_A^+ = u^+ + A \int_0^{u^+} \frac{du^+}{B(u^+)^p + 1} \tag{6.94}$$

Numerical methods were used to obtain A and B as functions of both N_{Re} and N_{Sc}, and p as a function of N_{Re} only. These functions are shown in Figures 6.14, 6.15, and 6.16. Plots of equations 6.93 and 6.94 confirm that at high Schmidt numbers most of the concentration gradient is located near the wall, with negligible gradient in the main bulk of the fluid.

Stanton numbers ($N^*_{\mathrm{Sh}}/N_{\mathrm{Re}}N_{\mathrm{Sc}}$) calculated as described above for N_{Re} of 10,000, 25,000, and 50,000 are plotted against N_{Sc} (or N_{Pr}) in Figure 6.17 for comparison with experimental heat- or mass-transfer data. The agreement is evidently good over a wide range of N_{Sc} (or N_{Pr}). Further data (shown in Figure 6.18) were in good agreement with predictions at $N_{\mathrm{Sc}} = 870$ in the low N_{Re} range of 1000 to 20,000. In this region agreement between experiment and theory was much better than that obtained by

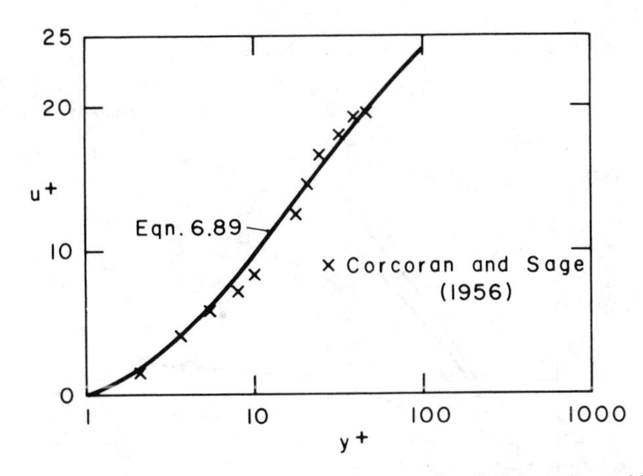

Figure 6.12. Generalized velocity distribution in smooth tubes at a Reynolds number of 2000 according to equations 6.89 to 6.92 (Gowariker and Garner, 1962).

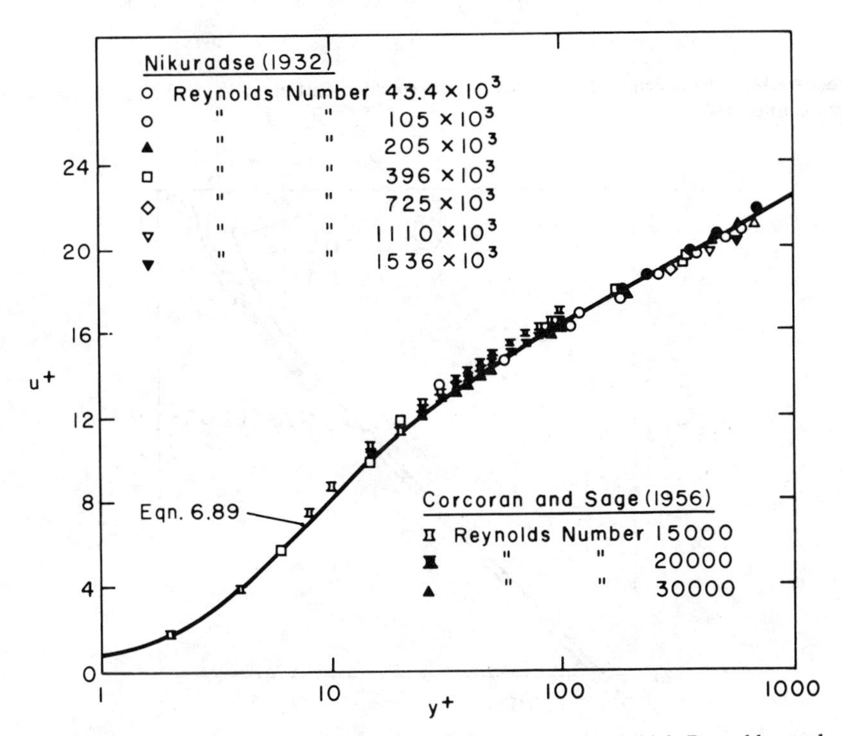

Figure 6.13. Generalized velocity distribution in smooth tubes at high Reynolds numbers according to equations 6.89 to 6.92 (Gowariker and Garner, 1962).

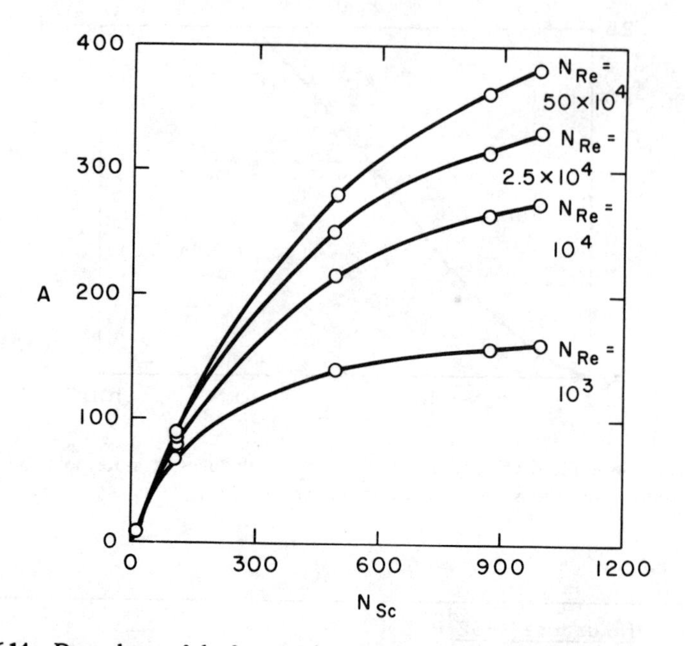

Figure 6.14. Dependence of the factor A from equation 6.94 upon N_{Sc} and N_{Re} (Gowariker and Garner, 1962).

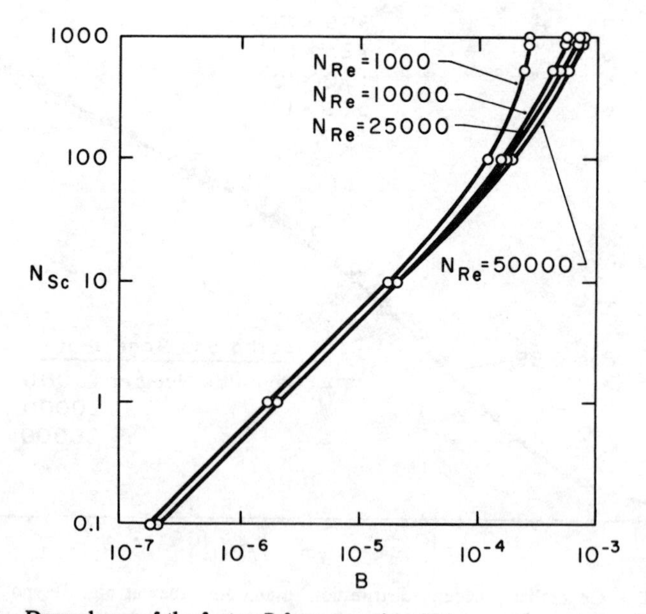

Figure 6.15. Dependence of the factor B from equation 6.94 upon N_{Sc} and N_{Re} (Gowariker and Garner, 1962).

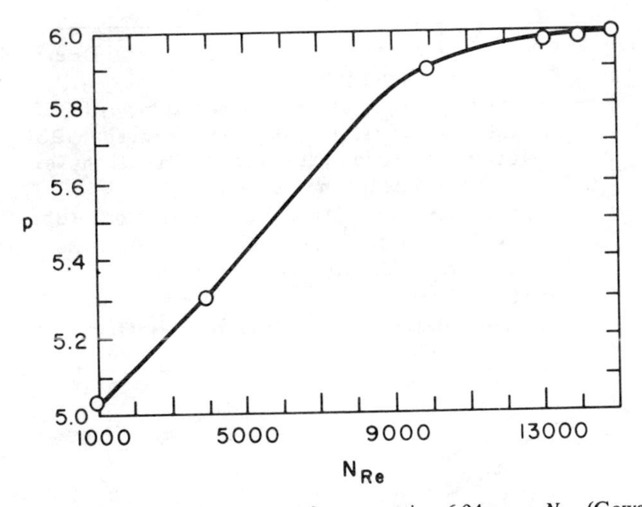

Figure 6.16. Dependence of the exponent p from equation 6.94 upon N_{Re} (Gowariker and Garner, 1962).

Deissler (1955) using the empirical equation

$$\frac{\epsilon_M}{\mu/\rho}=\frac{\epsilon_H}{\mu/\rho}=0.0154u^+y^+[1-\exp(-0.0154u^+y^+)], \qquad y^+\leqslant 26 \quad (6.95)$$

For $y^+>26$ Deissler's treatment links up with that of von Kármán.

A plot of Sherwood number (equation 6.50) against Reynolds number with Schmidt number as parameter is given by Gowariker and Garner in Figure 6.19.

The importance of achieving the correct representation of ϵ_D in the vicinity of the wall at high N_{Sc} has been shown from a comparison of the various analogies with experimental data. In the analogies of von Kármán, Lin et al., and Deissler, ϵ_D is assumed equal to ϵ_M and is given by the same expression in the turbulent core, obtained as follows. Combining equations 6.35 and 6.85,

$$\frac{\tau_W g_c}{\rho}\left(1-\frac{y}{r_t}\right)=(u^*)^2\left(1-\frac{y}{r_t}\right)=\left(\frac{\mu}{\rho}+\epsilon_M\right)\frac{du}{dy} \quad (6.96)$$

or

$$\epsilon_M=(u^*)^2\left(1-\frac{y}{r_t}\right)\frac{dy}{du}-\frac{\mu}{\rho} \quad (6.97)$$

and evaluating dy/du from equation 6.43,

$$\epsilon_M=\frac{\mu}{\rho}\left[\frac{y^+(1-y/r_t)}{2.5}-1\right], \qquad y^+>26 \text{ to } 33 \quad (6.98)$$

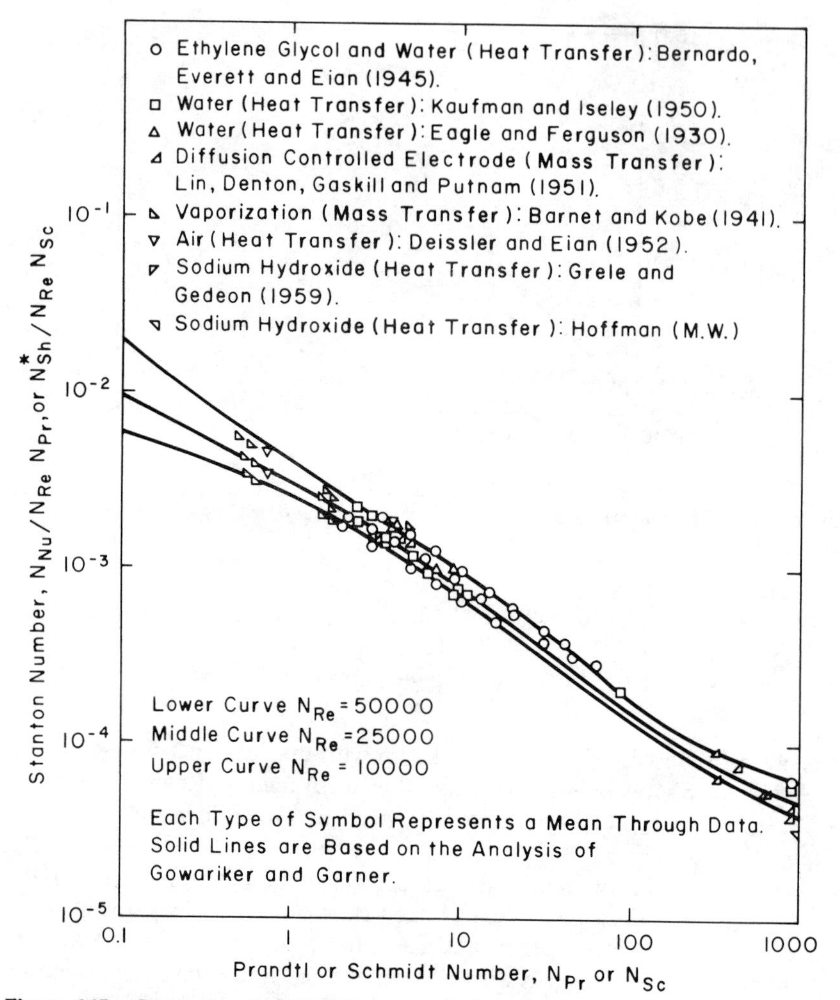

Figure 6.17. Comparison between experimental data and calculations according to the analogy of Gowariker and Garner (1962).

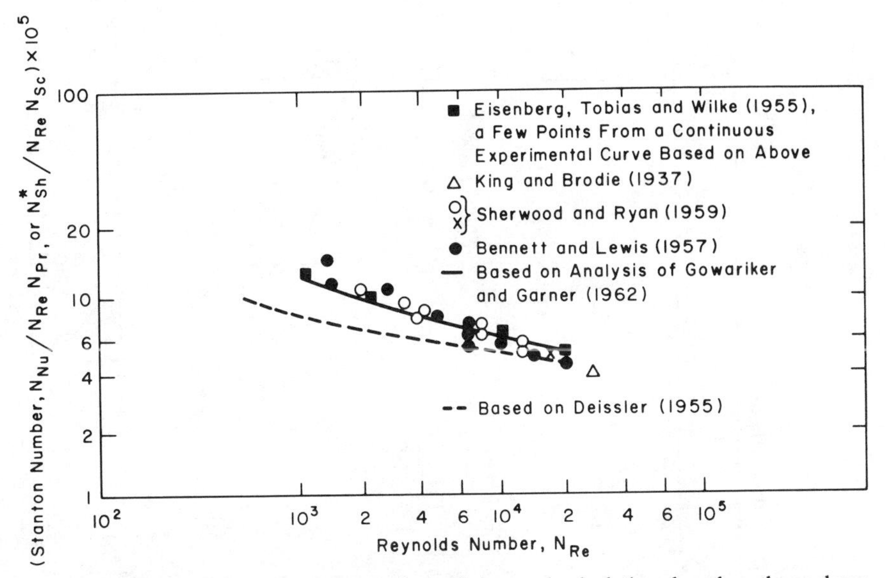

Figure 6.18. Comparison between experimental data and calculations based on the analyses of Gowariker and Garner (1962) and Deissler (1955).

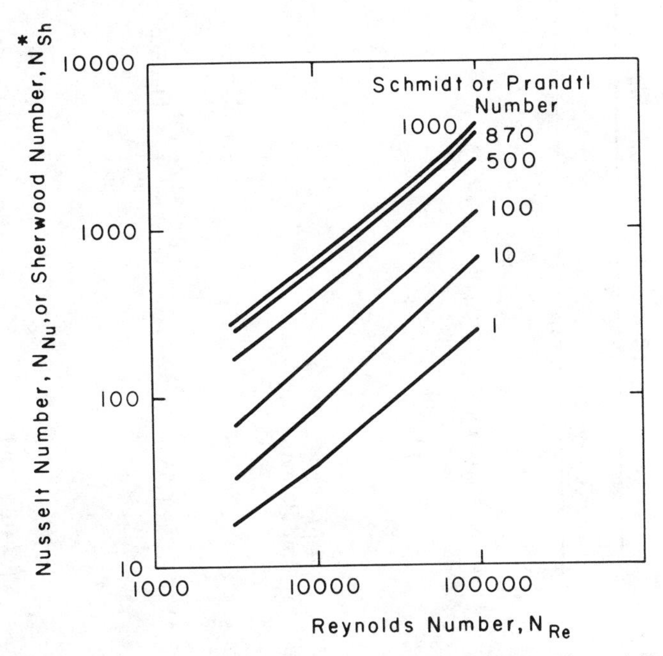

Figure 6.19. Dependence of the Sherwood (or Nusselt) number upon N_{Re} and N_{Sc} (or N_{Pr}) for turbulent flow in smooth tubes, according to the analogy of Gowariker and Garner (1962).

Table 6.1. Eddy diffusivities used in four different analogies.

Analogy	$y^+ \leqslant 5$	$5 \leqslant y^+ \leqslant y_b^+$	y_b^+	$y^+ > y_b^+$
von Kármán (1939)	$\epsilon_M = 0$	$\epsilon_M = \frac{\mu}{\rho}\left(\frac{y^+}{5} - 1\right)$	30	$\epsilon_M = \frac{\mu}{\rho}\left[\frac{y^+\left(1 - \frac{y}{r_t}\right)}{2.5} - 1\right]$
Lin, Moulton, and Putnam (1953)	$\epsilon_M = \frac{\mu}{\rho}\left(\frac{y^+}{14.5}\right)^3$	$\epsilon_M = \frac{\mu}{\rho}\left(\frac{y^+}{5} - 0.959\right)$	33	$\epsilon_M = \frac{\mu}{\rho}\left[\frac{y^+\left(1 - \frac{y}{r_t}\right)}{2.5} - 1\right]$
Deissler (1955)	$\epsilon_M = \frac{0.0154\mu u^+ y^+}{\rho}[1 - \exp(-0.0154 u^+ y^+)]$		26	$\epsilon_M = \frac{\mu}{\rho}\left[\frac{y^+\left(1 - \frac{y}{r_t}\right)}{2.5} - 1\right]$
Gowariker and Garner (1962)	← $\epsilon_M = \frac{\mu}{\rho}F_1(N_{\mathrm{Re}})\left[\frac{y^+}{u^+} - F_2(N_{\mathrm{Re}})\right]$ →			

Table 6.2. $\epsilon/(\mu/\rho)$ corresponding to analogies developed by von Kármán and by Lin, Moulton, and Putnam.

Distance y^+	$\epsilon/(\mu/\rho)$, von Kármán	$\epsilon/(\mu/\rho)$, Lin et al.
1	0	0.0003
10	1	1.04
30	5	5.04

The expressions proposed to represent eddy diffusivities in various regions of flow for four of the analogies referred to here are summarized in Table 6.1. Gowariker and Garner compared the values of $\epsilon/(\mu/\rho)$ predicted by the respective expressions of von Kármán and Lin et al. for three short distances from the wall, as shown in Table 6.2. The values at $y^+=30$ were obtained from the buffer-zone relationships.

Although the differences between $\epsilon/(\mu/\rho)$ from the two approaches seem slight, the consequence is that von Kármán's equation 6.69 is

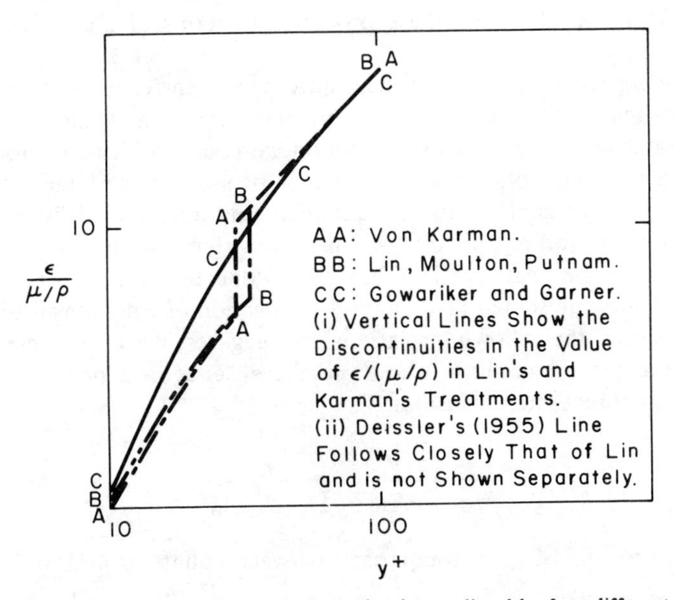

Figure 6.20. Comparison between values of $\epsilon/(\mu/\rho)$ predicted by four different analyses of turbulent transport at very high Reynolds numbers in smooth tubes (Gowariker and Garner, 1962).

applicable only up to N_{Sc} of 40, whereas the relationship of Lin et al. (equation 6.79) gave good correlation with experimental data up to N_{Sc} of 3200, as shown in Figures 6.7 and 6.8. These considerations underline the importance of obtaining the correct formulation of $\epsilon/(\mu/\rho)$ in the wall region at high N_{Sc}. Nevertheless, the first three models in Table 6.1 contain a considerable discontinuity in $\epsilon/(\mu/\rho)$ at y_b^+ (the outer edge of the turbulent core). This is shown for very high Reynolds numbers $[F_1(N_{Re}) \to 7, F_2(N_{Re}) \to 1]$ in Figure 6.20, after Gowariker and Garner, whose model (curve CC) avoids the discontinuity common to the others, The good results obtained by Lin et al. despite their discontinuity at y^+ of 33 are presumably related to their arbitrary selection of equation 6.71 so as to ensure agreement with mass-transfer data. Deissler's equation 6.95 has a comparable justification. It may be noted that single continuous equations, although again of complicated form, have been provided by Spalding (1961) for relating $\epsilon_M/(\mu/\rho)$ and y^+ to u^+. The reader is again referred for further study to the extensive list of analogy developments on page 232.

Analogies in Terms of j Factors

The empirical relationships for heat, mass, and momentum transfer were manipulated by Colburn (1933) and by Chilton and Colburn (1934) to obtain an analogy between these processes in terms of quantities designated as j factors.

In drawing analogies between heat and mass transfer it must be noted that most heat-transfer data are for zero mass transfer. Analogous mass-transfer rates are therefore closer to equimolal counter-diffusion processes, for which the net molal transfer is zero. Attempts have been made to "correct" transfer coefficients for unimolal unidirectional diffusion processes to allow for significant net mass transfer and therefore permit application of the analogy. This has been done in terms of the film or two-film theory, presented in Chapter 4. Inspection of equations 4.16, 4.19, and 4.25 shows the following relationship for gases, where a prime on k_ρ denotes equimolal counterdiffusion and the absence of a prime indicates unimolal unidirectional diffusion:

$$\frac{k_{\rho G}}{k'_{\rho G}} = \frac{P}{p_{BLM}} = \frac{1}{(1-y_A)_{LM}} \tag{6.99}$$

Similarly, for liquids, from comparison between equations 4.21 and 4.27,

$$\frac{k_{\rho L}}{k'_{\rho L}} = \frac{c}{c_{BLM}} = \frac{(\rho/M)_{av}}{(\rho_B/M_B)_{LM}} = \frac{1}{(1-x_A)_{LM}} \tag{6.100}$$

Evidently $k_{\rho G \text{ or } L} = k_\rho \to k_\rho^* \to k_\rho'$ as the concentration of A becomes increasingly dilute. Equations 6.99 and 6.100 are analogous to equation 5.214 and equation 6.160 given later for transfer with a high mass flux. It is considerations of this sort which account for the appearance of the term $k_\rho p_{BLM}/P$ in many common definitions of the mass-transfer j factor. The j-factor analogies will be presented here, however, in terms of k_ρ^*, corresponding to low concentrations and transfer rates for component A. Approximate correction for effects of a high net mass flux may be made by use of equations 6.99 and 6.100 above, setting $k_\rho' = k_\rho^*$, or by the methods presented later for high-mass-flux conditions.

Two *general* definitions may be written as

$$j_D = \frac{N_{\text{Sh}}^*}{N_{\text{Re}} N_{\text{Sc}}^{1/3}} \tag{6.101}$$

$$j_H = \frac{N_{\text{Nu}}}{N_{\text{Re}} N_{\text{Pr}}^{1/3}} \tag{6.102}$$

where j_D and j_H are the j factors for mass and heat transfer, respectively. The formulation of these quantities and of the analogy will be demonstrated first for the case of flow through tubes.

j Factors for Tubes

Consider a fluid in turbulent flow through a tube with mass transfer taking place between the fluid and the tube wall. The mass-transfer coefficient in the fluid may be assumed to be some function of the following variables:

$$k_\rho^* = f_1(d_t, V, \rho, \mu, D) \tag{6.103}$$

and by dimensional analysis,

$$\frac{k_\rho^* d_t}{D} = f_2\left[\left(\frac{d_t V \rho}{\mu}\right), \left(\frac{\mu}{\rho D}\right)\right] \tag{6.104}$$

or

$$N_{\text{Sh}}^* = f_2[N_{\text{Re}}, N_{\text{Sc}}] \tag{6.105}$$

Experimental rates of mass transfer were measured by Gilliland and Sherwood (1934) for evaporation of various liquids in a wetted-wall column with co- and countercurrent air in turbulent flow. Their results were expressed in terms of equation 6.105 as

$$(N_{\mathrm{Sh}})^*_{\mathrm{LM}} = 0.023 N_{\mathrm{Re}}^{0.83} N_{\mathrm{Sc}}^{0.44} \tag{6.106}$$

where the velocity is relative to the stationary tube wall and the ranges covered were $2000 \leqslant N_{\mathrm{Re}} \leqslant 35{,}000$, $0.6 \leqslant N_{\mathrm{Sc}} \leqslant 2.5$. The small range of Schmidt number cast uncertainty on the exponent on N_{Sc}, which was removed in a subsequent study by Linton and Sherwood (1950). In this work, mass transfer occurred to water flowing through tubes cast from such soluble substances as benzoic acid, cinnamic acid, and β-naphthol. Combination of these data with the previous evaporation studies increased the range of Schmidt-number coverage to $0.6 \leqslant N_{\mathrm{Sc}} \leqslant 2500$. The combined results of Sherwood and Gilliland and of Linton and Sherwood were correlated by the equation

$$(N_{\mathrm{Sh}})^*_{\mathrm{LM}} = 0.023 N_{\mathrm{Re}}^{0.83} N_{\mathrm{Sc}}^{1/3} \tag{6.107}$$

Division of this expression by $N_{\mathrm{Re}} N_{\mathrm{Sc}}$ and rearrangement yields

$$\frac{(N_{\mathrm{Sh}})^*_{\mathrm{LM}}}{N_{\mathrm{Re}} N_{\mathrm{Sc}}^{1/3}} N_{\mathrm{Re}}^{0.03} = 0.023 N_{\mathrm{Re}}^{-0.2} \tag{6.108}$$

The exponent on N_{Re} in equation 6.107 is often reduced slightly to 0.8, causing the factor $(N_{\mathrm{Re}})^{0.03}$ in equation 6.108 to become unity. This is justified by noting that, under conditions prevailing for equation 6.106, the liquid surface in the wetted-wall column would exhibit rippling and wave formation. This increases both the interface available for transfer and the transfer coefficient in the gas, with consequently greater dependence of k^*_ρ on N_{Re}. Such anomalies were absent in the case of the experiments with liquids flowing through soluble tubes. The result of this "rounding" of the exponent on N_{Re} is then

$$j_D = \frac{(N_{\mathrm{Sh}})^*_{\mathrm{LM}}}{N_{\mathrm{Re}} N_{\mathrm{Sc}}^{1/3}} = 0.023 N_{\mathrm{Re}}^{-0.2} \tag{6.109}$$

where the coefficient in $(N_{\mathrm{Sh}})^*_{\mathrm{LM}}$ is defined by

$$k^*_{\rho\mathrm{LM}} = \frac{(n_{AW})_{\mathrm{av}}}{(\rho_{AW} - \rho_{AB})_{\text{log mean}}} \tag{6.110}$$

In the case of heat transfer between a tube and a fluid in turbulent flow through it, the analogs of equations 6.103, 6.104, 6.105, and 6.110 are

$$h = f_3(d_t, V, \rho, \mu, c_p, k) \tag{6.111}$$

$$\frac{hd_t}{k} = f_4\left[\left(\frac{d_t V \rho}{\mu}\right), \left(\frac{c_p \mu}{k}\right)\right] \tag{6.112}$$

$$N_{\text{Nu}} = f_4[N_{\text{Re}}, N_{\text{Pr}}] \tag{6.113}$$

$$h = \frac{Q_W}{A_W(T_W - T_B)_{\text{log mean}}} \tag{6.114}$$

Colburn's (1933) correlation of experimental data in terms of equation 6.113 is

$$(N_{\text{Nu}})_f = 0.023(N_{\text{Re}})_f^{0.8}(N_{\text{Pr}})_f^{1/3} \tag{6.115}$$

where the subscript f indicates that the physical properties are evaluated at the arithmetic average of the wall and bulk temperatures along the tube. Dividing this expression by $(N_{\text{Re}})_f(N_{\text{Pr}})_f$ and rearranging yields

$$j_H = \frac{(N_{\text{Nu}})_f}{(N_{\text{Re}})_f(N_{\text{Pr}})_f^{1/3}} = 0.023(N_{\text{Re}})_f^{-0.2} \tag{6.116}$$

The friction factor for flow through a cylindrical tube is defined as

$$f = \frac{d_t \Delta P g_c / 4L}{\rho V^2/2} = \frac{\tau_W g_c}{\rho V^2/2} \tag{6.117}$$

For turbulent flow through smooth tubes the following empirical correlation is widely used over the range $30{,}000 < N_{\text{Re}} < 1{,}000{,}000$:

$$f = 0.046 N_{\text{Re}}^{-0.2} \tag{6.118}$$

Combination of equations 6.109, 6.116, and 6.118 shows that

$$j_D = j_H = \frac{f}{2} \tag{6.119}$$

The ranges of validity in the case of mass transfer are $2000 < N_{Re} < 300{,}000$, $0.6 < N_{Sc} < 2500$; and for heat transfer, $10{,}000 < N_{Re} < 300{,}000$, $0.6 < N_{Pr} < 100$. Equation 6.119 is the empirical Chilton-Colburn analogy between heat, mass, and momentum transfer in tube flow.

It is evident that this empirical analogy amounts to the substitution of the quantity $N_{Sc}^{2/3}$ for the entire denominator in the final expressions for either the Prandtl or von Kármán analogies (equations 6.55 and 6.69). The expressions reduce to the Reynolds analogy (equation 6.51) for N_{Pr} or N_{Sc} of unity.

Illustration 6.3

A smooth tube with an internal diameter of 1 in. is cast from solid naphthalene. Pure air enters the tube at a velocity of 30 ft/sec. If the average air pressure is 14.7 psia and the system is at 113°F, estimate the tube length required for the average concentration of naphthalene vapor in the air to reach a value of 5.6×10^{-5} lb-mass/ft³—that is to say, 25 percent of the saturation value.

Use all relevant relationships presented in this chapter in order to obtain minimum and maximum estimates of the required length, neglecting entrance effects.

Compare with Illustration 5.5 for laminar conditions.

SOLUTION. Relevant physical properties are obtainable from Illustrations 5.1 and 5.2. The naphthalene surface temperature will be taken to be 113°F (see Illustration 5.1).

$$N_{Re} = \frac{\frac{1}{12}(30)(3600)(0.0694)}{0.0457} = 13{,}650$$

The flow is therefore turbulent. The Schmidt number under the prevailing conditions has a value of 2.475 (Illustration 5.2), so that the Reynolds analogy between momentum and mass transfer (equation 6.51) is inapplicable. Attention is therefore confined to analogies represented by equations 6.55, 6.69, 6.70, and 6.79 through 6.81, and Figure 6.19. The result corresponding to equation 6.107 will also be examined.

The integration of equation 6.56 leads to

$$\ln \frac{\rho_{AW} - \rho_{AB1}}{\rho_{AW} - \rho_{AB2}} = \frac{4L}{Vd_t}\left(\frac{D}{d_t} N_{Sh}^*\right) = \frac{4L}{d_t} \frac{N_{Sh}^*}{N_{Re} N_{Sc}}$$

where $N_{Sh}^*/N_{Re}N_{Sc}$ is obtained from the equations cited above for the analogy under consideration. The friction factor will be calculated from the Blasius equation, valid for flow in smooth tubes over the Reynolds-number range 3000 to 100,000, as

$$f = 0.079 N_{Re}^{-0.25} = 0.079(13{,}650)^{-0.25} = 0.0073$$

From equation 6.55,

$$\frac{N_{Sh}^*}{N_{Re}N_{Sc}} = \frac{0.0073/2}{1+5\sqrt{0.0073/2}\;(2.475-1)} = 0.002505$$

$$\ln\frac{\rho_{AW}-\rho_{AB1}}{\rho_{AW}-\rho_{AB2}} = 2.303\log\frac{0.000224-0}{0.000224-0.000056} = 0.2876$$

$$L = \frac{d_t}{4}\frac{N_{Re}N_{Sc}}{N_{Sh}^*}\ln\frac{\rho_{AW}-\rho_{AB1}}{\rho_{AW}-\rho_{AB2}} = \frac{0.2876}{12(4)(0.002505)} = 2.39 \text{ ft.}$$

From equation 6.69,

$$\frac{N_{Sh}^*}{N_{Re}N_{Sc}} = \frac{0.0073/2}{1+5\sqrt{0.0073/2}\;\{2.475-1+\ln[1+\frac{5}{6}(2.475-1)]\}} = 0.00217$$

$$L = 2.39\left(\frac{0.002505}{0.00217}\right) = 2.76 \text{ ft.}$$

From equation 6.70 with $\epsilon_D/\epsilon_M = 1.6$,

$$\frac{N_{Sh}^*}{N_{Re}N_{Sc}} = \frac{(0.0073/2)1.6}{1+5\sqrt{0.0073/2}\;\{1.6(2.475)-1+\ln[1+\frac{5}{6}(1.6\times 2.475-1)]\}}$$

$$= 0.00258$$

$$L = 2.39\left(\frac{0.002505}{0.00258}\right) = 2.32 \text{ ft.}$$

Figure 6.6 shows that, when $N_{Sc} = 2.475$

$$\frac{F(\mu/\rho D)}{(\mu/\rho D)^{1/3}} = 1.0059$$

so

$$F\left(\frac{\mu}{\rho D}\right) = 1.0059(2.475)^{1/3} = 1.36$$

and in equation 6.80,

$$\phi_D = 1+\sqrt{\frac{0.0073}{2}}\left[\frac{14.5}{3}(2.475)^{2/3}(1.36) + 5\ln\frac{1+5.64(2.475)}{6.64[1+0.041(2.475)]} - 4.77\right] = 1.65$$

From equation 6.79,

$$\frac{N^*_{\mathrm{Sh}}}{N_{\mathrm{Re}}N_{\mathrm{Sc}}} = \frac{0.0073/2}{1.65} = 0.00221$$

$$L = 2.39\left(\frac{0.002505}{0.00221}\right) = 2.71 \text{ ft}$$

Graphical interpolation between points read from the vertical at $N_{\mathrm{Re}} = 13{,}650$ in Figure 6.19 gives $N^*_{\mathrm{Sh}} \doteq 68$ for $N_{\mathrm{Sc}} = 2.475$; then

$$\frac{N^*_{\mathrm{Sh}}}{N_{\mathrm{Re}}N_{\mathrm{Sc}}} = \frac{68}{13{,}650(2.475)} = 0.002015$$

$$L = 2.39\left(\frac{0.002505}{0.002015}\right) = 2.98 \text{ ft}$$

From equation 6.107,

$$k^*_{\rho\mathrm{LM}} = 0.023\left(\frac{0.2665}{1/12}\right)(13{,}650)^{0.83}(2.475)^{1/3} = 268.5 \text{ ft/hr}$$

A naphthalene balance over length L is

$$\frac{\pi}{4}\left(\tfrac{1}{12}\right)^2 30(3600)(5.6\times10^{-5}) = 268.5\pi\left(\tfrac{1}{12}\right)L\frac{5.6\times10^{-5}}{2.303\log[22.4\times10^{-5}/(22.4-5.6)\times10^{-5}]}$$

From which $L = 2.41$ ft. Thus, from the six estimates considered here,

$$2.32 \text{ ft} \leqslant L \leqslant 2.98 \text{ ft}.$$

The average value of L is 2.59 ft. The average flux at the wall over the first 2.59 ft along the tube is therefore

$$\frac{\pi d_t^2 V \Delta\rho_A}{4\pi d_t x} = \frac{\frac{1}{12}(30)(3600)(5.6\times 10^{-5})}{4(2.59)} = 0.0486 \text{ lb naphthalene/(ft}^2\text{)(hr)}$$

j Factors for a Flat Plate

The theoretical equations for laminar and turbulent boundary layers developed as equations 5.24 and 6.10 may be rearranged into the following forms, recalling equations 6.101 and 6.102:

$$j_D = j_H = \frac{C_{df}}{2} = 0.664 N_{\text{Re},L}^{-1/2} \qquad \text{(laminar)} \qquad (6.120)$$

$$j_D = j_H = \frac{C_{df}}{2} = 0.037 N_{\text{Re},L}^{-0.2} \qquad \text{(turbulent)} \qquad (6.121)$$

The coefficients k_ρ^* and h appearing respectively in j_D and j_H are defined as

$$k_\rho^* = \frac{(n_{A0})_{\text{av}}}{\rho_{A0} - \rho_{A\infty}} \qquad (6.122)$$

$$h = \frac{Q_0}{A_0(T_0 - T_\infty)} \qquad (6.123)$$

The accurate value 0.664 replaces the approximate 0.646 in equation 6.120, and equation 6.121 is found to provide an empirical extension of equation 6.10 to systems in which the Schmidt number differs from unity. The quantity C_{df} is the drag coefficient for the flat plate, which is oriented parallel to the direction of flow.

$$C_{df} = \frac{F_{df} g_c}{wL(\rho u_\infty^2/2)} = \frac{(\tau_0)_{\text{av}} g_c}{\rho u_\infty^2/2} \qquad (6.124)$$

Mass transfer begins at the leading edge of the plate, which has width w and length L in the direction of the stream. The total drag force on one side of the plate is F_{df}.

Equations 6.120 and 6.121 will apply for Reynolds numbers respectively below and above the region of transition from a laminar to a turbulent boundary layer. In considering the analogy $j_H = C_{df}/2$, McAdams (1954) indicates the transition region as lying in the range $8 \times 10^4 \leqslant N_{\mathrm{Re},L} \leqslant 5 \times 10^5$, although this is a little lower than the range customarily cited for the boundary-layer transition [e.g., Knudsen and Katz (1958), pp. 268–271; Schlichting (1961), p. 9-13].

Considerable experimental data have been accumulated, mostly on evaporation from a free liquid surface or from a wet solid surface into a controlled air stream. These data have been compared with equations 6.120 and 6.121 in graphical form by Sherwood and Pigford (1952) and show good general agreement, despite considerable scatter in the measurements, some deviation for a few liquid (water) systems, and a measure of uncertainty with regard to the location of transition. This last is to be expected, because the transition depends on the plate or surface roughness, the degree of turbulence in the free stream, heat transfer across the surface, and the amount of irregularity in the leading edge of the plate. Thus Dhawan (1952) has shown that the boundary layer may be turbulent over the entire plate when the leading edge is rough. Variations in each of these factors from one set of workers to another may largely account for the scatter observed in the data.

Illustration 6.4

Repeat Illustration 6.1(a), (b), (f), and (g), using appropriate j_D relationships so as to ensure suitable allowance for the influence of the non-unity Schmidt number. Compare the results with those in Illustration 6.1.

SOLUTION (a). Physical properties at these conditions are again available in Illustrations 5.1 and 5.2.

From equations 6.101 and 6.121,

$$k^*_{\rho m} = 0.037\left(\frac{0.2665}{1}\right)(6.28 \times 10^5)^{0.8}(2.475)^{1/3} = 582 \text{ ft/hr}$$

This is 37 percent greater than the value estimated in Illustration 6.1(a), where the deviation of N_{Sc} from unity was neglected.

SOLUTION (b).

$$q_{AW} = 582(1 \times 1)(0.000224 - 0) = 0.1303 \text{ lb naphthalene/hr}$$

SOLUTION (f). The total rate of transfer of component A from a plate of width w over length L, where $L > x_c$, may be written as

$$wL(n_{A0})_{\mathrm{av},x=L} = wL[(n_{A0})_{\mathrm{av},x=L}]_{\mathrm{turb}} - wx_c[(n_{A0})_{\mathrm{av},x=x_c}]_{\mathrm{turb}} + wx_c[(n_{A0})_{\mathrm{av},x=x_c}]_{\mathrm{lam}}$$

where it is assumed that the turbulent boundary layer beyond x_c is unaffected by whether conditions before x_c are laminar or turbulent. In this equation, $[(n_{A0})_{av,x=L}]_{turb}$ and $[(n_{A0})_{av,x=x_c}]_{turb}$ are the average fluxes of A over plate lengths L and x_c, respectively, for a wholly turbulent boundary layer, whereas $[(n_{A0})_{av,x=x_c}]_{lam}$ is the average flux of A over plate length x_c for a laminar boundary layer.

Equations 6.101, 6.120, and 6.121 may be combined with the definition of k_ρ^* given in equation 6.122 and inserted in the above expression to obtain

$$\frac{(n_{A0})_{av,x=L}}{\rho_{A0}-\rho_{A\infty}}\left(\frac{L}{D}\right)=\frac{k_{\rho m}^* L}{D}=N_{Sc}^{1/3}\left[0.664N_{Re,x_c}^{1/2}-0.037(N_{Re,x_c})^{0.8}+0.037(N_{Re,L})^{0.8}\right]$$

Aside from the slight revision of the numerical constants, as explained below equation 6.123, the principal difference between this expression and that obtained in Illustration 6.1(f) is in the factor $N_{Sc}^{1/3}$, which allows for the deviation of N_{Sc} from unity. When $N_{Re,x_c}=3.2\times10^5$,

$$\frac{k_{\rho m}^* L}{D}=0.037N_{Sc}^{1/3}[N_{Re,L}^{0.8}-15{,}190]$$

SOLUTION (g).

$$k_{\rho m}^*=0.037\left(\frac{0.2665}{1}\right)(2.475)^{1/3}\left[(6.28\times10^5)^{0.8}-15{,}190\right]=379\text{ ft/hr}$$

j Factors for Single Cylinders

The full analogy between momentum, heat, and mass transfer breaks down for flow around bluff bodies such as cylinders and spheres. This is because the total drag force, contained in the drag coefficient, consists of both form drag and skin friction. The full analogy still prevails, however, for the skin-friction *component* of the total drag, and in any event the more limited analogy between heat and mass transfer still holds—namely, $j_H=j_D$. This is well illustrated in Figure 6.21, taken from Sherwood and Pigford (1952, p. 70) for flow normal to the axis of single cylinders.

In Figure 6.21 the quantity f equals $(\tau_0)_{av}g_c/(\rho u_{av}^2/2)$ and is the skin-friction component of the total drag coefficient, isolated by a prescribed method. Additional results on the dissolution of solid cylinders in a turbulent water stream confirm Figure 6.21 and extend its approximate validity to the range $0.6\leqslant N_{Sc}\leqslant3000$. The coefficient in j_D is defined by equation 6.126.

j Factors for Single Spheres

Many extensive studies on heat and mass transfer from single spheres are reported in the scientific literature. Frössling (1940) showed theoretically

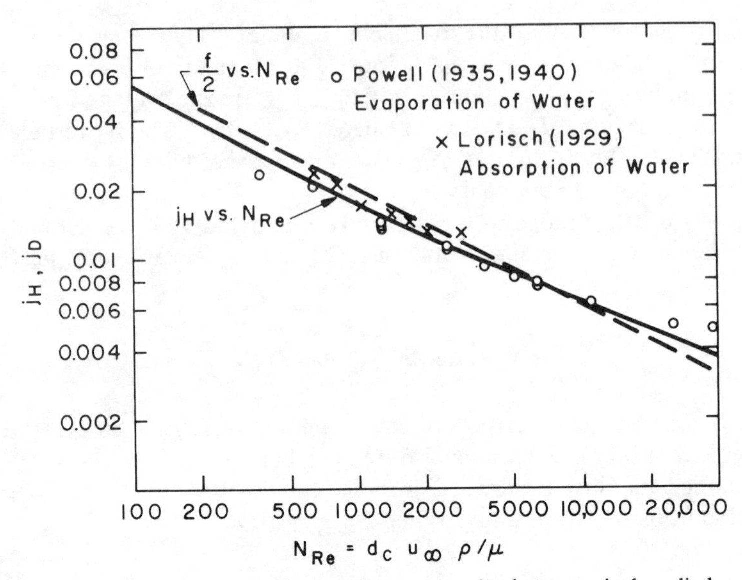

Figure 6.21. The j factors for heat and mass transfer between single cylinders and air streams flowing normal to the cylinder axis (Sherwood and Pigford, 1952).

that for nonangular bodies of revolution with their axes parallel to the direction of flow, the local values of the Sherwood number are proportional to the square root of the Reynolds number ($N_{Re} > 1$); this result was demonstrated theoretically for the region between the front stagnation point and the separation point at which the boundary layer leaves the surface. Theoretical predictions are lacking beyond the separation point. In the Stokesian region of flow ($N_{Re} < 1$) the exponents on N_{Re} and N_{Sc} are both $\frac{1}{3}$ (see Aksel'rud, 1953; Friedlander, 1957; and Bowman, Ward, Johnson, and Trass, 1961). The length dimension in N_{Sh} and N_{Re} is, of course, the diameter of the sphere, d_s.

A somewhat vexed question in transfer from immersed bodies is whether or not the contributions from molecular diffusion (N^*_{Sh0}), from natural convection ($N^*_{Sh\,nc}$), and from forced convection are additive. Several workers have correlated experimental values of Sherwood number by direct addition of terms representing transfer by purely molecular diffusion and by forced convection, in the form

$$N^*_{Sh} = N^*_{Sh0} + CN^m_{Re}N^{1/3}_{Sc} \tag{6.125}$$

where

$$k^*_\rho = \frac{D}{d_s}N^*_{Sh} = \frac{(n_{A0})_{av}}{\rho_{A0} - \rho_{A\infty}} \tag{6.126}$$

and N^*_{Sh0} assumes a value of 2, as shown in Table 2.2. Equations 6.127 to 6.132 in Table 6.3 show examples of correlation in these terms. Other investigators have omitted the molecular diffusion term N^*_{Sh0}, to obtain correlations such as equations 6.133 to 6.138 in Table 6.3. Additional relationships are tabulated by Griffith (1960).

The two groups of correlations in Table 6.3, those with and those without N^*_{Sh0}, both describe the data from which they were obtained in a satisfactory manner. The correlations of the first group, which include N^*_{Sh0}, generally cover smaller ranges of the Reynolds number with lower maximum values than the second group.

At low values of the Reynolds number, natural or free convection contributes to the rate of mass transfer, and the minimum rate attained may be considerably higher than that due to molecular diffusion alone. Ranz and Marshall (1952) measured rates of evaporation of spherical water drops to dry air at zero Reynolds number and obtained the following correlation:

$$N^*_{\text{Sh}} = 2 + 0.60 N_{\text{Gr}}^{1/4} N_{\text{Sc}}^{1/3} \tag{6.141}$$

Similar expressions have been presented by other investigators (e.g., Kyte, Madden, and Piret, 1953; Mathers, Madden, and Piret, 1957; Merk and Prins, 1953–1954). Garner and Keey (1958) measured dissolution rates of benzoic acid spheres in water and found that the minimum rate of mass transfer does not occur at $N_{\text{Re}} = 0$. In their studies water flowed successively upwards, downwards, and horizontally, so that natural convection should have in turn opposed, aided, and partially aided the forced convection flow patterns. They found that minimum k_ρ^* occurred for $20 \leqslant N_{\text{Re}} \leqslant 50$, and they reached the following conclusions:

1. Free or natural convection can aid or hinder mass-transfer rates in forced convection.
2. The true convective velocity cannot be obtained by the vectorial addition of the velocities associated with forced and free convection.
3. Free-convection effects are almost entirely absent when the Sherwood number attained by forced convection alone equals that attained in free convection.

Conclusion 1 above is consistent with quite different work by Oliver and Jenson (1964) on heat transfer to non-Newtonian fluids in tubes. From conclusion 3 (and also from equations 6.130 and 6.141), Garner and Keey consider that the effects of free or natural convection are negligible for Reynolds numbers satisfying the following expression:

$$N_{\text{Re}} \geqslant 0.4 N_{\text{Gr}}^{1/2} N_{\text{Sc}}^{-1/6} \tag{6.142}$$

Table 6.3. Some experimental correlations of forced-convection mass transfer from single spheres.[a]

Equation Number	Equation	Range of Variables	Reference
With N^*_{Sh0}			
6.127	$N^*_{Sh}=2+0.552N_{Re}^{1/2}N_{Sc}^{1/3}$	$2\leqslant N_{Re}\leqslant 800$	Frössling, (1938, 1940)
		$0.6\leqslant N_{Sc}\leqslant 2.7$	Maxwell and Storrow (1957)
6.128	$N^*_{Sh}=2+0.60N_{Re}^{1/2}N_{Sc}^{1/3}$	$2\leqslant N_{Re}\leqslant 200$	Ranz and Marshall (1952)
		$0.6\leqslant N_{Sc}\leqslant 2.5$	
6.129	$N^*_{Sh}=2+0.544N_{Re}^{1/2}N_{Sc}^{1/3}$	$50\leqslant N_{Re}\leqslant 350$	Hsu, Sato, and Sage (1954)
		$N_{Sc}=1$	
6.130	$N^*_{Sh}=2+0.95N_{Re}^{1/2}N_{Sc}^{1/3}$	$100\leqslant N_{Re}\leqslant 700$	Garner and Suckling (1958)
		$1200\leqslant N_{Sc}\leqslant 1525$	
6.131	$N^*_{Sh}=2+0.575N_{Re}^{1/2}N_{Sc}^{0.35}$	$1<N_{Re}$ $1\leqslant N_{Sc}$	Griffith (1960)
6.132	$N^*_{Sh}=2+0.79N_{Re}^{1/2}N_{Sc}^{1/3}$	$20\leqslant N_{Re}\leqslant 2000$	Rowe, Claxton, and Lewis (1965)
Without N^*_{Sh0}			
6.133	$N^*_{Sh}=0.43N_{Re}^{0.56}N_{Sc}^{1/3}$	$200\leqslant N_{Re}\leqslant 4\times 10^4$	Williams (1942) Maisel and Sherwood (1950)
		"air"$\leqslant N_{Sc}\leqslant$"water"	Linton and Sherwood (1950)

Table 6.3. Continued

Equation Number	Equation	Range of Variables	Reference
6.134	$N^*_{Sh} = 0.82 N_{Re}^{1/2} N_{Sc}^{1/3}$	$100 \leqslant N_{Re} \leqslant 3500$ $N_{Sc} = 1560$	Aksel'rud (1953)
6.135	$N^*_{Sh} = 0.582 N_{Re}^{1/2} N_{Sc}^{1/3}$	$300 \leqslant N_{Re} \leqslant 7600$ $N_{Sc} \doteq 1210$	Linton and Sutherland (1960)
6.136	$N^*_{Sh} = 0.692 N_{Re}^{0.514} N_{Sc}^{1/3}$	$500 \leqslant N_{Re} \leqslant 5000$	Pasternak and Gauvin (1960)
6.137	$N^*_{Sh} = 0.33 N_{Re}^{0.6} N_{Sc}^{1/3}$	$1500 \leqslant N_{Re} \leqslant 12000$	Evnochides and Thodos (1961)
6.138	$N^*_{Sh} = 0.74 N_{Re}^{1/2} N_{Sc}^{1/3}$	$120 \leqslant N_{Re} \leqslant 6000$ $N_{Sc} \doteq 2.44$	Skelland and Cornish (1963)
With $N^*_{Sh\,nc}$:			
6.139	$N^*_{Sh} = 44 + 0.48 N_{Re}^{1/2} N_{Sc}^{1/3}$	$20 \leqslant N_{Re} \leqslant 850$ $N_{Sc} = 1210$	Garner and Grafton (1954)
6.140	$N^*_{Sh} = N^*_{Sh\,nc} + 0.347 (N_{Re} N_{Sc}^{1/2})^{0.62}$	$1 \leqslant N_{Re} \leqslant 3 \times 10^4$ $0.6 \leqslant N_{Sc} \leqslant 3200$	Steinberger and Treybal (1960)
	$N^*_{Sh\,nc} = 2 + 0.569 (N_{Gr} N_{Sc})^{0.25}$	$N_{Gr} N_{Sc} < 10^8$	
	$N^*_{Sh\,nc} = 2 + 0.0254 (N_{Gr} N_{Sc})^{1/3} N_{Sc}^{0.244}$	$N_{Gr} N_{Sc} > 10^8$	

[a] For additional correlations see Griffith (1960). For graphical correlation up to $N_{Re} = 140{,}000$ see Steele and Geankoplis (1959).

This equation was substantiated by the findings of Garner and Hoffman (1960, 1961). Other evidence that the effects of free and forced convection are not additive is provided by Gaffney and Drew (1950); Dryden, Strang, and Withrow (1953); Bar-Ilan and Resnick (1957); and Acrivos (1958).

These considerations suggest that correlations of the form of equation 6.125—with or without $N^*_{\mathrm{Sh}0}$— can be used to describe forced-convection rates of mass transfer only when the effects of free or natural convection are negligible, as indicated by equation 6.142. Empirical correlations of data in the presence of natural convection are provided by equations 6.139 and 6.140 in Table 6.3.

The effects of free stream turbulence on heat and mass transfer from various immersed bodies have been examined in a number of studies (see Maisel and Sherwood, 1950; Sato and Sage, 1958; Brown, Sato, and Sage, 1958; Loyzantsky and Schwab, 1935; and Comings, Clapp, and Taylor, 1948). Briefly, the root-mean-square velocity fluctuation at a point in the fluid is a measure of the intensity of turbulence. Expressed as a percentage of the mainstream velocity, this gives α_T, the percent intensity of turbulence. The scale of turbulence is a measure of the magnitude of the turbulent eddies. Both scale and intensity can be measured with hot-wire anemometers.

Maisel and Sherwood (1950) found that the scale of turbulence had little effect on the rate of mass transfer from spheres, but the effect of the intensity of turbulence can be considerable. Thus at N_{Re} of 2440 a change in the intensity of turbulence from 3.5 to 23 percent changed k^*_ρ by 18 percent. Brown, Sato, and Sage (1958) measured rates of evaporation of n-heptane from porous ceramic spheres to air streams with turbulence intensities from 1.3 to 15.1 percent. Their results covered a range of Reynolds numbers from 200 to 7000 and are summarized in Figure 6.22. The relative Sherwood number appearing in this figure is defined as the Sherwood number at a turbulence intensity α_T divided by the Sherwood number at the same N_{Re} and N_{Sc} but at a turbulence intensity of zero. Evidently the effect of the intensity of turbulence increases with increasing Reynolds number.

The differences between the correlations found by various workers as shown in Table 6.3 may be due to differences in the intensity of turbulence in the fluid stream (Nienow et al., 1969), in the degree of surface roughness of the sphere (Mullin and Cook, 1965), in the manner in which the sphere was supported, in the way in which physical properties were determined (Rowe et al., 1965), and in the concentrations and corresponding bulk flow velocities normal to the solid surface (Nienow et al., 1969).

Equations 6.133 to 6.138 of Table 6.3 can be rearranged into customary

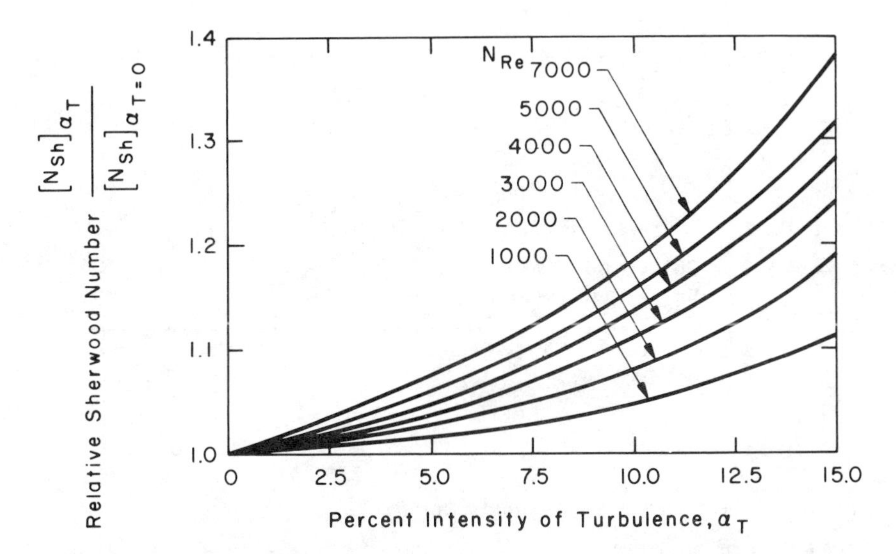

Figure 6.22. Relative Sherwood number for a sphere as a function of the intensity of turbulence (after Brown, Sato, and Sage, 1958).

j-factor form to obtain, in general,

$$j_D = \frac{N_{\mathrm{Sh}}^*}{N_{\mathrm{Re}} N_{\mathrm{Sc}}^{1/3}} = C N_{\mathrm{Re}}^{m-1} \tag{6.143}$$

Similar rearrangement of equations 6.127 to 6.132 yields a modified j factor as follows:

$$j_D' = \frac{N_{\mathrm{Sh}}^* - N_{\mathrm{Sh}0}^*}{N_{\mathrm{Re}} N_{\mathrm{Sc}}^{1/3}} = C N_{\mathrm{Re}}^{m-1} \tag{6.144}$$

Figure 6.23 shows a plot of j_D versus N_{Re} for the data of Skelland and Cornish (1963) and the correlations of Aksel'rud (1953), Linton and Sutherland (1960), and Evnochides and Thodos (1961). The modified j factor j_D' appears versus N_{Re} in Figure 6.24 for the data of Skelland and Cornish (1963) and the correlations of Frössling (1938), Maxwell and Storrow (1957), Ranz and Marshall (1952), and Garner and Suckling (1958).

Recommendation of one correlation rather than another from Table 6.3 and Figures 6.23 and 6.24 is somewhat difficult. Pasternak and Gauvin (1960) state that their correlation (equation 6.136) was obtained at an intensity of turbulence between 9 and 10 percent. Turbulence measure-

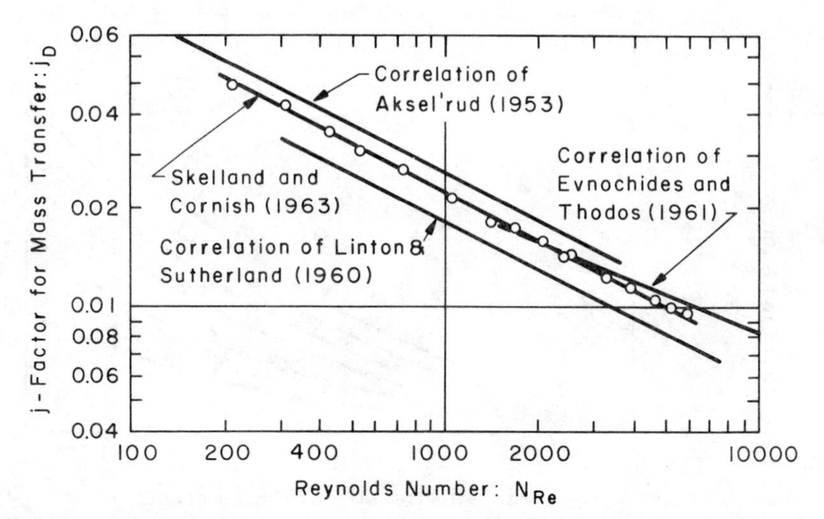

Figure 6.23 Comparison between *j*-factor correlations for forced-convection mass transfer from single spheres.

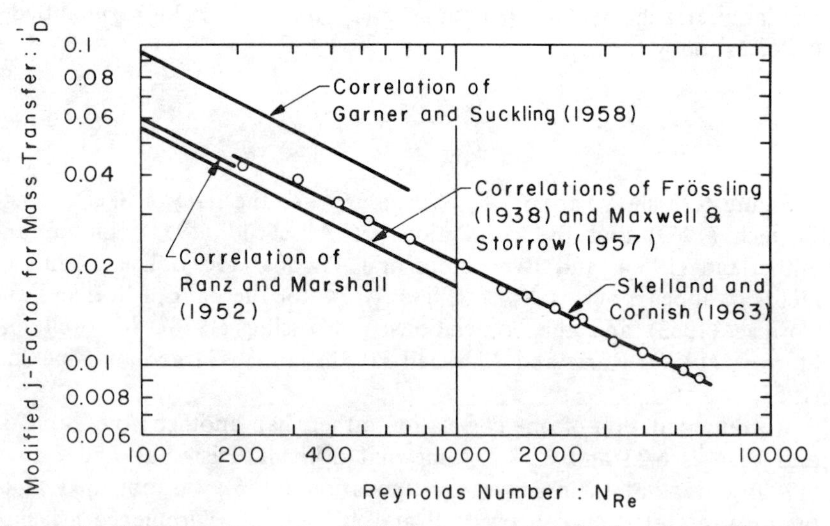

Figure 6.24. Comparison between modified *j*-factor correlations for forced-convection mass transfer from single spheres.

ments with a hot-wire anemometer showed that, according to Figure 6.22, the rates of transfer in the data of Skelland and Cornish were not increased by more than about 2 percent as a result of free-stream turbulence. In the absence, however, of information on intensity of turbulence, surface roughness, etc., it appears best to select the most conservative prediction for the problem at hand.

It is instructive to consider a few representative correlations for the rate of heat transfer between a sphere and a moving stream of fluid. A good correlation of experimental data was obtained by Drake (1961) with the following expression:

$$N_{\text{Nu}} = \frac{hd_s}{k} = 2 + 0.459 N_{\text{Re}}^{0.55} N_{\text{Pr}}^{1/3} \tag{6.145}$$

for $1 \leqslant N_{\text{Re}} \leqslant 70{,}000$, $0.6 \leqslant N_{\text{Pr}} \leqslant 400$.

For droplets evaporating in an air stream under conditions usually encountered in spray drying, Ranz and Marshall (1952) found

$$N_{\text{Nu}} = 2 + 0.60 N_{\text{Re}}^{1/2} N_{\text{Pr}}^{1/3} \tag{6.146}$$

for $2 \leqslant N_{\text{Re}} \leqslant 200$.

Evnochides and Thodos (1961) measured heat transfer to celite spheres from which nitrobenzene or water was evaporating to an air stream. They concluded that

$$N_{\text{Nu}} = 0.35 N_{\text{Re}}^{0.6} N_{\text{Pr}}^{1/3} \tag{6.147}$$

Comparison between equations 6.145 to 6.147 and Table 6.3 is sufficient to show the validity of the general analogy $j_H = j_D$ for the case of single spheres. An extension of the analogy to include half the total drag coefficient could not be expected to hold, as explained for the case of single cylinders. This is because of the presence of extensive form drag in the total drag force, contained in the drag coefficient.

Illustration 6.5

A solid sphere of benzoic acid has a diameter of $\frac{1}{2}$ in. and falls a distance of 10 ft through a stationary column of pure water. How much benzoic acid dissolves from the sphere during this fall?

How long would it take for the same amount of dissolution if the sphere were suspended in water entirely free from forced convection?

The system is at a temperature of 77°F in both cases.

SOLUTION. Relevant physical properties at 77°F are available from Illustration 5.6. The density of solid benzoic acid is 79.03 lb/ft^3, so that the terminal velocity of the sphere in water may be estimated by standard procedures, as described, for example, by W. L. McCabe and J. C. Smith (*Unit Operations of Chemical Engineering*, 2nd ed., McGraw-Hill, New York, 1967, pp. 167–168). The resulting terminal velocity u_t is 1.05 ft/sec, assuming that wall effects are insignificant. Calculations using the procedure of C. E. Lapple and C. B. Shepherd [*Ind. Eng. Chem.*, **32**, 605–617 (1940)], which are not shown here, indicate that the sphere attains velocities of 25, 50, 75, 90, and 99 percent of u_t in about 0.039, 0.084, 0.145, 0.203, and 0.264 sec, respectively, after release from rest. These values can be used in graphical integration of the expression $d(\text{distance}) = u\,dt$, to show that the sphere reaches 99 percent of its terminal velocity in a fall distance of only 2.1 in. The initial acceleration effects are therefore neglected, so that the sphere falls through 10 ft in 9.52 sec.

$$N_{\mathrm{Re}} = \frac{d_s u_t \rho}{\mu} = \frac{(0.5/12)1.05(3600)62.24}{2.16} = 4540$$

The difference between the densities of a saturated aqueous solution of benzoic acid and pure water at 77°F is 0.025 lb-mass/ft^3 (Garner and Keey, 1958, p. 221).

$$N_{\mathrm{Gr}} = \frac{(0.5/12)^3(4.17\times 10^8)(0.025)(62.24)^2}{62.265(2.16)^2} = 1.008\times 10^4$$

$$N_{\mathrm{Sc}} = 740 \qquad \text{(Illustration 5.6)}$$

In equation 6.142,

$$0.4N_{\mathrm{Gr}}^{1/2}N_{\mathrm{Sc}}^{-1/6} = \frac{0.4(1.008\times 10^4)^{1/2}}{(740)^{1/6}} = 13.4$$

This is much smaller than the Reynolds number, indicating—through equation 6.142—that natural convection effects are negligible.

The selection of a correlation for N_{Sh}^* from among those in Table 6.3 is somewhat arbitrary; although the expression due to Pasternak and Gauvin (1960) was developed for gas-phase systems (low N_{Sc}), comparison—in the original paper—with results from the aqueous systems of Linton and Sherwood demonstrated the effectiveness of equation 6.136 in this higher N_{Sc} range. Furthermore, in contrast to many of the other studies, the intensity of turbulence in the work by Pasternak and Gauvin is accurately known (9 to 10 percent), enabling allowance for this effect via Figure 6.22.

Substituting in equation 6.136,

$$N^*_{\mathrm{Sh}\,\alpha_T=10\%}=0.692(4540)^{0.514}(740)^{1/3}=476$$

From Figure 6.22, when $N_{\mathrm{Re}}=4540$ and $\alpha_T=10$ percent,

$$\frac{[N^*_{\mathrm{Sh}}]_{\alpha_T}}{[N^*_{\mathrm{Sh}}]_{\alpha_T=0}}=1.145$$

so that the Sherwood number corresponding to a turbulence intensity of zero—as in the stationary column of water—is

$$N^*_{\mathrm{Sh}}=\frac{476}{1.145}=416$$

Values of N^*_{Sh} calculated from equations 6.127 to 6.138 range from 332 to 580, with an average of 425. The above value of 416 for zero turbulence intensity differs by only about 2 percent from this average value and is used to complete the calculation.

$$(n_{A0})_{\mathrm{av}}=\frac{D}{d_s}N^*_{\mathrm{Sh}}(\rho_{A0}-\rho_{A\infty})=\frac{4.695\times10^{-5}}{(0.5/12)}(416)(0.213-0)$$

$$=0.1 \text{ lb } C_6H_5COOH/(\mathrm{ft}^2)(\mathrm{hr})$$

The sphere falls 10 ft in 9.52 sec, so that the amount dissolved during the fall is

$$(n_{A0})_{\mathrm{av}}\pi d_s^2\left(\frac{9.52}{3600}\right)=0.1\pi\left(\frac{0.5}{12}\right)^2\left(\frac{9.52}{3600}\right)=1.442\times10^{-6} \text{ lb } C_6H_5COOH$$

This represents a reduction in mass of the sphere of only 0.048 percent, showing the assumed constancy of d_s and u_t to be adequate.

If the sphere is suspended in water free from forced convection, then from equation 6.141,

$$N^*_{\mathrm{Sh}}=2+0.6(1.008\times10^4)^{1/4}(740)^{1/3}=56.3$$

It may be noted that alternative expressions to equation 6.141 have been proposed; Garner and Keey (1958), for example, use

$$N^*_{\mathrm{Sh}}=2+0.6(N_{\mathrm{Gr}})^{1/4}(N_{\mathrm{Sc}})^{1/4}$$

(It was this expression, equated to 6.130, that led to equation 6.142.) This alternative relationship gives

$$N_{\mathrm{Sh}}^{*}=2+0.6(1.008\times10^{4})^{1/4}(740)^{1/4}=33.3$$

This provides a more conservative (longer) estimate of the time for dissolution and is used to obtain the desired answer as

$$t=9.52\left(\frac{416}{33.3}\right)=119 \text{ sec}$$

j Factors for Single Oblate Spheroids

The oblate spheroid may be regarded as an idealized shape in many mass-transfer operations when one phase is dispersed in the other. In liquid-liquid extraction, for example, droplets of the disperse phase may approximate internally stagnant, non-oscillating oblate spheroids, owing to a particular combination of physical properties and the presence of trace quantities of surface-active impurities (see Garner and Hale, 1953; Garner and Skelland, 1955, 1956). At a given Reynolds number, spheroids of different major-to-minor-axis ratios (eccentricities) are subject to drag forces of different magnitudes (see Hughes and Gilliland, 1952; Malaika, 1949; Perry, 1950, p. 1018). This indicates that mass-transfer rates are also substantially dependent upon eccentricity.

Skelland and Cornish (1963) measured sublimation rates of oblate naphthalene spheroids in an airstream for $120\leqslant N_{\mathrm{Re}}\leqslant 6000$ and eccentricities between 1:1 (spheres) and 3:1. Attempts were made to characterize the geometry of the spheroids in terms of seven alternative dimensions. The correlation was clearly best using d_3, which is the total surface area of the body divided by the perimeter normal to flow. This represents the characteristic dimension for all bodies proposed by Pasternak and Gauvin (1960). The mass-transfer coefficient k_ρ^* was as defined by equation 6.126. It was based on the true surface of the spheroid, which, for an oblate spheroid of semimajor axis d and semiminor axis f, is given by

$$S_{\mathrm{os}}=2\pi d^2+\frac{\pi f^2 d}{(d^2-f^2)^{1/2}}\ln\left(\frac{d+(d^2-f^2)^{1/2}}{d-(d^2-f^2)^{1/2}}\right) \tag{6.148}$$

The volume of an oblate spheroid is

$$V_{\mathrm{os}}=\frac{4\pi f d^2}{3} \tag{6.149}$$

and the characteristic linear dimension d_3 is therefore

$$d_3 = d + \frac{1}{2}\left[\frac{f^2}{(d^2-f^2)^{1/2}} \ln\left(\frac{d+(d^2-f^2)^{1/2}}{d-(d^2-f^2)^{1/2}}\right)\right] \tag{6.150}$$

Measurements from 100 runs were correlated with an estimated standard deviation of 2.1 percent by the equation

$$j_D = \frac{N_{\mathrm{Sh}}^*}{N_{\mathrm{Re}} N_{\mathrm{Sc}}^{1/3}} = 0.74 N_{\mathrm{Re3}}^{-0.50} \tag{6.151}$$

where N_{Re3} denotes a Reynolds number using d_3. Correlations in terms of j_D and j_D' (equations 6.143 and 6.144) were statistically indistinguishable, so preference was given to the simpler j_D. Equation 6.142 demonstrated that, in this work, natural-convection effects could be neglected for $N_{\mathrm{Re}} > 15$, which is far below the minimum Reynolds number studied.

j Factors Generalized for All Shapes

A new general shape parameter was introduced by Pasternak and Gauvin (1960) to account for both body shape and orientation in equations for forced-convection transfer of heat and mass. The dimension, used as d_3 above for oblate spheroids, is defined as the total surface area of the body divided by the perimeter of the maximum projected area perpendicular to flow. The authors achieved good correlation of heat- and mass-transfer data for 20 shapes in different orientations, including spheres, cylinders with axes normal and parallel to flow, prisms, cubes in various orientations, and hemispheres situated with the flat section at the rear. Their expression, which correlated all these results with a "deviation" of 15 percent, is

$$j_D = j_H = 0.692 N_{\mathrm{Re}P,G}^{-0.486} \tag{6.152}$$

for $500 \leqslant N_{\mathrm{Re}P,G} \leqslant 5000$, and where $N_{\mathrm{Re}P,G}$ denotes a Reynolds number using the new general shape parameter of Pasternak and Gauvin. The intensity of turbulence was between 9 and 10 percent in all their studies; viscosity and density were evaluated at the average temperature of the fluid film around the body.

Heat- and mass-transfer rates were studied during evaporation of water from stationary bodies made from celite suspended in an air stream. Celite changes color from orange to pale yellow on drying, thereby yielding a

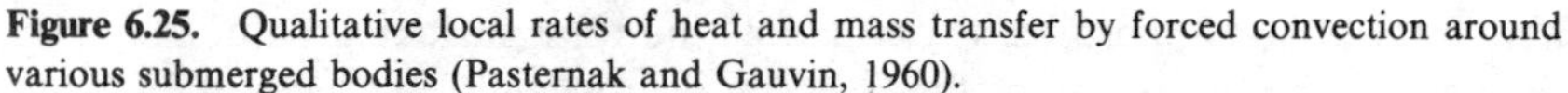

Figure 6.25. Qualitative local rates of heat and mass transfer by forced convection around various submerged bodies (Pasternak and Gauvin, 1960).

qualitative picture of *local* rates of mass transfer around the surface of the body. This information appears in Figure 6.25, where the distance between the body outline and the corresponding envelope at a particular point gives a measure of the relative rate of heat and mass transfer at that point. Shaded regions show areas of minimum transfer.

j Factors for Packed and Fluidized Beds

Packed and fluidized beds occur frequently in industrial processing activities, including the conduct of gaseous reactions catalyzed by solid surfaces, drying operations, heat transfer in granular beds and heat regenerators, and the adsorption and desorption of gases or liquids by solid particles.

Many of the studies made before 1963 used unduly shallow beds, with the result that end effects due to poor fluid distribution and undeveloped flow contributed excessively to the results. These conditions were reproduced by Gupta and Thodos (1963), who contrasted the findings obtained

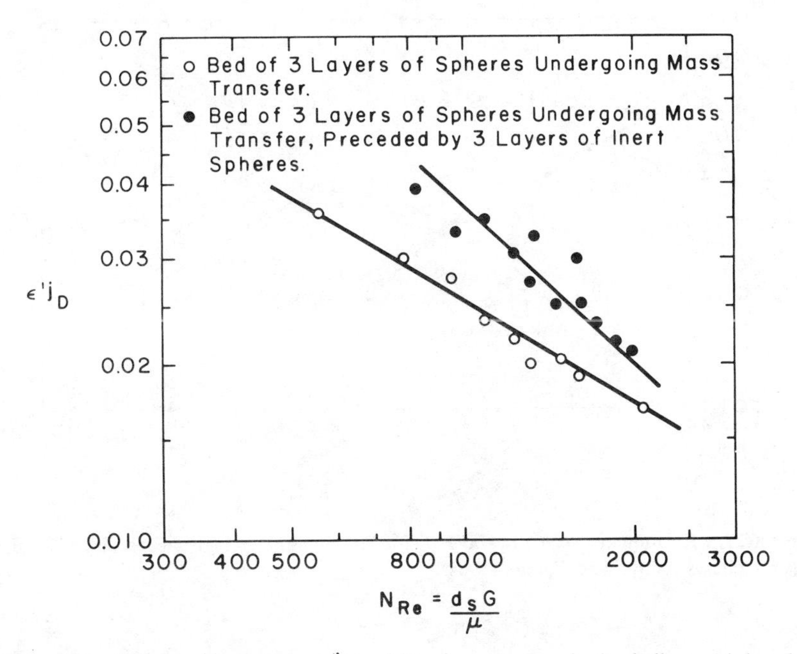

Figure 6.26. Relationships between $\epsilon' j_D$ and N_{Re} for mass transfer in shallow and deep beds of spheres (Gupta and Thodos, 1963).

with shallow and deep beds of spheres undergoing transfer of water vapor to an air stream, as shown in Figure 6.26.

Mass transfer between *liquids* and deep beds of packed spheres was studied by Wilson and Geankoplis (1966), who propose

$$j_D = \frac{1.09}{\epsilon' N_{\text{Re}}^{2/3}}; \quad N_{\text{Re}} = \frac{d_s G}{\mu} \tag{6.153}$$

for $0.0016 \leqslant N_{\text{Re}} \leqslant 55$, $165 \leqslant N_{\text{Sc}} \leqslant 70{,}600$, $0.35 \leqslant \epsilon' \leqslant 0.75$, and

$$j_D = \frac{0.25}{\epsilon' N_{\text{Re}}^{0.31}} \tag{6.154}$$

for $55 \leqslant N_{\text{Re}} \leqslant 1500$; $165 \leqslant N_{\text{Sc}} \leqslant 10{,}690$.

For mass or heat transfer between *gases* and deep beds of spheres, the

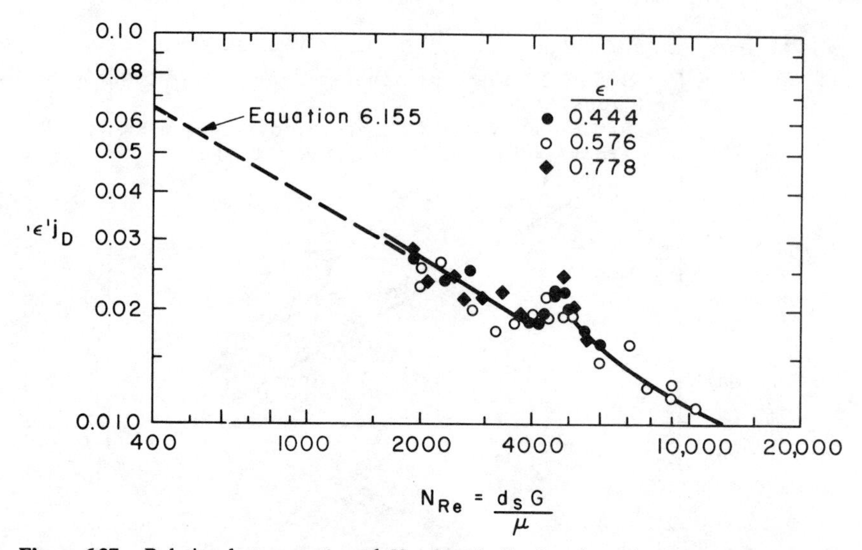

Figure 6.27. Relation between $\epsilon' j_D$ and N_{Re} for air flowing through packed and distended beds of spheres (Gupta and Thodos, 1964).

relationship recommended by Gupta and Thodos (1963) for low mass-transfer rates is

$$j_D = j_H = \frac{2.06}{\epsilon' N_{\text{Re}}^{0.575}}, \quad 95 \leqslant N_{\text{Re}} \leqslant 2453 \tag{6.155}$$

For N_{Re} between 1900 and 10,300, Gupta and Thodos (1964) present the correlation for high mass-transfer rates shown in Figure 6.27. In the absence of radiation effects these authors found j_H to exceed j_D by 5 percent for this range of N_{Re}.

Mass transfer in both gas and liquid fluidized beds of spheres has been correlated by Gupta and Thodos (1962b) with the equation

$$\epsilon' j_D = 0.010 + \frac{0.863}{N_{\text{Re}}^{0.58} - 0.483} \tag{6.156}$$

for $1 \leqslant N_{\text{Re}} \leqslant 2140$.

Approximate procedures for extending these relationships for spheres to particles of other shapes have been presented by Gupta and Thodos (1962a), although many of the data used in this development contain substantial entrance or end effects, as noted above.

In these expressions for packed and fluidized beds, G is the superficial mass velocity of the fluid in mass per unit time per unit cross section without packing. The coefficient in j_D is based on the average mass transfer per unit surface of the solid particles, which have known dimensions. The driving force for transfer is expressed as the logarithmic mean of the concentration differences at the inlet and outlet of the bed. Knowledge of the void fraction of the bed, ϵ', is usually available, and $N_{Re} = d_s G/\mu$.

Further information on fixed and fluidized beds is provided in the book by Kunii and Levenspiel (1969), and earlier studies are well reviewed by Leva (1959).

Mass Transfer in Noncircular Conduits

Heat transfer in turbulent flow through noncircular ducts has been successfully correlated by insertion of the "equivalent diameter" D_e in place of d_t in the appropriate circular tube correlation.

$$D_e = \frac{4(\text{Cross-sectional flow area})}{\text{Wetted perimeter}}$$

The procedure has been effective for annular, triangular, and rectangular conduits (Bennett and Myers, 1962, p. 340–341). This approach has not been well established in the case of mass transfer, but the generality of the relationship $j_H = j_D$ suggests that the use of D_e in such cases might constitute a reasonable extension of the analogy between heat and mass transfer. It appears, however, that some caution may be required in the use of D_e when the duct cross section has some sharp corners (Eckert and Gross, 1963, p. 141–142).

INTERFACIAL TURBULENCE

A spontaneous agitation of the interface between two immiscible liquids or between a gas and a liquid phase has sometimes been observed when a solute is passing from one phase to the other prior to the attainment of equilibrium. The effect appears to depend upon the direction of transfer of solute and to be associated with local variations in interfacial concentration and therefore in interfacial tension. The Marangoni effect, which refers to flow driven by surface tension variations, is apparently involved.

McBain and Woo (1937) observed this behavior when a layer of toluene containing 10 percent of methanol was gently contacted with a plane water layer. Water droplets were emulsified into the toluene phase while the water stayed clear. The reverse effect was observed upon increasing the methanol concentration in the toluene to 40 percent. No emulsification occurred in either phase when the methanol was located in the water instead of in the toluene. Ward and Brooks (1952) found that transfer of either valeric or butyric acid from water to toluene across a plane interface leads to spontaneous agitation of the toluene near the interface. The agitation is absent, however, when the solute is either acetic or propionic acid. A study of such interfacial phenomena was made by Wei (1955), who identified a variety of interfacial disturbances, ranging from rippling and twitching of the interface to spontaneous emulsification in the vicinity of the boundary between phases.

The effects are sometimes displayed in a dramatic manner when a droplet is suspended from a nozzle in another immiscible liquid with solute transferring from one phase to the other. Twitching and rippling of the interface may occur, perhaps with periodic pulsations of the entire drop, interspersed with periods of quiescence. Observations of this sort were reported by Lewis and Pratt (1953) and were filmed at about the same time by Garner, Kendrick, and Skelland (1953). Garner, Nutt, and Mohtadi (1955) subsequently reported the same phenomena for pendent drops of nitrobenzene, chlorobenzene, carbon tetrachloride, or chloroform, containing between 2 and 30 percent of either acetone, methanol, ethanol, or isopropanol. Distilled water formed the continuous phase.

Droplets undergoing mass transfer during free rise or fall through another immiscible liquid may exhibit comparable behavior, although this is most noticeable when the drops are moving very slowly. Observations of these effects have been made by Garner, Kendrick, and Skelland (1953); Lewis (1953); Sigwart and Nassenstein (1956); and Sherwood and Wei (1957).

A review of many reports of these phenomena is provided by Sternling and Scriven (1959). These authors present a simplified but extensive mathematical analysis of this type of interfacial instability in terms of the equations of motion and diffusion. Their results are in agreement with many of the observations, including the conversion of some systems from stable to unstable by reversal of the direction of solute transfer. Their analysis suggests that interfacial turbulence is usually promoted by the following eight factors:

1. Solute transfer out of the phase of higher viscosity.
2. Solute transfer out of the phase in which solute diffusivity is lower.
3. Large differences in kinematic viscosity (μ/ρ) and solute diffusivity between the two phases.
4. Large concentration gradients near the interface.
5. Large variation of interfacial tension with concentration.
6. Low viscosities and diffusivities in both phases.

7. Absence of surface-active agents.
8. Large interfacial area.

The importance of interfacial turbulence lies in the substantial increase it induces in the rates of mass transfer between two phases. Thus transfer rates may be much higher than predicted from a proper combination of single-phase rate coefficients on the assumption of a quiescent interface.

TRANSFER WITH A HIGH MASS FLUX

Mass transfer in a turbulent stream is modified by high-mass-flux conditions in a manner qualitatively similar to that outlined under this heading for laminar systems in Chapter 5.

Relationships are here developed which permit approximate allowance for the influence of high mass flux on the mass-transfer coefficients. Concentrations will be expressed in terms of mass fractions for consistency with the corresponding treatment given in Chapter 5 and because much of the published work in this area is in these terms.

Consider the transfer of the single component A from a rigid surface ($y=0$) into a turbulent stream of nontransferring component B. The total transfer of A is given by the molecular and eddy diffusive fluxes, assumed to be additive and caused by the concentration gradient, plus the convective flux associated with the bulk flow. If the flow immediately adjacent to the surface is assumed to be completely nonturbulent, it follows that the diffusive flux at that location occurs by a purely molecular mechanism. The total flux of component A at the surface is then described by equations 5.197 and 5.198 from Chapter 5. In fact, equations 5.197 to 5.209 are equally applicable to the turbulent conditions considered here and serve to define the driving force B, the coefficients k_w, k_w^0, and k_w^*, and the relationships between them.

The film theory is often used to interpret mass transfer in a turbulent stream near a phase boundary. In view of the hypothetical aspects of this procedure, however, a relation between coefficients for high- and low-mass-flux conditions will be estimated by a rather more rigorous analysis which is a modification of that given by Vivian and Behrmann (1965). For a given location x in the direction of flow, the local rate equation at a point between the surface and the bulk of the fluid is written as

$$n_{Ay}=\psi(n_{A0})_x=\frac{-\rho(D+\epsilon_D)}{(1-w_A)}\frac{\partial w_A}{\partial y} \tag{6.157}$$

where the proportionality factor ψ is a function of y, so that

$$(n_{A0})_x = \frac{-D\rho\left(1+\dfrac{1-\psi}{\psi}+\dfrac{\epsilon_D}{\psi D}\right)}{1-w_A}\frac{\partial w_A}{\partial y}$$

$$= -\frac{\rho(D+\epsilon_D')}{1-w_A}\frac{\partial w_A}{\partial y}, \qquad \epsilon_D' = \frac{D(1-\psi)+\epsilon_D}{\psi}$$

This equation may be integrated over the range $0 \leqslant y \leqslant \Delta_T$, where Δ_T is the distance from the surface to the point at which bulk conditions prevail. For steady-state conditions at the surface and in the bulk, the result is

$$(n_{A0})_x = \frac{w_{A0}-w_{A\infty}}{(1-w_A)_{\text{LM}}\displaystyle\int_0^{\Delta_T}\frac{dy}{\rho(D+\epsilon_D')}}$$

The average flux over the surface length 0 to x is given by

$$(n_{A0})_{\text{av}} = \frac{1}{x}\int_0^x \frac{w_{A0}-w_{A\infty}}{(1-w_A)_{\text{LM}}\displaystyle\int_0^{\Delta_T}\frac{dy}{\rho(D+\epsilon_D')}}\,dx$$

If w_{A0} and $w_{A\infty}$ are independent of x, then

$$(n_{A0})_{\text{av}} = \left[\frac{1}{x}\int_0^x \frac{dx}{\displaystyle\int_0^{\Delta_T}\frac{dy}{\rho(D+\epsilon_D')}}\right]\frac{w_{A0}-w_{A\infty}}{(1-w_A)_{\text{LM}}} \tag{6.158}$$

but

$$k_w = \frac{(n_{A0})_{\text{av}}}{w_{A0}-w_{A\infty}}; \qquad k_w^* = \lim_{w_A\to 0} k_w$$

so that

$$k_w(1-w_A)_{\text{LM}} = \frac{1}{x}\int_0^x \frac{dx}{\displaystyle\int_0^{\Delta_T}\frac{dy}{\rho(D+\epsilon_D')}} = k_w^* \tag{6.159}$$

Thus

$$\frac{k_w}{k_w^*} = \frac{1}{(1-w_A)_{LM}} \tag{6.160}$$

and with equations 5.200, 5.201, and 5.208,

$$\frac{k_w^0}{k_w^*} = \frac{1-w_{A0}}{(1-w_A)_{LM}} = \frac{\ln(1+B)}{B} \tag{6.161}$$

Equations 6.160 and 6.161 for a turbulent system are seen to be the same as equations 5.214 and 5.215 respectively. The latter results were derived for nonturbulent conditions from film theory.

Equation 6.161 is used to correct for the effects of high mass flux in turbulent systems partly because no other simple relationship is available, but also because some experimental confirmation of the expression has been obtained. Thus equation 6.161 was successful in correlating measurements by Humphrey and Van Ness (1957) on the aqueous dissolution of sodium thiosulfate pentahydrate crystals under turbulent conditions in an agitated vessel, using propeller and turbine agitation. Physical properties were evaluated at the arithmetic average of the surface and bulk compositions. Equations 6.160 and 6.161 were also verified by Vivian and Behrmann (1965) for the gas-phase mass-transfer coefficient in the aqueous absorption of ammonia in a short wetted-wall column. Physical properties were regarded as constant, and the average Reynolds number of the turbulent gas stream was 3,200.

Little attention has been given to the problem of composition-dependent physical properties under turbulent conditions. Hanna (1962) presented the following approximate correction for the case of isothermal binary diffusion in a perfect-gas mixture when the density is the only concentration-dependent property:

$$\frac{(k_w^0)_{\text{var}\,\rho}}{(k_w^0)_{\text{const}\,\rho}} = \frac{\ln(M_0/M_\infty)}{M_0/M_\infty - 1} \tag{6.162}$$

where M_0 and M_∞ are the molecular weights of the mixture at the wall and in the bulk of the fluid. As noted under this heading in Chapter 5, equation 6.162 is valid for both laminar and turbulent flow, regardless of geometry or Schmidt number.

Knuth and Dershin (1963) outlined a method for predicting transport rates in turbulent gaseous boundary layers with variable fluid properties. For this purpose they first developed a semiempirical correlation describing the dependence of the friction coefficient on the rate of mass injection at the wall for fluids with constant physical properties. Modified Reynolds analogies were next derived, relating mass and energy transfer coefficients to the friction coefficient. Physical properties in the mass-transfer expressions were tentatively evaluated at the reference composition previously established for laminar boundary layers (Knuth, 1963; see also Chapter 5). The experimental data were inadequate to test this approach.

In addition to a need for further theoretical study, it is evident that more experimental work is required to elucidate the phenomena of transfer with a high mass flux under turbulent conditions for a variety of geometries. For example, no attention has been given to turbulent liquid systems in which the viscosity and diffusivity are strongly dependent upon composition.

Illustration 6.6

Repeat Illustration 5.7 with the air velocity increased to 200 ft/sec.

(*Note*: Problems in which the temperature of the mass-transfer surface is *not* controlled by a separate heat source but instead adopts an adiabatically established steady-state value different from that of the gas stream may be handled as in Illustration 5.8. For turbulent flow situations the appropriate heat- and mass-transfer correlations would be substituted for those used to obtain k_ρ^* and h_G in that problem, and k_w^0/k_w^* would be evaluated from equation 6.161.)

SOLUTION. Bulk, average, and surface values of compositions and physical properties will be as in Illustration 5.7.

$$N_{\text{Re},x}=72{,}100\left(\frac{200}{20}\right)=721{,}000$$

Assuming usual levels of bulk turbulence and of roughness of the plate surface and leading edge, the boundary layer will be turbulent at the point of interest.

$$N_{\text{Sc}}=0.602 \qquad \text{(from Illustration 5.7)}$$

By a procedure analogous to that leading to equation 6.121, the combination of equations 6.9, 6.101, and 6.102 results in

$$j_{Dx}=j_{Hx}=0.0292N_{\text{Re},x}^{-0.2}$$

where j_{Dx} is the local value of j_D at x, containing the local coefficient of mass transfer, $k_\rho^* = n_{A0}/(\rho_{A0} - \rho_{A\infty})$. Then

$$k_\rho^* = 0.0292 \frac{(0.32)(3.88)}{0.75} (721{,}000)^{0.8} (0.602)^{1/3} = 1980 \text{ ft/hr}$$

Equation 5.206 gives

$$k_w^* = \rho k_\rho^* = 0.0621(1980) = 123 \text{ lb-mass/(ft}^2\text{)(hr)}$$

Illustration 5.7 shows that $w_{A0} = 0.1734$, so that according to equation 5.201,

$$B = \frac{w_{A0} - w_{A\infty}}{1 - w_{A0}} = \frac{0.1734 - 0}{1 - 0.1734} = 0.21$$

and from equation 6.161

$$(k_w^0)_{\text{const } \rho} = k_w^* \frac{\ln(1+B)}{B} = 123 \frac{2.303 \log(1+0.21)}{0.21}$$

$$= 111.6 \text{ lb-mass/(ft}^2\text{)(hr)}$$

The isothermal equation 6.162 is next applied:

$$(k_w^0)_{\text{var } \rho} = \frac{111.6(2.303)\log(26.23/29)}{26.23/29 \;\; - 1} = 118 \text{ lb-mass/(ft}^2\text{)(hr)}$$

The local value of the mass flux of water vapor is obtainable from equation 5.200 as

$$n_{A0} = 118(0.21) = 24.75 \text{ lb-mass/(ft}^2\text{)(hr)}$$

NOMENCLATURE

A, B	Components A and B.
A, B	Functions of N_{Re} and N_{Sc} in equation 6.94.
A_0	Area of surface, ft^2.

A_W	Wall area, ft^2.
B	A mass-transfer driving force defined by equation 5.201.
B_3, B_4	Coefficients in equations 6.19 and 6.20.
C_{df}	Drag coefficient for a flat plate, equation 6.124.
C_{dx}	Local drag coefficient, $2\tau_0 g_c/\rho u_\infty^2$.
c	Total concentration, total lb-mole/ft^3.
c_{BLM}	Logarithmic-mean concentration of component B in film of thickness z_{fL}, lb-mole/ft^3.
c_p	Specific heat, Btu/(lb-mass)(°F).
D	(Volumetric) molecular diffusivity, ft^2/hr.
D_e	Equivalent or hydraulic diameter, ft.
d	Semimajor axis of an oblate spheroid, ft.
d_c	Diameter of a cylinder, ft.
d_s	Diameter of a sphere, ft.
d_t	Tube diameter, ft.
d_3	(Total body surface)/(perimeter normal to flow): equation 6.150 for an oblate spheroid, ft.
F	Mechanical energy used to overcome friction in flow between points 1 and 2, (ft)(lb-force)/lb-mass.
F_{df}	Frictional drag force, lb-force.
$F_1(N_{Re})$, $F_2(N_{Re})$	Functions of the Reynolds number in equations 6.87, 6.90, and 6.91.
$F(\mu/\rho D)$	Function of the Schmidt number, defined in equation 6.81.
f	Fanning's friction factor, $\tau_w g_c/(\rho V^2/2)$.
f	Semiminor axis of an oblate spheroid, ft.
G	Superficial mass velocity, lb-mass/(ft^2)(hr).
g	Acceleration due to gravity, ft/hr^2.
g_c	Conversion factor, 32.174 (lb-mass)(ft)/(lb-force)(sec^2) *or* 4.17×10^8 (lb-mass)(ft)/(lb-force)(hr^2).
h	Individual heat-transfer coefficient, Btu/(hr)(ft^2)(°F).
i_{A0}	Mass flux of component A relative to the mass-average velocity and at the surface ($y=0$), lb-mass/(ft^2)(hr).

j_D	The j factor for mass transfer, equation 6.101.
j_D'	Modified j factor for mass transfer, equation 6.144.
j_{Dx}	Local value of j_D.
j_H	The j factor for heat transfer, equation 6.102.
(KE)	Average kinetic energy per unit mass, (ft)(lb-force)/lb-mass.
k	Thermal conductivity, Btu/(ft^2)(hr)(°F/ft).
k, k'	Constants in equation 6.42.
$k_{\rho G}, k_{\rho L}$	Individual mass-transfer coefficients, defined in equations 4.16 and 4.17.
$k'_{\rho G}, k'_{\rho L}$	Individual mass-transfer coefficients for equimolal counterdiffusion, defined in equations 4.14 and 4.15.
k_w	Individual mass-transfer coefficient for any concentration range, $n_A/\Delta w_A$, lb-mass/(ft^2)(hr).
k_w^0	Individual mass-transfer coefficient for any concentration range, $i_A/\Delta w_A$, lb-mass/(ft^2)(hr).
$k_w^*, k_\rho^*, k_{\rho LM}^*$	Individual mass-transfer coefficients for low concentrations and transfer rates: $n_A/\Delta w_A$, lb-mass/(ft^2)(hr); (n_A or $n_{A\text{av}})/\Delta\rho_A$, ft/hr; $n_{A\text{av}}/(\Delta\rho_A)_{\text{log mean}}$, ft/hr. (The quantities n_A and $\Delta\rho_A$ are defined locally for each application.)
$k_{\rho m}^*$	Integrated mean value of k_ρ^*, defined by equation 6.24.
L	Length; plate length in direction x, ft.
L_c	Characteristic length dimension, ft.
l	Height of control volume, ft.
M	Average molecular weight.
M_A, M_B	Molecular weights of components A and B.
M_0, M_∞	Molecular weight of mixture at the wall and in bulk of free stream.
m	Mass of fluid aggregate, lb-mass.
$N_{\text{Gr}}, N_{\text{Gr}D}$	Grashof numbers for mass transfer, in equations 6.140–142, $d_s^3 g(\rho_0-\rho_\infty)/(\mu/\rho)^2\rho_0$; $N_{\text{Gr}D}$ as in equation 5.44.

N_{Nu}, $(N_{Nu})_f$	Nusselt number for heat transfer; f indicates physical properties at arithmetic-mean film temperature, hd_t/k.
N_{Pr}, $(N_{Pr})_f$	Prandtl number for heat transfer; f indication as for $(N_{Nu})_f$, $c_p\mu/k$.
N_{Re}, $(N_{Re})_f$, N_{Ref}, $N_{Re,L}$, $N_{Re,P,G}$, $N_{Re,x}$, N_{Re3}	Reynolds number (d_s or d_t)$V\rho/\mu$; f indication as for $(N_{Nu})_f$; for a liquid film, $4Q\rho/w\mu$; $Lu_\infty\rho/\mu$; based on Pasternak and Gauvin's shape parameter; $xu_\infty\rho/\mu$; $d_3V\rho/\mu$.
N_{Sc}	Schmidt number, $\mu/\rho D$.
N^*_{Sh}, $(N_{Sh})^*_{LM}$, $(N_{Sh})^*_m$, $N^*_{Sh\,nc}$, N^*_{Sh0}, $(N_{Sh})^*_x$	Sherwood numbers kL_c/D in which the coefficient k is, respectively, k^*_ρ, $k^*_{\rho LM}$, $k^*_{\rho m}$; Sherwood number for natural convection from a sphere; for molecular diffusion from a sphere; local value at x.
n_{A0}, n_{AW}, n_{Ay}	Mass flux of component A relative to stationary coordinates at the surface ($y=0$); at the wall; in direction y, lb-mass/(ft^2)(hr).
P, ΔP, p_{BLM}	Total pressure; pressure drop due to friction in fully developed flow; logarithmic-mean partial pressure of component B in film of thickness z_{fG}, lb-force/ft^2 or atm.
p, p	Constant in equation 6.19; in equation 6.94.
Q	Volumetric flow rate, ft^3/hr.
Q_0, Q_W	Rate of heat transfer at the surface ($y=0$); at the wall, Btu/hr.
q	Constant in equation 6.20.
q_{AW}	Rate of transfer of A at the wall, lb-mass/hr.
q_W	Heat flux at the wall, Btu/(ft^2)(hr).
r, r_t	Radius; radius of a tube, ft.
S_{os}	Surface of an oblate spheroid, ft^2.
T^+, T_B, T_0, T_W, T_∞	Dimensionless temperature defined below equation 6.49; bulk or "mixing cup" temperature; at the surface ($y=0$); at the wall; in the bulk fluid or outside the thermal boundary layer, °R.
u, u^*, u^+, u_1, u_2	Time-averaged mass-average velocity in the x direction; friction velocity, $\sqrt{\tau_W g_c/\rho}$; u/u^*; u at levels y_1 and y_2, ft/hr.

$u_b, u_b^+, u_e, u_L, u_L^+, u_\infty$	Time-averaged velocity in the x direction at the outer edge of the buffer zone; u_b/u^*; mean eddy velocity; velocity at the outer edge of the laminar sublayer; u_L/u^*; velocity outside the momentum boundary layer, ft/hr.
V, V^+	Mean velocity in the x direction, ft/hr; V/u^*.
V_{os}	Volume of an oblate spheroid, ft^3.
v_0	Mass-average velocity in the y direction at the surface ($y=0$), ft/hr.
W_s'	Mechanical (shaft) work done by unit mass of fluid, (ft)(lb-force)/lb-mass.
w	Width normal to flow, ft.
$w_A, w_{A0}, w_{A\infty}$	Mass fraction of component A; at the surface ($y=0$); in the bulk of the stream or outside the concentration boundary layer.
x	Distance in the direction of flow, ft.
x_A	Mole fraction of component A.
y	Direction and distance normal to surface, ft.
y^+	$yu^*\rho/\mu$.
y_A	Mole fraction of component A.
y_b^+	Value of y^+ at the outer edge of the turbulent core.
Z	Vertical height above datum, ft.
α	Angle between an inclined plane and the horizontal.
α_T	Turbulence intensity, percent.
β_e	Volume expansion coefficient, equation 5.27.
β_v	Velocity gradient at the wall, hr^{-1}.
$\delta, \delta_c, \delta_f, \delta_L$	Thickness of the momentum and concentration boundary layers; film thickness; thickness of the laminar sublayer, ft.
$\epsilon, \epsilon_D, \epsilon_H, \epsilon_M$	Eddy diffusivity; eddy mass, heat, and momentum diffusivities, ft^2/hr.
ϵ'	Void fraction of a bed of particles,
θ	Angle between tube axis and the vertical.
Λ	Factor in equation 6.14, ft/hr.
μ	Viscosity, lb-mass/(ft)(hr).
$\rho, \rho_0, \rho_\infty$	Total density; at the surface ($y=0$); at the bulk concentration, lb-mass/ft^3.

ρ_A	Mass concentration of component A, lb-mass/ft^3.
ρ'_A, $\rho'_{A\infty}$	$\rho_A-\rho_{A0}$; $\rho_{A\infty}-\rho_{A0}$, lb-mass/ft^3.
ρ_{AB}, ρ_{Ab}, ρ_{A0}, ρ_{As}, ρ_{AW}, $\rho_{A\infty}$, ρ_{A1}, ρ_{A2}	Bulk average or "mixing cup" value of ρ_A; ρ_A at $y^+=30$; mass concentration of A at the surface ($y=0$); at the centerline; at the wall; in the bulk or outside the concentration boundary layer; at levels y_1 and y_2, lb-mass/ft^3.
ρ_A^+	A dimensionless concentration, $(\rho_{AW}-\rho_A)u^*/n_{AW}$.
$(\rho/M)_{av}$, $(\rho_B/M_B)_{LM}$	Mean value for the phase under consideration; see c_{BLM} above, lb-mole/ft^3.
$\tau_0=(\tau_{yx})_{y=0}$	Shear stress at the surface ($y=0$), lb-force/ft^2.
τ_{rx}	Shear stress in direction x on surface normal to r, lb-force/ft^2.
$\tau_W=(\tau_{rx})_{r=r_i}$	Shear stress at the wall, lb-force/ft^2.
ϕ_D	Defined by equations 6.79 and 6.80.

PROBLEMS

6.1 Water flows with a free-stream velocity of 3 ft/sec over a smooth flat plate of solid benzoic acid. The plate measures 1 ft$\times$2 ft and is oriented at zero incidence to the stream, with its longer sides parallel to the direction of flow. The system is at 77°F and the leading edge is rough enough to promote a turbulent boundary layer over the entire plate. Estimate the total rate of mass transfer from a single surface of the plate and the corresponding value of the average Sherwood number. Is the total transfer rate increased by reorienting the plate so that its shorter sides are parallel to the direction of flow? Explain your answer physically and support it with appropriate calculation. (See Illustration 5.6 for physical properties.)

6.2 Repeat all parts of Problem 6.1 for the case in which the plate is sufficiently smooth to sustain a laminar boundary layer along the leading portion of the surface. Assume that transition to a turbulent boundary layer begins at a critical Reynolds number $x_c u_\infty \rho/\mu$ of 3.11×10^5.

6.3 For the conditions of Problem 5.5, estimate the local values of the mass flux, boundary-layer thickness, and maximum velocity in the boundary layer on each side of the coated plate at a point 1 ft below the upper edge.

6.4 Pure ethanol flows with an average velocity of 4 ft/sec through a smooth, 1-in.-i.d. tube made from solid stearic acid. If the system is at 77°F, use all available expressions to obtain maximum and minimum estimates of the average concentration of stearic acid in solution at a cross section located 4 ft downstream from the tube inlet. Physical properties are given in Illustration 5.3.

6.5 Calculate and plot the radial concentration distribution of stearic acid at the cross section located 4 ft from the tube entrance in Problem 6.4. Use equations 6.82 through 6.84 to prepare two plots, using abscissas y^+ and y/d_t, respectively. Locate the boundaries of the wall layer and buffer zone on each plot.

6.6 The analogy between momentum and mass transfer represented by equation 6.54 may be adapted to the turbulent boundary layer on a flat plate by inserting equation 6.4 for $\tau_w g_c$, with δ replaced by equation 6.6 and u_∞ substituted for V. According to Eckert and Gross (1963, pp. 144, 299) the ratio u_L/V is then given by $1.305 N_{\text{Sc}}^{-1/6} N_{\text{Re},x}^{-1/10}$. The resulting expression gives the local Sherwood number; an average value is of course derived by integration along the surface.

Use these relationships to recalculate parts a, b, and c of Illustration 6.1, allowing for the non-unity N_{Sc}. Compare the results with those in Illustrations 6.1 and 6.4 as appropriate.

6.7 The following solid bodies made from naphthalene are suspended in turn in a pure air stream with an undisturbed velocity of 4 ft/sec. If the system is at 113°F, estimate the total rate of mass transfer from each of the bodies and arrange them in order of decreasing transfer rate:

(a) A cylinder 2 in. in diameter and 2 in. long, with axis parallel to the flow.

(b) The cylinder in (a), with axis normal to the flow.

(c) An oblate spheroid with a major axis of 2 in. and a volume of 2.095 in.^{3}, the major axis being normal to the flow.

(d) An oblate spheroid with a major axis of 2 in. and a volume of 1.397 in.^{3}, the major axis being normal to the flow.

(e) A hemisphere 2 in. in diameter, with the flat section at the rear.

(f) A cube with 2-in. sides, four sides being parallel to the direction of flow.

6.8 Derive equations 6.148 and 6.149 for the surface and volume of an oblate spheroid.

6.9 Air flows through a packed bed of naphthalene spheres with uniform diameters of 1/2 in. Entrance effects are eliminated by an inlet section which is packed with inert spheres of the same size to a height of about 4 in. The superficial velocity of the air, based on the cross section of bed without packing, is 7 ft/sec, and the average air pressure is 1 atm. If the void fraction ϵ' is 0.43 and the system is at 113°F, what depth of active packing would be required for the naphthalene to reach a concentration of 20 percent of saturation in the exit air? Physical properties are given in Illustration 5.1.

6.10 Prove that for a packed bed of uniformly sized spheres, the total surface of spheres per unit packed volume of bed is $6(1-\epsilon')/d_s$. What is the corresponding expression for cubes with side d_s and for cylinders with length and diameter equal to d_s?

6.11 A short, thin-walled tube has a smooth, thin coating of solid potassium chloride on its outer surface. The tube has an external diameter of 22 in. and a length of 1 ft, and is towed longitudinally at 5 ft/sec through water initially 5 percent saturated with KCl. The leading edge is sufficiently rough to ensure a turbulent boundary layer over the entire length of the tube. If the system is at 18.5°C, estimate the average mass-transfer coefficient over the outer surface. What error would be incurred by neglecting the effects of the high mass flux? Physical properties appear in Illustration 3.5.

6.12 A porous, solid cylinder has a diameter of 1 in. and is continuously supplied with liquid carbon tetrachloride to maintain a completely wet surface. The cylinder and CCl_4 are kept at 100°F by a separate source of heat. Pure air at 100°F and 1 atm flows over the cylinder at 5 ft/sec in a direction perpendicular to its axis. Estimate the total rate of evaporation of CCl_4 from a section of the cylinder that is 1 ft in length.

REFERENCES

Acrivos, A., *A. I. Ch. E. J.*, **4**, 285, (1958).

Aksel'rud, G. A., *Zh. fiz. khim.*, **27**, 1445, (1953).

Bar-Ilan, M., and W. Resnick, *Ind. Eng. Chem.*, **49**, 313, (1957).

Barnet, W. I., and K. A. Kobe, *Ind. Eng. Chem.*, **23**, 436, (1941).

Bennett, J. A. R., and J. B. Lewis, A. E. R. E. Rep., E/R. 1998, U.K. Atomic Energy Authority, (1957).

Bennett, C. O., and J. E. Myers, *Momentum, Heat, and Mass Transfer*, McGraw-Hill, New York, (1962).

Bernardo, Everett, and C. S. Eian, N.A.C.A. E-136, (1945).

Boelter, L. M. K., R. C. Martinelli, and F. Jonassen, *Trans. Am. Soc. Mech. Eng.*, **63**, 447–455, (1941).

Bogue, D. C., and A. B. Metzner, *Ind. Eng. Chem. Fundam.*, **2**, No. 2, 143–9, (1963).

Bowman, C. W., D. M. Ward, A. I. Johnson, and O. Trass, *Can. J. Chem. Eng.*, **39**, 9, (1961).

Brötz, W., *Chem. Ing. Tech.*, **26**, 470, (1954).

Brown, R. A. S., K. Sato, and B. H. Sage, *Chem. Eng. Data Ser.*, **3**, 263, (1958).

Chilton, T. H., and A. P. Colburn, *Ind. Eng. Chem.*, **26**, 1183, (1934).

Clapp, R. M., *International Developments in Heat Transfer*, Part III, Am. Soc. Mech. Eng., New York, (1961), 652-61, D-159, D-211–215.

Colburn, A. P., *Trans. A. I. Ch. E.*, **29**, 174–210, (1933).

Comings, E. W., J. T. Clapp, and J. F. Taylor, *Ind. Eng. Chem.*, **40**, 1076, (1948).

Corcoran, W. H., and B. H. Sage, *A. I. Ch. E. J.*, **2**, 251–258, (1956).

Deissler, R. G., N.A.C.A. Res. Memo. E52F05 (1952).

Deissler, R. G., N.A.C.A. Tech. Note 3145, (1954).

Deissler, R. G., N.A.C.A. Rep. 1210, (1955).

Deissler, R. G., and C. S. Eian, N.A.C.A. Tech. Note 2629, (1952).

Dhawan, S., N.A.C.A. Tech. Note 2567, (1952).

Drake, Jr., R. M., *J. Heat Transfer, Trans. Am. Soc. Mech. Eng.*, **83**, 170, (1961).

Dryden, C. E., D. A. Strang, and A. E. Withrow, *Chem. Eng. Prog.*, **49**, 191, (1953).

Eagle, A. E., and R. M. Ferguson, *Proc. Roy. Soc. (Lond.)*, Ser. A., **127**, 540, (1930).

Eckert, E. R. G., and R. M. Drake, Jr., *Heat and Mass Transfer*, McGraw-Hill, New York, (1959).

Eckert, E. R. G., and J. F. Gross, *Introduction to Heat and Mass Transfer*, McGraw-Hill, New York, (1963).

Eckert, E. R. G., and T. W. Jackson, N.A.C.A. Tech. Note 2207, (1950).

Eisenberg, M., C. W. Tobias, and C. R. Wilke, *Chem. Eng. Prog. Symp. Ser.*, **51**, No. 16, 1, (1955).

Evnochides, S., and G. Thodos, *A. I. Ch. E. J.*, **5**, 178, (1959).

Fage, A., and H. C. H. Townend, *Proc. Roy. Soc. (Lond.)*, **A135**, 656, (1932).

Friedlander, S. K., *A. I. Ch. E. J.*, **3**, 43, (1957).

Frössling, N., *Gerlands Beitr. Geophys.*, **52**, 170, (1938).

Frössling, N., *Lunds Univ. Årsskr. N. F.*, **36**, No. 4, (1940).

Gaffney, B. J., and T. B. Drew, *Ind. Eng. Chem.*, **42**, 1120, (1950).

Garner, F. H., and R. W. Grafton, *Proc. Roy. Soc.*, **A224**, 64, (1954).

Garner, F. H., and A. R. Hale, *Chem. Eng. Sci.*, **2**, 157, (1953).

Garner, F. H., and J. M. Hoffman, *A. I. Ch. E. J.*, **6**, 579, (1960).

Garner, F. H., and J. M. Hoffman, *A. I. Ch. E. J.*, **7**, 148, (1961).

Garner, F. H., and R. B. Keey, *Chem. Eng. Sci.*, **9**, 119, 218, (1958).

Garner, F. H., P. Kendrick, and A. H. P. Skelland, *Flow and Diffusion Patterns for Fluid Droplets*, a ciné film made in the Chemical Engineering Department of the University of Birmingham, (1953).

Garner, F. H., C. W. Nutt, and M. F. Mohtadi, *Nature*, **175**, 603, (1955).

Garner, F. H., and A. H. P. Skelland, *Chem. Eng. Sci.*, **4**, 149, (1955).

Garner, F. H., and A. H. P. Skelland, *Ind. Eng. Chem.*, **48**, 51, (1956).

Garner, F. H., and R. D. Suckling, *A. I. Ch. E. J.*, **4**, 114, (1958).

Gilliland, E. R., and T. K. Sherwood, *Ind. Eng. Chem.*, **26**, 516, (1934).

Gowariker, V. R., and F. H. Garner, A.E.R.E. R. 4197, Harwell, November, 1962.

Grele, M. D., and L. Gedeon, N.A.C.A. R.M. E53, 109, (1953).

Griffith, R. M., *Chem. Eng. Sci.*, **12**, 198, (1960).

Gupta, A. S., and G. Thodos, *Chem. Eng. Prog.*, **58**, 58–62, (July, 1962a); **58**, 62, (October, 1962a).

Gupta, A. S., and G. Thodos, *A. I. Ch. E. J.*, **8**, 608–610, (1962b).

Gupta, A. S., and G. Thodos, *A. I. Ch. E. J.*, **9**, 751–754, (1963).

Gupta, A. S., and G. Thodos, *Ind. Eng. Chem. Fund.*, **3**, 218–220, (1964).

Hanna, O. T., *A. I. Ch. E. J.*, **8**, 278–279, (1962).

Hanna, O. T., and O. C. Sandall, *A. I. Ch. E. J.*, **18**, 527–533, (1972).

Hoffman, E., *Forsch. Geb. Ingenieurwes.*, **11**, 159, (1940).

Hoffman, M. W., Oak Ridge Natl. Lab., 1370, Reactor Exper. Eng. Div. (Contract No. W-7405, 26).

Hsu, N. T., K. Sato, and B. H. Sage, *Ind. Eng. Chem.*, **46**, 870–876, (1954).

Hughes, R. R., and E. R. Gilliland, *Chem. Eng. Prog.*, **48**, 497, (1952).

Humphrey, D. W., and H. C. Van Ness, *A. I. Ch. E. J.*, **3**, 283–286, (1957).

Jackson, M. L., and N. H. Ceaglske, *Ind. Eng. Chem.*, **42**, 1188, (1950).

Jenkins, R., Proc. Ht. Trans. and Fl. Mech. Inst., Stanford Univ. Press, Stanford, California, pp. 147–158, (1951).

Johnstone, H. F., and R. L. Pigford, *Trans. Am. Inst. Chem. Eng.*, **38**, 25, (1942).

Kamei, S., and J. Oishi, *Mem. Fac. Eng., Kyoto Univ.*, **17**, 277, (1955).

Kármán, T. von, *Trans. Am. Soc. Mech. Eng.*, **61**, 705, (1939).

Kaufman, S. J., and F. D. Iseley, N.A.C.A.R.M. E-50 G. 31 (1950).

King, C. V., and S. S. Brodie, *South Am. Chem. Soc.*, **59**, 1375, (1937) [reproduced in Sherwood and Ryan, *Chem. Eng. Sci.*, **11**, 81, (1959)].

Knudsen, J. G., and D. L. Katz, *Fluid Dynamics and Heat Transfer*, McGraw-Hill, New York, (1958).

Knuth, E., *Int. J. Heat Mass Transfer*, **6**, 1–22, (1963).
Knuth, E., and Dershin, H., *Int. J. Heat Mass Transfer*, **6**, 999–1018, (1963).
Kramers, H., and P. J. Kreyger, *Chem. Eng. Sci.*, **6**, 42–48, (1956).
Kunii, D., and O. Levenspiel, *Fluidization Engineering*, Wiley, New York, (1969).
Kyte, J. R., A. J. Madden, and E. L. Piret, *Chem. Eng. Prog.*, **49**, 653, (1953).
Leva, M., *Fluidization*, McGraw-Hill, New York, (1959).
Lewis, J. B., *Trans. Inst. Chem. Eng. (Lond.)*, **31**, 323, 325, (1953).
Lewis, J. B., and H. R. C. Pratt, *Nature*, **171**, 1155, (1953).
Lin, C. S., E. B. Denton, H. L. Gaskill, and G. L. Putnam, *Ind. Eng. Chem.*, **43**, 2136, (1951).
Lin, C. S., R. W. Moulton, and G. L. Putnam, *Ind. Eng. Chem.*, **45**, 636–646, 1377, (1953).
Linton, W. H., and T. K. Sherwood, *Chem. Eng. Prog.*, **46**, 258, (1950).
Linton, M., and K. L. Sutherland, *Chem. Eng. Sci.*, **12**, 214, (1960).
Lorisch, W., *Mitt. Forsch.*, **322**, 46, (1929).
Loyzantsky, L. D., and B. A. Schwab, Central Aerodyn. Hydraul. Inst. (U.S.S.R.), Rep. 329, (1935).
Lyon, R. N., *Chem. Eng. Prog.*, **47**, 75, (1951).
Maisel, D. S., and T. K. Sherwood, *Chem. Eng. Prog.*, **46**, 131, 172, (1950).
Malaika, J., Ph.D. thesis, Department of Mechanics and Hydraulics, State University of Iowa, (1959).
Martinelli, R. C., *Trans. Am. Soc. Mech. Eng.*, **69**, 947, (1947).
Mathers, W. G., A. J. Madden, and E. L. Piret, *Ind. Eng. Chem.*, **49**, 961, (1957).
Maxwell, R. W., and J. A. Storrow, *Chem. Eng. Sci.*, **6**, 204, (1957).
McAdams, W. H., *Heat Transmission*, 3rd ed., McGraw-Hill, New York, (1954), p. 224, 249.
McBain, J. W., and T. Woo, *Proc. Roy. Soc. (Lond.)*, **A163**, 182, (1937).
Merk, H. J., and J. A. Prins, *Appl. Sci. Res.*, **A**, 4, 11, (1953–1954).
Metzner, A. B., and W. L. Friend, *Can. J. Chem. Eng.*, **36**, 235–240, (1958); *A. I. Ch. E. J.*, **4**, 393–402, (1958*a*).
Metzner, A. B., and P. S. Friend, *Ind. Eng. Chem.*, **51**, 879–882, (1959).
Mullin, J. W., and T. P. Cook, *J. Appl. Chem.*, **15**, 145, (1965).
Murphree, E. V., *Ind. Eng. Chem.*, **24**, 726, (1932).
Nienow, A. W., R. Unahabhokha, and J. W. Mullin, *Chem. Eng. Sci.*, **24**, 1655–1660, (1969).
Nikuradse, J., *Forschungsheft*, **361**, 1–22, (1933); *Pet. Eng.*, **11**, (6), 164; (8), 75; (9), 124; (11), 38; (12), 83, (1940).
Oliver, D. R., and V. G. Jenson, *Chem. Eng. Sci.*, **19**, 115–129, (1964).
Pasternak, I. S., and W. H. Gauvin, *Can. J. Chem. Eng.*, **38**, 35, (April 1960); *A. I. Ch. E. J.*, **7**, 254, (1961).

Perry, J. H., Ed., *Chemical Engineers' Handbook*, 3rd ed., McGraw-Hill, New York, (1950); 4th ed., (1963).

Powell, R. W., *Trans. Inst. Chem. Eng. (Lond.)*, **18**, 36, (1940).

Powell, R. W., and E. Griffiths, *Trans. Inst. Chem. Eng. (Lond.)*, **13**, 175, (1935).

Prandtl, L., *Z. Physik*, **11**, 1072, (1910).

Prandtl, L., *Z. Physik*, **29**, 487, (1928).

Prandtl, L., *Z. Ver. dtsch. Ing.*, **77**, 105, (1933).

Ranz, W. E., and W. R. Marshall, *Chem. Eng. Prog.*, **48**, 141, 173, (1952).

Reichardt, H., *Z. Angew. Math. Mech.*, **20**, 297, (1940); N.A.C.A. Tech. Memo. 1047, (1943); N.A.C.A. Tech. Memo. 1408, (1957) and N-41947, (1956).

Reichardt, H., *Z. Angew. Math. Mech.*, **31**, 2, 208–219, (1951).

Reynolds, O., *Proc. Manch. Lit. Phil. Soc.*, **8**, (1874); reprinted in *Scientific Papers of Osborne Reynolds*, Vol. II, Cambridge, London, (1901).

Rohsenow, W. M., and H. Y. Choi, *Heat, Mass, and Momentum Transfer*, Prentice-Hall, Englewood Cliffs, N.J., (1961).

Rowe, P. N., K. T. Claxton, and J. B. Lewis, *Trans. Inst. Chem. Eng. (Lond.)*, **43**, 14, (1965).

Sato, K., and B. H. Sage, *Trans. Am. Soc. Mech. Eng.*, **80**, 1380, (1958).

Schlichting, H., *Boundary Layer Theory*, McGraw-Hill, New York, (1955).

Schlichting, H., in *Handbook of Fluid Dynamics*, V. L. Streeter, Ed., McGraw-Hill, New York, p. **9**-13, (1961).

Schultz-Grunow, F., *Luftfahrt-Forsch.*, **17**, 239–246, (1940).

Seban, R. A., and T. T. Shimazaki, *Proc. Gen. Discuss. Heat Transfer*, Inst. Mech. Eng., London, and Am. Soc. Mech. Eng., New York, (1951), p. 122.

Sherwood, T. K., *Trans. A. I. Ch. E.*, **36**, 817–840, (1940).

Sherwood, T. K., *Chem. Eng. Prog. Symp. Ser.*, **55**, No. 25, 71–85, (1959).

Sherwood, T. K., and R. L. Pigford, *Absorption and Extraction*, McGraw-Hill, New York, (1952).

Sherwood, T. K., and J. M. Ryan, *Chem. Eng. Sci.*, **11**, 81, (1959).

Sherwood, T. K., and J. C. Wei, *Ind. Eng. Chem.*, **49**, 1030, (1957).

Sherwood, T. K., and B. B. Woertz, *Ind. Eng. Chem.*, **31**, 1034, (1939).

Sigwart, K., and H. Nassenstein, *Ver. deut. Ingenieruwes. Zeit.*, **98**, 453, (1956).

Skelland, A. H. P., *A. I. Ch. E. J.*, **12**, 69–75, (1966).

Skelland, A. H. P., *Non-Newtonian Flow and Heat Transfer*, Wiley, New York, (1967), Chapter 10.

Skelland, A. H. P., *J. Eng. Phys. (U.S.S.R.)*, **XIX**, No. 3, 385–394, (1970).

Skelland, A. H. P., and A. R. H. Cornish, *A. I. Ch. E. J.*, **9**, 73–76, (1963).

Spalding, D. B., Conf. Int. Dev. Heat Transfer, Am. Soc. Mech. Eng., Boulder, Colo., Part II, 439–446, (1961).

Steele, L. R., and C. J. Geankoplis, *A. I. Ch. E. J.*, **5**, 178, (1959).

Steinberger, R. L., and R. E. Treybal, *A. I. Ch. E. J.*, **6**, 227, (1960).

Sternling, C. V., and L. E. Scriven, *A. I. Ch. E. J.*, **5**, 514–523, (1959).

Taylor, G. I., *Brit. Adv. Comm. Aero. Rep. Mem.*, **272**, 31, 423–429, (1916).

Vivian, J. E., and W. C. Behrmann, *A. I. Ch. E. J.*, **11**, 656–661, (1965).

Ward, A. F. H., and L. H. Brooks, *Trans. Faraday Soc.*, **48**, 1124, (1952).

Wei, J. C., thesis, Massachusetts Institute of Technology, Cambridge, Mass., (1955).

Williams, G. C., Sc.D. thesis, Chemical Engineering, Massachusetts Institute of Technology, (1942).

Wilson, E. J., and C. J. Geankoplis, *Ind. Eng. Chem. Fund.*, **5**, 9–14, (1966).

7

Design of Continuous Columns from Rate Equations

Many types of mass-transfer operations are carried out in either continuous or stagewise columns. This chapter deals with the design of continuous columns, which are usually packed with Raschig rings, Pall rings, Berl saddles, Lessing rings, or other types of packing, to promote intimate contact between the two phases. Continuous contact is therefore maintained between the two countercurrent streams throughout the equipment, necessitating a differential type of treatment.

The preferred method of design involves determination of the number of transfer units (NTU) necessary to achieve the desired separation. Evaluation of the NTU requires preliminary construction of the equilibrium curve and the operating lines appropriate to the process on the x_A-y_A diagram. The chapter is accordingly divided into two main sections:

1. The first half of the chapter is concerned with the location of the necessary operating lines under a variety of circumstances for subsequent evaluation of the NTU.
2. The rest of the chapter is devoted to the formulation and evaluation of the corresponding NTU relationships.

A brief treatment of flooding precedes these more extensive topics. This is necessary because the prediction of the height of a transfer unit (HTU) involves knowledge of flooding conditions.

Capacity and Flooding

The capacity of a given packed column for two-phase contacting is limited by the approach of the system towards flooding conditions. In a gas-liquid system this refers to a state reached by increasing the gas rate at a fixed liquid rate until the gas-phase pressure drop begins to increase drastically with further increase in gas flow. This constitutes the flooding point in the column. Flooding is usually preceded by a so-called loading region in which the pressure gradient increases more gradually but at an increasing rate with gas velocity. Operable conditions require gas rates below the flooding state, and usually somewhat below the loading region or, alterna-

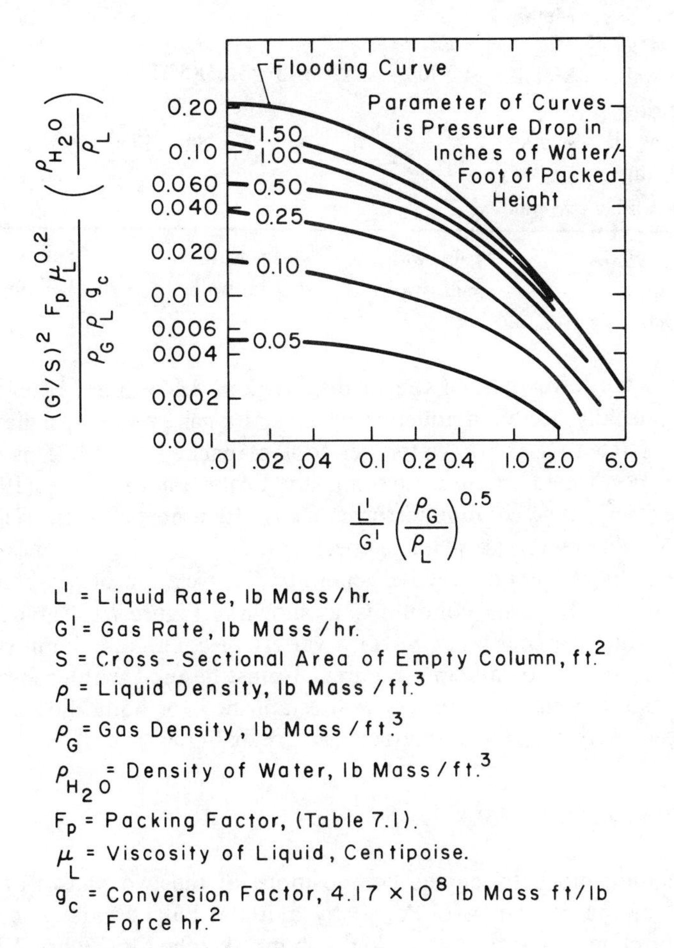

Figure 7.1. Generalized correlation of pressure drop in flow through packed beds (Eckert, 1963).

Table 7.1. Packing factors F_p (wet- and random-packed).[a]

		Nominal packing size, in.									
Type of packing	Material	$\frac{1}{4}$	$\frac{3}{8}$	$\frac{1}{2}$	$\frac{5}{8}$	$\frac{3}{4}$	1	$1\frac{1}{4}$	$1\frac{1}{2}$	2	3
Intalox saddles	Ceramic	600[b]		265		130	98		52	40	
Raschig rings	Ceramic	1000[b,c]	750[b,d]	640[e]	380[e]	255[e]	160[f]	125[g]	95[h]	65[h]	37[b,i]
Berl saddles	Ceramic	900[b]		380		170	110		65	45	
Pall rings	Plastic				97[b]		52		32	25	
Pall rings	Metal				71		48		28	20	
Raschig rings $\frac{1}{32}$-in. wall.	Metal	700[b]		300[b]	258	185[b]	115				
Raschig rings $\frac{1}{16}$-in. wall.	Metal			340	290	230	145	110	83	57	32[b]
Raschig rings $\frac{1}{8}$-in. wall.	Metal										38

[a] Eckert (1963).
[b] Extrapolated
[c] $\frac{1}{32}$-in. wall.
[d] $\frac{1}{16}$-in. wall.
[e] $\frac{3}{32}$-in. wall.
[f] $\frac{1}{8}$-in. wall.
[g] $\frac{3}{16}$-in. wall.
[h] $\frac{1}{4}$-in. wall.
[i] $\frac{3}{8}$-in. wall.

tively, in the lower part of the loading region. According to Leva (1953), loading usually occurs in aqueous systems at a gas pressure gradient in the region of 0.5 to 1 in. of water per foot of packed height. This range is roughly confirmed in other systems by Otake and Kimura (1953) and Eduljee (1960). The column performance is often most efficient (HTUs are lowest) in this region (see Hengstebeck, 1961).

Eckert (1963) has presented a generalized correlation of gas-phase pressure drop and flooding conditions, as shown in Figure 7.1. Table 7.1 gives values of the packing factor F_p for a variety of common column packings. The dimensions shown below Figure 7.1 must be used with each quantity.

Flooding correlations are less well established for liquid-liquid systems, but those available are summarized by Treybal (1963).

THE OPERATING LINE

The relationship between the compositions of the two phases at a given section of the column is obtained by material balance and is called the operating line. Consider the packed column sketched in Figure 7.2.

The reader should note carefully the location of sections 1 and 2, which is directionally the same as that used by some authors (Bird, Stewart, and Lightfoot, 1960, p. 694; Levenspiel, 1962, p. 401; McCabe and Smith, 1956,

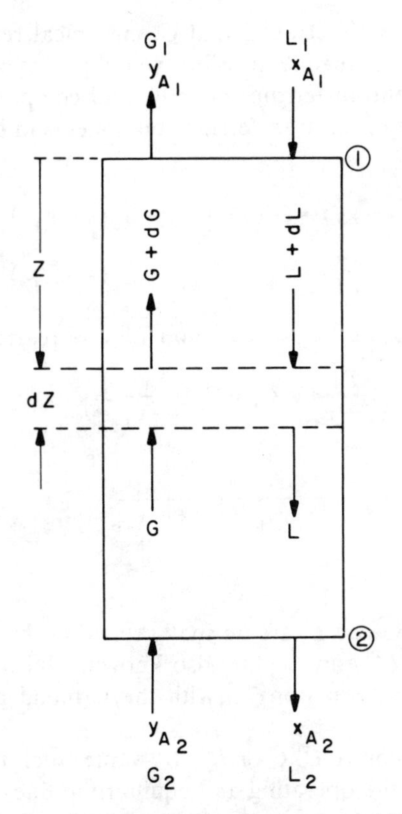

Figure 7.2. Terminology for a continuous column. A, total interfacial area, ft^2; G, L, flow rates of phases G and L, lb-mole/hr; Z, column height; S, cross-sectional area of empty column; x_A, y_A, concentration of component A in phases L and G, mole fraction.

p. 641; Oliver, 1966, p. 266; Larian, 1958, p. 177) and the reverse of that used by others (Bennett and Myers, 1962, p. 517; Foust et al., 1960, p. 271; Sherwood and Pigford, 1952, p. 118; Calderbank, 1967; Treybal, 1968).

A balance on component A over the differential volume $S\,dZ$ gives

$$d(Gy_A) = d(Lx_A) \tag{7.1}$$

Integrating between section 1 and any section within the column leads to

$$y_A = \frac{L}{G}x_A + \frac{G_1 y_{A1} - L_1 x_{A1}}{G} \tag{7.2}$$

This is the equation of the operating line, relating y_A to x_A at any section within the column. In the general case L and G may vary with location, giving a curved operating line. Equation 7.2 is effectively linear, however, in cases of very dilute streams for which composition changes due to mass transfer have a negligible effect on L and G. Graphical representation of the operating line is also readily possible when phases G and L each consist only of inert (nontransferring) material and component A. Thus if G'', L'' are the flow rates of nontransferring components in phases G and L (in lb-mole/hr), then

$$G'' = G(1-y_A) = G_1(1-y_{A1}) = G_2(1-y_{A2}) \tag{7.3}$$

$$L'' = L(1-x_A) = L_1(1-x_{A1}) = L_2(1-x_{A2}) \tag{7.4}$$

Substituting for L, L_1, G, and G_1 in equation 7.2 and rearranging,

$$\frac{y_A}{1-y_A} = C_1 + C_2\frac{x_A}{1-x_A} \tag{7.5}$$

where

$$C_1 = \frac{y_{A1}}{1-y_{A1}} - C_2\frac{x_{A1}}{1-x_{A1}} \tag{7.6}$$

$$C_2 = \frac{L''}{G''} \tag{7.7}$$

and x_{A1}, x_{A2}, y_{A1}, y_{A2}, G_2, and L_1 are normally stated in the conditions of a given problem, so that L'' and G'' are also known. The operating line is then readily plotted from equation 7.5 with the terminal points (x_{A1}, y_{A1}) and (x_{A2}, y_{A2}).

Operable conditions require L/G or L''/G'' values such that no point of contact occurs between the operating and equilibrium lines over the range y_{A1} to y_{A2}. For example, when equations 7.3 to 7.7 are appropriate and the equilibrium curve is below the operating line and concave upward, as in Figure 7.26, the minimum L''/G'' ratio is obtained from equation 7.5 by substituting $y_A = y_{A2}$ and $x_A = x_{AG2}$. The latter is the abscissa of the point on the equilibrium curve having the ordinate y_{A2}; an L'' greater than that obtained by this calculation is necessary for a column of finite height.

It will often be found that the two simplified methods just described for locating the operating line are not applicable. Procedures specific to a variety of situations in binary distillation will now be considered, followed by treatment of comparable cases in gas absorption, stripping, and liquid extraction, in which substantial curvature of the operating lines may be present. Illustration 7.4 involves a strongly curved operating line in a liquid extraction process without reflux.

Binary Distillation with Reflux

In the case of distillation without reflux, Figure 7.2 applies with L_1 as a liquid feed. In this case, however, stream G_1 cannot be richer in the more

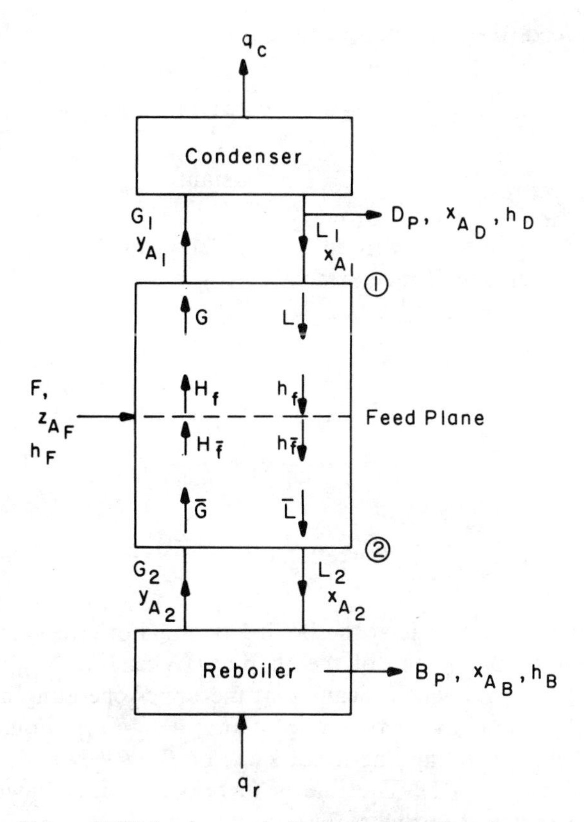

Figure 7.3. Continuous distillation in a packed column with reflux, L_1.

volatile component than the composition that would be in equilibrium with L_1—that is, the maximum value of y_{A1} is the equilibrium value corresponding to x_{A1}. This restriction on y_{A1} is removed by operating with reflux, as shown in Figure 7.3.

In many cases of distillation the heats of vaporization per mole are almost equal for all components, and sensible heat exchanges throughout the column, heats of mixing, and heat losses to the surroundings are all negligible. Under these conditions there is an equimolal exchange of the more and less volatile components between the phases. The condition is said to be one of "constant molal overflow." The operating line is then linear between consecutive inlet and outlet points on the column, because L and G are constants. In Figure 7.3 this means that:

Above the feed (the enriching section),

$$G_1 = G = \text{constant}$$
$$L_1 = L = \text{constant}$$

Below the feed (the stripping section),

$$G_2 = \bar{G} = \text{constant}$$

$$L_2 = \bar{L} = \text{constant}$$

Equation 7.2 may be written in the following form for the column section above the feed (since evidently $y_{A1} = x_{A1} = x_{AD}$ and $G_1 - L_1 = G - L = D_p$):

$$y_A = \frac{L}{G} x_A + \frac{D_p}{G} x_{AD} \tag{7.8}$$

$$y_A = \frac{R}{R+1} x_A + \frac{x_{AD}}{R+1} \tag{7.9}$$

$$R = \text{reflux ratio} = \frac{L_1}{D_p} = \frac{L}{D_p}$$

provided that the reflux is at the boiling point. (For effects of cold reflux, see Problem 7.5 at the end of the chapter.) When $x_A = x_{AD}$, equation 7.9 shows that $y_A = x_{AD}$, which means that the upper operating line intersects the 45° diagonal on a y_A-versus-x_A plot at $y_A = x_A = x_{AD}$. Equation 7.9 also shows that this operating line has a slope of $R/(R+1)$ and intersects the ordinate at $x_{AD}/(R+1)$. The line is therefore readily drawn on such a diagram, as sketched in Figure 7.4.

Equation 7.1 may next be integrated between section 2 and any section below the feed plane in Figure 7.3 to obtain

$$y_A = \frac{\bar{L}}{\bar{G}} x_A - \frac{\bar{L} x_{A2} - \bar{G} y_{A2}}{\bar{G}}$$

An A balance over the reboiler shows that

$$B_p x_{AB} = \bar{L} x_{A2} - \bar{G} y_{A2}$$

so

$$y_A = \frac{\bar{L}}{\bar{G}} x_A - \frac{B_p}{\bar{G}} x_{AB} \tag{7.10}$$

When $x_A = x_{AB}$, equation 7.10 shows that $y_A = x_{AB}$, which means that the lower operating line intersects the 45° diagonal at $y_A = x_A = x_{AB}$.

The locus of intersections of the operating lines above and below the feed, which are defined by equations 7.9 and 7.10, is found by subtracting 7.10 from 7.8:

$$y_A(G-\bar{G})=x_A(L-\bar{L})+D_p x_{AD}+B_p x_{AB} \tag{7.11}$$

An overall balance on A gives

$$Fz_{AF}=D_p x_{AD}+B_p x_{AB} \tag{7.12}$$

and from an overall balance around the feed plane

$$G-\bar{G}=F-(\bar{L}-L) \tag{7.13}$$

Combining equations 7.11 to 7.13,

$$y_A=\left(\frac{q}{q-1}\right)x_A-\frac{z_{AF}}{q-1} \tag{7.14}$$

where

$$q=\frac{\bar{L}-L}{F} \tag{7.15}$$

An enthalpy balance on a narrow section of the packed column which includes the feed plane is

$$Fh_F+Lh_f+\bar{G}H_{\bar{f}}=GH_f+\bar{L}h_{\bar{f}}$$

Since composition and temperature changes in the saturated liquid and saturated vapor streams are small in this narrow section,

$$h_f \doteq h_{\bar{f}}; \quad H_{\bar{f}} \doteq H_f$$

and

$$(\bar{L}-L)h_f=Fh_F+(\bar{G}-G)H_f \tag{7.16}$$

Combining equations 7.13 and 7.16,

$$\frac{\bar{L}-L}{F}=q=\frac{H_f-h_F}{H_f-h_f} \tag{7.17}$$

The quantity q is therefore obtained by dividing the latent heat of vaporization (H_f-h_f) into the enthalpy change involved in converting the feed into saturated vapor (H_f-h_F).

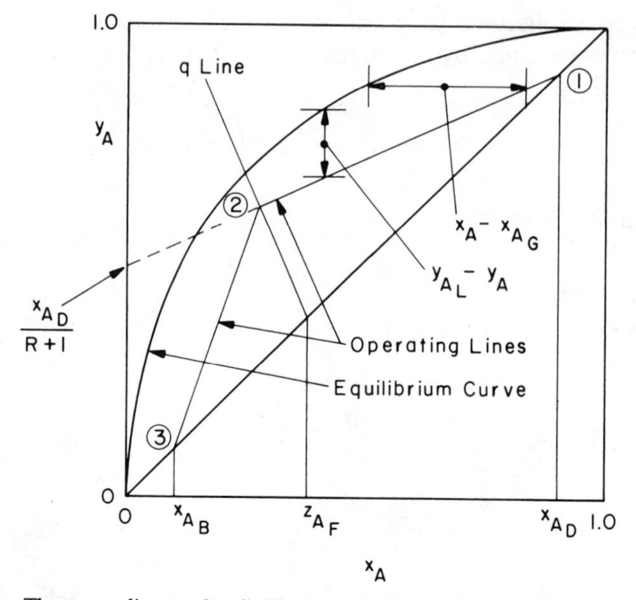

Figure 7.4. The x_A-y_A diagram for distillation with reflux, corresponding to Figure 7.3. The quantities x_A-x_{A_G} and y_{A_L}-y_A are used later in NTU evaluation.

It is now apparent that the graph of equation 7.14, which is the locus of intersections of the upper and lower operating lines, has a slope of $q/(q-1)$ and intersects the 45° diagonal at $y_A = x_A = z_{AF}$. It is therefore easily drawn on the x_A-y_A diagram, and is known as the q line, as shown in Figure 7.4.

In summary, the upper operating line intersects the 45° diagonal at $y_A = x_A = x_{AD}$, has a slope of $R/(R+1)$, and intersects the ordinate at $x_{AD}/(R+1)$. The lower operating line intersects the 45° diagonal at $y_A = x_A = x_{AB}$. It is inserted on the x_A-y_A diagram by joining the point $(y_A = x_{AB}, x_{AB})$ to the intersection of the upper operating line with the q line.

Minimum Reflux Ratio

When the operating lines on the x_A-y_A diagram either intersect with or become tangent to the equilibrium curve, the point at which this occurs corresponds to a state of equilibrium between the two adjacent phases in the column. The driving force causing mass transfer $(y_A - y_{AL})$ is zero at such a point, necessitating an infinitely tall column. Borrowing from equation 7.43, developed later in this chapter, it is evident that the $(\mathrm{NTU})_{OG}$ needed to reach this condition is infinite.

Suppose the reflux ratio is progressively decreased, resulting in a continual decrease in the slope of the upper operating line. When contact between this operating line and the equilibrium curve first occurs, the operating line will have a slope of $R_m/(R_m+1)$, where R_m is the minimum reflux ratio.

The more common equilibrium curves are concave downward for all x_A, and in such cases the point of contact referred to above occurs at the intersection of the q line and the equilibrium curve. The corresponding R_m is then obtained from the slope of the operating line through this point. In other cases the equilibrium curve may be concave upward in some region between x_{AD} and z_{AF}. The value of R_m may then be found from the slope of that operating line which is tangential to the equilibrium curve in the concave-upward region.

Total Reflux

A distillation column may be run under the limiting condition of total reflux during evaluation of a new type of packing. In such a case the flow rates of F, D_p, and B_p are all zero, and $q_r = q_c$ if heat losses are negligible. The reflux ratio ($R = L_1/D_p$) is infinite, so that the slope of the upper and lower operating lines is unity. The operating lines therefore coincide with the 45° diagonal on the x_A-y_A diagram for the binary system.

Side Streams and Multiple Feeds

Side streams are occasionally withdrawn from columns distilling binary mixtures, as shown in Figure 7.5. A material balance on the entire column above a horizontal plane between F and L_s leads to

$$y_A = \frac{\bar{L}}{\bar{G}} x_A + \frac{L_s x_{As} + D_p x_{AD}}{\bar{G}} \tag{7.18}$$

This is the operating line between the feed and the side stream; it intersects the 45° diagonal at

$$y_A = x_A = \frac{L_s x_{As} + D_p x_{AD}}{L_s + D_p} \tag{7.19}$$

The operating line for the column section above the side stream was given earlier as equation 7.8. Evidently $\bar{L} = L - L_s$ and $\bar{G} = G$, and equating the relationships in equations 7.8 and 7.18 shows that the two operating lines intersect on the vertical at $x_A = x_{As}$. The resulting construction of operating lines is shown on the x_A-y_A diagram of Figure 7.5.

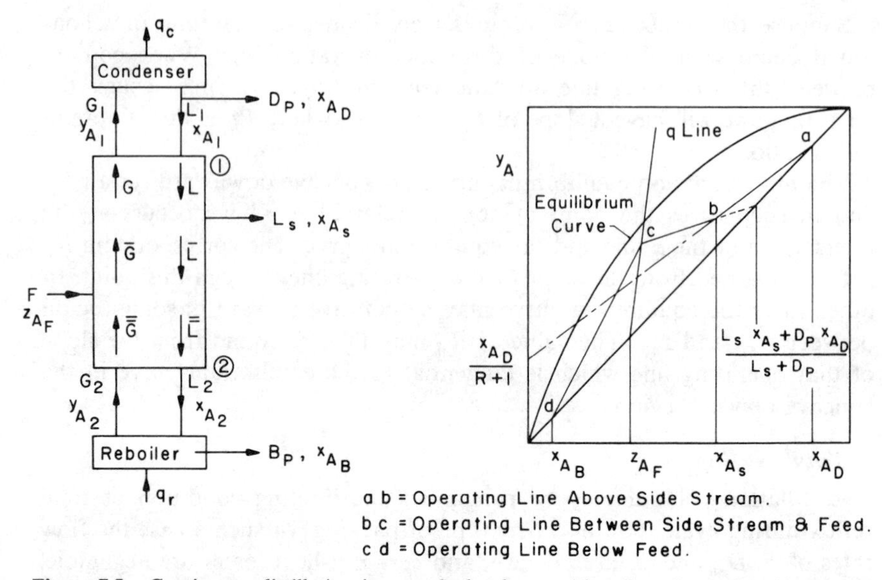

Figure 7.5. Continuous distillation in a packed column with reflux, L_1, and a side-stream product, L_s. Modifications to describe multiple feeds ($-L_s$) are detailed with equations 7.20 and 7.21. The x_A-y_A diagram is used to evaluate the NTU.

Instead of withdrawing a side-stream product, a second feed, of different composition from the first, may be added to the column by reversing the direction of L_s in Figure 7.5. The equation to the operating line between the two feeds is then obtained from a material balance as

$$y_A = \frac{\bar{L}}{\bar{G}} x_A + \frac{D_p x_{AD} - L_s x_{As}}{\bar{G}} \tag{7.20}$$

and this line intersects the 45° diagonal at

$$y_A = x_A = \frac{D_p x_{AD} - L_s x_{As}}{D_p - L_s} \tag{7.21}$$

It is clear that, for a boiling feed L_s, $\bar{L} = L + L_s$ and $\bar{G} = G$. Equating the relationships 7.8 and 7.20 shows that the upper and middle operating lines intersect on the vertical at $x_A = x_{As}$. The incorporation of these modifications will enable ready revision of the x_A-y_A diagram in Figure 7.5 to represent distillation with two feeds.

Open Steam

When distilling an aqueous solution in which water is the less volatile component, the heat required may be provided by direct injection of open steam into the bottom of the column. The reboiler is thereby eliminated, and the unit is as sketched in Figure 7.6. Design procedures are illustrated in the following worked example.

Illustration 7.1

Thirty lb-moles per hour of a mixture of isopropanol and water are to be concentrated in a packed distillation column to give a distillate containing 0.65 mole fraction isopropanol and a residue product containing 0.03 mole fraction isopropanol. The feed contains 0.25 mole fraction isopropanol and is at its bubble point.

The column is to be equipped with a total condenser and operated at atmospheric pressure. Heat will be supplied to the column by introducing

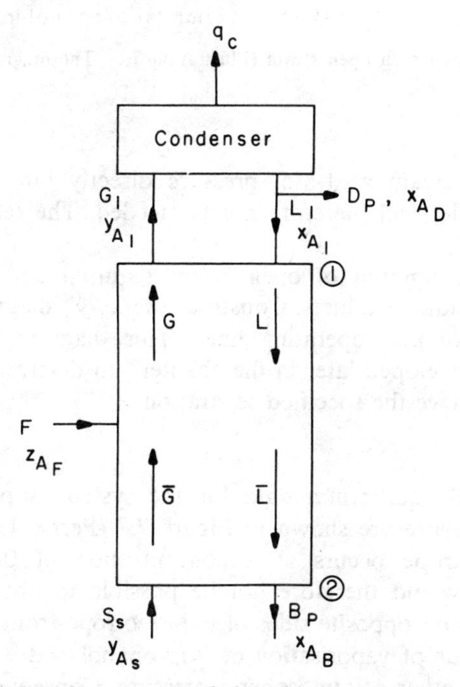

Figure 7.6. Distillation of an aqueous solution with open steam, S_s (Illustration 7.1).

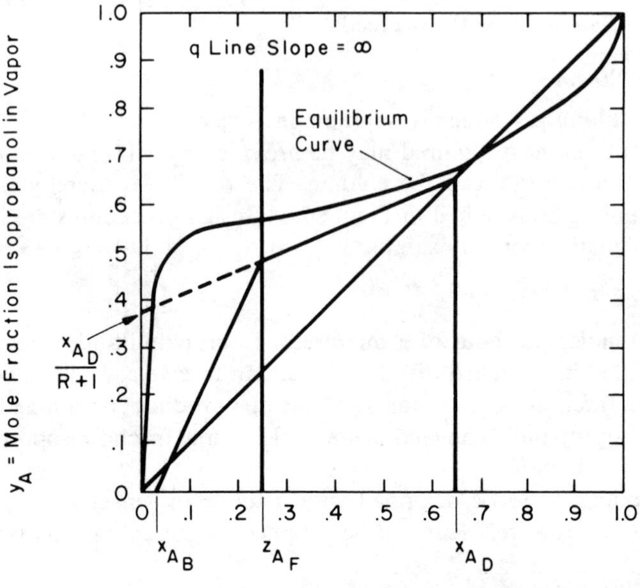

Figure 7.7. Distillation with open steam (Illustration 7.1). The diagram is used for NTU evaluation.

open, saturated steam at 1-atm pressure directly into the base of the column. A reboiler will therefore not be needed. The reflux ratio will be 0.75.

Determine the amount of open steam required and the amounts of distillate and residue products. Construct the x_A-y_A diagram, showing the equilibrium curve and operating lines. This diagram is for use—with equation 7.43, developed later in the chapter—in determining the NTU_{OG} necessary to achieve the specified separation.

Principles.

The vapor-liquid equilibrium data for the system isopropanol-water at atmospheric pressure are shown in Figure 7.7 (Perry, 1950, p. 574). Evidently an azeotrope occurs at a concentration of 0.7 mole fraction isopropanol. It would therefore not be possible to obtain distillate and residue products on opposite sides of the azeotrope from a given feed.

The molal heats of vaporisation of isopropanol and water are within 2 percent of each other at atmospheric pressure. Consequently all assumptions required for the use of straight operating lines are approximately satisfied, if the operation is adiabatic.

A material balance on the more volatile component in Figure 7.6 gives

$$Fz_{AF} + S_s y_{As} = B_p x_{AB} + D_p x_{AD}$$

Equations 7.13 and 7.15 show that

$$S_s = \bar{G} = G + (q-1)F = (L_1 + D_p) + (q-1)F = D_p(R+1) + (q-1)F$$

$$B_p = \bar{L} = L + qF = RD_p + qF$$

Substituting for S_s and B_p in the above material balance leads to

$$F[z_{AF} + (q-1)y_{As} - qx_{AB}] = D_p(Rx_{AB} + x_{AD} - y_{As} - Ry_{As})$$

This enables evaluation of D_p and hence B_p and S_s.

A material balance for the more volatile component over the lower section of the column below the feed gives

$$\bar{L}x_A + S_s y_{As} = \bar{G}y_A + B_p x_{AB}$$

and since $\bar{G} = S_s$ and $B_p = \bar{L}$,

$$\frac{\bar{L}}{\bar{G}} = \frac{B_p}{S_s} = \frac{y_A - y_{As}}{x_A - x_{AB}}$$

which shows that the lower operating line extends to the point (x_{AB}, y_{As}).

SOLUTION Substituting in the equation developed above,

$$30[0.25 + (1-1)(0) - (1)(0.03)] = D_p[0.75(0.03) + 0.65 - 0 - (0.75)(0)]$$

whence

$$D_p = 9.81 \text{ lb-mole/hr}$$

$$B_p = (0.75)(9.81) + (1)(30) = 37.35 \text{ lb-mole/hr}$$

$$S_s = 9.81(0.75 + 1) + (1-1)(30) = 17.16 \text{ lb-mole/hr}$$

The intersection of the operating line for the column section above the feed with the y_A axis is

$$\frac{x_{AD}}{R+1}=\frac{0.65}{0.75+1}=0.372$$

The upper operating line is drawn in Figure 7.7 by joining this intersection to the point (0.65, 0.65). Since the feed enters as a saturated liquid, q is unity, and the slope of the q line is infinite. The operating line for the section below the feed is obtained by joining the intersection of the upper operating line and the q line to the point $(x_{AB}, y_{As})=(0.03, 0.00)$.

The use of this diagram in evaluating the required NTU_{OG} is detailed later in the chapter.

The Leaking Condenser

As an aid towards the development of facility in the construction of operating lines it is often instructive to consider somewhat unorthodox but realistic situations. A case in point is provided by the following illustration, in which the interpretation of data from an aqueous distillation is complicated by leakage of condenser cooling water into the distillate product.

Illustration 7.2

A mixture of water and cellosolve is fed at 30 lb-mole/hr to an old, continuous distillation column, operating at atmospheric pressure. The feed is liquid at its bubble point and contains 0.3 mole fraction water and 0.7 mole fraction cellosolve (glycol monoethyl ether, $CH_2OHCH_2OC_2H_5$).

A distillate containing 0.85 mole fraction water and a residue product containing 0.05 mole fraction water are being obtained with a reflux ratio of 0.75. Tests have indicated that the condenser cooling water is leaking into the distillate product, to the extent of 0.1 mole of cooling water per mole of feed.

Establish the equilibrium curve and the operating lines on the x_A-y_A diagram under these conditions of faulty operation.

Principles

An overall material balance on the column shown in Figure 7.8 gives

$$F+L_c=D_p+B_p$$

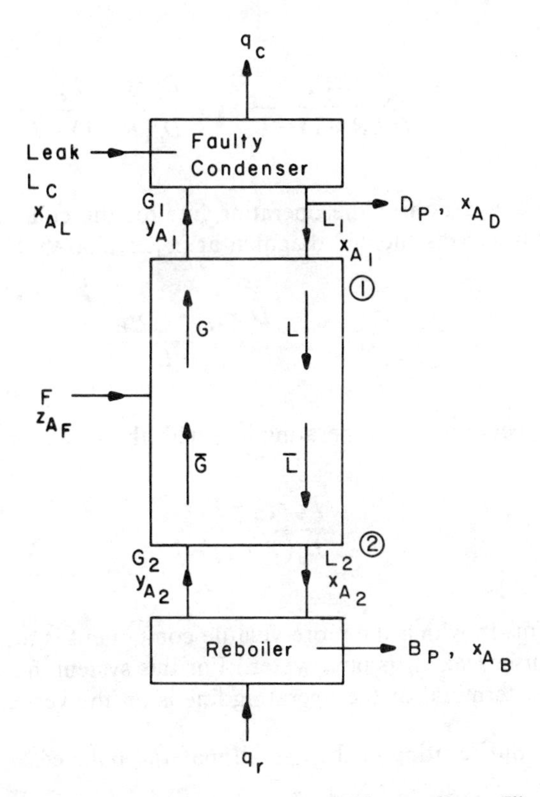

Figure 7.8. Distillation of an aqueous solution with a leaking condenser (Illustration 7.2).

and on the more volatile component (in this case water),

$$Fz_{AF}+L_c x_{AL}=D_p x_{AD}+B_p x_{AB}$$

A material balance on the upper section of the column above the feed gives

$$G=L+D_p-L_c=D_p(R+1)-L_c$$

and on the more volatile component,

$$Gy_A=Lx_A+D_p x_{AD}-L_c x_{AL}$$

Dividing throughout by G—expressed in the above form—and substituting

RD_p for L,

$$y_A = \frac{RD_p}{D_p(R+1) - L_c} x_A + \frac{D_p x_{AD} - L_c x_{AL}}{D_p(R+1) - L_c}$$

This is the equation of the operating line for the column section above the feed. It intersects the 45° diagonal at (x_A, x_A), so that

$$y_A = x_A = \frac{D_p x_{AD} - L_c x_{AL}}{D_p - L_c}$$

The intersection of this operating line with the y_A axis has the value

$$\frac{D_p x_{AD} - L_c x_{AL}}{D_p(R+1) - L_c}$$

For systems in which the more volatile component is not water, $x_{AL} = 0$. (The condenser leak L_c is pure water.) For this system, however, $x_{AL} = 1.0$. The upper terminal of the operating line is on the vertical at x_{AD}.

SOLUTION. Substituting in the overall material balance,

$$30 + 30(0.1) = D_p + B_p, \qquad B_p = 33 - D_p$$

and on the more volatile component (water),

$$30(0.3) + 3(1) = D_p(0.85) + (33 - D_p)(0.05)$$

$$D_p = 12.95 \text{ lb-mole/hr}$$

$$B_p = 20.05 \text{ lb-mole/hr}$$

The vapor-liquid equilibrium relationship for the system cellosolve-water at atmospheric pressure is given in Figure 7.9 (Chu et al., 1950). The intersection of the operating line for the column section above the feed

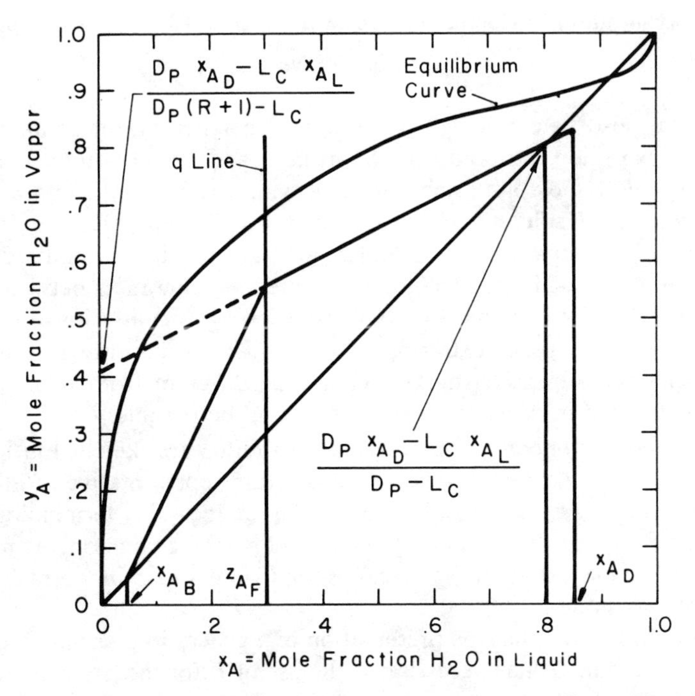

Figure 7.9. Aqueous distillation with a leaking condenser, corresponding to Figure 7.8. The diagram may be used to evaluate the NTU (Illustration 7.2).

with the 45° diagonal is at

$$y_A = x_A = \frac{D_p x_{AD} - L_c x_{AL}}{D_p - L_c} = \frac{12.95(0.85) - 3(1)}{12.95 - 3} = 0.805$$

This line has an intercept on the y_A axis of

$$\frac{D_p x_{AD} - L_c x_{AL}}{D_p(R+1) - L_c} = \frac{12.95(0.85) - 3(1)}{12.95(0.75+1) - 3} = 0.407$$

This enables the upper operating line to be drawn on Figure 7.9. Since the feed is a saturated liquid, the q line is vertical. The intersection of the upper operating line with the q line is joined to the point (0.05, 0.05) to obtain the operating line for the column section below the feed. The diagram is now ready for use in NTU evaluation, as described later.

Gas Absorption, Stripping or Desorption, and Liquid Extraction with Reflux

Reflux can also be employed in continuous countercurrent columns used for gas absorption, stripping or desorption, and liquid-liquid extraction. This will allow the solute-rich stream leaving the unit to attain a higher concentration of solute than would be in equilibrium with the feed. In considering the separation of a binary mixture it must be realized, however, that reflux will not enhance the degree of separation between components of the feed unless they both transfer significantly into the phase which absorbs, strips, or extracts the solute. (The proportions in which the two components transfer must of course be different from their proportions in the feed in order for the separation to be feasible.)

The individual operations performed with reflux are sketched in Figures 7.10, 7.11, and 7.12, and a generalized diagram representative of all three operations appears in Figure 7.13, which defines the terminology for algebraic purposes. In the most general case all three components may be present to some extent in each exit stream, but one will normally predominate, as shown.

Figure 7.13 shows that the preservation of a generalized terminology (the customary L and G) has necessitated the use of L for the gas phase and G for the liquid phase in the case of stripping. These symbols then automatically adopt their more conventional connotation for gas absorption. The direction of solute transfer is from G to L throughout, and the solute concentrations in these two phases are denoted by y_A and x_A, respectively. It may be noted that when treating gas absorption, some authors associate y_A with the phase losing the solute, whereas when dealing with liquid extraction, they associate y_A with the phase gaining the solute. One consequence of this on the usual x_A-y_A diagram is that the operating line lies above the equilibrium curve for gas absorption, but below the equilibrium curve for liquid extraction. In addition, the rate equations and corresponding NTU expressions for gas absorption require reversal of driving forces before they are valid for liquid extraction, when this inconsistent procedure is followed. Such anomalies are excluded by the present generalized nomenclature, in which y_A is consistently associated with G, the phase losing the solute. Reversal of the phases to which x_A and y_A refer would cause the operating line to lie below the equilibrium curve, as in distillation. The present system will be retained, however, since it may emphasize the differences between equimolal counterdiffusion and unimolal unidirectional diffusion and because it conforms to the conventional nomenclature for the major operation of gas absorption.

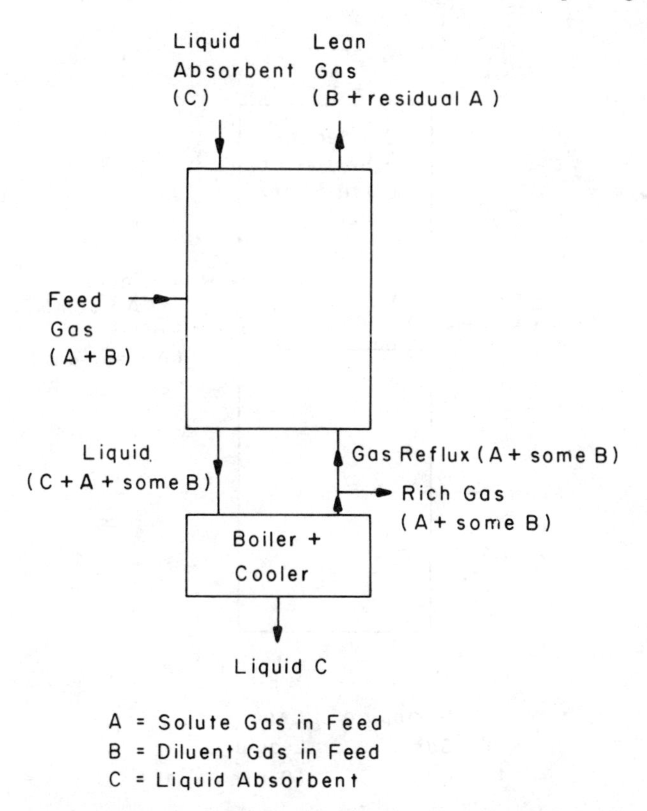

Figure 7.10. Gas absorption with reflux. In some cases, part of the A-rich liquid leaving the base of the column is removed as product and all of the gas leaving the boiler is refluxed to the column.

The usual appearance of phase equilibria on triangular coordinates may be sketched for each of the three operations under consideration. The use of these diagrams was outlined in Illustration 4.2; either mole or mass fractions are applicable, provided stream quantities are in corresponding units.

In gas absorption the triangular diagram is somewhat as shown in Figure 7.14 for the prevailing conditions of temperature and pressure.

When B is almost insoluble and C has negligible volatility, side AB becomes the locus of all possible gas-phase compositions and side AC the locus of all possible liquid-phase compositions. Side BC is the locus of nonexistent solutions of B in C, since only immiscible phases of pure B and pure C can occur in the absence of A. Even for this limiting system, however, the A apex still resembles that in the above diagram when

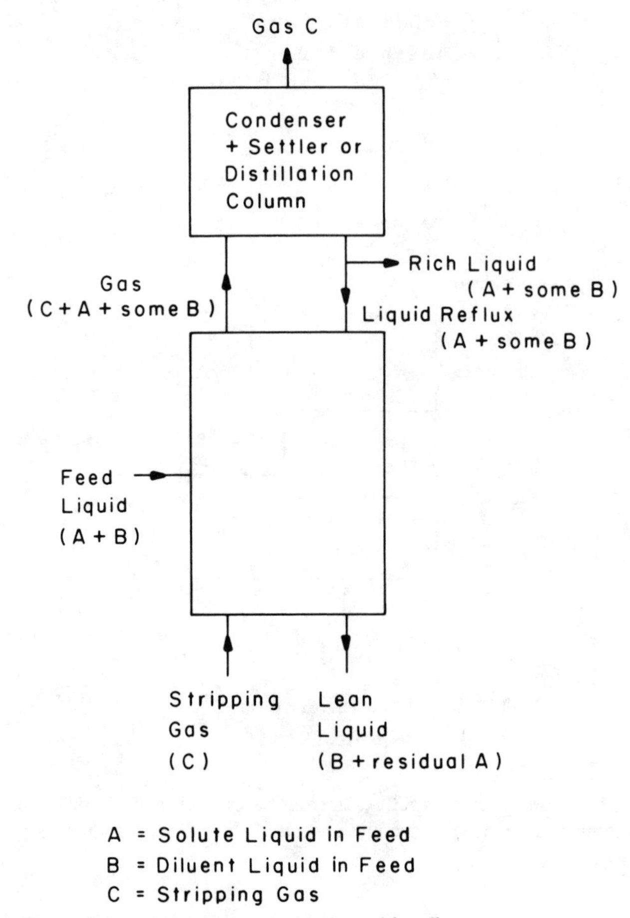

Figure 7.11. Stripping or desorption with reflux.

magnified sufficiently so that pure A is still correctly represented as a gas.

For stripping or desorption at a given temperature and pressure, the diagram is as sketched in Figure 7.15.

When B has negligible volatility and C is almost insoluble, side AB becomes the locus of all possible liquid-phase compositions, and side AC the locus of all possible gas-phase compositions.

Liquid extraction exhibits at least two common forms on the triangular diagram. Figure 7.16 shows both type-I and type-II systems, the former characterized by a plait point, as encountered in Illustration 4.2.

In type-I systems the tie lines linking equilibrium phases shrink to a

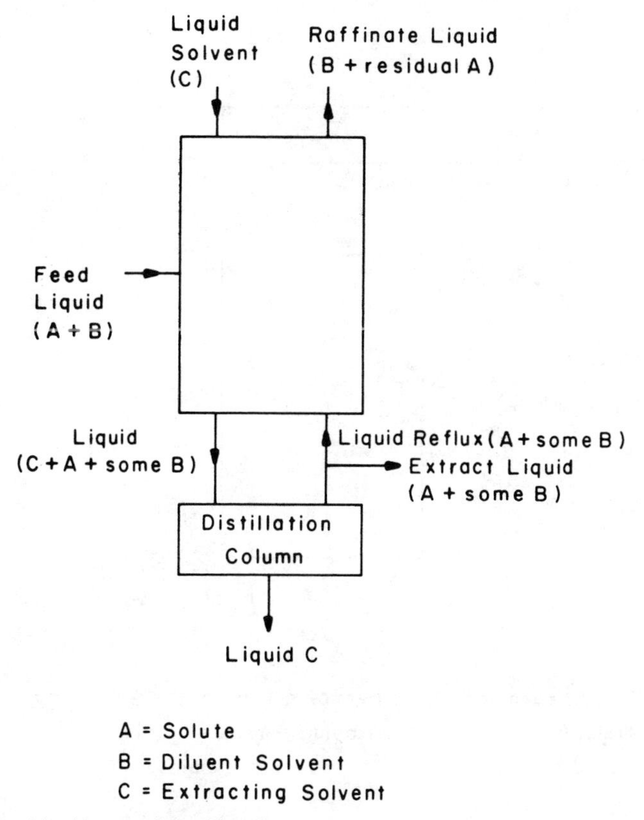

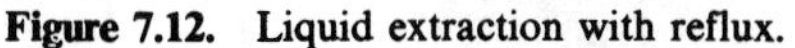
Figure 7.12. Liquid extraction with reflux.

point at the plait point, so that the two conjugate phases become identical.

When reflux is used in a type-II system the feed is theoretically separable into pure A and B, respectively, after solvent removal. This contrasts with type-I systems, for which, even with the use of reflux, a feed mixture is theoretically separable only into pure B at one end of the unit and a *mixture* of A and B at the other—after removal of solvent. Thus the most concentrated extract obtainable is E^x.

Any point in the two-phase region of Figures 7.14, 7.15, and 7.16 represents an overall mixture which separates into two phases linked, at equilibrium, by a tie line.

The Line-Ratio Principle on Triangular Diagrams

Consider the steady flow of two ternary streams into the mixer shown in Figure 7.17. The streams enter at rates of M and N mole/hr, respectively,

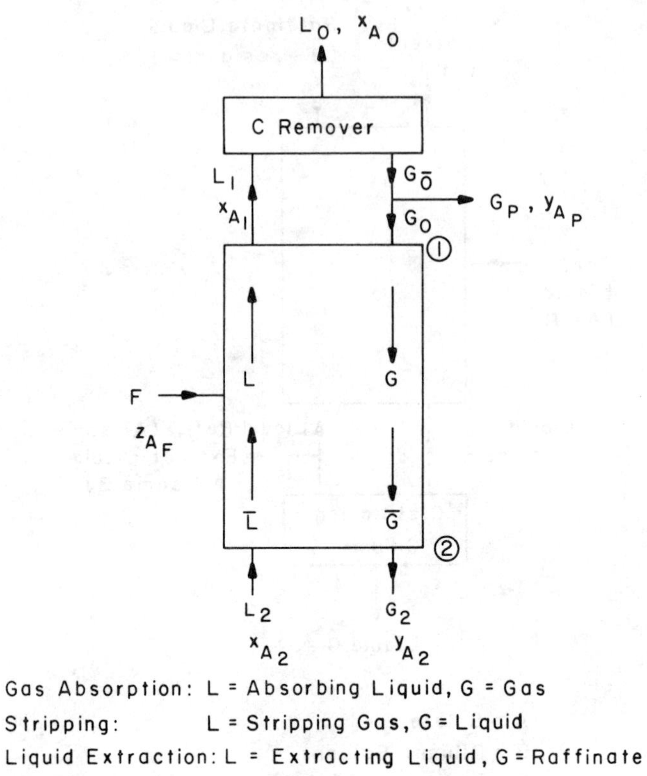

Figure 7.13. Generalized diagram for gas absorption, stripping, and liquid extraction with reflux.

and a third stream leaves at P mole/hr. In representing this process on the triangular diagram it is shown below that, when streams M and N are added, the point denoting the resultant stream P lies *on* the line MN such that

$$\frac{[\text{Moles (or mass) of } N]/\text{hr}}{[\text{Moles (or mass) of } M]/\text{hr}} = \frac{N}{M} = \frac{\text{length of line } MP}{\text{length of line } PN}$$

Let

x_{BM}, x_{BN}, x_{BP} = mole (or mass) fractions of B in streams M, N, and P.

x_{CM}, x_{CN}, x_{CP} = mole (or mass) fractions of C in streams M, N, and P.

M, N, P = moles (or masses) in streams M, N, and P per hour.

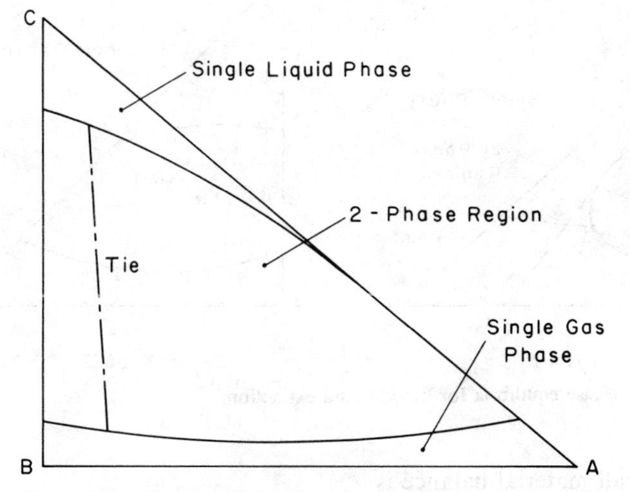

Figure 7.14. Phase equilibria for gas absorption.

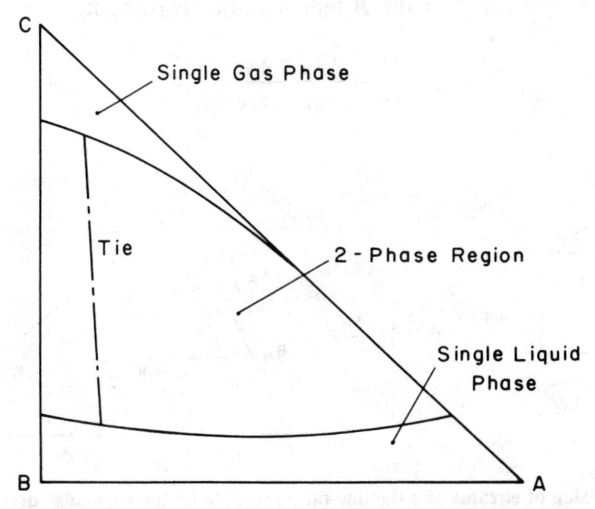

Figure 7.15. Phase equilibria for stripping or desorption.

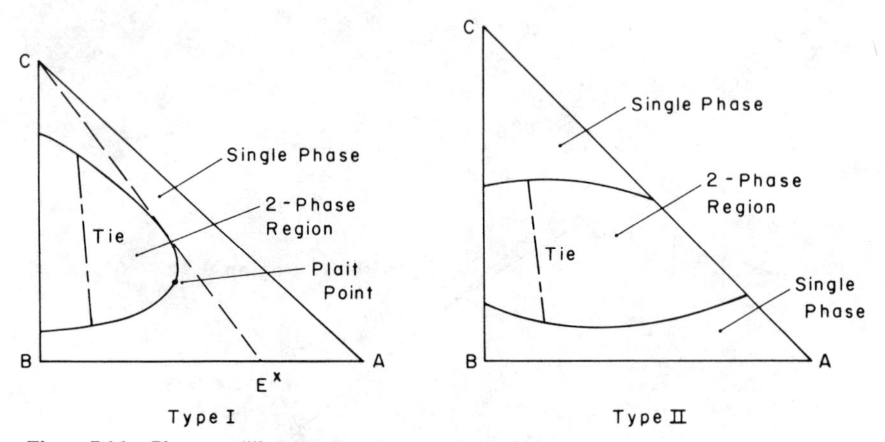

Figure 7.16. Phase equilibria for liquid-liquid extraction.

An overall material balance is

$$P = M + N$$

and for component B,

$$Px_{BP} = Mx_{BM} + Nx_{BN}$$

and for C,

$$Px_{CP} = Mx_{CM} + Nx_{CN}$$

Substituting $M + N$ for P in the B balance and rearranging,

$$\frac{N}{M} = \frac{x_{BM} - x_{BP}}{x_{BP} - x_{BN}}$$

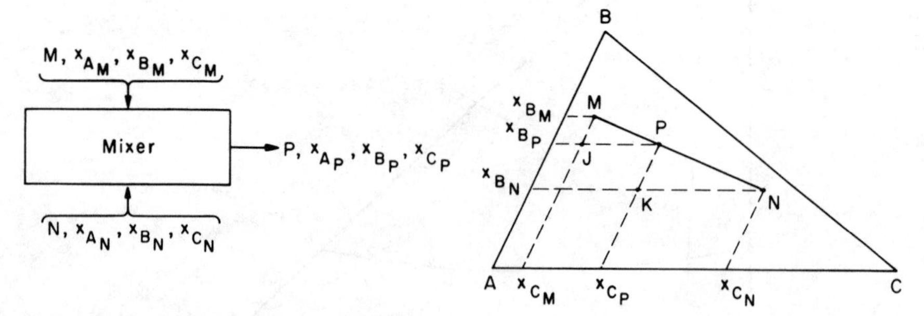

Figure 7.17. Mixing of streams and the line-ratio principle on the triangular diagram.

Similarly, from the C balance,

$$\frac{N}{M} = \frac{x_{CP} - x_{CM}}{x_{CN} - x_{CP}}$$

so that

$$\frac{x_{BM} - x_{BP}}{x_{BP} - x_{BN}} = \frac{x_{CP} - x_{CM}}{x_{CN} - x_{CP}}$$

This relationship can be satisfied only if the triangles MJP and PKN are similar (Figure 7.17), which in turn can be true only if P lies *on* the line MN. Furthermore, in the similar triangles MJP and PKN,

$$\frac{x_{BM} - x_{BP}}{x_{BP} - x_{BN}} = \frac{x_{CP} - x_{CM}}{x_{CN} - x_{CP}} = \frac{\text{line } MP}{\text{line } PN} = \frac{N}{M}$$

Since the triangle used in Figure 7.17 is neither equilateral nor isosceles, it follows that the graphical addition of streams and the line-ratio principle are independent of the shape of the triangle.

Material Balances on a Continuous Column Using Reflux

Consider the column shown in the generalized diagram of Figure 7.13. The C remover withdraws sufficient C from phase L_1 to convert it to a G phase. A material balance around the C remover leads to

$$L_1 - G_0 = L_0 + G_p = \Delta_1 \tag{7.22}$$

A similar balance over the C remover plus any portion of the column above the feed plane gives

$$L_1 - G_0 = L - G = \Delta_1 \tag{7.23}$$

Converting equation 7.23 into component B and C balances,

$$L_1(x_B)_{L_1} - G_0(y_B)_{G_0} = L(x_B) - G(y_B) = \Delta_1(x_B)_{\Delta_1} \tag{7.24}$$

$$L_1(x_C)_{L_1} - G_0(y_C)_{G_0} = L(x_C) - G(y_C) = \Delta_1(x_C)_{\Delta_1} \tag{7.25}$$

Equations 7.23, 7.24, and 7.25 demonstrate that the difference between adjacent streams in the column section above the feed plane is constant in amount and composition. This difference may therefore be represented by the single point Δ_1 on the triangular diagram. The "difference point" Δ_1 is located in Figure 7.18 from the following development in terms of the reflux ratio G_0/G_p.

Streams $G_{\bar{0}}$, G_p, and G_0 of Figure 7.13 are identical in composition and are represented by a common point in Figure 7.18. The compositions of L_0 and G_p will be specified and equation 7.22 shows that the points representing streams L_0, G_p, G_0, L_1, and Δ_1 all lie on the same straight line because of the graphical addition of streams and line-ratio principle. Then from equation 7.22,

$$\frac{G_p}{\Delta_1} = \frac{\text{line } L_0\Delta_1}{\text{line } L_0G_p}$$

$$\frac{G_0}{\Delta_1} = \frac{\text{line } L_1\Delta_1}{\text{line } L_1G_0}$$

The reflux ratio is known, and since

$$\frac{G_0}{G_p} = \frac{G_0}{\Delta_1}\cdot\frac{\Delta_1}{G_p} = \frac{\text{line } L_1\Delta_1}{\text{line } L_1G_0}\,\frac{\text{line } L_0G_p}{\text{line } L_0\Delta_1} \tag{7.26}$$

this enables location of Δ_1, because the line lengths L_1G_0, L_0G_p, and L_0L_1 are all readily measurable.

A material balance over the lower portion of the column below the feed plane in Figure 7.13 shows that

$$\bar{G} - \bar{L} = G_2 - L_2 = \Delta_2 \tag{7.27}$$

This equation may be converted into component-B and -C balances analogous to equations 7.24 and 7.25, to demonstrate that the difference between adjacent streams in the column section beneath the feed is constant in amount and composition. This quantity can therefore be represented by a second single difference point Δ_2 on the triangular diagram. Location of Δ_2 requires the following material balance about the feed plane:

$$F + \bar{L} + G = L + \bar{G}$$

$$\begin{aligned} F &= L - G + \bar{G} - \bar{L} \\ &= \Delta_1 + \Delta_2 \end{aligned} \tag{7.28}$$

The problem specifications will enable points F, G_2, and L_2 to be located on Figure 7.18. Equations 7.27 and 7.28 show that Δ_2 lies at the intersection of the extended lines L_2G_2 and Δ_1F, in accordance with the principle of the graphical addition of streams. It is clear that whether Δ_2 lies above or below the triangle merely depends upon the relative positions of points L_2, G_2, F, and Δ_1. Thus when Δ_2 lies above the diagram, equation 7.28 (together with equation 7.27) shows that $\Delta_1 - F = -\Delta_2$. If L_0 or L_2 contains no A or B, the corresponding point coincides with apex C.

The operating line on the distribution diagram (y_A versus x_A) may now be obtained from the triangular diagram of Figure 7.18 in the following manner. Corresponding x_A and y_A values for the two phases at a given column cross section are obtained from the intersection of random lines from the appropriate difference point (Δ_1 or Δ_2) with the upper and lower boundaries, respectively, of the two-phase region. The procedure is shown in detail in Illustration 7.4 for a column operating without reflux, in which case only one difference point is needed. Where reflux is used, as in Figures 7.13, 7.18, and 7.19, two difference points are required, and a discontinuity will occur between the two operating curves representing the two sections of the column. The discontinuity appears at the point on the distribution diagram corresponding to the intersection of the line $F\Delta_1\Delta_2$ with the upper and lower boundaries of the two-phase region on the triangular diagram. The procedure is sketched in Figure 7.19, which is not to scale.

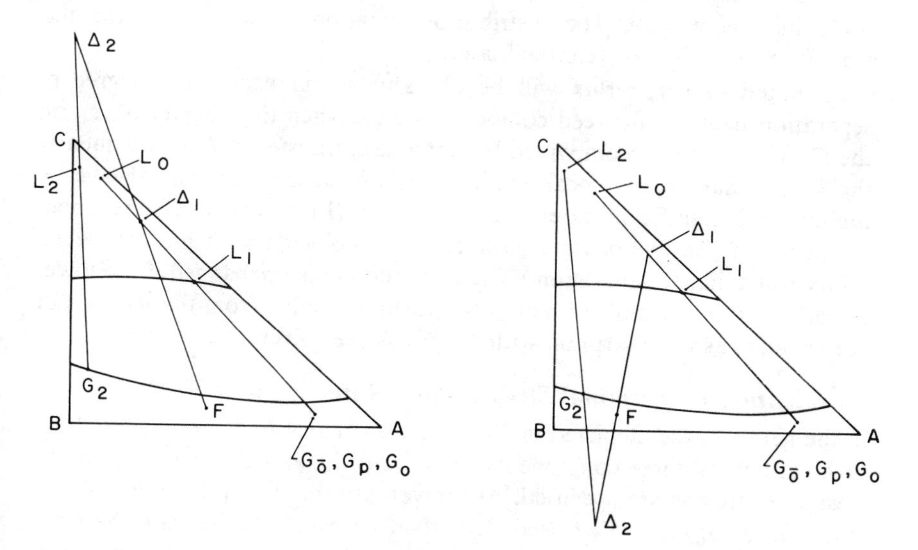

Figure 7.18. Representation of a continuous countercurrent column using reflux as in Figure 7.13, showing location of difference points Δ_1 and Δ_2 on the triangular diagram. This construction is used in locating operating lines on x_A-y_A coordinates for NTU evaluation.

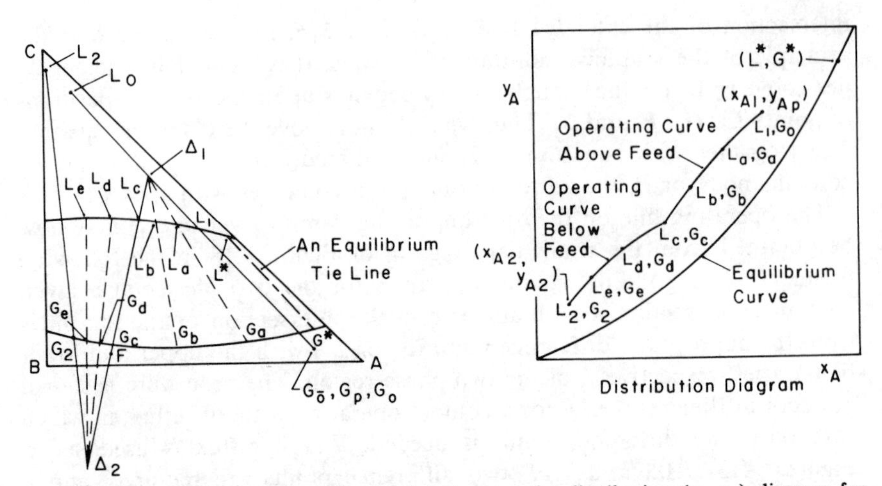

Figure 7.19. Construction of the operating lines on the distribution (x_A-y_A) diagram for NTU evaluation, using random lines from difference points Δ_1 and Δ_2 on the triangular diagram. The operation represented is shown in Figure 7.13.

The equilibrium curve on the distribution diagram may be obtained from the values of y_A and x_A read from the terminals of tie lines on the triangular diagram. One such point, (L^*, G^*), is shown on the equilibrium curve in Figure 7.19. The distribution diagram is now ready for the evaluation of NTU, as described later.

As noted earlier, reflux will be of value in increasing the degree of separation between the feed components only when the transfer of A into the C-rich phase is accompanied by significant transfer of B. Consequently the single-phase region near the C apex of Figure 7.18 will always be present to a significant extent in such cases. (In other words, the upper boundary of the two-phase region does not coincide with line AC when reflux will enhance separation.) The construction is unchanged if the lower boundary of the two-phase region is indistinguishable from the AB axis of the triangle, as in absorption with a nonvolatile solvent.

Evaluation of Terminal Stream Flow Rates

In the general case, depicted in Figure 7.13, streams F, L_2, L_0, G_p, and G_2 each contain all three components (A, B, and C), and the compositions of these five streams are specified. Also given are the flow rate of F and the reflux ratio, G_0/G_p, to be used. It is then necessary to estimate the flow rates of streams G_p, G_2, L_2, and L_0, the latter determining the size of the C remover needed for the operation. The estimation of these flow rates is conveniently performed on the triangular diagram. An overall material balance on Figure 7.13 gives

$$F+L_2=G_2+G_p+L_0=G_2+\Delta_1=\Sigma \tag{7.29}$$

This enables the location of Σ and other relevant points in Figure 7.20. Application of the line-ratio principle then allows the evaluation of both L_2 and Σ, since

$$L_2=F\left(\frac{\text{line}\,\Sigma F}{\text{line}\,\Sigma L_2}\right)$$

Next let

$$\frac{G_2}{\Delta_1}=N_1,$$

so

$$\Delta_1(1+N_1)=\Sigma$$

where

$$N_1=\frac{\text{line}\,\Sigma\Delta_1}{\text{line}\,G_2\Sigma}$$

is known from Figure 7.20. Hence Δ_1 and G_2 may be evaluated. From equation 7.22,

$$L_0+G_p=\Delta_1 \text{ (known)}$$

$$\frac{L_0}{G_p}=\frac{\text{line}\,\Delta_1 G_p}{\text{line}\,\Delta_1 L_0}=N_2 \text{ (known)}$$

$$L_0=N_2G_p, \quad \therefore G_p(1+N_2)=\Delta_1$$

Hence evaluate G_p and L_0.

Minimum Reflux Ratio

Suppose that, on the distribution diagram of Figure 7.19, the operating curve intersects or becomes tangent to the equilibrium curve at some point between the terminals of the column. At such a point the two adjacent phases have attained a state of equilibrium, and the driving force causing mass transfer $(y_A - y_{AL})$ has become zero. This, however, requires an infinitely tall column—borrowing from equation 7.54, developed later in this chapter, it follows that the NTU_{OG} needed to reach this condition is

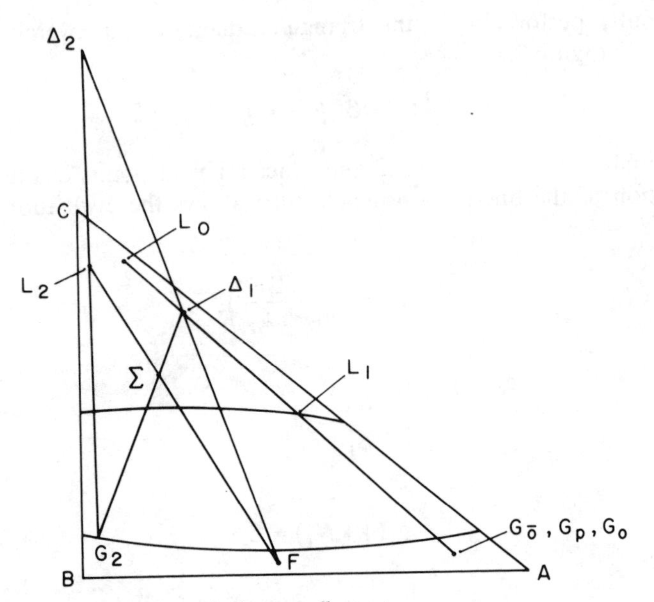

Figure 7.20. Evaluation of terminal stream flow rates.

infinite. Referring to the triangular diagram of Figure 7.19, the condition just described evidently corresponds to coincidence between a tie line and a random line from Δ_1 or Δ_2. This amounts to a location of Δ_1 or Δ_2 such that some tie line above or below the feed will extrapolate through the difference point appropriate to that section of the column. This situation must be avoided as follows to ensure a column of finite dimensions.

Equation 7.26 shows that Δ_1 approaches L_0 on the diagram as the reflux ratio increases. Thus the minimum reflux ratio corresponds to Δ_1 located at the closest intersection to L_0 of all extended tie lines in the column section above the feed with line L_0G_p. ("Above the feed" is intended in the sense of Figure 7.13; this could be the lower part of the column in the case of Figure 7.10 or in a liquid-extraction operation in which the extract phase was more dense than the raffinate phase.)

The point Δ_{1m} located in this manner is then used to find Δ_{2m} at the intersection of lines $F\Delta_{1m}$ and G_2L_2 extended, as shown in Figure 7.21.

If, however, any extended tie lines in the column section below the feed (Figure 7.13) intersect the extended line G_2L_2 between Δ_{2m} and L_2, then the intersection closest to L_2 finally determines Δ_{2m}. In this case, the intersection of the line $\Delta_{2m}F$ with the line L_0G_p finally determines Δ_{1m}, at a point farther from L_1 than found earlier by the extension of tie lines between G_p and F.

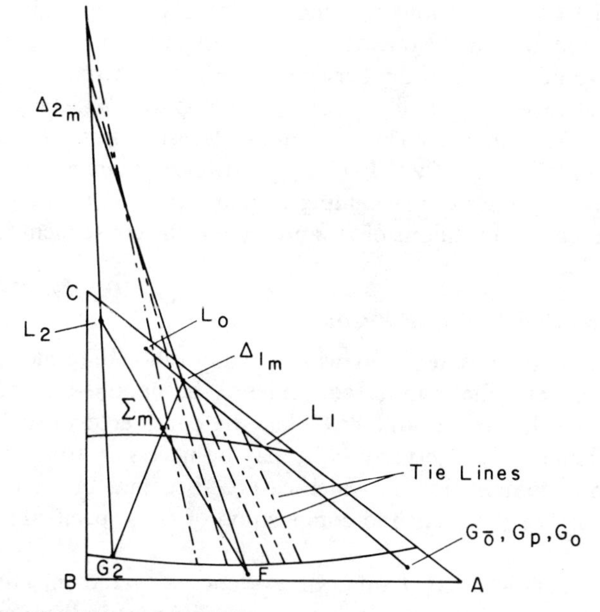

Figure 7.21. Evaluation of the minimum reflux ratio and of the minimum ratio of absorbent, stripping gas, or solvent to feed.

The minimum reflux ratio is then obtainable from equation 7.26 and Figure 7.21 as

$$\left(\frac{G_0}{G_p}\right)_{\min} = \frac{\text{line}\, L_1\Delta_{1m}}{\text{line}\, L_1 G_0}\,\frac{\text{line}\, L_0 G_p}{\text{line}\, L_0\Delta_{1m}} \tag{7.30}$$

The corresponding minimum ratio of absorbent, stripping gas, or solvent to feed is found from the relationship

$$\left(\frac{L_2}{F}\right)_{\min} = \frac{\text{line}\, \Sigma_m F}{\text{line}\, \Sigma_m L_2} \tag{7.31}$$

This follows from equation 7.29.

Total Reflux

Another limiting condition in column operation is that of total reflux, in which the flow rates of feed and product streams G_2 and G_p are all zero, and $L_2 = L_0 = \Delta$ in amount and composition. The result is shown in triangular coordinates in Figure 7.22. The intersection of random lines from Δ with the upper and lower bounds of the two-phase region would provide points on the single operating curve on an x_A-y_A diagram. No discontinuity would appear in the curve, since F equals zero.

In the special case of L_2 and L_0 consisting of pure C, these points of course coincide with apex C. This condition corresponds to the conventional definition of total reflux. Evidently, however, there is an infinite number of operating lines corresponding to total reflux, depending on the particular identical compositions of L_2 and L_0 and the consequent location of Δ.

Side Streams and Multiple Feeds

Occasionally a product is required with a composition intermediate between G_p and G_2 and that cannot conveniently be obtained by blending with the feed. A side-stream product G_s may be considered in such cases, as shown in Figure 7.23. Alternatively, it may be necessary to process two feeds of differing composition, in which case G_s becomes the second feed into the column. The flow rate and composition of G_s is specified in either case.

The column is now divided into three sections by the injection and withdrawal of streams. Since the flow rates are different in these sections, three difference points will be required on the triangular diagram: Δ_1 for the column section between the top and G_s, Δ_2 between G_s and F, and Δ_3 between F and the base of the column at 2. These difference points will be identified by the following balances. On the C-remover:

$$L_1 - G_0 = L_0 + G_p = \Delta_1 \tag{7.32}$$

On the entire column above a horizontal plane between F and G_s:

$$\bar{L} - \bar{G} = G_s + L_0 + G_p = G_s + \Delta_1 = \Delta_2 \tag{7.33}$$

On the column section below F:

$$G_2 - L_2 = \bar{\bar{G}} - \bar{\bar{L}} = \Delta_3 \tag{7.34}$$

Around the feed plane at F:

$$\begin{aligned} F &= \bar{L} - \bar{G} + \bar{\bar{G}} - \bar{\bar{L}} \\ &= \Delta_2 + \Delta_3 \end{aligned} \tag{7.35}$$

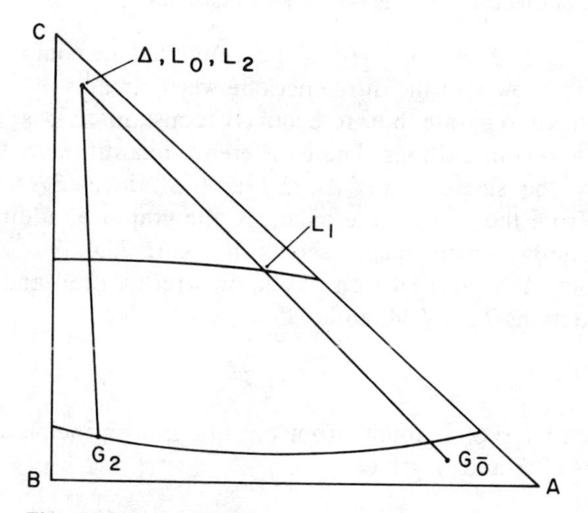

Figure 7.22. Total reflux.

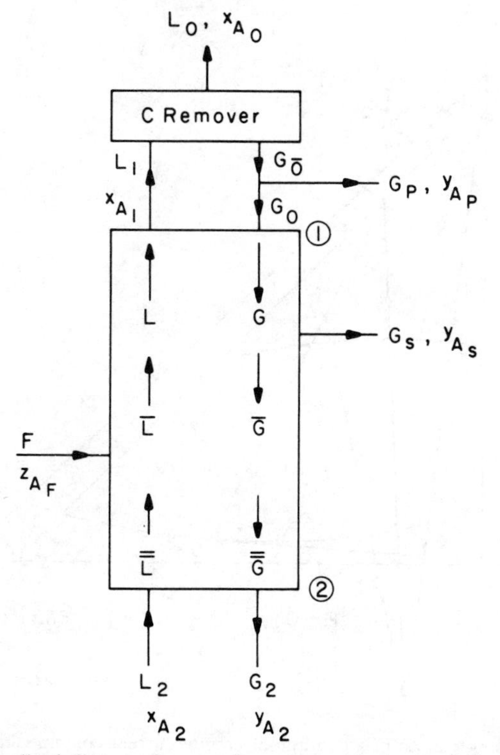

Figure 7.23. Generalized diagram for gas absorption, stripping, and liquid extraction with reflux and a side-stream product (positive G_s) or two feeds (negative G_s).

Equations 7.32 to 7.35 may all be converted to component-B and -C balances to show that the differences between streams in the three column sections have separate but respectively constant values with regard to amounts and compositions. These difference quantities are therefore represented by the single points Δ_1, Δ_2, and Δ_3, located on the triangular diagram from the appropriate balances, the graphical addition of streams, and the line-ratio principle, as shown in Figure 7.24.

The point Δ_1 is located from the known reflux ratio and equation 7.26. From equations 7.33, 7.34, and 7.35,

$$\Delta_3 = (F - G_s) - \Delta_1 = G_2 - L_2 \tag{7.36}$$

The point $F - G_s$ is found from the line-ratio principle and the known flow rates of F and G_s, since

$$\text{line}\, F(F - G_s) = (\text{line}\, FG_s)\frac{G_s}{F - G_s}$$

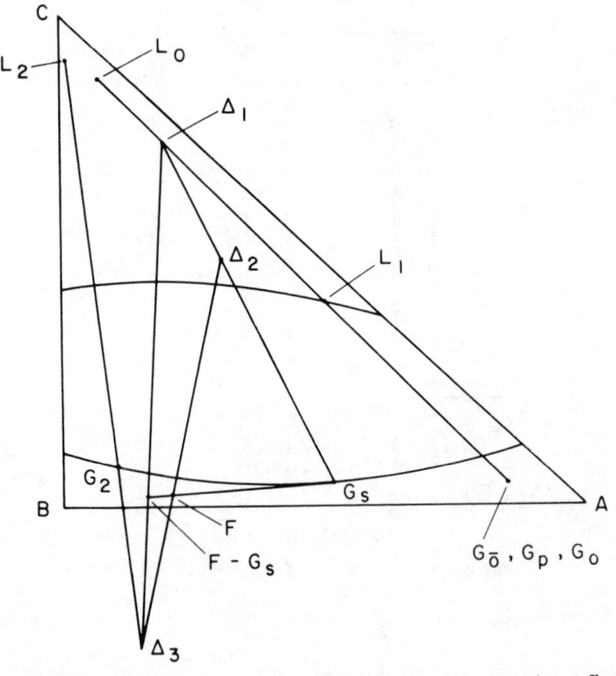

Figure 7.24. Representation of a continuous countercurrent column using reflux and with a side-stream product G_s as in Figure 7.23, showing location of difference points Δ_1, Δ_2, and Δ_3 on the triangular diagram. This construction is used in locating operating lines on x_A-y_A coordinates for NTU evaluation.

This permits location of Δ_3 at the intersection of the extended lines L_2G_2 and $\Delta_1(F-G_s)$. The point Δ_2 is then found, in accordance with equations 7.33 and 7.35, at the intersection of the extended line Δ_3F with the line Δ_1G_s.

Suppose that, instead of a side-stream product, G_s is a second feed into the column, having a composition different from F. Equations 7.33 and 7.36 are then revised to

$$\Delta_2=\Delta_1-G_s \tag{7.37}$$

$$\Delta_3=(F+G_s)-\Delta_1=G_2-L_2 \tag{7.38}$$

The construction appears in Figure 7.25, where point $F+G_s$ is located from the following, since rates of F and G_s are known:

$$\text{line}\, G_s(F+G_s)=(\text{line}\, FG_s)\frac{F}{F+G_s}$$

The position of Δ_3 is found at the intersection of extended lines L_2G_2 and $\Delta_1(F+G_s)$, and Δ_2 lies at the intersection of lines Δ_3F and $G_s\Delta_1$, as prescribed by equations 7.35 and 7.37.

The operating lines for either side-stream or two-feed operation are then constructed on the distribution diagram in a manner analogous to that shown in Figure 7.19. Random lines are drawn from the difference points Δ_1, Δ_2, and Δ_3, and the coordinates of their intersection with the upper and

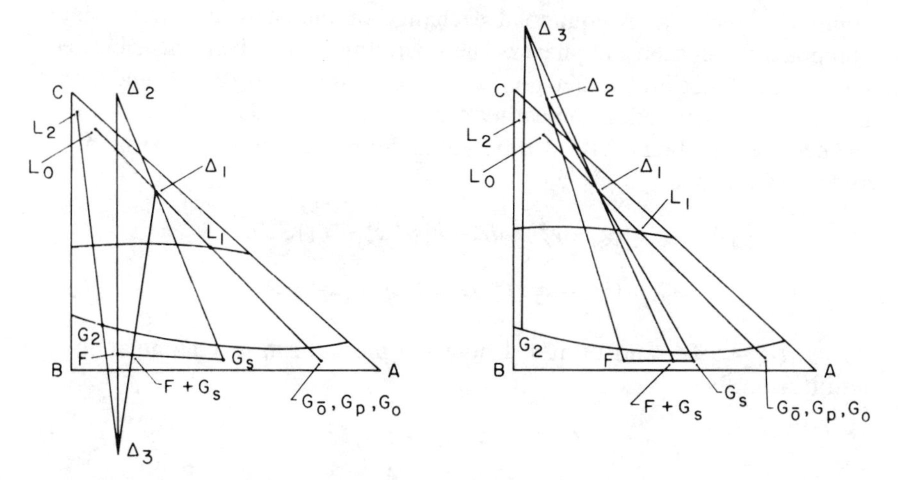

Figure 7.25. Representation of a continuous countercurrent column using reflux and two feeds (F and G_s) as in Figure 7.23, showing location of difference points Δ_1, Δ_2, and Δ_3 on the triangular diagram. This construction is used in locating operating lines on x_A-y_A coordinates for NTU evaluation.

lower bounds of the two-phase region provide x_A and y_A values, respectively, for the operating curve. Two discontinuities occur in the operating curve, corresponding to the intersections of the lines $\Delta_2 G_s$ and $\Delta_2 F$ with the upper and lower boundaries of the two-phase region on the triangular diagram. The NTU may then be evaluated as described later.

The Futility of Raffinate-Type Reflux

Most of the earlier treatments of this subject showed a portion of the stream G_2 being returned to the column after preliminary mixing with the incoming stream L_2 in Figures 7.13 and 7.23. Comparison with Figure 7.12 suggests that this might be called raffinate-type reflux, as distinct from the extract-type reflux commonly employed at the other end of the column. It was shown by Skelland (1961), however, that raffinate-type reflux is of no value either when accompanied by extract-type reflux or when used alone. It involves the needless use of a mixer, reflux dividing equipment, and possibly an auxiliary pumping unit for the refluxed raffinate-type material.

TRANSFER UNITS IN EQUIMOLAL COUNTERDIFFUSION

Equimolal counterdiffusion is the prevailing mechanism in distillation, provided that the heats of vaporization per mole are equal for all components and that sensible heat exchanges throughout the column, heats of mixing, and heat losses to the surroundings are all negligible. Under these conditions there is an equimolal exchange of the more and less volatile components between the phases. The operating line is then linear between consecutive inlet and outlet points on the column, because L and G are constants, as detailed in the earlier section on binary distillation.

The rate equations 4.10 in differential form are written in terms of x_A and y_A as follows:

$$d(N_A A) = k'_y a(y_A - y_A^*) S\,dZ = k'_x a(x_A^* - x_A) S\,dZ$$
$$= K'_y a(y_A - y_{AL}) S\,dZ = K'_x a(x_{AG} - x_A) S\,dZ \tag{7.39}$$

where $dV_v = S\,dZ$, and primes denote the mechanism of equimolal counterdiffusion. Now

$$d(N_A A) = d(G y_A) = d(L x_A) \tag{7.40}$$

and because of the constancy of L and G for this mechanism

$$d(N_A A) = G\,dy_A = L\,dx_A \tag{7.41}$$

Combining equation 7.39 with 7.41 and integrating,

$$\int_0^Z dZ = G\int_{y_{A1}}^{y_{A2}} \frac{dy_A}{k_y'aS(y_A - y_A^*)} = L\int_{x_{A1}}^{x_{A2}} \frac{dx_A}{k_x'aS(x_A^* - x_A)}$$

$$= G\int_{y_{A1}}^{y_{A2}} \frac{dy_A}{K_y'aS(y_A - y_{AL})} = L\int_{x_{A1}}^{x_{A2}} \frac{dx_A}{K_x'aS(x_{AG} - x_A)} \tag{7.42}$$

If the process is interpreted in terms of the two-film theory, equations 4.18 and 4.20 show that the individual and overall mass-transfer coefficients are independent of the concentration of component A when D_G and D_L do not vary. (This also requires constant m in the case of the overall K's.) Although the coefficients are dependent on the flow rates of the two phases, L and G have been shown to be constants for this mechanism. The capacity coefficients are accordingly treated as constants (if necessary, as averaged values over the column length of interest) and removed from the integral signs to give

$$Z = \frac{G}{k_y'aS}\int_{y_{A1}}^{y_{A2}} \frac{dy_A}{y_A - y_A^*} = \frac{L}{k_x'aS}\int_{x_{A1}}^{x_{A2}} \frac{dx_A}{x_A^* - x_A}$$

$$= \frac{G}{K_y'aS}\int_{y_{A1}}^{y_{A2}} \frac{dy_A}{y_A - y_{AL}} = \frac{L}{K_x'aS}\int_{x_{A1}}^{x_{A2}} \frac{dx_A}{x_{AG} - x_A} \tag{7.43}$$

If A denotes the more volatile component in distillation, then the operating line lies below the equilibrium curve on a plot of y_A vs x_A. The negative nature of $y_A - y_{AL}$ is of course countered by the fact that, in this case, $y_{A2} < y_{A1}$. Each integral represents the total change in composition of a given phase between sections 1 and 2 of Figure 7.2 divided by the available driving force causing the transfer. Each integral is therefore a measure of the difficulty of separation, and has been defined by Chilton and Colburn (1935) as the *number of transfer units* (NTU). Clearly the ratio Z/NTU may then be called the *height of a transfer unit* (HTU) and is given by the quantity outside each integral in equation 7.43. The latter relationship shows that there is an individual and an overall NTU expression for each of the G and L phases, combined with *corresponding* HTU expressions as follows:

$$Z = (\text{HTU})_G(\text{NTU})_G = (\text{HTU})_L(\text{NTU})_L$$

$$= (\text{HTU})_{OG}(\text{NTU})_{OG} = (\text{HTU})_{OL}(\text{NTU})_{OL} \tag{7.44}$$

By setting the NTU equal to unity, it is evident from equation 7.43 that the HTU is the column height necessary to effect a change in phase composition equal to the average driving force in the region under consideration. The evaluation of the NTU and HTU to determine the column height needed to obtain a specified separation will be considered shortly.

TRANSFER UNITS IN UNIMOLAL UNIDIRECTIONAL DIFFUSION

This mechanism is approximated in such operations as gas absorption, stripping or desorption, and liquid-liquid extraction. The transfer of component A from one phase to the other is not accompanied by any transfer in the reverse direction, so that L and G are no longer constant between sections 1 and 2 of Figure 7.2. The rate equations for this case are as follows:

$$d(N_A A) = k_y a(y_A - y_A^*) S\,dZ = k_x a(x_A^* - x_A) S\,dZ$$
$$= K_y a(y_A - y_{AL}) S\,dZ = K_x a(x_{AG} - x_A) S\,dZ \qquad (7.45)$$

The absence of primes on the coefficients compared with equation 7.39 signifies the difference in mechanism. Equation 7.40 remains valid, but equation 7.41 no longer holds. If it is assumed either that the solute (A) is the only component being transferred or that the solute transfer is accompanied by an equimolal countertransfer of the respective solvents (non-A) between phases, then

$$dG = d(Gy_A) = G\,dy_A + y_A\,dG$$
$$dG = d(Gy_A) = \frac{G\,dy_A}{1 - y_A} \qquad (7.46)$$

and

$$dL = d(Lx_A) = \frac{L\,dx_A}{1 - x_A} \qquad (7.47)$$

Combining equations 7.40, 7.45, 7.46, and 7.47, and integrating,

$$\int_0^Z dZ = \int_{y_{A1}}^{y_{A2}} \frac{G}{k_y aS} \frac{dy_A}{(1 - y_A)(y_A - y_A^*)} = \int_{x_{A1}}^{x_{A2}} \frac{L}{k_x aS} \frac{dx_A}{(1 - x_A)(x_A^* - x_A)}$$
$$= \int_{y_{A1}}^{y_{A2}} \frac{G}{K_y aS} \frac{dy_A}{(1 - y_A)(y_A - y_{AL})} = \int_{x_{A1}}^{x_{A2}} \frac{L}{K_x aS} \frac{dx_A}{(1 - x_A)(x_{AG} - x_A)} \qquad (7.48)$$

If the process is interpreted in terms of the two-film theory, equations 4.24 and 4.26 show that the coefficients are dependent on the concentration of component A because of the term $(1-y_A)_{\mathrm{LM}}$ or $(1-x_A)_{\mathrm{LM}}$ in the denominator. Accordingly, from equations 4.24 and 4.26, if D_G and D_L do not vary, the quantities $k_y(1-y_A)_{i\mathrm{LM}}$, $k_x(1-x_A)_{i\mathrm{LM}}$, $K_y(1-y_A)_{o\mathrm{LM}}$, and $K_x(1-x_A)_{o\mathrm{LM}}$ should be independent of concentration (assuming constant m in the case of the overall K's), where

$$k_y(1-y_A)_{i\mathrm{LM}}=\frac{k_y(y_A-y_A^*)}{\ln[(1-y_A^*)/(1-y_A)]} \tag{7.49}$$

$$k_x(1-x_A)_{i\mathrm{LM}}=\frac{k_x(x_A^*-x_A)}{\ln[(1-x_A)/(1-x_A^*)]} \tag{7.50}$$

$$K_y(1-y_A)_{o\mathrm{LM}}=\frac{K_y(y_A-y_{AL})}{\ln[(1-y_{AL})/(1-y_A)]} \tag{7.51}$$

$$K_x(1-x_A)_{o\mathrm{LM}}=\frac{K_x(x_{AG}-x_A)}{\ln[(1-x_A)/(1-x_{AG})]} \tag{7.52}$$

Equation 7.48 may now be multiplied and divided throughout by either $(1-y_A)_{\mathrm{LM}}$ or $(1-x_A)_{\mathrm{LM}}$, to obtain

$$\begin{aligned}
Z &= \int_{y_{A1}}^{y_{A2}} \frac{G}{k_y aS(1-y_A)_{i\mathrm{LM}}}\,\frac{(1-y_A)_{i\mathrm{LM}}\,dy_A}{(1-y_A)(y_A-y_A^*)} \\
&= \int_{x_{A1}}^{x_{A2}} \frac{L}{k_x aS(1-x_A)_{i\mathrm{LM}}}\,\frac{(1-x_A)_{i\mathrm{LM}}\,dx_A}{(1-x_A)(x_A^*-x_A)} \\
&= \int_{y_{A1}}^{y_{A2}} \frac{G}{K_y aS(1-y_A)_{o\mathrm{LM}}}\,\frac{(1-y_A)_{o\mathrm{LM}}\,dy_A}{(1-y_A)(y_A-y_{AL})} \\
&= \int_{x_{A1}}^{x_{A2}} \frac{L}{K_x aS(1-x_A)_{o\mathrm{LM}}}\,\frac{(1-x_A)_{o\mathrm{LM}}\,dx_A}{(1-x_A)(x_{AG}-x_A)}
\end{aligned} \tag{7.53}$$

Various experimental correlations for mass-transfer coefficients show that k_y or k_x is proportional to $(G/S \text{ or } L/S)^{0.8}$, and this relationship—at least in terms of mass velocities—has been extended to include the capacity coefficients $k_y a$ and $k_x a$ (Brown et al., 1950, pp. 529–530; Badger and Banchero, 1955, pp. 446–449). It is therefore customary to consider the capacity coefficients as varying roughly with the first power of the flow rate of the corresponding phase between sections 1 and 2. The considerations following equation 7.48 to this point have provided some justification for regarding quantities such as $G/k_y aS(1-y_A)_{i\text{LM}}$ as constant for a particular situation—generally at an average of the values at sections 1 and 2 of Figure 7.2. These quantities are therefore removed from the integral signs in equation 7.53 to give

$$Z=\left[\frac{G}{k_y aS(1-y_A)_{i\text{LM}}}\right]_{\text{av}}\int_{y_{A1}}^{y_{A2}}\frac{(1-y_A)_{i\text{LM}}\,dy_A}{(1-y_A)(y_A-y_A^*)}$$

$$=\left[\frac{L}{k_x aS(1-x_A)_{i\text{LM}}}\right]_{\text{av}}\int_{x_{A1}}^{x_{A2}}\frac{(1-x_A)_{i\text{LM}}\,dx_A}{(1-x_A)(x_A^*-x_A)}$$

$$=\left[\frac{G}{K_y aS(1-y_A)_{o\text{LM}}}\right]_{\text{av}}\int_{y_{A1}}^{y_{A2}}\frac{(1-y_A)_{o\text{LM}}\,dy_A}{(1-y_A)(y_A-y_{AL})}$$

$$=\left[\frac{L}{K_x aS(1-x_A)_{o\text{LM}}}\right]_{\text{av}}\int_{x_{A1}}^{x_{A2}}\frac{(1-x_A)_{o\text{LM}}\,dx_A}{(1-x_A)(x_{AG}-x_A)} \tag{7.54}$$

Each integral again defines an NTU expression, while the quantities in brackets outside each integral constitute the *corresponding* HTU. Individual and overall expressions for each of the L and G phases are identified by associating equation 7.44, term by term, with equation 7.54. For example, the individual G-phase transfer-unit relationships are

$$(\text{HTU})_G=\left[\frac{G}{k_y aS(1-y_A)_{i\text{LM}}}\right]_{\text{av}},$$

$$(\text{NTU})_G=\int_{y_{A1}}^{y_{A2}}\frac{(1-y_A)_{i\text{LM}}\,dy_A}{(1-y_A)(y_A-y_A^*)}$$

It may be noted that, although the rate equations 7.45 could be integrated to obtain Z using mass-transfer coefficients, the procedure in terms of transfer units is preferable. This is because the coefficients are strongly dependent on flow rate and would therefore vary with position. In contrast, the HTU has been shown to be less dependent on both flow rate and composition changes in a given application, and this greater stability renders it more suitable for design.

Approximate Expressions for NTU in Unimolal Unidirectional Diffusion

In many cases the evaluation of the integrals in equation 7.54 is facilitated by use of the arithmetic mean in place of the logarithmic mean $(1-y_A)_{\text{LM}}$ or $(1-x_A)_{\text{LM}}$, incurring only a small error (Wiegand, 1940). Thus in the case of $(\text{NTU})_G$,

$$(1-y_A)_{i\text{LM}} \doteq \frac{(1-y_A^*)+(1-y_A)}{2} \tag{7.55}$$

Insertion in the first integral of equation 7.54 leads to

$$(\text{NTU})_G = \int_{y_{A1}}^{y_{A2}} \frac{dy_A}{y_A - y_A^*} + \tfrac{1}{2}\ln\frac{1-y_{A1}}{1-y_{A2}} \tag{7.56}$$

The acceptability of equation 7.55 must be considered in any given case. The application of this approximation to the remaining three integrals in equation 7.54 results in

$$(\text{NTU})_L = \int_{x_{A1}}^{x_{A2}} \frac{dx_A}{x_A^* - x_A} + \tfrac{1}{2}\ln\frac{1-x_{A2}}{1-x_{A1}} \tag{7.57}$$

$$(\text{NTU})_{OG} = \int_{y_{A1}}^{y_{A2}} \frac{dy_A}{y_A - y_{AL}} + \tfrac{1}{2}\ln\frac{1-y_{A1}}{1-y_{A2}} \tag{7.58}$$

$$(\text{NTU})_{OL} = \int_{x_{A1}}^{x_{A2}} \frac{dx_A}{x_{AG} - x_A} + \tfrac{1}{2}\ln\frac{1-x_{A2}}{1-x_{A1}} \tag{7.59}$$

The reader should perhaps be cautioned regarding the incorrect forms of equations 7.57 and 7.59 which are often seen in the literature, where the logarithmic terms are erroneously inverted.

EVALUATION OF THE NUMBER OF TRANSFER UNITS

The integrals in equations 7.43, 7.54, and 7.56 to 7.59 are often evaluated numerically by graphical integration. Information for this procedure is obtained from the equilibrium-curve–operating-line plot on (x_A, y_A) coordinates, as sketched in Figure 7.4 for equimolal counterdiffusion (binary distillation) and Figure 7.26 for unimolal unidirectional diffusion (gas absorption, etc.), the latter without reflux.

It is often inconvenient to determine interfacial compositions (x_A^*, y_A^*) corresponding to each point on the operating line, so overall NTU values are frequently determined in preference to the individual ones. In evaluating the number of G-phase transfer units, values of $1-y_A$ and y_A-y_{AL} are readily obtained to enable evaluation of the integrand for a series of y_A values between y_{A1} and y_{A2}. (The quantity y_A-y_{AL} is evidently the vertical distance between the operating line and the equilibrium curve at a given y_A.) The quantity $(1-y_A)_{oLM}$ for use in NTU_{OG} is defined by equation 7.51. A plot such as that sketched in Figure 7.27 is then prepared, and the NTU_{OG} is given by the area under the curve between the limits of integration.

The evaluation could alternatively be performed with a digital computer. This would require equations for the equilibrium and operating curves similar to those in the computer program shown in Table 8.5 of Chapter 8. Numerical integration would then be effected using Simpson's rule, for which a standard subroutine is usually to be found in Fortran subroutine libraries—alternatively, see Conte (1965).

It should be noted that when reflux is used, the change in L and G below the feed plate necessitates the determination of separate NTU values for the column segments above and below the feed—namely, between points 1 and 2 and between points 2 and 3 for the distillation operation in Figure 7.4. Similarly, three NTU values are determined when operating with side streams or multiple feeds—for example, between points a and b, b and c, and c and d in Figure 7.5.

The evaluation of $(\text{NTU})_{OL}$ may be performed in an analogous manner, noting that $x_{AG}-x_A$ is the horizontal distance between the operating line and the equilibrium curve at a given x_A.

The selection of NTU expression for computation is arbitrary, but it is normally considered appropriate to use that relationship corresponding to the phase offering the greater resistance to mass transfer. This is the liquid phase in the case of absorption of rather insoluble gases, and the gas phase when highly soluble gases are being absorbed.

With regard to distillation, an empirical correlation has been presented by Yu and Coull (1950) which enables fairly quick estimation of the

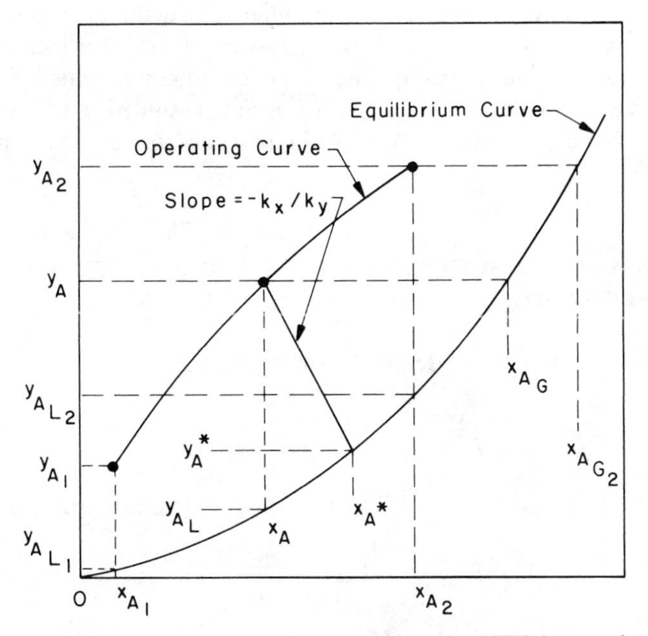

Figure 7.26. Evaluation of components in expressions for NTU in gas absorption etc., without reflux, corresponding to Figure 7.2.

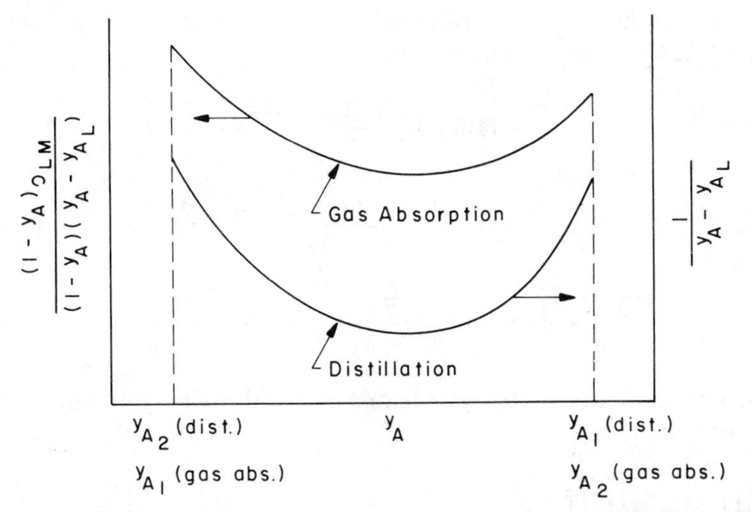

Figure 7.27. Graphical integration for $(NTU)_{OG}$ for equimolal counterdiffusion, typified by distillation, and for unimolal unidirectional diffusion, typified by gas absorption.

$(\text{NTU})_{OG}$ from a knowledge of the minimum reflux, the operating reflux, and the minimum $(\text{NTU})_{OG}$. A relationship was provided for evaluation of the latter quantity in terms of the relative volatility, which must be constant over the column. The correlation also assumed a boiling-liquid feed. It showed a maximum deviation of 6 percent from a large number of data.

THE RELATIONSHIP BETWEEN OVERALL AND INDIVIDUAL HEIGHTS OF TRANSFER UNITS

Equation 4.11 may be rewritten as follows:

$$\frac{G}{K_y' aS} = \frac{G}{k_y' aS} + \frac{mG}{L}\frac{L}{k_x' aS} \tag{7.60}$$

so that from equations 7.43 and 7.44 for equimolal counterdiffusion,

$$(\text{HTU})_{OG} = (\text{HTU})_G + \frac{mG}{L}(\text{HTU})_L \tag{7.61}$$

Similarly, from equation 4.12,

$$(\text{HTU})_{OL} = (\text{HTU})_L + \frac{L}{m'G}(\text{HTU})_G \tag{7.62}$$

In the case of unimolal unidirectional diffusion, the HTU expressions in equation 7.54 may be rearranged to give

$$\frac{1}{K_y} = (\text{HTU})_{OG}\frac{aS(1-y_A)_{o\text{LM}}}{G} \tag{7.63}$$

$$\frac{1}{k_y} = (\text{HTU})_G\frac{aS(1-y_A)_{i\text{LM}}}{G} \tag{7.64}$$

$$\frac{1}{k_x} = (\text{HTU})_L\frac{aS(1-x_A)_{i\text{LM}}}{L} \tag{7.65}$$

Inserting these expressions in equation 4.7 and multiplying throughout by $G/aS(1-y_A)_{o\text{LM}}$,

$$(\text{HTU})_{OG} = (\text{HTU})_G\frac{(1-y_A)_{i\text{LM}}}{(1-y_A)_{o\text{LM}}} + \frac{mG}{L}(\text{HTU})_L\frac{(1-x_A)_{i\text{LM}}}{(1-y_A)_{o\text{LM}}} \tag{7.66}$$

A parallel development using equation 4.8 yields

$$(\mathrm{HTU})_{OL}=(\mathrm{HTU})_L\frac{(1-x_A)_{i\mathrm{LM}}}{(1-x_A)_{o\mathrm{LM}}}+\frac{L}{m'G}(\mathrm{HTU})_G\frac{(1-y_A)_{i\mathrm{LM}}}{(1-x_A)_{o\mathrm{LM}}} \tag{7.67}$$

When the controlling resistance to transfer lies in the G phase,

$$(1-y_A)_{i\mathrm{LM}}\doteq(1-y_A)_{o\mathrm{LM}}$$

If the solutions are also dilute,

$$(1-x_A)_{i\mathrm{LM}}\doteq(1-y_A)_{o\mathrm{LM}}$$

In this special case equation 7.66 reduces to equation 7.61, so that the form of expression for the two transfer mechanisms becomes identical. Similarly, when the L phase is controlling,

$$(1-x_A)_{i\mathrm{LM}}\doteq(1-x_A)_{o\mathrm{LM}}$$

If the solutions are also dilute,

$$(1-y_A)_{i\mathrm{LM}}\doteq(1-x_A)_{o\mathrm{LM}}$$

and equation 7.67 reduces to 7.62. The individual HTU's are not obtainable, however, as the slope and intercept of a plot of $(\mathrm{HTU})_{OL}$ against $L/m'G$ [or of $(\mathrm{HTU})_{OG}$ against mG/L]. This is because $(\mathrm{HTU})_G$ and $(\mathrm{HTU})_L$ are not independent of the flow rates (Colburn, 1943; Garner, Ellis, and Fosbury, 1953).

For the operations of gas absorption, stripping, and distillation the following empirical correlations of experimental data have been presented by Cornell, Knapp, Close, and Fair (1960, 1960).

For Raschig rings, Berl saddles, and spiral tile,

$$(\mathrm{HTU})_L=\phi_1(N_{\mathrm{Sc},L})^{0.5}(C')\left(\frac{Z_r}{10}\right)^{0.15} \tag{7.68}$$

For Raschig rings,

$$(\mathrm{HTU})_G=\frac{\psi(N_{\mathrm{Sc},G})^{0.5}}{\left(\frac{L'}{S}f_1f_2f_3\right)^{0.6}}\left(\frac{d_c}{12}\right)^{1.24}\left(\frac{Z_r}{10}\right)^{1/3} \tag{7.69}$$

For Berl saddles,

$$(\mathrm{HTU})_G = \frac{\psi (N_{\mathrm{Sc},G})^{0.5}}{\left(\frac{L'}{S} f_1 f_2 f_3\right)^{0.5}} \left(\frac{d_c}{12}\right)^{1.11} \left(\frac{Z_r}{10}\right)^{1/3} \tag{7.70}$$

where ϕ_1 is obtained from Figure 7.28 or 7.29, C' from Figure 7.30, and ψ from Figure 7.31 or 7.32. The necessary flooding velocities are obtainable from Figure 7.1 and Table 7.1, as described earlier in this chapter. $N_{\mathrm{Sc},L}$ and $N_{\mathrm{Sc},G}$ are the liquid- and gas-phase Schmidt numbers, d_c is the column diameter in inches, and Z_r is the packed height between redistributors in feet.

$$f_1 = \left(\frac{\mu_L \text{ in centipoises}}{1.005}\right)^{0.16}$$

$$f_2 = \left(\frac{1}{\rho_L \text{ in g/ml}}\right)^{1.25}$$

$$f_3 = \left(\frac{72.8}{\sigma \text{ in dyn/cm}}\right)^{0.8}$$

In the application of equations 7.61, 7.68, 7.69, and 7.70 to distillation, the following relationship was used to obtain m:

$$m = \frac{\alpha}{[1 + (\alpha - 1)\bar{x}_A]^2}$$

where α is relative volatility and $\bar{x}_A$ is the average liquid mole fraction in the range of application. The correlation for Raschig rings is considered to be good for other types of ring except Pall rings, for which it would be conservative (Eckert, 1963). The Berl-saddle correlation was also found to fit data for McMahon packing and Intalox saddles. Goodloe and Stedman packings are not included in the correlations.

No such relationships have been established for liquid-liquid extraction. The design is effected in such cases by the use of HTU measurements made in a pilot plant in which the system, packing, and flow rates are the same as those to be used on the full scale.

Illustration 7.3

A mixture of chloroform and benzene is to be fed at 30 lb-mole/hr to a packed distillation column operating at atmospheric pressure. This feed

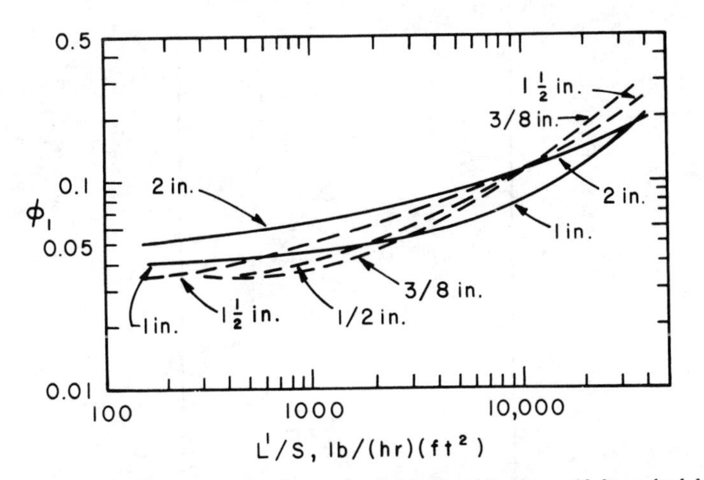

Figure 7.28. $(HTU)_L$ correlation for various-size Raschig rings: 10-ft packed height; less than 50 percent of flooding (Cornell et al., 1960).

contains 0.45 mole fraction chloroform; a distillate containing 0.95 mole fraction chloroform and a residue or bottom product containing 0.15 mole fraction chloroform are required.

The feed will be a subcooled liquid at 155°F, a total condenser will be used, and reflux returned to the column as a saturated (boiling) liquid at a reflux ratio of 4 to 1.0. The column will be randomly packed with 1.5-in. ceramic Raschig rings.

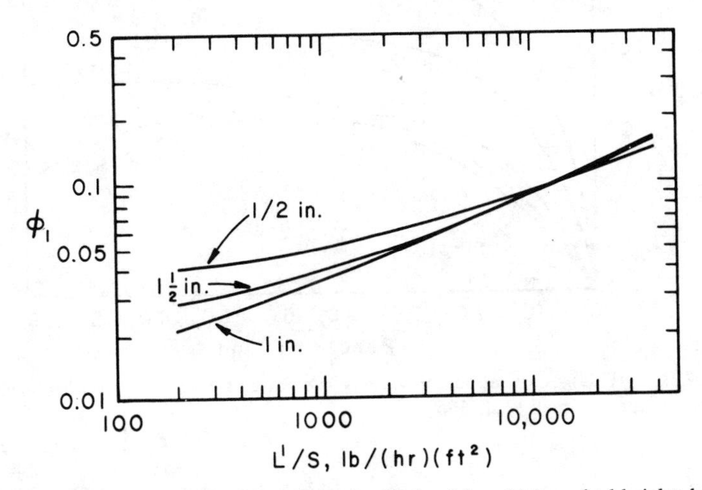

Figure 7.29. $(HTU)_L$ correlation for various-size Berl saddles: 10-ft packed height; less than 50 percent of flooding (Cornell et al., 1960).

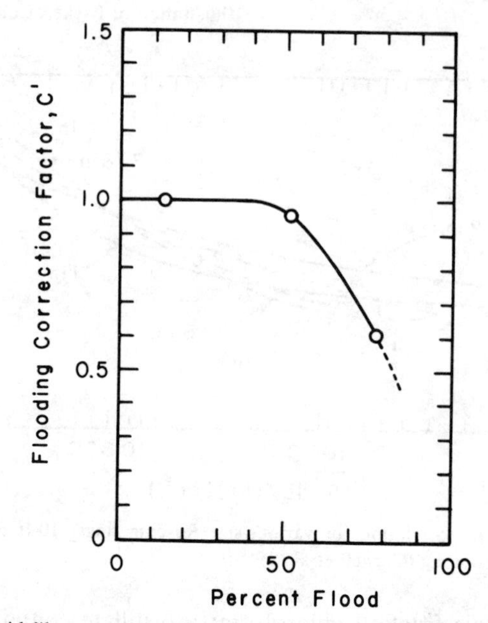

Figure 7.30. Liquid-film correction factor for operation at high percentage of flooding (Cornell et al., 1960).

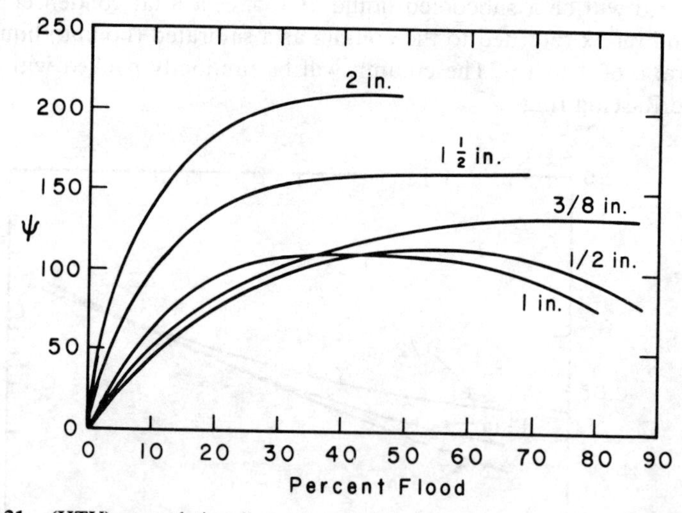

Figure 7.31. $(HTU)_G$ correlation for various-size Raschig rings: 10-ft packed height; 1-ft column diameter (Cornell et al., 1960).

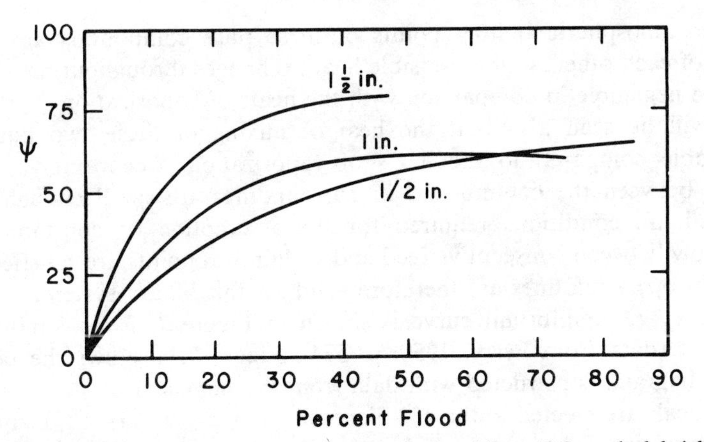

Figure 7.32. $(HTU)_G$ correlation for various-size Berl saddles: 10-ft packed height; 1-ft column diameter (Cornell et al., 1960).

Determine the following:

(a) The amounts of products obtained per hour.

(b) The number of overall gas-phase transfer units, $(NTU)_{OG}$, required for the separation.

(c) The flow rates of the liquid and vapor streams within the column, above and below the feed.

(d) A suitable column diameter.

(e) The approximate $(HTU)_{OG}$ in the enriching and stripping sections of the column, located respectively above and below the feed.

(f) The necessary column height.

(g) The location of the feed point.

(h) The condenser and reboiler heat loads.

SOLUTION (a). An overall material balance gives

$$F = B_p + D_p, \text{ or } D_p = 30 - B_p$$

and a balance on chloroform,

$$30(0.45) = B_p(0.15) + (30 - B_p)(0.95)$$

$$B_p = 18.75 \text{ lb-mole/hr}, \quad D_p = 11.25 \text{ lb-mole/hr}$$

SOLUTION (b). The molal latent heats of vaporization at atmospheric pressure are 12,700 Btu/lb-mole for chloroform and 13,250 Btu/lb-mole for benzene (Perry, 1950, p. 218)—i.e., within 4.5 percent of each other.

The atmospheric boiling points of these pure compounds are within 35°F of each other, so that sensible heat exchanges throughout the column will be negligible in comparison with the heats of vaporization.

It will be seen also that the heat of mixing of these two liquids is negligible compared to the heats of vaporization. Accordingly, if heat losses between the column and its surroundings are small enough to be ignored, all conditions required for the assumption of constant molal overflow between consecutive feed and withdrawal points are satisfied, and straight operating lines are therefore valid for this binary system.

The x_A-y_A equilibrium curve is shown in Figure 7.33 for atmospheric pressure (data from Perry, 1950, p. 574). Figure 7.34 shows the boiling-point diagram, constructed with data from the same source.

Verticals are erected to the 45° diagonal from x_{AB}, x_{AF} $(=z_{AF})$, and x_{AD}. The operating lines for the column sections above and below the feed intersect on the q line, which passes through the point (x_{AF}, x_{AF}) and has a slope of $q/(q-1)$, where, from equation 7.17,

$$q=\frac{H_f-h_F}{H_f-h_f}$$

CALCULATION OF q. Many tabulated data are in mass units instead of molal units; accordingly,

Molecular weight of chloroform = 119.39

Molecular weight of benzene = 78.11

Calculation of h_F. From Figure 7.34 (Perry, 1950, p. 574) the bubble point of the feed is 167.5°F, and its dew point is 170.5°F. Taking 65°F as an enthalpy datum, the enthalpy of the entering feed is

$$h_F=c_{pF}(t_F-65°)+\Delta H_{SF}$$

where c_{pF} is the average specific heat of the liquid feed between 65 and 155°F, and ΔH_{SF} is the measured heat of solution of chloroform in benzene at the feed composition and 65°F (ΔH_{SF} is negative when heat is evolved on mixing). The value of c_{pF} is interpolated from International Critical Tables (abbreviated here to I.C.T.) V, 126, (1929) as 36.1 Btu/(lb-mole of feed)(°F). The value of ΔH_{SF} is interpolated from I.C.T. V, 155, to be -139.8 Btu/lb-mole of feed. Substituting,

$$h_F=36.1(155-65)-139.8=3105.2 \text{ Btu/lb-mole of feed}$$

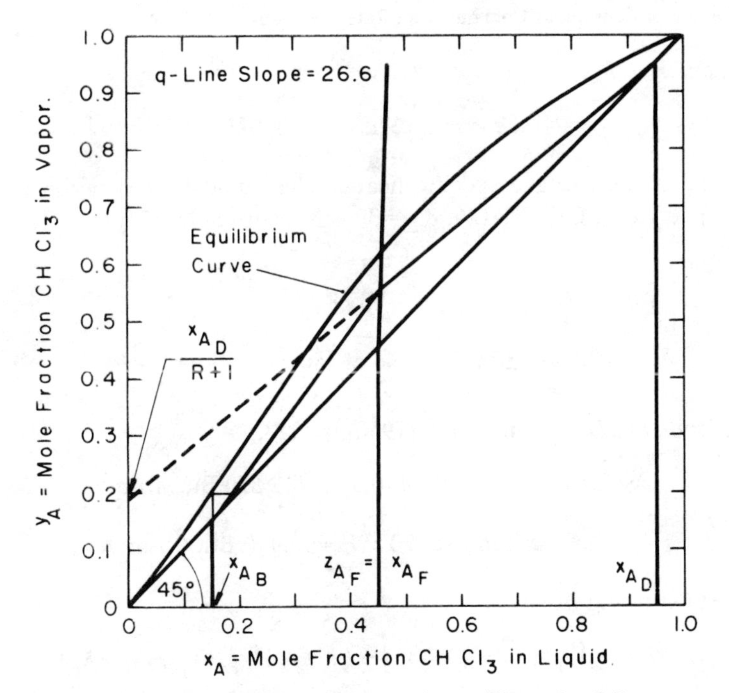

Figure 7.33. Distillation of a chloroform-benzene mixture (Illustration 7.3).

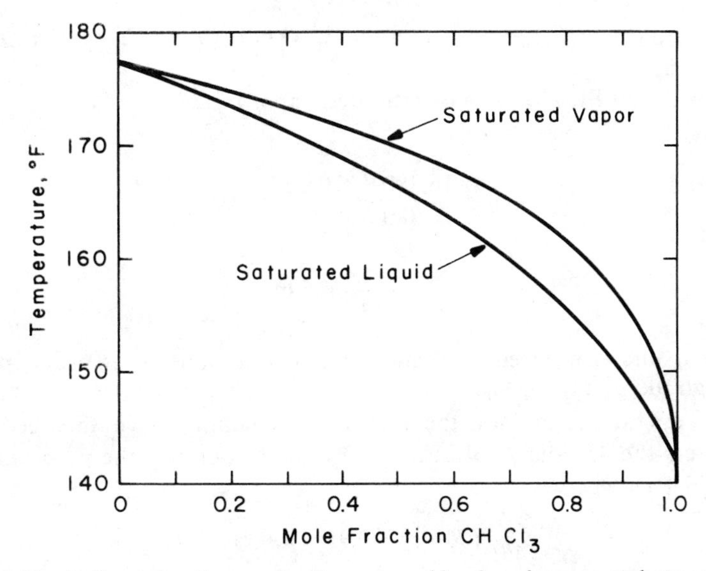

Figure 7.34. Boiling-point diagram for the system chloroform-benzene at 1-atm pressure (Illustration 7.3).

Calculation of h_f.

$$h_f = c_{pL}(167.5 - 65) + \Delta H_{SF}$$

Here c_{pL} is the average specific heat of the liquid feed between 65° and 167.5°F. From I.C.T. V, 126, $c_{pL} = 36.4$ Btu/(lb-mole of feed)(°F). Substituting,

$$h_f = 36.4(167.5 - 65) - 139.8$$

$$= 3590.2 \text{ Btu/lb-mole of saturated-liquid feed.}$$

Calculation of H_f. From Perry (1950), p. 218,

$$\lambda_A \text{ for chloroform at } 170.5°\text{F} = 12{,}570 \text{ Btu/lb-mole}$$

$$\lambda_B \text{ for benzene at } 170.5°\text{F} = 13{,}490 \text{ Btu/lb-mole}$$

Thus substituting in

$$H_f = y_A[c_{pA}(t_G - t_0) + \lambda_A] + (1 - y_A)[c_{pB}(t_G - t_0) + \lambda_B]$$

where c_{pB} and c_{pA} are the average specific heats of liquid benzene and chloroform between 65° and 170.5°F (Perry, 1950, p. 228),

$$H_f = 0.45[28.7(170.5 - 65) + 12{,}570] + 0.55[33.6(170.5 - 65) + 13{,}490]$$

$$= 16{,}380 \text{ Btu/lb-mole of saturated-vapor feed.}$$

Thus

$$q = \frac{16{,}380 - 3105.2}{16{,}380 - 3590.2} = 1.039$$

$$\frac{q}{q-1} = 26.6$$

The q line is inserted in Figure 7.33 with a slope of 26.6 and passing through the point (x_{AF}, x_{AF}).

The operating line for the section of column above the feed plate intersects the 45° diagonal at $x_{AD} = 0.95$ and intersects the y_A ordinate at $x_{AD}/(R+1)$, where

$$x_{AD}/(R+1) = 0.95/(4+1) = 0.19$$

First the upper and then the lower operating line is drawn, starting at the points (x_{AD}, x_{AD}) and (x_{AB}, x_{AB}), respectively, and intersecting on the q line.

From equation 7.43 the NTU_{OG} below and above the feed are obtained as

$$(NTU)_{OG,\text{ below feed}} = \int_{y_{A2}}^{y_{Af}} \frac{dy_A}{y_{AL} - y_A}$$

$$(NTU)_{OG,\text{ above feed}} = \int_{y_{Af}}^{y_{A1}} \frac{dy_A}{y_{AL} - y_A}$$

where y_{Af} is the vapor composition at the level at which the feed is introduced.

Values of y_A and y_{AL} are read from Figure 7.33 for various values of x_A between 0.185 and 0.95, where 0.185 is the composition of the liquid leaving the base of the column and entering the reboiler. This is evident from Figure 7.33, where the "step" represents the reboiler in the manner of McCabe and Thiele (see e.g. McCabe and Smith, 1967, pp. 558–559). The composition of the condensed liquid distillate is 0.95. When a total condenser is used, this is also the composition of the vapor leaving the top of the column. Table 7.2 shows some of the required values.

The quantity $1/(y_{AL} - y_A)$ is plotted against y_A in Figure 7.35. From Figure 7.33, y_{Af} is 0.55, assuming that the feed is introduced at the level corresponding to the intersection of the q line and operating lines. The area under the curve between $y_{A2} = 0.2$ and $y_{Af} = 0.55$ gives

$$(NTU)_{OG,\text{ below feed}} = 5.5$$

Table 7.2. Quantities used in evaluating $(NTU)_{OG}$ in Illustration 7.3.

x_A	y_A	y_{AL}	$y_{AL} - y_A$	$\frac{1}{y_{AL} - y_A}$
0.185	0.200	0.254	0.054	18.52
0.3	0.350	0.415	0.065	15.4
0.4	0.480	0.550	0.070	14.3
0.455	0.550	0.611	0.061	16.4
0.6	0.670	0.755	0.085	11.78
0.8	0.830	0.904	0.074	13.5
0.95	0.950	0.981	0.031	32.3

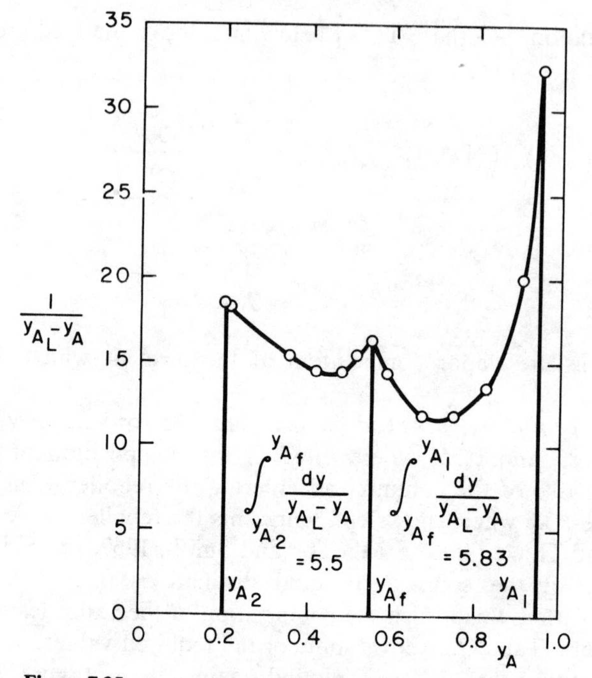

Figure 7.35. Evaluation of $(\text{NTU})_{OG}$, Illustration 7.3.

Similarly the area under the curve between $y_{Af} = 0.55$ and $y_{A1} = 0.95$ gives

$$(\text{NTU})_{OG,\ \text{above feed}} = 5.83$$

The total number of overall gas-phase transfer units needed is $5.5 + 5.83 = 11.33$.

SOLUTION (c). From the definition of the reflux ratio R and equations 7.13 and 7.17,

$$L = RD_p = 4(11.25) = 45 \text{ lb-mole/hr}$$

$$G = D_p(R+1) = 11.25(4+1) = 56.25 \text{ lb-mole/hr}$$

$$\bar{L} = RD_p + qF = 4(11.25) + 1.039(30) = 76.17 \text{ lb-mole/hr}$$

$$\bar{G} = D_p(R+1) - F(1-q) = 11.25(4+1) - 30(1-1.039)$$

$$= 57.42 \text{ lb-mole/hr}$$

SOLUTION (d). The molal flow rates of L and G are greater below the feed than above, as are the mass flow rates of L. The mass flow rate of G, however, is greater at the top of the column than at the bottom. The approach to flooding will therefore be checked at both the top and bottom of the column for a selected diameter. Figure 7.33 shows that $x_{A2}=0.185$ and $y_{A2}=0.2$. Thus the vapor G_2 entering the column consists of

$$57.42(0.2)119.39 = 1370 \text{ lb/hr } CHCl_3$$

$$57.42(0.8)78.11 = \underline{3585 \text{ lb/hr}}\ C_6H_6$$

$$\text{Total} = 4955 \text{ lb/hr} = G'_2$$

The boiling-point diagram in Figure 7.34 shows that the temperature in the reboiler is 174.2°F. The volume of G_2 entering the column per hour is then

$$57.42(359)\left(\frac{460+174.2}{492}\right) = 26{,}550 \text{ ft}^3/\text{hr}.$$

so that

$$\rho_{G2} = \frac{4955}{26{,}550} = 0.187 \text{ lb/ft}^3$$

$$L_2 = 76.17 \text{ lb-mole/hr}$$

consisting of

$$76.17(0.185)119.39 = 1685 \text{ lb/hr } CHCl_3$$

$$76.17(0.815)78.11 = \underline{4855 \text{ lb/hr}}\ C_6H_6$$

$$\text{Total} = 6540 \text{ lb/hr} = L'_2$$

The boiling-point diagram shows the temperature of L_2 to be 173.5°F. The density is estimated as

$$\rho_{L2} = 62.9 \text{ lb/ft}^3.$$

Then

$$\frac{L'_2}{G'_2}\left(\frac{\rho_{G2}}{\rho_{L2}}\right)^{0.5} = \frac{6540}{4955}\left(\frac{0.187}{62.9}\right)^{0.5} = 0.0719$$

Figure 7.1 shows that flooding occurs for this abscissa when the ordinate reaches a value of 0.16; i.e.,

$$\frac{\left(\overline{G}'_{\text{flooding}}/S\right)^2 F_p \mu_L^{0.2}}{\rho_G \rho_L g_c}\left(\frac{\rho_{\text{H}_2\text{O}}}{\rho_L}\right)=0.16$$

At 173.5°F, μ_{L2} is estimated to be 0.33 cP (Perry, 1963, p. 3-199–200). For 1.5-in. Raschig rings, Table 7.1 shows that $F_p = 95$. Therefore

$$\frac{\overline{G}'_{\text{flooding}}}{S}=\left[\frac{0.16(0.187)(62.9)^2(4.17\times 10^8)}{95(0.33)^{0.2}(62.4\times 0.972)}\right]^{1/2}=3270\,\text{lb}/(\text{hr})(\text{ft}^2)$$

where 0.972 is the specific gravity of water at 173.5°F. For operation at 60 percent of the flooding rate,

$$\frac{G'_2}{S}=0.6(3270)=1962\ \text{lb}/(\text{hr})(\text{ft}^2)$$

The cross section of the column would then be $4955/1962 = 2.525\ \text{ft}^2$, and the column diameter

$$\frac{d_c}{12}=\left(\frac{4(2.525)}{\pi}\right)^{0.5}=1.795\ \text{ft}$$

If a 2-ft-diameter column is chosen, then the column cross section is π ft^2, and

$$\frac{G'_2}{S}=\frac{4955}{\pi}=1579\ \text{lb}/(\text{hr})(\text{ft}^2)$$

This is (1579/3270)(100) = 48.3 percent of flooding.

Entirely analogous calculations for the top of the column show that a diameter of 2 ft will lead to 36 percent of flooding in that region.

A 2-ft-diameter column will be selected. Figure 7.1 shows that the resulting gas-phase pressure gradient will correspond to operation somewhat below the lower limit of the loading region, at least in the section below the feed. This is a common operating condition and allows for some future increase in capacity of the column, while being in the vicinity of most efficient operation, namely, near loading conditions. The ratio of packing diameter to column diameter will be 1.5/24 or $\frac{1}{16}$, which is in the range recommended by Eckert (1961, 1963), who proposes that

d_{packing}/d_c should not be greater than $\frac{1}{15}$ for good liquid distribution. Most packed distillation columns are small in diameter, because of difficulties in maintaining effective liquid distribution and in avoiding channeling. Thus Teller, Miller, and Scheibel (Perry, 1963, p. **18**-49) indicate that diameters usually do not exceed 2 ft, whereas McCabe and Smith (1967, p. 601) state that they are usually 3 ft or less.

SOLUTION (e). HTU values may vary along the column with variations in physical properties, mass flow rates, and percentage of flooding. Conservative estimates of column height should utilize maximum estimates for the HTU. Since the present example is largely vapor-phase controlling, attention will be focused on column regions where the HTU_G is highest. From equation 7.69,

$$\frac{(\text{HTU}_G)_{\text{bottom of section}}}{(\text{HTU}_G)_{\text{top of section}}} \doteq \frac{[\rho_L^{1.25}/L']^{0.6}_{\text{bottom}}}{[\rho_L^{1.25}/L']^{0.6}_{\text{top}}}$$

since, for more than 35 percent of flooding, ψ is approximately constant for 1.5-in. Raschig rings (Figure 7.31), and variations in μ_L and σ are minor.

Calculations analogous to those given in part (d) show that, at the top of the stripping section, $\rho_L = 74.1\ \text{lb/ft}^3$ and $\bar{L}'/S = 2360\ \text{lb/(hr)(ft}^2)$. Substitution in the above expression shows that the ratio of the HTU_G at the top to that at the bottom of the stripping section is approximately unity. A similar result is found for the enriching section of the column. HTU calculations may accordingly be made using conditions at the foot of each section.

For the stripping section:

$$f_1 = \left(\frac{0.33}{1.005}\right)^{0.16} = 0.837$$

$$f_2 = \left(\frac{1}{1.008}\right)^{1.25} = 0.986$$

$$f_3 = \left(\frac{72.8}{22}\right)^{0.8} = 2.6$$

$$N_{\text{Sc},G}^{0.5} \doteq 1.0$$

$$\psi = 160.5 \qquad \text{(Figure 7.31 at 48.3 percent flooding).}$$

Provision for the redistribution of liquid is made at intervals of 5 column diameters along the column, so that $Z_r = 10$ ft. Substituting in equation 7.69,

$$(\mathrm{HTU})_G = \frac{160.5(1.0)}{[(6540/\pi)(0.837)(0.986)2.6]^{0.6}}\left(\frac{24}{12}\right)^{1.24}$$

$$= 2.44 \text{ ft.}$$

To obtain $(\mathrm{HTU})_L$, we need

$\phi_1 = 0.0625$ (Figure 7.28 at $L_2'/S = 6540/\pi$)

$C' = 0.95$ (Figure 7.30 at 48.3 percent flooding)

$Z_r = 10;$

$$N_{\mathrm{Sc},L} = \frac{\mu_L}{\rho_L D};$$

$\mu_L \doteq 0.0033$ Poises;

$\rho_L \doteq 1.008$ gm/cm^3

D will be estimated by the Lusis and Ratcliff correlation (equation 3.20) as follows:

From Perry (1963, p. **14**-20), molal volumes are obtained as

$CHCl_3 = A$:

Carbon		14.8
Hydrogen		3.7
Chlorine	$3 \times 24.6 =$	73.8
		92.3 cm^3/gm-mole

$C_6H_6 = B$:

Carbon	$6 \times 14.8 =$	88.8
Hydrogen	$6 \times 3.7 =$	22.2
Benzene ring		-15
		96 cm^3/gm-mole

Also, μ_B at 173.5°F (=78.6°C) is 0.32 cP; $T = 273 + 78.6 = 351.6$°K. Substituting in equation 3.20,

$$D_{AB}^0 = \frac{8.52 \times 10^{-8}(351.6)}{0.32(96)^{1/3}}\left[1.4\left(\frac{96}{92.3}\right)^{1/3} + \frac{96}{92.3}\right] = 5.04 \times 10^{-5}\ \text{cm}^2/\text{sec}$$

This diffusivity applies to very dilute solutions. Although a correction for variation with concentration could be made by the methods of Chapter 3, this will not be done here because D_{AB} appears to the fractional exponent 0.5 in equation 7.68, and particularly because this process is largely vapor-phase controlled. Therefore

$$N_{\text{Sc},L} = \frac{0.0033 \times 10^5}{1.008(5.04)} = 65$$

From equation 7.68,

$$(\text{HTU})_L = 0.0625(65)^{0.5}(0.95) = 0.477\ \text{ft}$$

$$m = \frac{\alpha}{[1 + (\alpha - 1)\bar{x}_A]^2}$$

where, for the stripping section,

$$\bar{x}_A = 0.5(0.185 + 0.455) = 0.32$$

The relative volatility is defined as

$$\alpha = \alpha_{AB} = \frac{y_A/x_A}{y_B/x_B} = \frac{y_A(1 - x_A)}{x_A(1 - y_A)}$$

where x_A and y_A are equilibrium values. From the equilibrium curve in Figure 7.33, when $\bar{x}_A = 0.32$,

$$\alpha = \frac{0.445(1 - 0.32)}{0.32(1 - 0.445)} = 1.704$$

$$m = \frac{1.704}{[1 + 0.704(0.32)]^2} = 1.136$$

It may be noted here that the use of m in this context has been questioned by Hengstebeck (1961, pp. 235, 246, 249).

Substitution in equation 7.61 yields the $(\mathrm{HTU})_{OG}$ for the stripping section as

$$(\mathrm{HTU})_{OG} = 2.44 + \frac{1.136(57.42)}{76.17}(0.477) = 2.85 \text{ ft}$$

Calculations analogous to those in parts (d) and (e) above give the following results for the enriching section of the column. Values are those at the foot of this section, denoted by subscript *e*.

$$x_{Ae} = 0.455, \quad y_{Ae} = 0.55$$

$$G'_e/S = 1810 \text{ lb}/(\text{hr})(\text{ft}^2), \quad L'_e/S = 1390 \text{ lb}/(\text{hr})(\text{ft}^2),$$

$$(\text{percent flooding})_e \doteq 41.6 \text{ percent}$$

$$t_{L_e} \doteq 167°\text{F}, \ \rho_{L_e} \doteq 74.1 \text{ lb/ft}^3,$$

$$f_1 = 0.845, \quad f_2 = 0.804, \quad f_3 = 2.56,$$

$$(N_{\mathrm{Sc},G})^{0.5} \doteq 1.0, \quad \psi \doteq 160.0,$$

$$\phi_1 = 0.055, \quad C' \doteq 1.0, \quad \bar{x}_A = 0.7025,$$

$$\alpha = 2.145, \quad m = 0.66$$

$$(\mathrm{HTU})_G = 3.50 \text{ ft}; \ (\mathrm{HTU})_L = 0.44 \text{ ft}; \ (\mathrm{HTU})_{OG} = 3.86 \text{ ft}.$$

SOLUTION (f).

$$\text{Height of the enriching section} = 5.83(3.86) = 22.5 \text{ ft}$$

$$\text{Height of the stripping section} = 5.5(2.85) = \underline{15.7 \text{ ft}}$$

$$\text{Total column height} = 38.2 \text{ ft}$$

Although possible errors of up to 75 percent were noted in the prediction of $(\mathrm{HTU})_G$, the correlation was rather good in the case of Raschig rings (see Cornell, Knapp, Close, and Fair, 1960, 1960).

SOLUTION (g). The feed is introduced at a distance of 22.5 ft from the top of the column—that is, $(22.5/38.2) \times 100 = 58.9$ percent of the way down.

SOLUTION (h). An enthalpy balance on the condenser in Figure 7.3 gives

$$G_1H_1 = L_1h_D + D_ph_D + q_c$$

but

$$G_1 = L_1 + D_p = D_p(R+1)$$

$$q_c = D_p(R+1)(H_1 - h_D)$$

If heat losses to the surroundings are negligible, an overall enthalpy balance on Figure 7.3 yields

$$q_r = B_ph_B + D_ph_D - Fh_F + q_c$$

All terms for substitution in these expressions will now be calculated.

Calculation of h_D. From Figure 7.34, the bubble point of the distillate D_p is 146.5°F.

The average specific heat of a liquid mixture of chloroform and benzene between 65 and 146.5°F and at composition D_p, is interpolated from I.C.T. V, 126, as 28.8 Btu/(lb-mole)(°F).

The heat of solution ΔH_{SD}, at 65°F and the composition of D_p, is interpolated (I.C.T. V, 155) as −34.4 Btu/lb-mole of D_p. Therefore,

$$h_D = 28.8(146.5 - 65) - 34.4 = 2312.6 \text{ Btu/lb-mole of } D_p.$$

Calculation of H_1. From Figure 7.34, the dew point of G_1 is 152°F.

From Perry (1950), p. 218,

$$\lambda_A \text{ for chloroform at } 152°\text{F} = 12{,}900 \text{ Btu/lb-mole}$$

$$\lambda_B \text{ for benzene at } 152°\text{F} = 13{,}800 \text{ Btu/lb-mole}$$

The average specific heats of liquid chloroform and benzene between 65° and 152°F are (Perry, 1950, p. 228)

$$c_{pA} = 28.4 \text{ Btu/(lb-mole)(°F)}$$

$$c_{pB} = 33.4 \text{ Btu/(lb-mole)(°F)}$$

Substituting in the expression given earlier above for H_f,

$$H_1 = 0.95[28.4(152-65) + 12{,}900] + 0.05[33.4(152-65) + 13{,}800]$$

$$= 15{,}435 \text{ Btu/lb-mole of } G_1.$$

Substituting in the equation for q_c, the condenser heat load,

$$q_c = 11.25(4+1)(15{,}435 - 2312.6)$$

$$= 739{,}000 \text{ Btu/hr.}$$

Calculation of h_B. From Figure 7.34, the bubble point of the residue product B_p is 174.2°F.

The average specific heat of a liquid mixture of chloroform and benzene between 65 and 174.2°F and at composition B_p is interpolated from I.C.T. V, 126, as 38.65 Btu/(lb-mole)(°F).

The heat of solution ΔH_{SB}, at 65°F and the composition of B_p, is interpolated (I.C.T. V, 155) to be -58.1 Btu/lb-mole of B_p. Thus

$$h_B = 38.65(174.2 - 65) - 58.1 = 4166.9 \text{ Btu/lb-mole of } B_p.$$

Substituting in the equation for q_r, the reboiler heat load,

$$q_r = 18.75(4166.9) + 11.25(2312.6) - 30(3105.2) + 739{,}000 = 750{,}000 \text{ Btu/hr}$$

Illustration 7.4. An aqueous solution of acetone is to be extracted countercurrently at a rate of 300 lb/hr with 100 lb/hr of ethyl propionate at 86°F. The solution initially contains 0.40 mass fraction acetone, which is to be reduced to a finished (solvent-free) concentration of 0.15 (mass fraction).

Determine the number of overall G phase (raffinate) transfer units, $(\text{NTU})_{OG}$, required in a packed extraction column to achieve this degree of extraction.

If the extraction is obtained in a packed pilot-plant column that is 7 ft high, what is the height of an overall G-phase transfer unit, $(\text{HTU})_{OG}$, and also the height equivalent to a theoretical stage (H.E.T.S.) under these operating conditions?

Transfer is from the G to the L phase.

SOLUTION. Equation 7.54 shows that

$$(\text{NTU})_{OG} = \int_{y_{A1}}^{y_{A2}} \frac{(1-y_A)_{o\text{LM}}\, dy_A}{(1-y_A)(y_A - y_{AL})}$$

but from equation 7.51,

$$(1-y_A)_{o\text{LM}} = \frac{y_A - y_{AL}}{\ln[(1-y_{AL})/(1-y_A)]}$$

Combination with the above expression for $(\mathrm{NTU})_{OG}$ gives

$$(\mathrm{NTU})_{OG}=\int_{y_{A1}}^{y_{A2}}\frac{dy_A}{(1-y_A)\times 2.303\log\left(\dfrac{1-y_{AL}}{1-y_A}\right)}$$

This expression is now used to evaluate the number of overall transfer units in the G (aqueous) phase, which is where the principal resistance to transfer lies for this system.

The operating (distribution) diagram of Figure 7.36 must first be constructed. The tie-line data and plait point of Figure 4.5, Illustration 4.2, have been converted to mole fractions of acetone in each phase and plotted as the equilibrium curve.

The operating line cannot be plotted from equation 7.5, because the two solvents are partially miscible under these extraction conditions, which are far from dilute. To obtain the operating curve, the terminal streams G_1, L_1, and G_2 are located on the triangular diagram, which was first obtained in Illustration 4.2, and is now reproduced in Figure 7.37. A material balance on Figure 7.2 gives

$$G_2+L_1=G_1+L_2=\Sigma$$

so that

$$\frac{\text{Rate }(G_2)}{\text{Rate }(G_2+L_1)}=\frac{300\text{ lb/hr}}{400\text{ lb/hr}}=\frac{\text{Length of line }L_1\Sigma}{\text{Length of line }L_1G_2}$$

This enables the location of Σ; the point L_2 then lies at the intersection of the binodal curve and the extended line $G_1\Sigma$. In accordance with an overall balance in Figure 7.2, showing $G_1-L_1=G-L=G_2-L_2=\Delta$, the lines L_2G_2 and L_1G_1 are extended to intersect at the difference point Δ. The fixed location of Δ for the entire column is readily demonstrated by successive balances on any two of the three components involved. Random lines from Δ intersect the binodal curve at w_{AB} on the water-rich side and at w_{AC} on the ethyl propionate–rich side. These intersections, when converted to mole fractions, give points (x_A, y_A) on the operating curve of Figure 7.36.

Values of y_{AL} and y_A are read from the equilibrium and operating curves, respectively, at various values of x_A in Figure 7.36, and Table 7.3 is constructed. The final column of this table is plotted against y_A in Figure 7.38, and the area under the curve between $y_{A1}=0.0513$ and $y_{A2}=0.171$ is measured to be 3.63 units, so that

$$(\mathrm{NTU})_{OG}=3.63$$

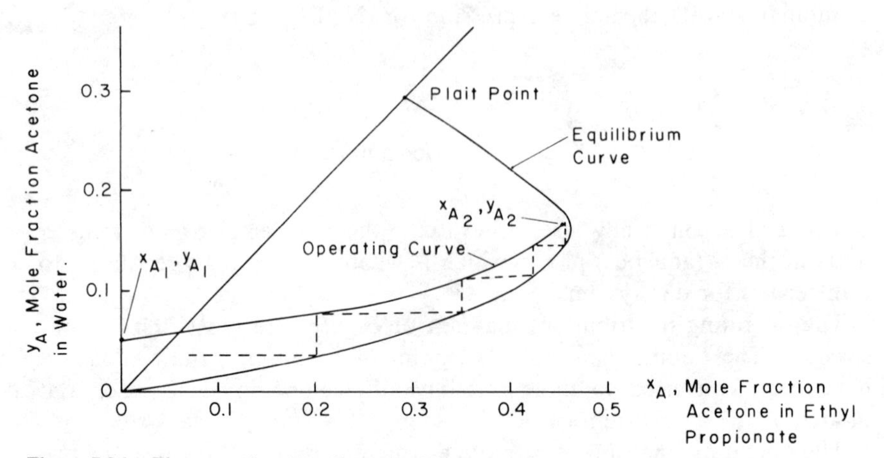

Figure 7.36. The operating (distribution) diagram for the liquid extraction process without reflux in Illustration 7.4.

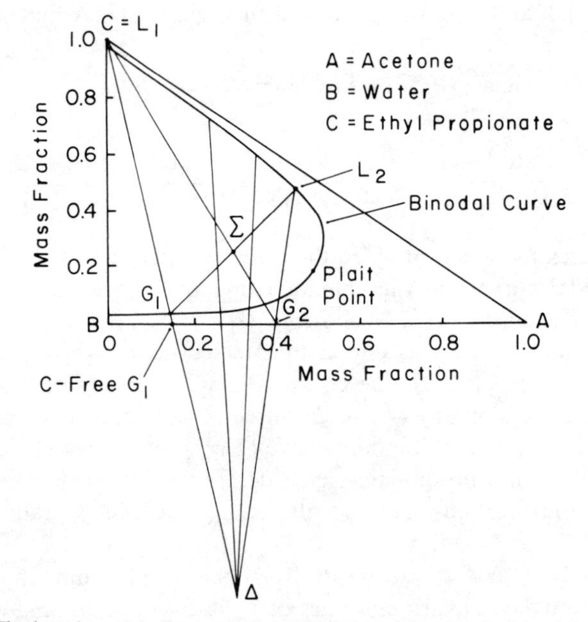

Figure 7.37. The location of Δ, from which random lines intersect the binodal curve to give points on the operating curve in Figure 7.36 (Illustration 7.4).

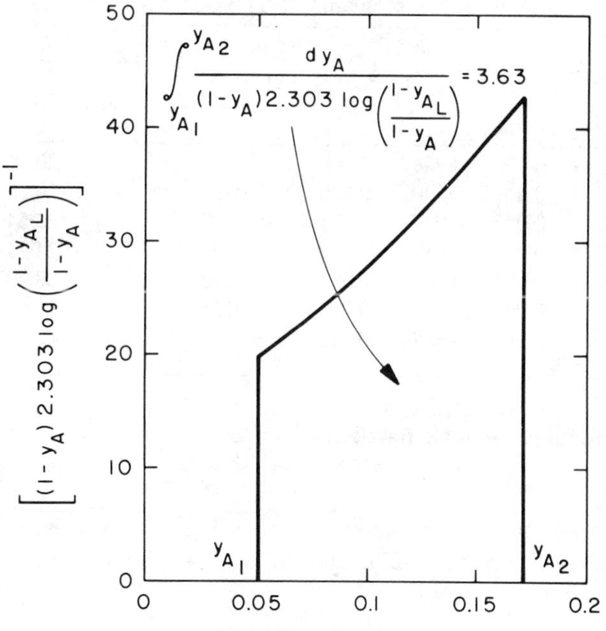

Figure 7.38. Evaluation of $(NTU)_{OG}$, Illustration 7.4.

For a column height of Z ft,

$$(\mathrm{HTU})_{OG} = \frac{Z}{(\mathrm{NTU})_{OG}} = \frac{7}{3.63} = 1.93 \text{ ft}$$

under these particular conditions of operation.

In contrast with continuous contactors, stagewise contactors utilize intermittent contact between the phases. The stages often take the form of horizontal plates or trays of varied design, arranged vertically above each other in a column. The two phases enter a stage from opposite directions in countercurrent flow, mix together to facilitate transfer, and then separate and leave the stage. When the two phases leave in a state of equilibrium the stage is said to be an "ideal" or "theoretical" one. This concept has been extended to packed columns by defining the height (of packing) equivalent to a theoretical stage (HETS) such that the streams leaving this section are in equilibrium.

The number of theoretical stages to which this packed column is equivalent is given by the broken-line stepwise construction between the

Table 7.3. Quantities used in evaluating $(NTU)_{OG}$ in Illustration 7.4.[a]

x_A	y_A	y_{AL}	$1-y_A$	$1-y_{AL}$	$\log\left(\frac{1-y_{AL}}{1-y_A}\right)$	$\frac{1}{(1-y_A)2.303\log\left(\frac{1-y_{AL}}{1-y_A}\right)}$
0.0	0.0513	0.00	0.9487	1.000	0.0229	20
0.075	0.06	0.01	0.940	0.990	0.0222	20.85
0.150	0.07	0.025	0.930	0.975	0.0207	22.54
0.225	0.0825	0.041	0.9175	0.959	0.0191	24.73
0.300	0.099	0.062	0.901	0.938	0.0170	28.40
0.375	0.1225	0.090	0.8775	0.910	0.0155	31.95
0.458	0.171	0.148	0.829	0.852	0.0123	42.65

[a] All concentrations are mole fractions.

operating and equilibrium curves shown in Figure 7.36 (see, e.g., McCabe and Smith, 1967, pp. 526–527). This shows that about 3.6 ideal stages are present, so that

$$\text{HETS} = \frac{7}{3.6} = 1.95 \text{ ft.}$$

Although in this particular case the HETS and $(HTU)_{OG}$ have almost the same values, this is not generally true. It has resulted on this occasion because the equilibrium and operating curves of Figure 7.36 are nearly parallel (Perry, 1950, p. 550).

The HETS varies much more widely than the HTU with flow rates, type of system, solute concentration, and packing. This is because it represents the imposition of a fictitious stagewise mechanism upon what is actually a continuous, differential process. Design in terms of HTU rather than HETS is desirable for this reason.

NOMENCLATURE

A, B, C — Components A, B, and C.

A — Area; total interfacial area, ft^2.

a — Interfacial area per unit volume, ft^2/ft^3.

B_p — Bottom product, lb-mole/hr.

C' — Flooding correction factor for $(HTU)_L$ at high gas rates, equation 7.68.

C_1, C_2	Constants in equations 7.5 to 7.7.
c_{pA}, c_{pB}	Average specific heats of components A and B, Btu/(lb-mole)(°F).
c_{pF}	Specific heat of feed, Btu/(lb-mole)(°F).
c_{pL}	Average specific heat of a liquid phase, Btu/(lb-mole)(°F).
D, D_{AB}, D_{AB}^0	(Volumetric) molecular diffusivity; of A in B; at "infinite" dilution, ft^2/hr or cm^2/sec.
D_G, D_L	Molecular diffusivities in the gas and liquid phases, ft^2/hr.
D_p	Distillate product, lb-mole/hr.
d_c	Column diameter, inches.
F	Feed, lb-mole/hr.
F_p	Packing factor.
f_1	Ratio of liquid viscosity under column conditions to viscosity of water at 20°C.
f_2	Ratio of density of water at 20°C to that of liquid under column conditions.
f_3	Ratio of surface tension of water at 20°C to that of liquid under column conditions.
G, $\overline{G}$, $\overline{\overline{G}}$	Flow rate of phase G; between L_s and F planes *or* below F plane; below F plane, lb-mole/hr.
G'	Mass flow rate of phase G, lb-mass/hr.
G''	Defined by equation 7.3.
G^*	A phase in equilibrium with L^*.
G_p	A-rich product, lb-mole/hr.
G_0, $G_{\overline{0}}$	Reflux; stream that splits to give G_0 and G_p, lb-mole/hr.
G_s	Side-stream product *or* second feed, lb-mole/hr.
G_1, G_2	Flow rate of phase G at sections 1 and 2, lb-mole/hr.
g_c	Conversion factor, 4.17×10^8 (lb-mass)(ft)/(lb-force)(hr^2).
H_f, $H_{\bar{f}}$	Enthalpy of gas phase at points just above and just below the feed plane, Btu/lb-mole.
ΔH_{SB}, ΔH_{SD}, ΔH_{SF}	Heats of solution at 65°F and the compositions of B_p, D_p, and F, respectively, Btu/lb-mole of solution.

h_B, h_D, h_F	Enthalpies of B_p, D_p, and F, Btu/lb-mole.
h_f, $h_{\bar{f}}$	Enthalpy of liquid phase at points just above and just below the feed plane, Btu/lb-mole.
HETS	Height (of packing) equivalent to a theoretical stage, ft.
$(\mathrm{HTU})_G$, $(\mathrm{HTU})_L$	Heights of individual G-phase and L-phase transfer units, ft.
$(\mathrm{HTU})_{OG}$, $(\mathrm{HTU})_{OL}$	Heights of overall G-phase and L-phase transfer units, ft.
K_x, K_y, k_x, k_y	Overall and individual mass-transfer coefficients based on Δx and Δy, as defined in equation 7.45, lb-mole/(ft²)(hr).
K'_x, K'_y, k'_x, k'_y	Overall and individual mass-transfer coefficients based on Δx and Δy for equimolal counterdiffusion, defined in equation 7.39, lb-mole/(ft²)(hr).
L, $\bar{L}$, $\bar{\bar{L}}$	Flow rate of phase L; between L_s and F planes *or* below F plane; below F plane, lb-mole/hr.
L'	Mass flow rate of phase L, lb-mass/hr.
L''	Defined by equation 7.4.
L^*	A phase in equilibrium with G^*.
L_c	Leakage rate of cooling water in faulty condenser, lb-mole/hr.
L_0	Flow rate of stream leaving C remover (not $G_{\bar{0}}$), lb-mole/hr.
L_s	Side-stream product *or* second feed, lb-mole/hr.
L_1, L_2	Flow rate of phase L at sections 1 and 2, lb-mole/hr.
M	Flow rate of stream M, lb-mole/hr *or* lb-mass/hr.
m, m'	Slope of equilibrium curve; see below equation 4.3 in Chapter 4.
N	Flow rate of stream N, lb-mole/hr *or* lb-mass/hr.
N_A	Molal flux of component A relative to stationary coordinates, lb-mole/(ft²)(hr).

$N_{\mathrm{Sc}\,G}$, $N_{\mathrm{Sc},L}$	Schmidt number ($\mu/\rho D$) for a gas; for a liquid.
N_1, N_2	G_2/Δ_1; L_0/G_p.
$(\mathrm{NTU})_G$, $(\mathrm{NTU})_L$	Number of individual G-phase and L-phase transfer units.
$(\mathrm{NTU})_{OG}$, $(\mathrm{NTU})_{OL}$	Number of overall G-phase and L-phase transfer units.
P	Flow rate of stream P, lb-mole/hr or lb-mass/hr.
q	$(\bar{L}-L)/F=(H_f-h_F)/(H_f-h_f)$; defined below equation 7.17.
q_c, q_r	Condenser and reboiler heat loads, Btu/hr.
R	Reflux ratio, L_1/D_p; G_0/G_p.
S	Cross-sectional area of empty column, ft^2.
S_s	Flow rate of open steam, lb-mole/hr.
T	Absolute temperature, °K.
t_F, t_G, t_O	Temperature of the feed; gas temperature; datum temperature, °F.
V_v	Contacting volume, ft^3.
x_A, $\bar{x}_A$, x_{AB}, x_{AD}, x_{AL}, x_{AO}, x_{As}, x_{A1}, x_{A2}	Mole fraction of component A in the L phase; average over a composition range; in B_p; in D_p; in L_c; in L_0; in L_s; in L_1 and L_2.
x_A^*	Local equilibrium concentration in the L phase at the interface, mole fraction.
x_{AG}	L-phase concentration which would be in equilibrium with existing G-phase concentration, mole fraction.
x_B, x_{BM}, x_{BN}, x_{BP}	Mole fraction of component B; in streams M, N, and P.
x_C, x_{CM}, x_{CN}, x_{CP}	Mole fraction of component C; in streams M, N, and P.
$(x_B)_{L1}$, $(x_C)_{L1}$	Mole fractions of B and C in L_1.
$(x_B)_{\Delta 1}$, $(x_C)_{\Delta 1}$	Mole fractions of B and C in Δ_1.
$(1-x_A)_{i\mathrm{LM}}$, $(1-x_A)_{\mathrm{oLM}}$	Defined by equations 7.50 and 7.52.
y_A, y_{Af}, y_{AP}, y_{As}, y_{A1}, y_{A2}	Mole fraction of component A in the G phase; at feed plane; in G_p; in G_s; in G_1 and G_2.

y_A^*	Local equilibrium concentration in the G phase at the interface, mole fraction.
y_{AL}	G-phase concentration that would be in equilibrium with existing L-phase concentration, mole fraction.
y_B, y_C	Mole fraction of components B and C in the G phase.
$(y_B)_{G0}, (y_C)_{G0}$	Mole fractions of B and C in G_0.
$(1-y_A)_{i\text{LM}}, (1-y_A)_{o\text{LM}}$	Defined by equations 7.49 and 7.51.
Z	Column height, ft.
Z_r	Packed height between redistributors, ft.
z_{AF}	Mole fraction of A in feed.
α	Relative volatility, defined in Illustration 4.1.
$\Delta, \Delta_1, \Delta_2, \Delta_3$	Difference between flow rates of adjacent streams at any horizontal plane in a column; for the column portion between section 1 and either the F or G_s plane; between the G_s and F planes *or* between the F plane and section 2; between the F plane and section 2, lb-mole/hr.
Δ_{1m}, Δ_{2m}	Δ_1 and Δ_2 for minimum reflux ratio, lb-mole/hr.
λ_A, λ_B	Latent heats of vaporization for A and B, Btu/lb-mole.
μ_B	Viscosity of solvent B (as in equation 3.20), centipoises.
μ_L	Liquid viscosity, centipoises.
$\rho_G, \rho_{H_2O}, \rho_L$	Densities of gas, water, and liquid, lb-mass/ft^3 when used in Figure 7.1; when used in f_2, ρ_L is in gm/cm^3.
Σ, Σ_m	Sum, as defined in equation 7.29; corresponding to conditions of minimum reflux ratio, lb-mole/hr.
σ	Liquid surface tension, dyn/cm.
ϕ_1	Factor in equation 7.68.
ψ	Factor in equations 7.69 and 7.70.

PROBLEMS

Binary Distillation

7.1 A chloroform-benzene mixture is to be fed at 30 lb-mole/hr to the top of a continuous, packed distillation column operating at atmospheric pressure and with a total condenser. For the column exclusive of the reboiler it is known that $(NTU)_{OG} = 5$ under the proposed conditions. The feed is a boiling liquid and contains 0.55 mole fraction chloroform. If a residue product containing 0.15 mole fraction chloroform is obtained, what will be the composition of the distillate if no reflux is used?

Physical properties are given in Illustration 7.3.

7.2 A boiling liquid mixture of chloroform and benzene is fed at 30 lb-mole/hr to the reboiler of a packed distillation column operating at atmospheric pressure. The residue product, feed, and distillate contain 0.25, 0.3, and 0.6 mole fraction chloroform, respectively, with a reflux ratio of 6. Determine the $(HTU)_{OG}$ if the packed height is 12 ft.

7.3 A mixture containing 0.45 mole fraction chloroform and 0.55 mole fraction benzene is to be fed to a continuous, packed distillation column operating at atmospheric pressure. Find the minimum reflux ratio for the following three feed conditions for a distillate containing 0.95 mole fraction chloroform.

(a) Feed as a boiling liquid.
(b) Feed as 50 mole percent vapor, 50 mole percent liquid.
(c) Feed as a saturated vapor.

If the column operation is converted to total reflux, what is the $(NTU)_{OG}$ in the column if the reflux and reboiler liquids contain 0.95 and 0.15 mole fraction chloroform, respectively?

7.4 Suppose that in Illustration 7.3 the total condenser is replaced by a partial condenser, of capacity sufficient to provide the same amount of liquid reflux while delivering a saturated-vapor product containing 0.95 mole fraction chloroform. This vapor product is subsequently condensed in a total condenser. The feed will again enter at 155°F, and it will be assumed that the vapor product and liquid reflux leaving the partial condenser are in equilibrium. Determine the $(NTU)_{OG}$ needed within the column.

7.5 Suppose that the conditions of Illustration 7.3 are repeated, except that the condenser is so large that the reflux returns to the column as a subcooled liquid at 100°F instead of at its bubble point (146.5°F). The reflux ratio is again 4. Show that the equation for the operating line above the feed in this case is

$$y_A = \frac{R\beta}{R\beta+1} x_A + \frac{x_{AD}}{R\beta+1}; \qquad \beta = \frac{H-h_{L1}}{H-h_L}$$

where h_{L1} is the enthalpy of the cold liquid reflux, and H and h_L are the enthalpies of the saturated vapor and liquid streams close to the top of the column—approximated for present purposes as H_1 and h_D from Illustration 7.3. Plot the equilibrium curve and operating lines for these conditions, and estimate whether the $(NTU)_{OG}$ is greater or less than was needed in Illustration 7.3 for the same separation with boiling reflux.

7.6 A mixture of chloroform and benzene is to be fed at 30 lb-mole/hr to a continuous, packed distillation column operating at atmospheric pressure. The mixture contains 0.45 mole fraction chloroform, and the feed will be a boiling liquid. A total condenser will be used, with a reflux ratio of 5. Three products are required, comprising a distillate, side stream, and residue containing 0.95, 0.68, and 0.15 mole fraction chloroform respectively. The mole ratio of distillate to side-stream product will be 1 to 1. If the column is randomly packed with 1.5-in. ceramic Berl saddles, determine the following:

(a) The amounts of products obtained per hour.
(b) The $(NTU)_{OG}$ required for the separation.
(c) The flow rates of the liquid and vapor streams in all three sections of the column.
(d) A column diameter such that the maximum approach to flooding is 45 percent.
(e) The $(HTU)_{OG}$ in each section of the column.
(f) The required column height.
(g) The location of the feed and side-stream take-off points.

Physical properties are given in Illustration 7.3.

Gas Absorption and Stripping

7.7 For the SO_2 absorption conditions described in Problem 4.1, estimate

(a) The percentage approach to flooding conditions.
(b) The average individual and overall G- and L-phase HTU values using the expressions given for k_La and k_Ga with equation 7.54.

7.8 Pure water will be used to absorb SO_2 from dry air by countercurrent contact in a column packed with 1-in. ceramic Berl saddles. The gas mixture will enter the column at a rate of 900 lb/hr, and the partial pressure of SO_2 in the inlet and outlet gas streams will be 0.080 and 0.0015 atm, respectively. The water flow rate is to be 35,000 lb/hr, and the column will operate at 45 percent of the gas flooding velocity. If the system is at 68°F and 1 atm, estimate the diameter and height of the column necessary to achieve the specified absorption.

Equilibrium data for this system are given in Perry (1963, p. **14**-6).

7.9 The desorption or stripping of oxygen from water into air in a countercurrent column packed with 1-in. Raschig rings has been studied by R. P. Whitney and J. E. Vivian [*Chem. Eng. Prog.*, **45**, 323–337 (1949)]. The column diameter and packed height were 8 and 24.5 in., respectively, and the system was at 62°F and 1 atm. The following measurements are taken from the authors' run 4:

$$\frac{L'}{S} = 4200 \text{ lb}/(\text{hr})(\text{ft}^2)$$

$$\frac{G'}{S} = 355 \text{ lb}/(\text{hr})(\text{ft}^2)$$

O_2 concn in entering water

$$= 6.97 \times 10^{-5} \text{ lb-mole/ft}^3 \text{ of soln}$$

O_2 concn in leaving water

$$= 2.80 \times 10^{-5} \text{ lb-mole/ft}^3 \text{ of soln}$$

If the air entering the column was saturated with water vapor, estimate the $(\text{HTU})_{OL}$ under these conditions. Consider dry air to contain 0.21 mole fraction O_2, and obtain equilibrium data for the system from Perry (1963, p. **14**-6).

Liquid Extraction

7.10 A solution of cyclohexane in *n*-heptane is being extracted countercurrently at 300 lb/hr in a packed column with aniline at 77°F, using extract reflux which is effectively solvent free. The feed solution contains 0.6 mass fraction cyclohexane, and the finished (solvent-free) extract and raffinate products contain 0.80 and 0.30 mass fraction cyclohexane, respectively, when the extract reflux ratio is 3.5. Determine

(a) The $(\mathrm{NTU})_{OG}$ corresponding to this operation.
(b) The amounts of finished raffinate and extract products.
(c) The aniline feed rate.

Equilibrium and saturation data for this system are given by T. G. Hunter and T. Brown, *Ind. Eng. Chem.*, **39**, 1343–1345, (1947).

7.11 Estimate the minimum reflux ratio for the extraction conditions of Problem 7.10.

7.12 Two separate aqueous solutions, F_1 and F_2, are to be extracted countercurrently in a packed column at 86°F, using ethyl propionate as extracting solvent. The stream F_1 is fed to one end of the column at 300 lb/hr and contains 0.60 mass fraction acetone, while F_2 is introduced part way along the column at 400 lb/hr and contains 0.25 mass fraction acetone. Both F_1 and F_2 are solvent free. Ethyl propionate will be fed at 233 lb/hr to obtain a final solvent-free raffinate containing 0.15 mass fraction acetone. Determine the $(\mathrm{NTU})_{OG}$ required in the packed column.
Physical data for this system are provided in Illustration 4.2.

REFERENCES

Badger, W. L., and J. T. Banchero, *Introduction to Chemical Engineering*, McGraw-Hill, New York, (1955).

Bennett, C. O., and J. E. Myers, *Momentum, Heat, and Mass Transfer*, McGraw-Hill, New York, (1962).

Bird, R. B., W. E. Stewart, and E. N. Lightfoot, *Transport Phenomena*, Wiley, New York, (1960).

Brown, G. G., et al., *Unit Operations*, Wiley, New York, (1950).

Calderbank, P. H., in *Mixing*, Vol. II, V. W. Uhl and J. B. Gray, Eds., Academic Press, New York, (1967), p. 49.

Chilton, T. H., and A. P. Colburn, *Ind. Eng. Chem.*, **27**, 255, (1935).

Chu, J. C., R. J. Getty, L. F. Brennecke, and R. Paul, *Distillation Equilibrium Data*, Reinhold, New York, (1950), p. 218.

Colburn, A. P., discussion in *A.I.Ch.E. Trans.*, **39**, 383, (1943).

Conte, S. D., *Elementary Numerical Analysis*, McGraw-Hill, New York, (1965), pp. 136–137.

Cornell, D., W. G. Knapp, and J. R. Fair, *Chem. Eng. Prog.*, **56** (7), 68–74, (1960).

Cornell, D., W. G. Knapp, H. J. Close, and J. R. Fair, *Chem. Eng. Prog.*, **56** (8), 48–53, (1960).

Eckert, J. S., *Chem. Eng. Prog.*, **57** (9), 54, (1961).

Eckert, J. S., *Chem. Eng. Prog.*, **59** (5), 76, (1963); see also "Tower Packings," Bulletin TP 54, The U. S. Stoneware Co., Akron, Ohio.

Eduljee, H. E., *Brit. Chem. Eng.*, **5**, 330, (1960).

Foust, A. S., L. A. Wenzel, C. W. Clump, L. Maus, and L. B. Andersen, *Principles of Unit Operations*, Wiley, New York, (1960).

Garner, F. H., S. R. M. Ellis, and D. W. Fosbury, *Trans. Inst. Chem. E. (Lond.)*, **31**, 348–362, (1953).

Hengstebeck, R. J., *Distillation: Principles and Design Procedures*, Reinhold, New York, pp. 274–275, (1961).

International Critical Tables (I.C.T.), McGraw-Hill, New York, (1929).

Larian, M. G., *Fundamentals of Chemical Engineering Operations*, Prentice-Hall, Englewood Cliffs, N. J., (1958).

Leva, M., *Tower Packings and Packed Tower Design*, 2nd ed., U. S. Stoneware Co., Akron, Ohio, (1953).

Levenspiel, O., *Chemical Reaction Engineering*, Wiley, New York, (1962).

McCabe, W. L., and J. C. Smith, *Unit Operations of Chemical Engineering*, McGraw-Hill, New York, (1956).

Oliver, E. D., *Diffusional Separation Processes*, Wiley, New York, (1966).

Otake, and Kimura, *Chem. Eng. (Japan)*, **17**, 261, (1953).

Perry, J. H., Ed., *Chemical Engineers' Handbook*, 3rd ed. McGraw-Hill, New York, (1950); 4th ed., (R. H. Perry, C. H. Chilton, and S. D. Kirkpatrick, Eds.), (1963).

Sherwood, T. K., and R. L. Pigford, *Absorption and Extraction*, McGraw-Hill, New York, (1952).

Skelland, A. H. P., *Ind. Eng. Chem.*, **53**, 799–800, (1961).

Teller, A. J., S. A. Miller, and E. G. Scheibel, in *Chemical Engineers' Handbook*, Perry, R. H., C. H. Chilton, and S. D. Kirkpatrick, Eds., 4th ed., McGraw-Hill, New York, (1963).

Treybal, R. E., *Liquid Extraction*, 2nd ed., McGraw-Hill, New York, (1963), pp. 491–492.

Treybal, R. E., *Mass Transfer Operations*, 2nd ed., McGraw-Hill, New York, (1968), p. 245.

Wiegand, J. H., *Trans. A. I. Ch. E.*, **36**, 679, (1940).

Yu, R. T., and J. Coull, *Chem. Eng. Prog.*, **46**, 89, (1950).

8

Design of Stagewise Columns from Rate Equations

Stagewise columns achieve contact between two phases in a discontinuous manner in stages which may, for example, take the form of bubble-cap plates or perforated plates. Both types of plate are widely used in gas-liquid contacting, such as distillation and gas absorption. In liquid-liquid extraction, the lower density difference between phases and the lower interfacial tensions cause bubble-cap plates to be ineffective, but perforated plates are effective and have been widely used.

Skelland and Cornish (1965) have presented a procedure for the design of perforated-plate extraction columns which is intended to eliminate the need for experimental determination of stage efficiencies, because these are normally obtained at substantial cost in time, effort, and money. Furthermore, the applicability of such efficiencies measured on small-pilot-plant to large-scale equipment is always an uncertain matter. Skelland and Cornish's procedure consists essentially in using rate equations for mass transfer during droplet formation, free rise (or fall), and coalescence on each plate, to locate a pseudo-equilibrium curve. This curve is used in place of the true equilibrium relationship when stepping off the desired number of actual stages between the pseudo-equilibrium and operating curves on the x_A-y_A diagram. Although the treatment is in principle valid for both gas-liquid and liquid-liquid systems, it is evident that greater success is to be anticipated in its application to liquid-liquid systems. This is because the much smaller density difference between phases and the substantially higher viscosity of the disperse phase cause the flow pattern to be less turbulent and more nearly predictable for liquid-liquid than for gas-liquid systems.

An outline and extension of the approach of Skelland and Cornish will now be given, although it should be recognized that expressions currently available for some stages of the procedure must be regarded as provisional only. On the other hand, the method may go a considerable way towards unifying the various aspects of hydrodynamic and mass-transfer study on drops into a coherent approach to design from a phenomenological point of view.

Correlations for plate efficiency in vapor-liquid systems will not be considered here. For such treatments, including in particular the A.I.Ch.E. correlation, the reader is referred to the *Bubble-Tray Design Manual*, A.I.Ch.E., New York, (1958), and to the surveys by Smith (1963, Chapter 16) and Oliver (1966, pp. 333–346).

Whereas the two phases in contact have previously been denoted by G and L regardless of which is dispersed, it is convenient in this chapter to refer instead to the continuous and the dispersed phases, identified as $\mathfrak{C}$ and $\mathfrak{D}$, respectively, where $\mathfrak{C}$ and $\mathfrak{D}$ are the flow rates of the continuous and dispersed phases through the entire cross section of column, total lb-mole/hr.

THE PSEUDO-EQUILIBRIUM CURVE

Consider the nth stage of a perforated-plate column, as shown in Figure 8.1.

For transfer into the disperse phase and with $y_{A\mathfrak{C}}$ constant for a given stage because of the mixing provided by the moving droplets, the rate of mass transfer in the nth stage is

$$q = K_{df}A_f(y_{A\mathfrak{C},n} - y_{Af})_{\mathrm{LM}} + K_{dr}A_r(y_{A\mathfrak{C},n} - y_{Ar})_{\mathrm{LM}} + K_{dc}A_c(y_{A\mathfrak{C},n} - y_{Ac})_{\mathrm{LM}} \tag{8.1}$$

which is the sum of the transfer rates during droplet formation on plate n,

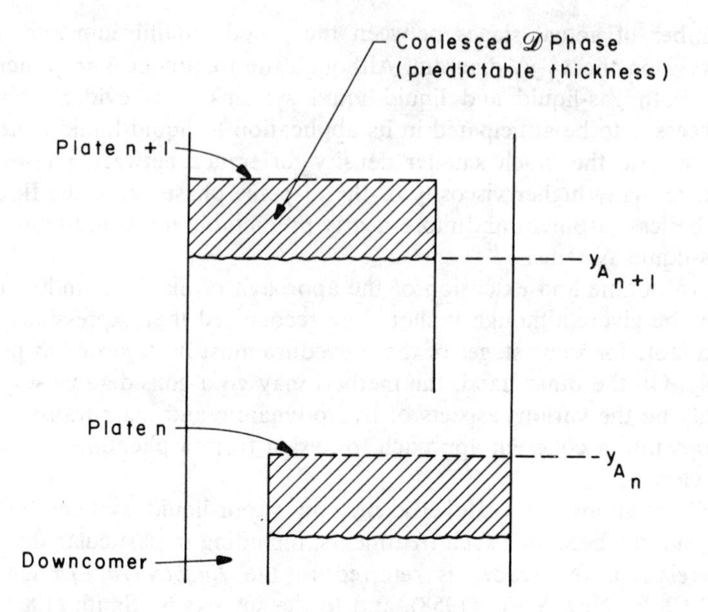

Figure 8.1. Plates $n+1$ and n in a perforated-plate column.

free rise, and coalescence beneath plate $n+1$. Now

$$(y_{A\mathfrak{C},n}-y_{Af})_{\mathrm{LM}} \doteq (y_{A\mathfrak{C},n}-y_{An}) \tag{8.2}$$

$$(y_{A\mathfrak{C},n}-y_{Ac})_{\mathrm{LM}} \doteq (y_{A\mathfrak{C},n}-y_{A\,n+1}) \tag{8.3}$$

and if $\mathfrak{D}$ does not vary significantly over stage n, then

$$(y_{A\mathfrak{C},n}-y_{Ar})_{\mathrm{LM}} \doteq \frac{(y_{A\mathfrak{C},n}-y_{An})-(y_{A\mathfrak{C},n}-y_{A\,n+1})}{\ln\left[(y_{A\mathfrak{C},n}-y_{An})/(y_{A\mathfrak{C},n}-y_{A\,n+1})\right]} \tag{8.4}$$

These approximations may be inserted in equation 8.1, giving

$$q=K_{df}A_f(y_{A\mathfrak{C},n}-y_{An})+K_{dr}A_r\left[\frac{(y_{A\mathfrak{C},n}-y_{An})-(y_{A\mathfrak{C},n}-y_{A\,n+1})}{\ln\left[(y_{A\mathfrak{C},n}-y_{An})/(y_{A\mathfrak{C},n}-y_{A\,n+1})\right]}\right]$$

$$+K_{dc}A_c(y_{A\mathfrak{C},n}-y_{A\,n+1}) \tag{8.5}$$

If it is assumed that only solute (A) is transferred or that solute transfer is accompanied by equimolal countertransfer of solvents between phases, then

$$q = \mathfrak{D}_{n+1} y_{A\,n+1} - \mathfrak{D}_n y_{An} \tag{8.6}$$

and from material balances, if $\mathfrak{D}$ enters the column at section 2,

$$\mathfrak{D}_n = \mathfrak{D}_2 \frac{1 - y_{A2}}{1 - y_{An}} \tag{8.7}$$

$$\mathfrak{D}_{n+1} = \mathfrak{D}_2 \frac{1 - y_{A2}}{1 - y_{A\,n+1}} \tag{8.8}$$

Suppose that A_f, A_r, A_c, K_{df}, K_{dr}, and K_{dc} can all be predicted. A trial-and-error procedure can then be used for estimating $y_{A\,n+1}$ corresponding to a given pair of $y_{A\,n}$ and $y_{A\mathcal{C},n}$ values in the following manner:

1. Assume a value of $y_{A\,n+1}$ corresponding to a selected pair of y_{An} and $y_{A\mathcal{C},n}$ values in Figure 8.2.
2. Calculate $\mathfrak{D}_n$ and $\mathfrak{D}_{n+1}$ corresponding to y_{An} and the assumed $y_{A\,n+1}$.
3. Calculate q from equations 8.5 and 8.6.

This process is repeated, if necessary, until the two estimates of q are in agreement, signifying that the assumed value of $y_{A\,n+1}$ is correct. The

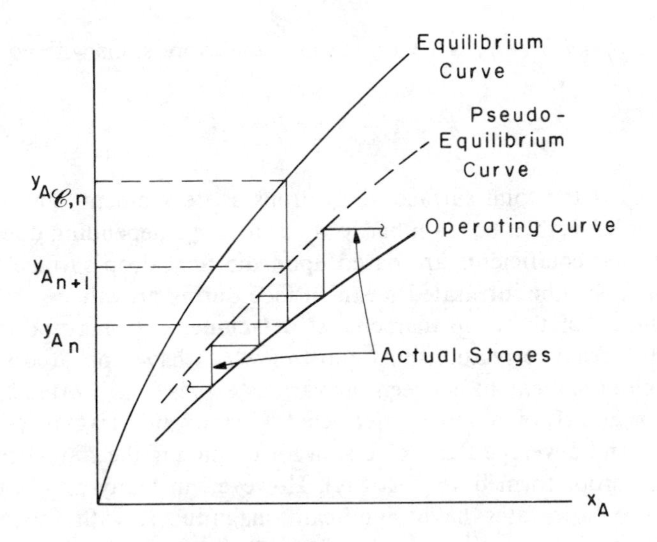

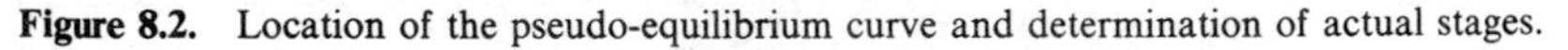
Figure 8.2. Location of the pseudo-equilibrium curve and determination of actual stages.

pseudo-equilibrium curve may be constructed in this way and used with the operating curve to step off the number of actual perforated plates necessary to achieve the desired composition change from y_{A2} to y_{A1}. (The operating curve is located in the manner described in Chapter 7.)

Relationships available for estimating the various interfacial areas and coefficients in equation 8.5 will now be reviewed for the case of liquid-liquid systems.

ESTIMATION OF INTERFACIAL AREA DURING DROPLET FORMATION

If a drop is assumed to grow as a sphere, the total integrated time-average surface during formation, with n_0 perforations per plate, is

$$A_{av} = \frac{n_0}{t_f} \int_0^{t_f} a\, dt \qquad (8.9)$$

If d_o and d_t are the diameters of the perforation and of the drop at time t, and u_o is the average velocity through the perforation, then a at t is πd_t^2, and the drop volume at t is $\pi d_o^2 u_o t/4 = \pi d_t^3/6$. Therefore,

$$dt = \frac{2d_t^2}{d_o^2 u_o} dd_t, \qquad t_f = \frac{2}{3} \frac{d_p^3}{d_o^2 u_o}$$

in which $d_t = d_p$ when $t = t_f$. Combining these expressions with equation 8.9,

$$A_{av} = \tfrac{3}{5} n_0 \pi d_p^2 = \tfrac{3}{5} A_d \qquad (8.10)$$

where A_d is the total surface of n_0 drops at detachment. The term A_f in equation 8.5 is then equal either to A_d or to $\frac{3}{5} A_d$, depending upon whether the selected coefficients are based upon the final drop surface at detachment or upon the integrated mean surface during growth.

Estimates of the drop diameter at detachment, d_p, may be made from several alternative relationships, most of which have appeared since 1950, although disagreement between the various expressions is often substantial. Harkins and Brown (1919) neglected kinetic and drag forces in their analysis and developed an expression for predicting the static drop volume (i.e., for drops formed very slowly). However, in many practical applications, the flow rates have significant magnitudes, with the result that kinetic and drag forces may become appreciable. Correlations under such conditions have been presented by Hayworth and Treybal (1950); Siemes (1956); Ueyama (1957); Null and Johnson (1958); Poutanen and Johnson (1960) (for gas bubbles in liquids); Rao, Kumar, and Kuloor (1966); Scheele and Meister (1968); Miss Anjali Basu (1970); Heertjes, de Nie, and de Vries (1971); de Chazal and Ryan (1971); Izard (1972); and Skelland and Raval (1972) (for power-law non-Newtonian systems). An excellent review of the available literature on droplet and bubble formation has been published by Kumar and Kuloor (1970).

Unfortunately, in spite of these extensive investigations, the correlations show areas of disagreement, and the expression presented by one investigator often fails to fit the experimental data of another. Thus the equation of Hayworth and Treybal fitted their 639 measurements with an average deviation of 7.5 percent, but predicted drop sizes larger than those found by Ruby and Elgin (1955) in multinozzle systems undergoing mass transfer and with countercurrent flow. Furthermore, when compared with 24 sets of measurements by Null and Johnson (1958), Hayworth and Treybal's equation revealed deviations exceeding 100 percent for one-third of the sets, with a maximum deviation of 377 percent. A maximum error of 94 percent was also found by Null and Johnson (1958) between their measured values and their own model. Neither the equation of Hayworth and Treybal (1950) nor the procedure of Null and Johnson (1958) fitted the measurements of Rao, Kumar, and Kuloor (1966), who accordingly provided their own trial-and-error procedure. Miss Basu (1970) compared her extensive measurements of drop size in a variety of systems with the correlations of Hayworth and Treybal (1950); Null and Johnson (1958); Rao, Kumar, and Kuloor (1966); and Scheele and Meister (1968). She concluded that none of these relationships adequately fitted her experimental data. de Chazal and Ryan (1971) found their measurements to be overestimated by more than 50 percent in some regions by Scheele and Meister's correlation.

It may be noted that Hayworth and Treybal (1950) varied the interfacial tension by use of the hydrophobic surfactant Alkaterge C, while Rao, Kumar, and Kuloor (1966) used the hydrophilic agent Lissapol. Skelland and Caenepeel (1972), however, have shown in related mass-transfer studies that the influence of surfactants during drop formation is highly specific and more complex than previously realized. Apparently the use of surface-active agents in drop formation measurements *could* lead to correlations suited only to the systems studied.

The correlation of drop size by Scheele and Meister (1968) is as follows

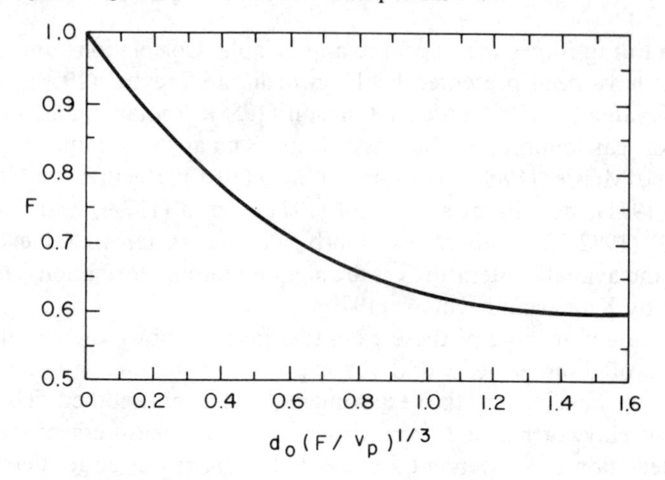

Figure 8.3. The Harkins and Brown correction factor F foı use in equation 8.11 (Scheele and Meister, 1968).

for short nozzles in the nonjetting region ($u_o < u_{oj}$):

$$v_p = F\left\{\frac{\pi\sigma g_c d_o}{g\Delta\rho} + \frac{5\pi\mu_c d_o^3 u_o}{d_p^2 g\Delta\rho} - \frac{\pi\rho_d d_o^2 u_o^2}{4g\Delta\rho} + 4.5\left[\left(\frac{\pi d_o^3 u_o}{4g\Delta\rho}\right)^2 \rho_d \sigma g_c\right]^{1/3}\right\} \quad (8.11)$$

The drag term (containing μ_c) may be neglected when μ_c is less than 10 cP, and the Harkins and Brown correction factor F is read from Figure 8.3. Equation 8.11 showed an average error of 11 percent in correlating measurements from 15 liquid-liquid systems. The data covered a range of interfacial tension from 1.8 to 45.4 dyn/cm and included two systems in which mass transfer was occurring.

At nozzle velocities somewhat greater than those for which equation 8.11 was developed a "jetting velocity" is reached, above which drops form from the break-up of a jet issuing from the nozzle. Scheele and Meister (1968) present the following equation for prediction of the "jetting velocity" in a given system:

$$u_{oj} = 2\left[\frac{\sigma g_c}{\rho_d d_o}\left(1 - \frac{d_o}{d_p}\right)\right]^{1/2} \quad (8.12)$$

The authors propose a brief iterative procedure in which a preliminary estimate of d_p is given by $d_p = (6F\sigma g_c d_o / g\Delta\rho)^{1/3}$; this is used for a first estimate of u_{oj} from equation 8.12; d_p is then recomputed from equation 8.11 using $u_o = u_{oj}$; finally u_{oj} is recalculated using this second estimate of d_p.

The number of perforations per plate is given by

$$n_0 = \frac{4Q_{d2}}{\pi d_o^2 u_o} \tag{8.13}$$

where Q_{d2} is the volumetric flow rate of disperse phase entering the column (ft^3/hr); $u_o/3600$ is normally selected to be between $\frac{1}{2}$ and 1.0 ft/sec; and $12d_o$, the perforation diameter, is usually chosen as $\frac{1}{8}$ to $\frac{1}{4}$ in., with the smaller diameters used for higher-interfacial-tension systems. The perforations are commonly located on either a square or a triangular pitch of $\frac{1}{2}$ to $\frac{3}{4}$ in.

ESTIMATION OF INTERFACIAL AREA DURING DROPLET RISE

The interfacial area A_r associated with the rising drops between two consecutive plates, may be estimated from the expression

$$A_r = \frac{(A_0 - A_D)(H - h_c)\phi_d}{\text{volume per drop}}(\text{surface area per drop}) \tag{8.14}$$

where ϕ_d is the disperse phase hold-up, A_D and A_0 are the cross-sectional areas of the downcomer and of the entire column, respectively, H is the vertical distance between plates, and h_c is the thickness of the coalesced layer. Procedures are now described for the evaluation of each of the terms in equation 8.14.

Disperse-Phase Holdup ϕ_d

The disperse-phase holdup ϕ_d is the fraction of column active volume occupied by the disperse phase, exclusive of the coalesced layer, and with the active volume defined to exclude the downcomer. The value of ϕ_d is obtainable from a finding by Beyaert, Lapidus, and Elgin (1961) and by Weaver, Lapidus, and Elgin (1959) that, for all fluid-particle systems in vertical motion, whether gas-solid, liquid-solid, or liquid-liquid,

$$\frac{u_s}{u_t} = f_s(\phi_d) \tag{8.15}$$

This indicates that the ratio of the slip velocity u_s to the terminal velocity of a single particle in a quiescent fluid, u_t, is a single or unique function of disperse-phase holdup. The slip velocity is the net linear velocity between the two phases; it is the relative velocity which would prevail if each phase were constrained to flow through that fraction of column cross section equal to the volume fraction of the phase under consideration. In a countercurrent spray column the slip velocity is

$$u_s = \frac{u_d}{\phi_d} + \frac{u_c}{1-\phi_d} \tag{8.16}$$

where the superficial velocities u_c and u_d for the continuous and disperse phases are based on the empty-column cross section. In the case of a perforated-plate column, u_c is taken to be zero because the continuous phase flows horizontally across the plate and does not affect the holdup. The value of u_d is then based on the column cross-section excluding the downcomer.

It was found that the function in equation 8.15 can be established from the correlation given by Zenz (1957) for the fluidization of solids, as shown in Figure 8.4.

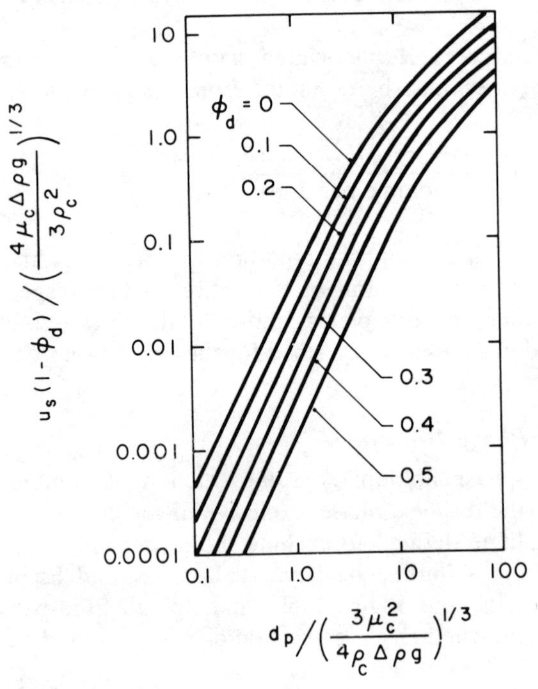

Figure 8.4. Correlation for fluidized solids by Zenz (1957).

The procedure for estimating ϕ_d may be itemized as follows:

1. Estimate the equivalent drop diameter d_p from equation 8.11 and evaluate the abscissa in Figure 8.4.
2. Obtain the ordinate in Figure 8.4 corresponding to an assumed ϕ_d and hence calculate u_s appropriate to a solid particle of the same density and diameter as the disperse liquid droplets.
3. Determine u_t ($=u_s$ at $\phi_d=0$) for the same hypothetical solid particle from Figure 8.4.
4. Calculate u_s/u_t corresponding to the assumed ϕ_d.
5. Evaluate u_t for a droplet of disperse phase with the diameter estimated in step 1 above. This is done with the Johnson and Braida or the Klee and Treybal correlation, to be described.
6. Compute u_s for the two liquid phases corresponding to the assumed ϕ_d by combining the results of steps 4 and 5 above.
7. Calculate $u_s\phi_d$ from step 6.
8. Repeat steps 2, 4, 6, and 7 above for other assumed values of ϕ_d, and plot $u_s\phi_d$ against ϕ_d.
9. From equation 8.16 with $u_c=0$, obtain $u_s\phi_d=u_d$, which is known. This permits the evaluation of ϕ_d from the plot in step 8; u_s may then be computed.

A simpler, but supposedly more approximate, procedure utilizes a relationship proposed by Thornton (1956):

$$u_s=u_t(1-\phi_d) \tag{8.17}$$

Combination with equation 8.16 ($u_c=0$ for a perforated-plate column) gives

$$u_t(1-\phi_d)=\frac{u_d}{\phi_d} \tag{8.18}$$

Whether to use the simpler procedure for the estimation of ϕ_d depends on the accuracy desired. In any event, Garner and Skelland (1955) showed that minute amounts of certain surface-active impurities can halve the terminal velocity of a given droplet in another immiscible liquid. Such impurities may often be present in recycled solvent, and since both methods of calculation of ϕ_d require an estimation of u_t, it is by no means clear which procedure conforms more closely to a real situation.

Droplet Terminal Velocity u_t

The droplet terminal velocity u_t is needed for evaluating ϕ_d and may be obtained from the correlation due to Hu and Kintner (1955) as modified by Johnson and Braida (1957) for continuous phase viscosities up to about 30 cP. Their empirical relationship appears in Figure 8.5 as a plot of

$$\frac{C_D N_{We} N_P^{0.15}}{(\mu_c/\mu_{H_2O})^{0.14}} \text{ versus } \frac{N_{Re}}{N_P^{0.15}}$$

where

$$N_{Re} = \frac{d_p u_t \rho_c}{\mu_c} \tag{8.19}$$

$$N_P = \frac{4N_{Re}^4}{3C_D N_{We}^3} = \frac{\rho_c^2 \sigma^3 g_c^3}{g \mu_c^4 \Delta\rho} \tag{8.20}$$

$$C_D = \text{drag coefficient} = \frac{4\Delta\rho d_p g}{3\rho_c u_t^2} \tag{8.21}$$

$$N_{We} = \frac{d_p u_t^2 \rho_c}{\sigma g_c} \tag{8.22}$$

An alternative set of equations for evaluating u_t is obtainable from the expressions of Klee and Treybal (1956), which may be written in lb-mass-ft-hr units as

$$u_t = \frac{53{,}000 \Delta\rho^{0.58} d_p^{0.70}}{\rho_c^{0.45} \mu_c^{0.11}}, \qquad d_p < d_{pt} \tag{8.23}$$

$$u_t = \frac{577 \Delta\rho^{0.28} \mu_c^{0.10} (\sigma g_c)^{0.18}}{\rho_c^{0.55}}, \qquad d_p > d_{pt} \tag{8.24}$$

$$d_{pt} = 7.25 \sqrt{\frac{\sigma g_c}{g \Delta\rho N_P^{0.15}}} \tag{8.25}$$

The onset of oscillations with larger droplets causes a maximum to appear in the plot of u_t versus d_p in a given liquid-liquid system. This occurs at the transition value of the drop diameter given by equation 8.25 and also at the abscissa value indicated in Figure 8.5. These relationships enable the estimation of u_t with an accuracy within about 10 percent for $d_p < d_{pt}$ and about 15 percent for $d_p > d_{pt}$, provided that the system is free from surface-active agents, which normally reduce the terminal velocity.

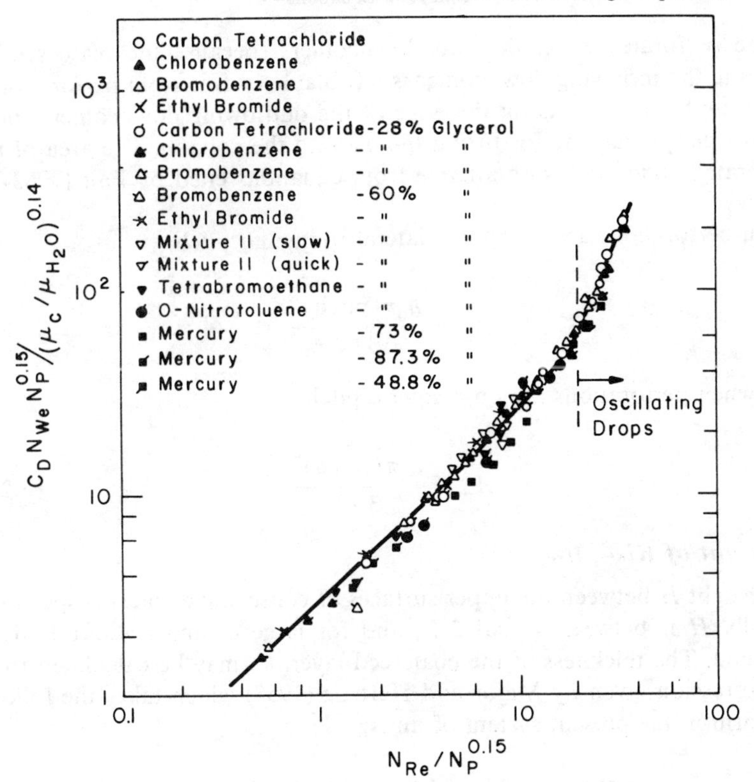

Figure 8.5. Correlation of terminal velocities of single droplets in liquid-liquid systems. This modified Hu and Kintner plot was presented by Johnson and Braida (1957) and shows data on organic-solvent and mercury drops falling in water and in aqueous glycerol solutions.

Downcomer and Column Cross Sections, A_D and A_0

The average velocity in the downcomer, u_D, is customarily set equal to the terminal velocity of some arbitrary-size, small droplet of the disperse phase, say $12d_p = \frac{1}{16}$ in., so that u_D may be calculated from equation 8.23. The cross-sectional area of the downcomer, A_D, is then found from the equation

$$A_D = \frac{Q_{c1}}{u_D} \tag{8.26}$$

where Q_{c1} is the volumetric flow rate of the continuous phase entering the column, in ft^3/hr.

The area of the entire column cross section, A_0, is the sum of the areas

of the perforated zone, the two downcomers (because the plate section opposite the inflowing downcomer is left blank, or free from perforations), a narrow blank strip along the edge of the outflowing downcomer, and a narrow peripheral ring for fitting the plate to the column. The area of the perforated zone, A_{pz}, is obtainable from equations cited by Fair (1963) as follows:
When perforations are on an equilateral triangular pitch,

$$A_{pz} = \frac{n_0 \pi (\text{pitch})^2}{3.62} \tag{8.27}$$

and when perforations are on a square pitch,

$$A_{pz} = \frac{n_0 \pi (\text{pitch})^2}{3.14} \tag{8.28}$$

Height of Rise, $H\text{-}h_c$

The height H between the upper surfaces of consecutive plates is specified. Usually H is between $\frac{1}{2}$ and 2 ft, and for large columns about $1\frac{1}{2}$ ft is common. The thickness of the coalesced layer, h_c, may be calculated from an expression given by Major and Hertzog (1955) which takes the following form in the present system of units:

$$h_c = \frac{0.521 \times 10^{-9} \sigma g_c \mu_d^{0.4} \mu_c^{0.2}}{\Delta\rho d_o^{1.4}} + \frac{u_o^2 \rho_d}{2g\Delta\rho}\left(1 \quad \frac{0.71}{\log N_{\text{Re},o}}\right)^{-2} + \frac{0.592 \times 10^{-8} u_D^2 \rho_c}{\Delta\rho} \tag{8.29}$$

where h_c is measured from the top surface of the plate when the lighter liquid is dispersed—in other words, h_c includes the plate thickness. The first term in equation 8.29 makes empirical allowance for the influence of interfacial tension, and the second and third terms represent the orifice effect and frictional effects in the downcomer, respectively. An alternative procedure for the estimation of h_c has been provided by Bussolari, Schiff, and Treybal (1953).

Droplet Surface Area

The volume per drop is needed in equation 8.14, and it may be estimated from equation 8.11. During free fall, non-oscillating droplets assume a shape which may be approximated by an oblate spheroid with the major axis oriented horizontally. [Sustained oscillations are apparently initiated when the Weber number $d_p u_t^2 \rho_c / \sigma g_c$ reaches a value of 3.58 (Hu and Kintner, 1955; Basu, 1970, p. 86).] The surface area of an oblate spheroid may be conveniently expressed in terms of eccentricity and droplet volume. The relationship is provided by Heertjes, Holve, and Talsma (1954) as

$$S_{os} = 2\pi \left(\frac{3v_p E}{4\pi} \right)^{2/3} \left[1 + \frac{1}{E\sqrt{E^2 - 1}} \ln \left(E + \sqrt{E^2 - 1} \right) \right] \tag{8.30}$$

The eccentricity of the non-oscillating droplets is obtainable from the correlation given by Wellek, Agrawal, and Skelland (1966):

$$E = 1.0 + 0.093 \left(\frac{d_p u_s^2 \rho_c}{\sigma g_c} \right)^{0.98} \left(\frac{\mu_c}{\mu_d} \right)^{0.07} \tag{8.31}$$

This relationship fitted 198 experimental measurements on 45 systems free from surface-active agents with an average absolute deviation of 6 percent over a droplet Reynolds-number range of 6 to 1354. The ratio of the area of an oblate spheroid to that of a sphere of equal volume is (Garner and Tayeban, 1960, p. 479)

$$\frac{S_{os}}{S_s} = \tfrac{1}{2} E^{2/3} + \frac{1}{2E^{1/3}\sqrt{E^2 - 1}} \ln \left(E + \sqrt{E^2 - 1} \right) \tag{8.32}$$

At eccentricities of 1.5, 2, 2.5, and 3, the above ratio is approximately 1.031, 1.095, 1.17, and 1.26, respectively. It is evident, therefore, that realistic distortion of the drops leads to significant increases in the surface between phases during free rise.

The term A_r is calculated from equation 8.14 using either S_s or S_{os}, depending upon whether the selected coefficients are based on the surface of a sphere having the same volume as the drop or on the surface of an oblate spheroid.

ESTIMATION OF INTERFACIAL AREA DURING DROPLET COALESCENCE

The interfacial area available for mass transfer during droplet coalescence, A_c, will be defined as the plane interface between the continuous phase and the coalesced disperse phase; A_c is accordingly expressed as

$$A_c = A_0 - A_D \tag{8.33}$$

ESTIMATION OF THE OVERALL TRANSFER COEFFICIENT DURING DROPLET FORMATION

In an early approach, Licht and Pansing (1953) simplified the drop surface to a plane, and considered the process to be continuous-phase controlled. Their resulting relationship has been written by Treybal (1963) as

$$K_{df}^{*} = \frac{0.805}{m}\left(\frac{\rho}{M}\right)_{\text{av},d}\sqrt{\frac{D_c}{t_f}} \tag{8.34}$$

where K_{df}^{*} is based on the integrated average surface during drop formation, given by equation 8.10. The overall mass-transfer coefficient during formation is, of course, compounded from the individual coefficients for the disperse and continuous phases by means of equation 4.7. This equation assumes local equilibrium at the interface; m is the slope of the equilibrium curve, dy_A/dx_A, and may vary with composition. Measurements by Garner and Skelland (1954) exceeded equation 8.34 by an average of 50 percent, but the data included the effects of drop detachment. In this regard, Popovich et al. (1964) contend from experiment that negligible mass transfer occurs during drop detachment. This conflicts with Licht and Conway (1950), who concluded from their measurements that the effects of formation and detachment on mass transfer are about equal. The question is currently unresolved.

Relationships for the Individual Disperse-Phase Coefficient

In one approach the droplet surface is considered to grow by the addition of fresh elements to the surface. Analyses along these lines have been made by Groothuis and Kramers (1955), Beek and Kramers (1962), Heertjes and de Nie (1966), Calderbank and Patra (1966), and Skelland and Hemler (1969). In another approach, the droplet surface is thought to increase by the stretching of existing elements already at the surface. Developments on this basis are given by Beek and Kramers (1962); Angelo, Lightfoot, and Howard (1966); Heertjes and de Nie (1966); and Angelo and Lightfoot (1968).

Heertjes and de Nie (1966) compared theoretical relationships corresponding to each mechanism with their experimental results and concluded that the expressions based on growth of the surface by the addition of fresh elements fitted their data best. In contrast, Angelo and Lightfoot (1968) considered their results to be well represented by expressions resulting from the surface-stretch model. Other relationships have appeared, and these have been reviewed by Skelland and Cornish (1965).

Heertjes, Holve, and Talsma (1954) applied Higbie's (1935) penetration theory to molecular diffusion into a droplet growing uniformly and spherically about a point and at a rate much greater than the rate of diffusion. Their expression has been given as (Perry, 1963)

$$k_{df}^{*}=\frac{24}{7}\left(\frac{\rho}{M}\right)_{\mathrm{av}}\sqrt{\frac{D_d}{\pi t_f}} \tag{8.35}$$

Equation 8.35 fitted measurements by Johnson and Hamielec (1960), but not those by Heertjes et al. (1954) or by Popovich, Jervis, and Trass (1964), who found Ilkovic's (1934) expression to agree best with their results, as follows:

$$k_{df}^{*}=1.31\left(\frac{\rho}{M}\right)_{\mathrm{av}}\sqrt{\frac{D_d}{\pi t_f}} \tag{8.36}$$

Additional expressions are found in the references cited above, and also in the papers of Sawistowski and Goltz (1963), Zheleznyak (1967), and Coulson and Skinner (1952). As an example of an analysis based on the assumption that the surface of a growing droplet is extended by the addition of fresh surface elements, one may consider the following treatment bý Skelland and Hemler (1969). It will be assumed that:

1. The drop grows spherically about a point source.
2. The concentration driving force remains constant throughout drop formation at Δy_{Af}.
3. The fresh surface elements continually added to the growing surface are subject to mass transfer in accordance with the penetration theory.

The solute transferred during the formation of a single drop may be expressed as

$$q_{sf}=k_{df}^{*}\,\pi d_p^2\Delta y_{Af}t_f=\Delta y_{Af}\int_0^{\pi d_p^2}k_d^{*}(t_f-t)\;da$$

so that the average mass-transfer coefficient during drop formation, based on the drop surface at detachment, is

$$k_{df}^{*}=\frac{1}{\pi d_p^2 t_f}\int_0^{\pi d_p^2}k_d^{*}\;(t_f-t)\;da \tag{8.37}$$

From assumption 3,

$$k_d^* = 2\left(\frac{\rho}{M}\right)_{\text{av}} \sqrt{\frac{D_d}{\pi(t_f - t)}} \tag{8.38}$$

where each fresh element arriving at the surface at time t is exposed for a time $t_f - t$. Relationships stated below equation 8.9 show that

$$d_t = d_p\left(\frac{t}{t_f}\right)^{1/3}, \quad a = \pi d_p^2\left(\frac{t}{t_f}\right)^{2/3} \tag{8.39}$$

Combining equations 8.37, 8.38, and 8.39 leads to

$$k_{df}^* = \frac{4}{3}\left(\frac{\rho}{M}\right)_{\text{av}} \sqrt{\frac{D_d}{\pi t_f}} \int_0^{t_f} \left(\frac{t_f - t}{t_f}\right)^{1/2} \left(\frac{t}{t_f}\right)^{-1/3} \frac{dt}{t_f}$$

Defining $z = t/t_f$,

$$k_{df}^* = \frac{4}{3}\left(\frac{\rho}{M}\right)_{\text{av}} \sqrt{\frac{D_d}{\pi t_f}} \int_0^{1.0} (1-z)^{1/2} z^{-1/3}\, dz \tag{8.40}$$

The definite integral in equation 8.40 is a standard form of the beta function, which is in turn composed of gamma functions in the following general manner (Jenson and Jeffreys, 1963, pp. 154, 157):

$$\beta(p,q) = \frac{\Gamma(p)\Gamma(q)}{\Gamma(p+q)} = \int_0^{1.0} (1-z)^{q-1} z^{p-1}\, dz \tag{8.41}$$

where p and q are positive. Comparison between equations 8.40 and 8.41 shows that $p = \frac{2}{3}$, $q = \frac{3}{2}$, and therefore,

$$\int_0^{1.0} (1-z)^{1/2} z^{-1/3}\, dz = \beta\left(\tfrac{2}{3}, \tfrac{3}{2}\right) = \frac{\Gamma\left(\frac{2}{3}\right)\Gamma\left(\frac{3}{2}\right)}{\Gamma\left(\frac{13}{6}\right)} \tag{8.42}$$

The right-hand side of equation 8.42 was evaluated from tabulated values of the gamma function in the manner described by Jenson and

Jeffreys (1963, p. 152). The result may be stated in the following forms:

$$k_{df}^{*} = 1.48\left(\frac{\rho}{M}\right)_{av}\sqrt{\frac{D_d}{\pi t_f}} = \frac{24}{7}(0.4317)\left(\frac{\rho}{M}\right)_{av}\sqrt{\frac{D_d}{\pi t_f}} \tag{8.43}$$

The second form of equation 8.43 facilitates comparison with the relationship of Heertjes, Holve, and Talsma in equation 8.35. It would appear that the above development is applicable to either the disperse or the continuous phase when appropriate physical properties are used.

The coefficients k_{df}^{*} in equations 8.35 to 8.43 are based on A_d rather than A_{av}, and do not include effects of drop detachment from the nozzle.

Correlation of Experimental Data

Mass-transfer rates during drop formation were measured by Skelland and Minhas (1971) in one binary and two ternary systems which were disperse-phase controlled. The systems were ethyl acetate–water, acetic acid–chlorobenzene–water, and acetic acid–carbon tetrachloride + Nujol–water, with water as the continuous phase in the ternary systems and the disperse phase in the binary system. Solute diffused into the drops in the first system and out in the other two. Drops formed simultaneously on three thin-walled glass nozzles set on $\frac{3}{4}$-in. triangular pitch. The measured mass-transfer rates—which included the effects of droplet detachment—were higher than predicted by the theoretical models of Heertjes et al., Skelland and Hemler, Coulson and Skinner, Groothuis and Kramers, and Ilkovic, and were empirically correlated using least-squares statistical techniques by the equation

$$k_{df}^{*} = 0.0432\frac{d_p}{t_f}\left(\frac{\rho}{M}\right)_{av}\left(\frac{u_o^2}{d_p g}\right)^{0.089}\left(\frac{d_p^2}{t_f D_d}\right)^{-0.334}\left(\frac{\mu_d}{\sqrt{\rho_d d_p \sigma g_c}}\right)^{-0.601} \tag{8.44}$$

This expression correlated 23 measurements with an average absolute deviation of about 26 percent. The coefficient is based on A_d.

Relationships for the Individual Continuous-Phase Coefficient

For the continuous-phase coefficient during droplet formation, an expression has been obtained by assuming the droplet to grow away from a fixed orifice, instead of around a fixed point. The result, due to Michels and

Pigford, is in the form of a coefficient based on A_d and has been stated as (Perry, 1963)

$$k_{cf}^{*}=4.6\left(\frac{\rho}{M}\right)_{\text{av}}\sqrt{\frac{D_c}{\pi t_f}} \tag{8.45}$$

This relationship, however, gives a rate roughly 3 times greater than that found both experimentally and theoretically for gas bubbles by Calderbank and Patra (1966). Their analysis was in terms of growth of the bubble surface by the addition of fresh surface elements.

Correlation of Experimental Data

A study similar to that resulting in equation 8.44 was performed by Skelland and Hemler (1969), but using three ternary systems in which mass transfer was continuous-phase controlled. The systems were as follows, in the order *solute–disperse phase–continuous phase*: acetic acid–water–toluene, benzoic acid–chlorobenzene–water, and acetic acid–water–benzyl isopentyl ether. Transfer was from the disperse to the continuous phase throughout. Three nozzles were used, as in the work of Skelland and Minhas (1971), and their 20 data points were correlated with an average absolute deviation of 11 percent using statistical techniques by the equation

$$k_{cf}^{*}=0.386\left(\frac{\rho}{M}\right)_{\text{av}}\left(\frac{D_c}{t_f}\right)^{0.5}\left(\frac{\rho_c\sigma g_c}{\Delta\rho g t_f \mu_c}\right)^{0.407}\left(\frac{g t_f^2}{d_p}\right)^{0.148} \tag{8.46}$$

The coefficient in this expression is based on A_d and includes the effects of drop detachment from the nozzle.

ESTIMATION OF THE OVERALL TRANSFER COEFFICIENT DURING DROPLET RISE

Relationships are here presented for the disperse- and continuous-phase coefficients in turn, corresponding to a variety of hydrodynamic conditions and with various distributions of mass-transfer resistance.

Relationships for the Individual Disperse Phase Coefficient

The coefficients here are for use with driving forces expressed as mole fractions, to be consistent with equations 8.5 and 8.6. Consider a balance on component A diffusing into a spherical droplet that is rising vertically during time dt:

$$k_{dr}\pi d_p^2 (y_A^* - y_A)\, dt = \frac{\pi d_p^3}{6}\left(\frac{\rho}{M}\right)_{\text{av}} dy_A \tag{8.47}$$

Integration is facilitated by considering y_A^* constant at its average value between sections a and b, giving

$$k_{dr} = -\frac{d_p}{6t}\left(\frac{\rho}{M}\right)_{\text{av}} \ln\left[\frac{y_A^* - y_{Aa}}{y_A^* - y_{Aa}} - \frac{y_{Ab} - y_{Aa}}{y_A^* - y_{Aa}}\right]$$

or

$$k_{dr} = -\frac{d_p}{6t}\left(\frac{\rho}{M}\right)_{\text{av}} \ln(1 - E_f) \tag{8.48}$$

where E_f is the fractional extraction in time t. Analytical expressions have been developed for E_f for a variety of situations, namely, when the droplets are internally stagnant, internally circulating because of the frictional drag of the continuous phase, and oscillating with partial internal circulation. The expressions have been developed with and without finite resistance in the continuous phase in each case, and the analytical results have sometimes been approximated by simpler empirical equations. The results appear in Table 8.1, where equation 8.49, for example, arises from combination of equation 8.48 with E_f obtained from equation 2.86.

The penetration theory has been applied by Rose and Kintner (1966) and Angelo et al. (1966) to the stretching and contracting surfaces of oscillating drops. Patel and Wellek (1967), however, contend that the results of these analyses are not significantly different from equations 8.56 and 8.57 for oscillating conditions in Table 8.1.

Correlation of Experimental Data

Skelland and Wellek (1964) made extensive measurements of k_{dr}^* for nonoscillating and oscillating droplets in four binary liquid-liquid systems where the continuous-phase resistance was virtually zero. The values were correlated by statistical analysis after equations 8.49, 8.53, and 8.56 were shown to be inadequate for fitting the data. The discrepancy between these equations and the experimental results may relate to the simplifying assumption of spherical drops and to the unknown degree of internal circulation or oscillation in the droplets (Garner and Skelland, 1951, 1954, 1955; Garner, Skelland, and Haycock, 1954).

Table 8.1. Theoretical expressions for k^*_{dr}.

Droplet Condition[a]	Eq. no.	Equation	Reference
Stagnant, $R_c=0$	8.49	$k^*_{dr}=-\frac{d_p}{6t}\left(\frac{\rho}{M}\right)_{av}\ln\left\{\frac{6}{\pi^2}\sum_{n=1}^{\infty}\frac{1}{n^2}\exp\left[\frac{-D_d n^2\pi^2 t}{(d_p/2)^2}\right]\right\}$	Eq. 8.48 and Eq. 2.86
	8.50	or $k^*_{dr}=-\frac{d_p}{6t}\left(\frac{\rho}{M}\right)_{av}\ln\left[1-\frac{\pi D_d^{1/2}t^{1/2}}{d_p/2}\right]$ when $E_f<0.5$[b]	Vermeulen (1953), Johnson et al. (1958)
Stagnant, $R_c>0$	8.51	$k^*_{dr}=-\frac{d_p}{6t}\left(\frac{\rho}{M}\right)_{av}\ln\left\{6\sum_{n=1}^{\infty}B_n\exp\left[-\lambda_n^2\frac{D_d t}{(d_p/2)^2}\right]\right\}$, $B_n,\lambda_n=$ functions of R_c	Grober (1925)
	8.52	or $k^*_{dr}=-\frac{d_p}{6t}\left(\frac{\rho}{M}\right)_{av}\ln\left[1-0.905\left(\frac{D_d\pi^2 t}{(d_p/2)^2}\right)^{1/2}-0.0189\right]$ when $E_f<0.5$[b]	Johnson and Hamielec (1960)

Circulating, $R_c = 0$	8.53	$k^*_{dr} = -\frac{d_p}{6t}\left(\frac{\rho}{M}\right)_{av} \ln\left\{\frac{3}{8}\sum_{n=1}^{\infty} B_n^2 \exp\left[-\lambda_n \frac{16 D_d t}{(d_p/2)^2}\right]\right\}$	Kronig and Brink (1950)
	8.54	or $k^*_{dr} = -\frac{d_p}{6t}\left(\frac{\rho}{M}\right)_{av} \ln\left[1 - \frac{R^{1/2}\pi D_d^{1/2} t^{1/2}}{d_p/2}\right]$,	Johnson et al. (1958)
		$R \doteq 2.25$, when $E_f < 0.5$[b]	
Circulating, $R_c > 0$		Eq. 8.53 with B_n, λ_n functions of R_c;	Elzinga and Banchero (1960)
	8.55	also $k^*_{dr} = 2\left(\frac{\rho}{M}\right)_{av} \sqrt{\frac{D_d}{\pi t_e}}, \quad t_e = \frac{d_p}{u_s}$	Higbie (1935)
Oscillating, $R_c = 0$	8.56	$k^*_{dr} = -\frac{d_p}{6t}\left(\frac{\rho}{M}\right)_{av} \ln\left\{2\sum_{n=1}^{\infty} B_n \exp\left[\frac{-\lambda_n D_d t N'_{Pe}}{128 d_p^2}\right]\right\}$	Handlos and Baron (1957)
Oscillating, $R_c > 0$	8.57	$k^*_{dr} = \left(\frac{\rho}{M}\right)_{av} \frac{\lambda_1 u_s}{768(1 + \mu_d/\mu_c)}$, λ_1 a function of R_c,	Wellek and Skelland (1965)
		$\lambda_1 = 2.88$ when $R_c = 0$	

[a] R_c = continuous-phase resistance to mass transfer.

[b] An empirical approximation when $E_f < 0.5$.

For *nonoscillating drops with internal circulation*, Skelland and Wellek (1964) correlated their 115 measurements on four binary systems with an average deviation of ± 34 percent as follows:

$$k_{dr}^{*} = 31.4 \frac{D_d}{d_p} \left(\frac{\rho}{M} \right)_{\text{av}} \left(\frac{4 D_d t}{d_p^2} \right)^{-0.34} \left(\frac{\mu_d}{\rho_d D_d} \right)^{-0.125} \left(\frac{d_p u_s^2 \rho_c}{\sigma g_c} \right)^{0.37} \tag{8.58}$$

For *oscillating drops* their results were statistically correlated with the equation

$$k_{dr}^{*} = 0.32 \frac{D_d}{d_p} \left(\frac{\rho}{M} \right)_{\text{av}} \left(\frac{4 D_d t}{d_p^2} \right)^{-0.14} \left(\frac{d_p u_s \rho_c}{\mu_c} \right)^{0.68} \left(\frac{\sigma^3 g_c^3 \rho_c^2}{g \mu_c^4 \Delta \rho} \right)^{0.10} \tag{8.59}$$

Equation 8.59 was subsequently shown by Brunson and Wellek (1970) to be the best of 12 alternative relationships for oscillating droplets, because it fitted 110 data points from three systems with an average absolute deviation of 15.6 percent. The measurements corresponded to an interfacial tension range of 3.5 to 21 dyn/cm.

Table 8.1 and equations 8.58 and 8.59 provide coefficients k_{dr}^{*} which are based on the surface area of a sphere having the same volume as the drop.

Relationships for the Individual Continuous-Phase Coefficient

The coefficients will be written in forms suitable for use with driving forces expressed as mole fractions to ensure consistency with equations 8.5 and 8.6. Expressions are again available for stagnant, circulating, and oscillating drops. Theoretical developments are well reviewed by Calderbank (1967), but k_{cr}^{*} equations will be presented here in empirical terms to conform more closely with experiment.

Correlation of Experimental Data

It seems likely that k_{cr}^{*} will be modified more than k_{dr}^{*} by interaction with adjacent droplets, but information on such effects is fragmentary. Empirical correlations of experimental k_{cr}^{*} values are given in Table 8.2 for various droplet conditions and degrees of interaction. Equation 8.60 is a special case of 8.61, which showed an average deviation of less than ± 2 percent from 100 measured values. The maximum deviation between equation 8.64 and 24 data points from three systems was ± 12 percent. No error estimates are available for equations 8.62 and 8.63.

Table 8.2. Empirical correlations for k_{cr}^*.

Droplet condition	Eq. no.	Correlation	Reference
Stagnant, single, spherical[a]	8.60	$k_{cr}^* = 0.74 \frac{D_c}{d_p} \left(\frac{\rho}{M}\right)_{av} \left(\frac{d_p u_s \rho_c}{\mu_c}\right)^{1/2} \left(\frac{\mu_c}{\rho_c D_c}\right)^{1/3}$	Skelland and Cornish (1963)
Stagnant, single, oblate spheroidal[b]	8.61	$k_{cr}^* = 0.74 \frac{D_c}{d_3} \left(\frac{\rho}{M}\right)_{av} \left(\frac{d_3 u_s \rho_c}{\mu_c}\right)^{1/2} \left(\frac{\mu_c}{\rho_c D_c}\right)^{1/3}$	Skelland and Cornish (1963)
Circulating, single drop stream	8.62	$k_{cr}^* = 0.6 \frac{D_c}{d_p} \left(\frac{\rho}{M}\right)_{av} \left(\frac{d_p u_s \rho_c}{\mu_c}\right)^{1/2} \left(\frac{\mu_c}{\rho_c D_c}\right)^{1/2}$, low σ	Garner and Tayeban (1960)
Circulating, swarms of drops	8.63	$k_{cr}^* = 0.725 \left(\frac{\rho}{M}\right)_{av} \left(\frac{d_p u_s \rho_c}{\mu_c}\right)^{-0.43} \left(\frac{\mu_c}{\rho_c D_c}\right)^{-0.58} u_s(1-\phi_d)$, low σ	Ruby and Elgin (1955), Treybal (1963)
Oscillating, single drop stream	8.64	$k_{cr}^* = \frac{D_c}{d_p} \left(\frac{\rho}{M}\right)_{av} \left[50 + 0.0085 \left(\frac{d_p u_s \rho_c}{\mu_c}\right)^{1.0} \left(\frac{\mu_c}{\rho_c D_c}\right)^{0.7}\right]$, low σ	Garner and Tayeban (1960)

[a] For other expressions, see Table 6.3.

[b] d_3 is given by equation 6.150, after Pasternak and Gauvin (1960).

With the exception of the oblate spheroidal drop, the coefficients k_{cr}^* in Table 8.2 are based on the surface area of a sphere with the same volume as the drop.

ESTIMATION OF THE OVERALL TRANSFER COEFFICIENT DURING DROPLET COALESCENCE

Although mass transfer during droplet coalescence has received only preliminary theoretical attention, two recent correlations of experimental data are available for k_{dc}^* and k_{cc}^* respectively.

Relationships for the Individual Disperse Phase Coefficient

In a preliminary analysis, Johnson and Hamielec (1960) regarded each drop as spreading over the entire coalescence surface in a uniform layer, to which Higbie's (1935) penetration theory may be applied, giving

$$k_{dc}^* = 2\left(\frac{\rho}{M}\right)_{av}\sqrt{\frac{D_d}{\pi t_f}} \tag{8.65}$$

The term t_f in equation 8.65 assumes transient transfer between the arrival of consecutive drops.

Correlation of Experimental Data

Mass-transfer rates during drop coalescence were measured by Skelland and Minhas (1971) and were in general lower than predicted by equation 8.65. The results, which were obtained with the same three systems and triple nozzle equipment used to develop equation 8.44, were statistically correlated by the expression

$$k_{dc}^* = 0.173\frac{d_p}{t_f}\left(\frac{\rho}{M}\right)_{av}\left(\frac{\mu_d}{\rho_d D_d}\right)^{-1.115}\left(\frac{\Delta\rho g d_p^2}{\sigma g_c}\right)^{1.302}\left(\frac{u_s^2 t_f}{D_d}\right)^{0.146} \tag{8.66}$$

The average absolute deviation between equation 8.66 and 23 data points was 25 percent.

Relationships for the Individual Continuous-Phase Coefficient

Correlation of Experimental Data

Skelland and Hemler (1969) performed an experimental study similar to that leading to equation 8.66. The triple-nozzle equipment and the three ternary systems investigated were the same as those used to develop equation 8.46. Statistical correlation of their 20 measurements was achieved with an average absolute deviation of 22 percent by the equation

$$k_{cc}^* = 5.959 \times 10^{-4} \left(\frac{\rho}{M} \right)_{av} \left(\frac{D_c}{t_f} \right)^{0.5} \left(\frac{\rho_c u_s^3}{g \mu_c} \right)^{0.332} \left(\frac{d_p^2 \rho_c \rho_d u_s^3}{\mu_d \sigma g_c} \right)^{0.525} \tag{8.67}$$

EFFECTS OF SURFACE-ACTIVE CONTAMINATION

Trace amounts of surface-active impurities, unknown in structure and concentration, are frequently present in commercial equipment. This leads to difficulties in interpreting the performance of such plant in terms of experimental and theoretical studies on drops. Garner and Skelland (1956), Garner and Hale (1953), and others have shown the rate of mass transfer to be very substantially reduced by the presence of such impurities, because they accumulate at the interface between the disperse and continuous phases. This inhibits circulation within the drops, changes the pattern of droplet oscillation, sets up mechanical barriers to transfer across the interface, and modifies the shape of the drops.

The formulation of generalized expressions to account for these effects is prevented by their specific dependence upon the structure and concentration of the surface-active contaminant. In a preliminary and tentative attempt to find some guidance on this matter, however, one may note that in several experimental studies (Garner and Skelland, 1956; Garner and Hale, 1953; and Lindland and Terjesen, 1956), k_{dr}^* was not reduced by various specific surface-active agents below half the value predicted for stagnant spheres (equations 8.49 to 8.52). Skelland and Caenepeel (1972) found, from measurements over a range of concentration of cationic or anionic surfactants, that their values of k_{df}^* and k_{cf}^* showed average absolute deviations of 34 percent (±) and 44 percent (−), respectively, from equation 8.35; k_{cr}^* deviated by an average of 38 percent (−) from equation 8.61; and k_{dc}^* and k_{cc}^* values revealed average absolute deviations of 59 percent (+) and 28 percent (±), respectively, from equation 8.65. [Physical properties corresponding to either the disperse or the continuous phase were substituted in these equations as appropriate. The + and − signs in parentheses after each percent error show that the model in question respectively overestimated (+), underestimated (−), or ran roughly through the middle (±) of most of the points in a given set of experimental data.] It must be emphasized that these equations did not correlate the various aspects of mass transfer in the presence of surface-active agents. The purpose of this comparison is simply to indicate the extent of the average deviation between the measurements and those equations which came closest to the data in each instance.

A notable feature of these experimental studies is the frequent occurrence of a minimum in the coefficient at surfactant concentrations which are much less than the bulk values corresponding to interfacial saturation in static systems. This phenomenon is tentatively linked (Davies, 1969; Skelland and Caenepeel, 1972) to the optimum concentration of soluble surface-active agent often found to give the maximum damping of waves on a free liquid surface (Davies and Vose, 1965). In contrast, Skelland and Caenepeel found that the continuous-phase coefficient during coalescence was substantially increased by surfactants in their systems, with maxima at intermediate concentrations in the case of the cationic agent. Their explanation was in terms of extended surface due to an observed retardation of droplet coalescence in these cases.

The selection of the appropriate correlations for the disperse and continuous phase coefficients during free rise or fall requires a knowledge of whether drops of the relevant size are internally stagnant, circulating, or oscillating. In a detailed review of the influence of surface-active contaminants on the hydrodynamic and mass-transfer behavior of drops, Davies and Rideal (1963) note that internal circulation is inhibited in commercial systems by a reduction in drop size and by the use of nonpolar solvents, because of traces of strongly adsorbed impurities. These impurities are less strongly adsorbed at the interface with polar solvents, which therefore tend to give circulating drops. It was remarked that even large drops of commercial benzene are always stagnant, and that circulation is reduced in drops 0.5 cm in diameter by protein concentrations of only 0.0005 percent when the interfacial tension exceeds 30 dyn/cm. The authors noted that the addition of a few percent of a short-chain alcohol or acetic acid to the dispersed solvent will often displace the adsorbed impurity from the interface, thereby restoring the transfer rate to that corresponding to circulating droplets. This remedy seems most likely to prove effective in the case of interfacial films that are only weakly adsorbed. It has been noted earlier that sustained oscillations of the drops apparently begin when the Weber number $d_p u_t^2 \rho_c / \sigma g_c$ reaches a value of 3.58 (Hu and Kintner, 1955; Basu, 1970, p. 86). The influence of surface-active contamination upon this criterion requires investigation. A qualitative determination of the droplet condition in a given system may be achieved in a preliminary glassware experiment in which small amounts of aluminum particles are inserted in the disperse phase and the system is then observed with reflected light (Garner and Skelland, 1956).

The disperse phase in liquid extraction is normally selected as that having the larger volume, so as to maximize the interface. It may be better, however, to select the disperse phase as that offering the least resistance to transfer, as revealed by the distribution coefficient. The stopping of circulation currents within the drop by surface-active contaminants may then be less serious in terms of the overall resistance to transfer, since the process is continuous-phase controlling.

A COMPUTER PROGRAM FOR THE PROVISIONAL DESIGN PROCEDURE

The provisional design procedure described throughout much of this chapter has been written in Fortran IV language for digital computer application by W. L. Conger.

Table 8.3. Expressions used to calculate quantities in the computer program for the provisional design of a perforated-plate extraction column.[a]

Quantity	Expressions
A_f	A_f(AF) = A_d(AD, 8.10); n_0(PN, 8.13); u_0(UO) < u_{0j}(8.12); d_p(DP) $=(6v_p/\pi)^{1/3}$; v_p(VOLP, 8.11 + Figure 8.3).
A_r	A_r(AR, 8.14); A_0{AO $=A_{pz}+2A_D+$(peripheral band of width w_b); when $[(4/\pi)(A_{pz}+2A_D)]^{1/2}\leqslant 0.75$ ft, $w_b=\frac{1}{2}$(pitch), otherwise $w_b=$ pitch; A_{pz} (APZ) from 8.27 or 8.28}; A_D(ADOWN, 8.26); u_D[UDOWN, 8.23 + ($d_p=5.208\times10^{-3}$ ft)]; h_c(HC, 8.29); ϕ_d[PHID $=0.5-0.5(1-4u_d/u_t)^{0.5}$, 8.18]; u_d[UD $=Q_{d2}/(A_0-A_D)$]; u_t(UT, 8.23 or 8.24); d_{pt}(DPT, 8.25); N_p(ANP, 8.20).
A_c	A_c(AC, 8.33).
K^*_{df}	K^*_{df}(KDF, 4.7); k^*_{df}(CDF, 8.44); k^*_{cf}(CCF, 8.46); t_f(TF $=n_0v_p/Q_{d2}$).
K^*_{dr}	K^*_{dr}(KDR, 4.7); k^*_{dr}(CDR, 8.50, 8.58, or 8.59); k^*_{cr}(CCR, 8.60, 8.63, or 8.64); u_s(US, 8.17).
K^*_{dc}	K^*_{dc}(KDC, 4.7); k^*_{dc}(CDC, 8.66); k^*_{cc}(CCC, 8.67).

q(QA, 8.5; QB, 8.6); $\mathfrak{D}_n$(DN, 8.7); $\mathfrak{D}_{n+1}$(DN1, 8.8).

[a] The term denoting the quantity in the computer program is given in parentheses, followed by the number of the equation used for its estimation.

Table 8.4. Data required to use the computer program in the provisional design of a perforated plate extraction column for a specific separation.[a]

MOLSC, MOLSD (molecular weights of continuous and disperse phases), DENC (ρ_c), VISC (μ_c), DC (D_c), DEND (ρ_d), VISD (μ_d), DD (D_d), TEN (σ), QC1 (Q_{c1}), QD2 (Q_{d2}), D2 ($\mathfrak{D}_2$), OD (d_o), UO (u_o), PITCH (distance between perforations), HTH(H), HD (number denoting type of drop; 1 = oscillating, 2 = circulating, 3 = stagnant), HH (number indicating direction of transfer; 1 = disperse to continuous, 2 = continuous to disperse), MP (number specifying pitch geometry; 1 = triangular, 2 = square), NT (number showing whether a new system is to be used on the next run; 1 = same system, any other number = a new system), ND (number of points to be calculated along the pseudo-equilibrium curve), YA1 (y_{A1}), YA2 (y_{A2}), XA1 (x_{A1}), XA2 (x_{A2}), C1, C2 (constants for the equilibrium curve, $y_A = \text{C1}\, x_A^{\text{C2}}$), $B(1)$, $B(2)$, $B(3)$,... (constants describing the operating curve).

---

[a]Terms are as they appear in the program; the definitions in parentheses have meaning and units as in the nomenclature at the end of the chapter.

The expressions used for calculating various quantities in the program are listed in Table 8.3, and the data required for its application to the provisional design of a column for a specified separation are assembled in Table 8.4. The computer program as prepared by Conger appears in Table 8.5; it contains many labeled segments to facilitate extension or replacement of individual sections by improved relationships as they become available from further research. The computer printout gives the number of real plates required for a prescribed separation, the number of perforations per plate, the column diameter, and the cross-sectional area of the downcomers.

COMPARISONS BETWEEN PREDICTIONS AND PUBLISHED DATA

Skelland and Conger (1973) have applied the provisional design method in Tables 8.3 to 8.5 to all the appropriate published results on perforated-plate extraction columns. Criteria determining whether or not published data were appropriate for comparison are noted in Table 8.6. These considerations led to the elimination of some—or occasionally all—of the

Table 8.5. FORTRAN IV computer program for the provisional design of a perforated-plate extraction column (terms are defined in Tables 8.3 and 8.4 and at the foot of this table; v_p is calculated omitting the term containing μ_c as noted below equation 8.11).

```
C
C
C      THIS PROGRAM CALCULATES THE PSEUDO-EQUILIBRIUM CURVE AND THE
C      NUMBER OF STAGES FOR LIQUID-LIQUID EXTRACTION IN PERFORATED
C      PLATE COLUMNS
C
C
       REAL M,KDF,KDR,KDC,MOLSC,MOLSD
       INTEGER HD,HH
       DIMENSION YAR(90),M(90),KDF(90),KDR(90),KDC(90),YA(90),YN1(90),
      1X(90),XX(90),B(5),TITLE(18),DFV(9),FTR(9)
C
C
C      THE INTEGER ND DESIGNATES THE NUMBER OF POINTS TO BE CALCULATED
C      -MAXIMUM NUMBER IS 50
C
C
     9 CONTINUE
C
C
C      THIS SECTION TO STATEMENT 10 READS IN BASIC EXTRACTION SYSTEM &
C      PLATE DATA
C
C
       READ(5,196)ND
       WRITE(6,197)ND
       READ(5,200) MOLSC,DENC,VISC,DC
       WRITE(6,299)
       WRITE(6,300)MOLSC,DENC,VISC,DC
       READ(5,200) MOLSD,DEND,VISD,DD
       WRITE(6,298)
       WRITE(6,301)MOLSD,DEND,VISD,DD
       READ(5,201) TEN
       WRITE(6,302)TEN
C
C
C      THE C1,C2 ARE CONSTANTS IN THE EQUILIBRIUM EQUATION
C      THE EQUATION IS OF THE FORM    YAR=C1*X**C2
C
C
       READ(5,203)C1,C2
       WRITE(6,304)C1,C2
       READ(5,216)OD,HTH,PITCH
    10 CONTINUE
C
C
C      THIS SECTION THROUGH STATEMENT 11 READS IN MATERIAL BALANCE, FLOW
C      RATE, & OTHER SYSTEM DATA ASSOCIATED WITH A SINGLE RUN
C
C
       READ(5,199)TITLE
       WRITE(6,198)TITLE
       READ(5,400)NT,HD,HH,MP
       WRITE(6,401)NT,HD,HH,ND,MP
C
C
C      NT IS AN INTEGER WHICH DENOTES WHETHER OR NOT THE NEXT RUN WILL
C      USE THE SAME BASIC EXTRACTION SYSTEM & PLATE DATA AS THE PRESENT
C      RUN
C            NT=1 NEW SYSTEM FOR NEXT RUN
```

```
C             NT=ANY OTHER NUMBER THAN 1 , SAME SYSTEM NEXT RUN
C
C
C       THE INTEGER HD DESIGNATES THE TYPE OF DROPLET IN THE COLUMN
C             HD=1  DROPLETS ARE OSCILLATING
C             HD=2  DROPLETS ARE CIRCULATING
C             HD=3  DROPLETS ARE STAGNANT
C
C
C       THE INTEGER HH DESIGNATES THE DIRECTION OF TRANSFER OF SOLUTE
C             HH = 1  TRANSFER FROM DISPERSE TO CONTINUOUS PHASE
C             HH = 2  TRANSFER FROM CONTINUOUS TO DISPERSE PHASE
C
C
C       THE INTEGER MP DESIGNATES THE PITCH GEOMETRY
C             MP=1  TRIANGULAR PITCH
C             MP=2  SQUARE PITCH
C
C
        READ(5,402)QC1,QD2,UO,D2,YA2,YA1,XA2,XA1
C
C
C       THE B(I) ARE CONSTANTS IN THE OPERATING LINE EQUATION- THE
C       EQUATION IS OF THE FORM--
C       YA=B(1)+B(2)*X+B(3)*X**2+B(4)*X**3+B(5)*X**4
C
C
        READ(5,217)(B(I),I=1,5,1)
        WRITE(6,305)B(1),B(2),B(3),B(4),B(5)
        WRITE(6,306)DD,UO,HTH,PITCH
     11 WRITE(6,303)QC1,QD2,YA2,D2,YA1,XA2,XA1
C
C
C       THIS MATRIX OF NUMBERS THROUGH STATEMENT 12 IS USED IN CALCULATING
C       "F" FOR THE SCHEELE & MEISTER DETERMINATION OF THE VOLUME OF A
C       DROPLET
C
C
        FTR(1)=1.0
        FTR(2)=0.884
        FTR(3)=0.782
        FTR(4)=0.708
        FTR(5)=0.663
        FTR(6)=0.63
        FTR(7)=0.61
        FTR(8)=0.6
        FTR(9)=0.6
        DFV(1)=0.0
        DFV(2)=0.2
        DFV(3)=0.4
        DFV(4)=0.6
        DFV(5)=0.8
        DFV(6)=1.0
        DFV(7)=1.2
        DFV(8)=1.4
     12 DFV(9)=1.6
        PART=(ABS(XA1-XA2))/(ND-5)
        IF(XA1.LT.XA2)GO TO 5
        XX(1)=XA2-2.0*PART
        GO TO 6
      5 XX(1)=XA1-2.0*PART
```

```
    6 DO 7 I=2,ND,1
    7 XX(I)=XX(I-1)+PART
      DENDIF=ABS(DENC-DEND)
C
C
C     THIS SECTION THROUGH STATEMENT 334 DETERMINES THE VOLUME &
C     DIAMETER OF THE DROPLETS
C
C
      VOOF=((3.14159*TEN*OD/DENDIF)-(3.14159*DEND*(OD**2.0)*(UO**2.0)/
     1(4.0*DENDIF*4.17E8))+4.5*(((3.14159*(OD**3.0)*UO/(4.0*4.17E8*
     2DENDIF))**2.0)*DEND*TEN*4.17E8)**0.333)
      VF=OD*(1.0/VOOF)**0.333
      F=0.0
      COMP=0.0
      DO 332 I=1,9,1
      IF(VF.EQ.DFV(I))GO TO 331
      IF(COMP.NE.0.0)GO TO 332
      IF(DFV(I).LT.VF)GO TO 332
      COMP=DFV(I)-VF
      IF(COMP.LT.0.001)GO TO 331
      FRACT=(COMP/(DFV(I)-DFV(I-1)))
      F=FTR(I)-FRACT*(FTR(I)-FTR(I-1))
      GO TO 332
  331 F=FTR(I)
  332 CONTINUE
      IF(F.NE.0.0)GO TO 333
      F=0.6
  333 VOLP=VOOF*F
  334 DP=((6.0*VOLP)/3.14159)**0.3333
C
C
C     NEXT, VARIOUS QUANTITIES ,SUCH AS THE AREA OF THE COLUMN, ARE
C     CALCULATED
C
C
      PN=(4.0*QD2)/(3.14159*UO*OD**2)
      AD=PN*3.14159*DP**2
      AF=AD
      UDOWN=(53000*(    DENDIF)**0.58*5.208E-3**0.70)/(DENC**.45*VISC
     1**0.11)
      ADOWN=QC1/UDOWN
      WRITE(6,403)ADOWN,PN
      IF(MP.NE.1)GO TO 15
      APZ=(PN*3.14159*PITCH**2.0)/3.62
      GO TO 16
   15 APZ=(PN*3.14159*PITCH**2.0)/3.14
   16 CONTINUE
      AT=APZ+2.0*ADOWN
      DB=(4.0*AT/3.14159)**0.5
      IF(DB.GT.0.75)GO TO 17
      DIA=DB+PITCH
      GO TO 18
   17 DIA=DB+2.0*PITCH
   18 CONTINUE
      WRITE(6,307)DIA
      AO=(3.14159*DIA**2)/4.0
      WRITE (6,207)
      SURD=3.14159*DP**2
      REYO=OD*UO*DEND/VISD
C
```

```
C
C     THE HEIGHT OF THE COALESCED LAYER IS CALCULATED
C
C
      HC=0.2172*TEN*(VISD**0.4)*(VISC**0.2)/((    DENDIF)*(OD**1.4))
     2+(UO**2.0*DEND)/(8.34E8*(    DENDIF))/((1.0-0.71/ALOG(REYO))**2.0)
     3+(0.592E-8*UDOWN**2*DENC)/(DENDIF)
      ANP=DENC**2*TEN**3*4.17E8**3/(4.17E8*VISC**4*(DENDIF))
      DPT=7.25*(TEN/((    DENDIF)*ANP**0.15))**0.5
C
C
C     THE TERMINAL VELOCITY OF THE DROPLETS IS DETERMINED
C
C
      IF(DP-DPT)20,20,30
   20 UT=53000*(    DENDIF)**0.58*DP**0.70/(DENC**0.45*VISC**0.11)
      GO TO 31
   30 UT=577*(    DENDIF)**0.28*VISC**0.10*(TEN*4.17E8)**0.18/
     4(DENC**0.55)
   31 UD=QD2/(AO-ADOWN)
      PHID=(1.0-(1.0-4.0*UD/UT)**0.5)/2.0
      AF=(AO-ADOWN)*(HTH-HC)*PHID*SURD/VOLP
      AC=AO-ADOWN
      US=UT*(1-PHID)
      TF=PN*VOLP/QD2
C
C
C     THIS SECTION, THROUGH STATEMENT 57,CALCULATES THE INDIVIDUAL
C     COEFFICIENTS OF MASS TRANSFER FOR BOTH PHASES & FOR DROPLET
C     FORMATION, FREE RISE (OR FALL), & COALESCENCE
C
C
      CDF=(0.0432*DP*DEND)/(TF*MOLSD)*((UO**2)/(DP*4.17E8))**0.089*
     5((TF*DD)/(DP**2.0))**0.334*((DEND*DP*TEN*4.17E8)**0.5/VISD)**0.601
      CCF=0.386*DENC/MOLSC*(DC/TF)**0.5*(DENC*TEN*4.17E8/((DENDIF)*
     64.17E8*TF*VISC))**0.407*(4.17E8*TF**2/DP)**0.148
      T=(HTH-HC)/US
      IF(HD.EQ.1)GO TO 53
      IF(HD.EQ.2)GO TO 50
      CDR=-DP/(6.0*T)*DEND/MOLSD*ALOG(1.0-(3.14159*DD**0.5*T**0.5/
     7(DP/2.0)))
   49 CCR=0.74*DC/DP*DENC/MOLSC*(DP*US*DENC/VISC)**0.5*
     1(VISC/(DENC*DC))**0.3333
      GO TO 56
   50 CDR=31.4*DD/DP*DEND/MOLSD*(DD*DEND/VISD)**0.125*(DP*US**2*DENC/
     2(TEN*4.17E8))**0.37*(DP**2/(4.0*DD*T))**0.34
   51 CCR=0.725*DENC/MOLSC*(VISC/(DP*US*DENC))**0.43*(DENC*DC/VISC)
     3**0.58*US*(1.0-PHID)
      GO TO 56
   53 CDR=0.32*DD/DP*DEND/MOLSD*(DP**2.0/(4.0*DD*T))**0.14*
     4(DP*US*DENC/VISC)**0.68*((TEN**3*4.17E8**3*DENC**2)/(4.17E8*
     5VISC**4*DENDIF))**0.1
   54 CCR=DC/DP*DENC/MOLSC*(50.0+0.0085*(DP*US*DENC/VISC)*(VISC/(DENC*
     6DC))**0.7)
   56 CONTINUE
      CDC=0.173*DP/TF*DEND/MOLSD*(DEND* DD/VISD)**1.115*(((DENDIF)
     1*4.17E8*DP**2.0)/(TEN*4.18E8))**1.302*(US**2*TF/DD)**0.146
   57 CCC=5.959E-4*DENC/MOLSC*(DC/TF)**.5*((DENC*US**3)/(4.17E8*VISC))**
     10.332*((DP**2*DENC*DEND*US**3)/(VISD*TEN*4.17E8))**0.525
      I=0
      DO 100 J=1,ND,1
```

```
      IF(XX(J).LT.0.0)GO TO 100
      I=I+1
      X(I)=XX(J)
  500 YAR(I)=C1*X(I)**C2
C
C
C     THE SLOPE OF THE EQUILIBRIUM LINE (M) IS DETERMINED FROM THE FIRST
C     DERIVATIVE OF THE EQUILIBRIUM LINE EQUATION
C
C
      M(I)=C1*C2*X(I)**(C2-1.0)
C
C
C     THE OVERALL COEFFICIENTS FOR MASS TRANSFER ARE NOW CALCULATED
C
C
      KDF(I)=(CCF*CDF)/(CCF+M(I)*CDF)
      KDR(I)=(CCR*CDR)/(CCR+M(I)*CDR)
      KDC(I)=(CCC*CDC)/(CCC+M(I)*CDC)
      YA(I)=B(1)+B(2)*X(I)+B(3)*X(I)**2+B(4)*X(I)**3+B(5)*X(I)**4
  100 CONTINUE
      IF(HH.EQ.2)GO TO 520
      IF(YA2.LT.YA(I))GO TO 520
      I=I+1
      X(I)=X(I-1)+PART
      GO TO 500
  520 CONTINUE
      NN=I
C
C
C     IN THIS SECTION, THROUGH STATEMENT 110, THE RATE OF MASS TRANSFER
C     AT INTERVALS THROUGHOUT THE COLUMN LENGTH IS CALCULATED BY TWO
C     DIFFERENT MEANS & THE COMPOSITION IN THE DISPERSE PHASE IS
C     ADJUSTED UNTIL THE TWO CALCULATIONS DIFFER BY A VERY SMALL
C     (ACCEPTABLE) AMOUNT
C
C
      DO 110 I=1,NN,1
      YN1(I)=(YAR(I)+YA(I))/2
  101 QA=KDF(I)*AF*(YAR(I)-YA(I))+KDR(I)*AR*((YN1(I)-YA(I))/ALOG((YAR(I)
     1-YA(I))/(YAR(I)-YN1(I))))+KDC(I)*AC*(YAR(I)-YN1(I))
      DN=D2*(1.0-YA2)/(1.0-YA(I))
      DN1=D2*(1.0-YA2)/(1.0-YN1(I))
      QB=DN1*YN1(I)-DN*YA(I)
      IF(QA-QB)102,109,104
  102 IF(ABS(QA-QB).LE.0.001)GO TO 109
      YN1(I)=YN1(I)-0.0001
      GO TO 101
  104 IF(ABS(QA-QB).LE.0.001)GO TO 109
      YN1(I)=YN1(I)+0.00011
      GO TO 101
  109 IF(QA-QB)105,108,106
  105 IF(ABS(QA-QB).LE.0.0001)GO TO 108
      YN1(I)=YN1(I)-0.00001
      GO TO 101
  106 IF(ABS(QA-QB).LE.0.0001)GO TO 108
      YN1(I)=YN1(I)+0.000011
      GO TO 101
  108 IF(QA-QB)1105,1108,1106
 1105 IF(ABS(QA-QB).LE.0.00001)GO TO 1108
      YN1(I)=YN1(I)-0.000001
```

```
      GO TO 101
 1106 IF(ABS(QA-QB).LE.0.00001)GO TO 1108
      YN1(I)=YN1(I)+0.0000011
      GO TO 101
 1108 IF(QA-QB)2105,2108,2106
 2105 IF(ABS(QA-QB).LE.0.000001)GO TO 2108
      YN1(I)=YN1(I)-0.0000001
      GO TO 101
 2106 IF(ABS(QA-QB).LE.0.000001)GO TO 2108
      YN1(I)=YN1(I)+0.00000011
      GO TO 101
 2108 WRITE(6,208)X(I),YN1(I),YAR(I),YA(I),M(I),QA,QB
  110 CONTINUE
      IF(HH.NE.1)GO TO 160
C
C
C     IF HH=1 PLATE TO PLATE CALCULATIONS ARE MADE FOR TRANSFER FROM
C     DISPERSE TO CONTINUOUS PHASE. SECTION ENDS BEFORE STATEMENT 160
C
C
      YN=YA2
      MM=0
C
C
C     MM IS COUNTER FOR STAGES
C
C
  126 JJ=1
      DO 140 J=1,NN,1
      IF(JJ.NE.1)GO TO 140
      IF(YA(J).LT.YN)GO TO 140
      IF(YA(J).EQ.YN)GO TO 135
      COMP=YA(J)-YN
      IF(COMP.LT.0.0000000001)GO TO 135
      FRACT=COMP/(YA(J)-YA(J-1))
      XA=X(J)-FRACT*(X(J)-X(J-1))
      JJ=JJ+1
      GO TO 140
  135 XA=X(J)
      JJ=JJ+1
  140 CONTINUE
  141 MM=MM+1
      IF(MM.GE.50)GO TO 171
      WRITE(6,213)MM,YN,XA
      JJ=1
      DO 150 J=1,NN,1
      IF(JJ.NE.1)GO TO 150
      IF(X(J).LT.XA)GO TO 150
      IF(X(J).EQ.XA)GO TO 145
      COMP=X(J)-XA
      IF(COMP.LT.0.0000000001)GO TO 145
      FRACT=(X(J)-XA)/(X(J)-X(J-1))
      YN=YN1(J)-FRACT*(YN1(J)-YN1(J-1))
      JJ=JJ+1
      GO TO 150
  145 YN=YN1(J)
      JJ=JJ+1
  150 CONTINUE
      IF(YN.GT.YA1)GO TO 126
      WRITE(6,214)MM,YN,YA1
      GO TO 171
```

```
  160 XA=XA1
C
C
C     IF HH=2 PLATE TO PLATE CALCULATIONS ARE MADE FOR TRANSFER FROM
C     CONTINUOUS TO DISPERSE PHASE.  SECTION ENDS AT STATEMENT 171
C
C
      MM=0
C
C
C     MM IS COUNTER FOR STAGES
C
C
  161 YN=B(1)+B(2)*XA+B(3)*XA**2+B(4)*XA**3+B(5)*XA**4
      JJ=1
      DO 170 J=1,NN,1
      IF(JJ.NE.1)GO TO 170
      IF(YN1(J).LT.YN)GO TO 170
      IF(YN1(J).EQ.YN)GO TO 165
      COMP=YN1(J)-YN
      IF(COMP.LT.0.0000000001)GO TO 165
      FRACT=(YN1(J)-YN)/(YN1(J)-YN1(J-1))
      XA=X(J)-FRACT*(X(J)-X(J-1))
      JJ=JJ+1
      GO TO 170
  165 XA=X(J)
      JJ=JJ+1
  170 CONTINUE
      MM=MM+1
      IF(MM.GE.50)GO TO 171
      WRITE(6,213)MM,YN,XA
      IF(XA.GT.XA2)GO TO 161
      WRITE(6,215)MM,XA,XA2
  171 IF(NT.EQ.1)GO TO 9
      GO TO 10
C
C
C     MANY OF THE FORMAT STATEMENTS USED IN THIS PROGRAM USE "H" FORMAT
C     CODE & MAY BE CHANGED TO A LITERAL FORMAT CODE AS THE USER DESIRES
C
C
  196 FORMAT(I5)
  197 FORMAT('X',9H     ND = ,I5)
  198 FORMAT(///,20X,18A4)
  199 FORMAT(18A4)
  200 FORMAT(3F10.4,F20.10)
  201 FORMAT (F10.7)
  203 FORMAT(2F10.5)
  207 FORMAT(///,69H     XA         YN1         YAF          YA            M
     1     QA          QB)
  208 FORMAT('X',5F10.5,2F10.7)
  213 FORMAT ('X',24HCONCENTRATIONS ON STAGE ,I3/6H YN = ,F10.7, 9H     X
     1A = ,F10.7)
  214 FORMAT ('X',25HTOTAL NUMBER OF STAGES = ,I5, 9H     YN = ,F10.7,
     1   6HYA1 = ,F10.7)
  215 FORMAT ('X',25HTOTAL NUMBER OF STAGES = ,I5, 9H     XA = ,F10.7,
     1   6HXA2 = ,F10.7)
  216 FORMAT(3F10.7)
  298 FORMAT('0',19HDISPERSE PHASE DATA)
  217 FORMAT(5F10.5)
  299 FORMAT('0',21HCONTINUOUS PHASE DATA)
```

```
300 FORMAT('X',34H      MOLECULAR WEIGHT,     MOLSC = ,F10.4,24H     POUND
   1S PER POUND MOLE,/24H       DENSITY,      DENC = ,F10.4,29H    POUNDS M
   2ASS PER CUBIC FOOT,/26H      VISCOSITY,       VISC = ,F10.4,28H    POUN
   3DS MASS PER FOOT HOUR,/36H       MOLECULAR DIFFUSIVITY,      DC = ,F20
   4.10,23H    SQUARE FEET PER HOUR)
301 FORMAT('X',34H      MOLECULAR WEIGHT,     MOLSD = ,F10.4,24H     POUND
   1S PER POUND MOLE,/24H       DENSITY,      DEND = ,F10.4,29H    POUNDS M
   2ASS PER CUBIC FOOT,/26H      VISCOSITY,       VISD = ,F10.4,28H    POUN
   3DS MASS PER FOOT HOUR,/36H       MOLECULAR DIFFUSIVITY,      DD = ,F20
   4.10,23H    SQUARE FEET PER HOUR)
302 FORMAT('X',30HINTERFACIAL TENSION,     TEN = ,F10.7,24H    POUNDS FO
   1RCE PER FOOT)
303 FORMAT('X',30HFLOW RATE AND COMPOSITION DATA,/43H      VOL FLOW RAT
   1E CONT PHASE IN,      QC1 = ,F10.3,22H    CUBIC FEET PER HOUR,/43H
   2  VOL FLOW RATE DISP PHASE IN,      QD2 = ,F10.3,22H    CUBIC FEET PE
   3R HOUR,/50H       MOLE FRACTION IN DISPERSE PHASE IN,      YA2 = ,F10.
   47,/41H       FLOW RATE OF DISP PHASE IN,      D2 = ,F10.5,23H    POUND
   5MOLES PER HOUR,/47H       MOLE FRACTION IN DISP PHASE OUT,      YA1 =
   6,F10.7,/47H       MOLE FRACTION IN CONT PHASE OUT,      XA2 = ,F10.7,/
   746H       MOLE FRACTION IN CONT PHASE IN,      XA1 = ,F10.7)
304 FORMAT('X',30HEQUILIBRIUM EQUATION CONSTANTS,/10H        C1 = ,F10.5,
   1/10H      C2 = ,F10.5)
305 FORMAT('X',33HOPERATING LINE EQUATION CONSTANTS,/12H       B(1) = ,F
   110.5,/12H      B(2) = ,F10.5,/12H       B(3) = ,F10.5,/12H       B(4) =
   2 ,F10.5,/12H      B(5)  = ,F10.5)
306 FORMAT('X',11HCOLUMN DATA,/29H       DIA OF ORIFICE,      OD = ,F10.7,
   17H    FEET,/34H       VEL THROUGH ORIFICE,      UO = ,F10.3,16H    FEET
   2PER HOUR,/67H      HEIGHT BETWEEN UPPER SURFACES OF CONSECUTIVE PLA
   3TES,      HTH = ,F10.5,7H    FEET,/13H       PITCH = ,F10.7, 7H    FEET)
307 FORMAT('X',31H       COLUMN DIAMETER,      DIA = ,F10.7,7H    FEET)
400 FORMAT(4I5)
401 FORMAT('X',5HNT = ,I5,/6H HD = ,I5,/6H HH = ,I5,/6H ND = ,I5,/6H M
   1P = ,I5)
402 FORMAT(4F10.5,4F10.7)
403 FORMAT('0',30HRESULTS OF DESIGN CALCULATIONS,/26H       AREA OF DOW
   1NCOMER = ,F20.10,/41H       NUMBER OF PERFORATIONS PER PLATE = ,F10
   2.3)
    RETURN
    END
```

The following terms not defined in Tables 8.3 and 8.4 are also used in the program: COMP, comparison value—used when comparing calculated quantities with tabulated quantities; DENDIF, absolute value of the difference in densities of the two phases; DN, DN1, values of the disperse-phase flow rate at plates n and $n+1$; DPT, estimate of the diameter of a droplet; F, the Harkins and Brown correction factor; FRACT, interpolation factor; FTR (I), DFV (I), values from Harkins and Brown's F Chart (Fig. 8.3); M(I), slope of equilibrium line from equation; MM, stage counter; PART, increment for calculating values of XX(I); REYO, Reynolds number based on orifice diameter; SURD, surface area of a droplet; T, time of rise (or fall) of droplet; VOOF, estimate of volume of a droplet before applying the Harkins and Brown correction factor F; VF, estimate of DFV; VOLP, volume of a droplet; XA, continuous-phase concentration on the plate; X(I), concentration in continuous phase, eliminating any negative values calculated in XX(I); XX(I) concentration (may contain negative values) in continuous phase; YA(I), disperse-phase concentration as calculated by operating-line equation; YAR(I), disperse-phase concentration in equilibrium with continuous-phase concentration X(I); YN, disperse-phase concentration on the plate; YN1, disperse-phase concentration as calculated using pseudo-equilibrium curve.

Table 8.6. Requirements for the applicability of the provisional design program.

1. No interfacial turbulence phenomena.
2. No interference from hydrogen bonding between solute and raffinate as described by Licht and Conway (1950) and Garner, Ellis, and Hill (1955).
3. No surface-active contamination.
4. No jetting or streaming of the disperse phase from the perforations.
5. All perforations operating.
6. All relevant physical properties known.
7. All relevant operating conditions and equipment dimensions known.
8. Drops coalesce normally beneath each plate.
9. Operation not erratic, as when near flooding.
10. Plate wetting characteristics ensure good dispersion.
11. Contactor exclusively of the perforated-plate type.
12. Interfacial tension high (⩾25 dyn/cm).

runs in the following papers. Criteria which were unfulfilled for the runs concerned are indicated in parentheses in each case:

Allerton, Strom, and Treybal, 1943 (4,5); Garner, Ellis, and Fosbury, 1953 (4,8); Garner, Ellis, and Hill, 1955 (2,10); Garner, Ellis, and Hill, 1956 (10,12); Goldberger and Benenati, 1959 (7,11); Mayfield and Church, 1952 (4,12); Moulton and Walkey, 1944 (4,6,9); Nandi and Ghosh, 1950 (2,4,5,6); Pyle, Colburn, and Duffey, 1950 (3,6,11); Row, Koffolt, and Withrow, 1941 (7,9,11); and Treybal and Dumoulin, 1942 (4,5,9).

The scarcity of appropriate data compelled the use of some measurements which were of very dubious acceptability in terms of the tabulated requirements. In consequence, the results may be classified into four groups—*A*, *B*, *C*, and *D*—the last three of which did *not* satisfy all the criteria of Table 8.6. This means that only Group *A* can be regarded as providing fully eligible data, but examination of results for the "ineligible" Groups *B*, *C*, and *D* is nevertheless instructive.

Group A comprised high-interfacial-tension systems which largely conformed to the specifications of Table 8.6. The group consists of 65 runs corresponding to eight sets of data from six different papers. Results for group *A* are summarized in Table 8.7 and plotted as individual and averaged values, respectively, in Figure 8.6. The overall average error

Table 8.7. Summarized comparison between actual and predicted plates required for the 65 separations in group A.[a]

Reference	System[b]	No. of perforations per plate	Plate spacing, in.	Perforation diam., in.	Downcomer diam., in.	Column diam., in.	No. of runs used	No. of plates used	Av. no. of plates predicted
Allerton, Strom, and Treybal (1943)	Kerosene–benzoic acid–water	51	4.75	0.1875	2×0.485	3.63	10	11	9.9
Garner, Ellis, and Fosbury (1953)	Toluene-diethylamine-water	59	6	0.125	1	4	10	8.627[c]	7.3
Garner, Ellis, and Hill (1955)	Water-diethylamine-toluene	59	6	0.125	1	4	11	8	7.1
Goldberger and Benenati (1959)	Toluene–benzoic acid–water	15		0.046	0.497	2.718	4	1	1
Row, Koffolt, and Withrow (1941)	Toluene–benzoic acid–water	465	6	0.125	1	8.75	7	10	8.43

Treybal and Dumoulin (1942)	Toluene–benzoic acid–water	17	3	0.1875	0.406	3.56	7	17	10.3
Treybal and Dumoulin (1942)	Toluene–benzoic acid–water	17	6	0.1875	0.406	3.56	7	9	5.86
Treybal and Dumoulin (1942)	Toluene–benzoic acid–water	17	9	0.1875	0.406	3.56	9	6	4.66

[a]*Note*: The droplets were assumed to be stagnant and transfer was from the organic phase throughout. From the last three columns, the overall average error $= \frac{100}{65} \sum (n_{\text{predicted}} - n_{\text{actual}}) / n_{\text{actual}} = -18.6\%$.

[b]In the order *disperse-phase–solute–continuous-phase*.

[c]0.627 = no. of nozzles on toluene distributor/no. of perforations per plate.

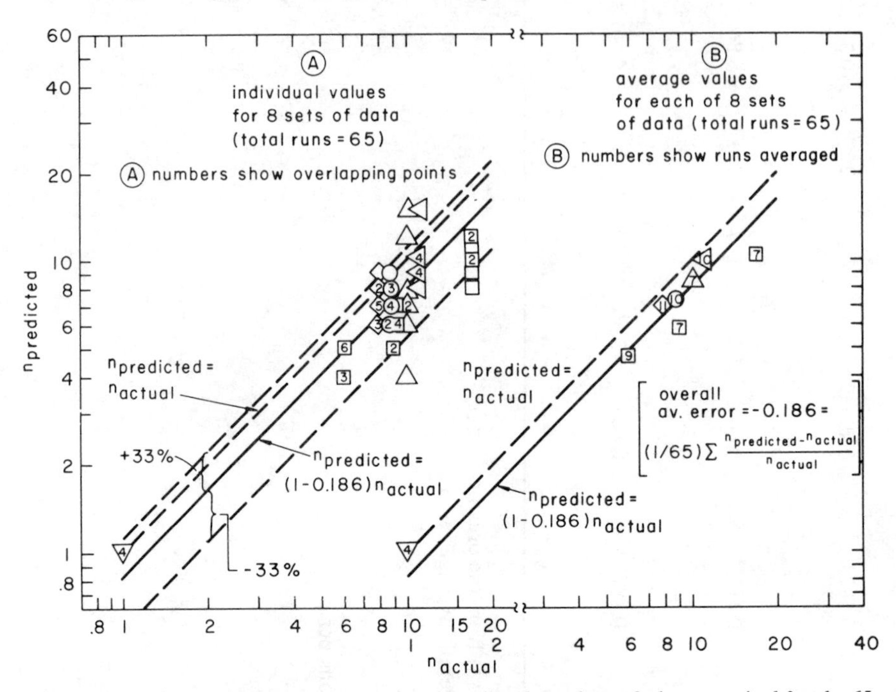

Figure 8.6. Comparison between actual and predicted numbers of plates required for the 65 separations in Group A: ◁, data of Allerton, Strom, and Treybal (1943); ○, data of Garner, Ellis, and Fosbury (1953); ◇, data of Garner, Ellis, and Hill (1955); ▽, data of Goldberger and Benenati (1959); △, data of Row, Koffolt, and Withrow (1941); □, data of Treybal and Dumoulin (1942). Table 8.7 gives some relevant details.

between the number of perforated plates predicted by the program and the number actually used to achieve the measured separation is -18.6 percent. The corresponding average absolute error without regard to algebraic sign is 22.4 percent. Figure 8.6 shows that about 90 percent of the data are within ± 33 percent of the relationship $n_{\text{actual}} = n_{\text{predicted}}/(1-0.186)$.

The fact that the program *underestimates* the actual number of plates used by an average of 18.6 percent is consistent with the probable presence of trace amounts of surface-active impurities in the six experimental studies which provided these 65 runs. Garner, Ellis, and Fosbury (1953), for example, attributed their decline in extraction with continued recycling of raffinate to the accumulation of surface-active contamination in the system.

Group B consisted of systems which in the present context exhibited "imperfect" operation. About 83 percent of these data also involved mild hydrogen bonding between raffinate and solute. The group comprises 54 runs with dilute solute by Nandi and Ghosh (1950) in a nine-plate, 1.75-in.-i.d. column with 36 or 72 holes per plate. About 83 percent of the runs transferred acetone from water to benzene or to kerosene of unspecified molecular weight. The remainder transferred benzoic acid from benzene to water. The authors indicate that some perforations were not functioning in much of the study, and there are implications of jetting or streaming in some runs.

The interference from hydrogen bonding referred to here was described by Licht and Conway (1950) and by Garner, Ellis, and Hill (1955). They postulate that transfer of solute out of a solvent to which it adheres by hydrogen bonding occurs much less readily than from a solvent to which it is linked by the weaker Van der Waals forces. Thus, when transferring diethylamine in alternate directions between toluene and water, Garner, Ellis, and Hill (1955) found the transfer rate from toluene to be severalfold greater than from water. The hydrogen bond between acetone and water is probably considerably weaker than that between diethylamine and water (Pimentel, 1972; Pimentel and McClellan, 1971), but is certainly significant.

The overall average error between the actual and predicted numbers of plates for the 54 runs of group *B* was −30.4 percent, and the average absolute error was 36.6 percent. The corresponding average errors obtained for the combined 119 runs of groups *A* and *B* together were −23.9 percent and 28.8 percent, respectively.

Group C comprises 19 runs by Garner, Ellis, and Hill (1955) on a system with strong hydrogen bonding between raffinate and solute. In this study diethylamine was transferred from water to toluene at the relatively slow rate described above. The column dimensions were as listed in Table 8.7. The number of plates was consistently underpredicted for these 19 runs, the overall average error being −62 percent. The explanation by the authors in terms of retarded transfer due to strong hydrogen bonding between diethylamine and water may be retained here.

Group D consisted of 40 runs on a low interfacial tension system in which adipic acid was transferred from methyl isobutyl ketone to water. In this study by Garner, Ellis, and Hill (1956), each phase was dispersed in turn, using the column described in Table 8.7. The overall average error and the average absolute error between actual and predicted numbers of plates were +68.5 and 69 percent, respectively, corresponding to an overestimation of the number of plates required. The direction of this result suggests that the effects of interfacial tension may be inadequately represented by the present formulation.

A further limitation on the present provisional program, then, is its restriction to high-interfacial-tension systems. This is perhaps related to the fact that two-thirds of the data leading to the expressions for k_{df}^* and k_{dc}^* were for high σ, while the correlations for k_{cf}^* and k_{cc}^* were obtained exclusively from high interfacial tension systems. In this regard, 83 percent of the runs in the fairly successful group A were continuous-phase controlled, and the continuous-phase resistance was significant in the other 17 percent.

It may be that, as in the case of circulating and noncirculating drops during free rise, different expressions are needed for the individual coefficients during formation and coalescence, depending upon whether σ is high or low. In any event, the restriction to high σ is certainly in the range of much industrial interest. This is because high-interfacial-tension systems are considered desirable to facilitate phase separation and the avoidance of stable emulsions (Treybal, 1963, p. 131; Treybal, 1968, p. 413-4; Perry, 1963, p. 14-41).

CONCLUSIONS

It must not be thought that the design procedure outlined here for sieve-plate extraction columns is presented as the final form of treatment. Indeed, many of the simplifications involved become apparent from an examination of the review by Olney and Miller (1963). The program must be regarded as provisional with respect to expressions for some phases of the process. A framework, however, has been delineated, and areas needing refinement are apparent. For design purposes, relationships based on the correlation of experimental data are probably to be preferred for some time to come. Although such empirical expressions leave many questions about mechanisms unanswered, they do represent measurements against which further theoretical developments can be tested.

More quantitative design information is clearly needed on such phenomena as droplet detachment after growth, drop size distribution, droplet coalescence and redispersion during free rise or fall, interfacial turbulence, non-Newtonian properties, interactions such as those affecting k_{cr} for swarms of stagnant and oscillating drops, coalescence mechanisms at the plane coalescence interface, surface-active contamination, effects associated with the direction of transfer, and the influence of plate wetting characteristics on the effectiveness of dispersion.

NOMENCLATURE

A_{av}	Total integrated average surface of n_0 drops during formation, ft^2.

A_c	Total interfacial area between two consecutive plates during droplet coalescence, ft^2.
A_D	Cross-sectional area of downcomer, ft^2.
A_d	Total surface of n_0 drops at detachment, ft^2.
A_f	Total interfacial area between two consecutive plates during drop formation, ft^2.
A_0	Cross-sectional area of entire column, ft^2.
A_{pz}	Area of perforated zone per plate, ft^2.
A_r	Total interfacial area between two consecutive plates during droplet rise, ft^2.
a	Surface area of a single growing drop at time t, ft^2.
B_n	Coefficient for the nth term in a series.
$\mathcal{C}$	Flow rate of continuous phase through entire cross section of column, lb-mole/hr.
C_D	Drag coefficient.
$\mathfrak{D}$	Flow rate of disperse phase through entire cross section of column, lb-mole/hr.
$\mathfrak{D}_n, \mathfrak{D}_{n+1}, \mathfrak{D}_2$	Values of $\mathfrak{D}$ at plates n and $n+1$ and entering the column, respectively, lb-mole/hr.
D_c, D_d	Molecular diffusivity of solute in continuous and disperse phases, ft^2/hr.
d_H	Major (horizontal) axis of an oblate spheroid, ft.
d_o	Orifice, nozzle, or perforation diameter, ft.
d_p	Droplet diameter (assumed spherical), ft.
d_{pt}	Transition value of d_p, equation 8.25, ft.
a_t	Diameter of growing drop at time t, ft.
d_V	Minor (vertical) axis of an oblate spheroid, ft.
E	Eccentricity, d_H/d_V.
E_f	Fractional extraction.
F	Harkins and Brown correction factor.
g	Acceleration due to gravity, ft/hr^2.
g_c	Conversion factor, 4.17×10^8 (lb-mass)(ft)/(lb-force)(hr^2).
H	Height between upper surfaces of consecutive plates, ft.
h_c	Thickness of coalesced layer, including plate thickness, ft.

K_{df}, K_{dr}, K_{dc}	Overall coefficients of mass transfer based on Δy_A during droplet formation, free rise (or fall), and coalescence, lb-mole/(hr)(ft^2).
k^*_{cf}, k^*_{cr}, k^*_{cc}	Individual continuous-phase coefficients of mass transfer for low concentrations and transfer rates during droplet formation, free rise (or fall), and coalescence, lb-mole/(hr)(ft^2).
k^*_{df}, k^*_{dr}, k^*_{dc}	Individual disperse-phase coefficients of mass transfer for low concentrations and transfer rates during droplet formation, free rise (or fall), and coalescence, lb-mole/(hr)(ft^2).
M	Mean molecular weight of phase under consideration.
m	Slope of the equilibrium curve, dy_A/dx_A.
N_P	Dimensionless group defined by equation 8.20.
N'_{Pe}	Modified Peclet number, $(d_p u_t/D_d)/(1+\mu_d/\mu_c)$.
N_{Re}	Reynolds number, $d_p u_t \rho_c/\mu_c$.
$N_{\mathrm{Re},o}$	Orifice Reynolds number, $d_o u_o \rho_d/\mu_d$.
N_{Sc}	Schmidt number, $\mu/\rho D$.
N_{We}	Weber number, $d_p u_t^2 \rho_c/\sigma g_c$.
n	Stage number n, *or* an integer (in Table 8.1).
n_0	Number of perforations per plate.
Q_{c1}	Volumetric flow rate of continuous phase entering column, ft^3/hr.
Q_{d2}	Volumetric flow rate of disperse phase entering column, ft^3/hr.
q	Rate of mass transfer, lb-mole/hr.
q_{sf}	Solute transferred during formation of a single drop, lb-mole.
R	Ratio of effective diffusivity to molecular diffusivity.
S_{os}	Surface area of an oblate spheroid, ft^2.
S_s	Surface area of a sphere, ft^2.
t	Time of free rise (or fall) of droplet, $(H-h_c)/u_s$, hr.
t_e	Time of exposure in equation 8.55.
t_f	Time of formation of a single drop, hr.
u_c, u_d	Superficial velocities of continuous and disperse phases, based on the empty-column cross section, ft^3/(hr)(ft^2 column cross section) = ft/hr.
u_D	Average velocity in the downcomer, ft/hr.

u_o	Velocity through orifice, nozzle, or perforation, ft/hr.
u_{oj}	Nozzle velocity at which a jet first forms, ft/hr.
u_s	Slip velocity, ft/hr.
u_t	Terminal velocity of a single drop, ft/hr.
v_p	Volume of a single drop, ft^3.
x_A	Mole fraction of component A in continuous phase.
y_A	Mole fraction of component A in disperse phase.
y_{Aa}, y_{Ab}	Mole fraction of component A in the disperse phase at sections a and b.
$y_{An}, y_{A\,n+1}$	Mole fraction of component A in the disperse phase on plates n and $n+1$.
y_A^*	Local equilibrium concentration of A in the $\mathfrak{D}$ phase at the interface, mole fraction.
$y_{A\mathfrak{C},n}$	Disperse-phase concentration of A that would be in equilibrium with existing continuous-phase concentration in stage n, mole fraction.
y_{Af}, y_{Ar}, y_{Ac}	Mole fraction of component A in disperse phase during droplet formation, free rise (or fall), and coalescence.
y_{A1}, y_{A2}	Mole fraction of component A in the disperse phase at outlet and inlet to column.
z	t/t_f.
$\beta(p,q)$	Beta function, equation 8.41.
$\Gamma(p), \Gamma(q)$	Gamma functions, equation 8.41.
λ_1, λ_n	Eigenvalues for first and nth terms in a series.
μ_c, μ_d	Viscosities of continuous and disperse phases, lb-mass/(ft)(hr).
μ_{H_2O}	Viscosity of water, lb-mass/(ft)(hr).
ρ	Mean density of phase under consideration, lb-mass/ft^3.
$\rho_c, \rho_d, \Delta\rho$	Densities of continuous and disperse phases; absolute value of $\rho_c - \rho_d$, lb-mass/ft^3.
$(\rho/M)_{av}$	Mean value of ρ/M for the phase under consideration, lb-mole/ft^3.
σ	Interfacial tension, lb-force/ft.
ϕ_d	Disperse-phase holdup, i.e., fraction of column active volume occupied by disperse phase.

Subscripts:

A	Component A.

c, d	Continuous and disperse phases.
LM	Logarithmic mean.

PROBLEMS

8.1 A droplet is growing on the tip of a vertical nozzle. Derive the average time of surface exposure during formation, $\bar{t}_e$, expressed as a fraction of the time of formation, t_f. The time-average droplet surface during formation was shown to be $3\pi d_p^2/5$ in equation 8.10. For a single growing drop, show whether the drop surface reaches $3\pi d_p^2/5$ in a time less than, equal to, or greater than $\bar{t}_e$.

Assume that the drop grows at a constant volumetric rate and that it remains truly spherical throughout growth.

8.2 A spray column is used in a liquid extraction process which has a low interfacial tension and is continuous-phase controlled. Use the form of equation 8.14 corresponding to a spray column plus equations 8.17 and 8.63 to show whether an optimum disperse-phase holdup is indicated for a maximum rate of mass transfer during free rise of the drops. Neglect variations in drop size and concentration driving force with holdup.

8.3 Nitrobenzene containing a few percent of acetic acid is to be extracted by spraying it in droplet form down a water-filled column 4 ft high and 0.0225 ft^2 in cross section. The nitrobenzene feed rate is 0.5 ft^3/hr, and it is withdrawn as soon as it reaches the foot of the column. The water rate through the column is zero, and the temperature is 68°F. The measured values of the droplet diameter and velocity of fall in stationary water are, respectively, 0.01475 ft and 0.39 ft/sec, and the interfacial tension is about 20 dyn/cm. The equilibrium distribution ratio for the prevailing conditions may be taken to be 16, expressed as lb-moles of acetic acid per ft^3 of aqueous phase/lb-mole of acetic acid per ft^3 of nitrobenzene phase. Calculate

(a) the overall mass-transfer coefficient during free fall of the droplets, K_{dr}^*, assuming that (i) the droplets are stagnant; (ii) the droplets contain maximum-speed internal circulation currents due to frictional drag of the water; (iii) the droplets are oscillating. Compare the results of (i), (ii), and (iii) with the experimental value of 0.503 lb-mole/(hr)(ft^2) to ascertain the hydrodynamic condition of the droplets.

(b) Compute the total interfacial area available for mass transfer in the column at any instant during free fall of the drops.

(c) Evaluate the capacity coefficient K_{dr}^*a for cases (i) to (iii) in (a) above, where a is as defined in Chapter 4.

8.4 Apply the computer program in Table 8.5 to predict the column requirements (n, n_0, A_D, and column diameter) for the extraction described in run H7 of Garner, Ellis, and Hill (1955). Use $u_o = 2150$ ft/hr, HD = 3 (i.e., stagnant drops), ND = 50, and TEN(σ) = 0.00172 lb-force/ft (Treybal, 1963, p. 498). The remaining data in Table 8.4 may be assembled from the original reference and by appropriate prediction. Compare the computed column characteristics with those used by the authors.

8.5 Perform computations similar to those in Problem 8.4, but for runs I2, H4, and H6 of Garner, Ellis, and Hill (1955). Use the following respective values for u_o, calculated from the recorded data as $u_o = 4Q_{d2}/\pi d_o^2 n_0$:

$$\text{I2: } u_o = 1612 \text{ ft/hr};$$

$$\text{H4: } u_o = 1195 \text{ ft/hr};$$

$$\text{H6: } u_o = 1340 \text{ ft/hr}.$$

8.6 Use the computer program in Table 8.5 to characterize the necessary column dimensions (n, n_0, A_D, and column diameter) for the aqueous extraction of benzoic acid from kerosene, as performed by Allerton, Strom, and Treybal (1943). Their runs are unnumbered, but the four to be considered here may be identified on their pp. 374–375 as having V_K = 153.8, 133.9, 94.0, and 133.0 ft^3 kerosene/(hr)(ft^2), respectively. The corresponding u_o values are 1131, 985, 693, and 979 ft/hr; also HD = 3, ND = 50, and TEN(σ) = 0.00206 lb-force/ft. Other data in Table 8.4 may be compiled from the original paper and by suitable prediction. (It appears that benzoic acid exists largely as a dimer in kerosene.) Compare the computed results with the column characteristics used in the original study.

REFERENCES

Allerton, J., B. O. Strom, and R. E. Treybal, *Trans. Am. Inst. Chem. Eng.*, **39**, 361–384, (1943).

Angelo, J. B., and E. N. Lightfoot, *A. I. Ch. E. J.*, **14**, 531, (1968).

Angelo, J. B., E. N. Lightfoot, and D. W. Howard, *A. I. Ch. E. J.*, **12**, 751–760, (1966).

Basu, Miss Anjali, *Momentum and Mass Transfer of Droplets,* Ph.D. thesis in chemical engineering, Indian Institute of Technology, Kharagpur, India, (1970).

Beek, W. J., and H. Kramers, *Chem. Eng. Sci.,* **17**, 909–921, (1962).

Beyaert, B. O., L. Lapidus, and J. C. Elgin, *A. I. Ch. E. J.,* **7**, 46, (1961).

Brunson, R. J., and R. M. Wellek, *Can. J. Chem. Eng.,* **48**, 267–274, (1970).

Bussolari, R., S. Schiff, and R. E. Treybal, *Ind. Eng. Chem.,* **45**, 2413, (1953).

Calderbank, P. H., in *Mixing: Theory and Practice,* Vol. 2, V. W. Uhl and J. B. Gray, Eds., Academic Press, New York, (1967), pp. 53–59.

Calderbank, P. H., and R. P. Patra, *Chem. Eng. Sci.,* **21**, 719–721, (1966).

Coulson, J. M., and S. J. Skinner, *Chem. Eng. Sci.,* **1**, 197, (1952).

Davies, J. T., private communication, (December 9, 1969).

Davies, J. T., and E. K. Rideal, *Interfacial Phenomena,* 2nd ed., (1963), pp. 333–335, 330–339.

Davies, J. T., and R. W. Vose, *Proc. Roy. Soc.,* **A286**, 218–234, (1965).

de Chazal, L. E. M., and J. T. Ryan, *A. I. Ch. E. J.,* **17**, 1226, (1971).

Elzinga, E. R., and J. T. Banchero, *Chem. Eng. Prog. Symp. Ser.,* **55** (No. 29), 149, (1960).

Fair, J. R., in *Design of Equilibrium Stage Processes,* B. D. Smith, McGraw-Hill, New York, (1963), p. 557.

Garner, F. H., S. R. M. Ellis, and D. W. Fosbury, *Trans. Inst. Chem. Eng.,* (*Lond.*), **31**, 348–362, (1953).

Garner, F. H., S. R. M. Ellis, and J. W. Hill, *A. I. Ch. E. J.,* **1**, 185–192, (1955).

Garner, F. H., S. R. M. Ellis, and J. W. Hill, *Trans. Inst. Chem. Eng.,* (*Lond.*), **34**, 223–234, (1956).

Garner, F. H., and A. R. Hale, *Chem. Eng. Sci.,* **2**, 157, (1953).

Garner, F. H., and A. H. P. Skelland, *Trans. Inst. Chem. Eng.,* (*Lond.*), **29**, 315, (1951).

Garner, F. H., and A. H. P. Skelland, *Ind. Eng. Chem.,* **46**, 1255, (1954).

Garner, F. H., and A. H. P. Skelland, *Chem. Eng. Sci.,* **4**, 149, (1955).

Garner, F. H., and A. H. P. Skelland, *Ind. Eng. Chem.,* **48**, 51, (1956).

Garner, F. H., A. H. P. Skelland, and P. J. Haycock, *Nature,* **173**, 1239, (1954).

Garner, F. H., and M. Tayeban, *An. Real Soc. Esp. Fis. Quim., Ser.* B—*Quim.,* **LVI(B),** 479, (1960).

Goldberger, W. M., and R. F. Benenati, *Ind. Eng. Chem.,* **51**, 641–644, (1959).

Grober, H., *Z. Ver. dtsch. Ing.,* **69**, 705, (1925).

Groothuis, H., and H. Kramers, *Chem. Eng. Sci.,* **4**, 17–25, (1955).

Handlos, A. E., and T. Baron, *A. I. Ch. E. J.,* **3**, 127, (1957).

Harkins, W. D., and F. E. Brown, *J. Am. Chem. Soc.,* **41**, 499, (1919).

Hayworth, C. B., and R. E. Treybal, *Ind. Eng. Chem.,* **42**, 1174, (1950).

Heertjes, P. M., and L. H. de Nie, *Chem. Eng. Sci.*, **21**, 755–768, (1966).

Heertjes, P. M., L. H. de Nie, and H. J. de Vries, *Chem. Eng. Sci.*, **26**, 441, (1971).

Heertjes, P. M., W. A. Holve, and H. Talsma, *Chem. Eng. Sci.*, **3**, 122, (1954).

Higbie, R., *Trans. Am. Inst. Chem. Eng.*, **31**, 365, (1935).

Hu, S., and R. C. Kintner, *A. I. Ch. E. J.* **1**, 42, (1955).

Ilkovic, D., *Coll. Czech. Chem. Commun.*, **6**, 498, (1934).

Izard, J. A. *A. I. Ch. E. J.*, **18**, 634–638, (1972).

Jenson, V. G., and G. V. Jeffreys, *Mathematical Methods in Chemical Engineering*, Academic Press, London and New York, (1963).

Johnson, A. I., and L. Braida, *Can. J. Chem. Eng.*, **35**, 165, (1957).

Johnson, A. I., and A. E. Hamielec, *A. I. Ch. E. J.*, **6**, 145, (1960).

Johnson, A. I., A. E. Hamielec, D. Ward, and A. Golding, *Can. J. Chem. Eng.*, **36**, 221, (1958).

Klee, A. J., and R. E. Treybal, *A. I. Ch. E. J.*, **2**, 444, (1956).

Kronig, R., and J. C. Brink, *Appl. Sci. Res.*, **A-2**, 142 (1950).

Kumar, R., and N. R. Kuloor, *Adv. in Chem. Eng.*, **8**, T. B. Drew, G. R. Cokelet, J. W. Hoopes Jr., and T. Vermeulen, Eds., Academic Press, New York, **8** 256–368, (1970).

Licht, W., and J. B. Conway, *Ind. Eng. Chem.*, **42**, 1151, (1950).

Licht, W., and W. F. Pansing, *Ind. Eng. Chem.*, **45**, 1885, (1953).

Lindland, K. P., and S. G. Terjesen, *Chem. Eng. Sci.*, **5**, 1–12, (1956).

Major, C. J., and R. R. Hertzog, *Chem. Eng. Prog.*, **51**, No. 1, 17-J, (1955).

Mayfield, F. D., and W. L. Church, *Ind. Eng. Chem.*, **44**, 2253–2260, (1952).

Moulton, R. W., and J. E. Walkey, *Trans. Am. Inst. Chem. Eng.*, **40**, 695–707, (1944).

Nandi, S. K., and S. K. Ghosh, *J. Indian Chem. Soc., Ind. and News Ed.*, **13**, 93–107, (1950).

Null, H. R., and H. F. Johnson, *A. I. Ch. E. J.*, **4**, 273, (1958).

Oliver, E. D., *Diffusional Separation Processes,* Wiley, New York, (1966).

Olney, R. B., and R. S. Miller, "Liquid-Liquid Extraction," in *Modern Chemical Engineering,* A. Acrivos, Ed., **1**, 89–139, Reinhold, New York, (1963).

Pasternak, I. S., and W. H. Gauvin, *Can. J. Chem. Eng.*, **38**, 35, (1960).

Patel, J. M., and R. M. Wellek, *A. I. Ch. E. J.*, **13**, 384–386, (1967).

Perry, R. H., C. H. Chilton, and S. D. Kirkpatrick, Eds., *Chemical Engineers Handbook,* 4th ed., McGraw-Hill, New York, **21**-25; **14**-64–65; (1963).

Pimentel, G. C., private communication, (June 26, 1972).

Pimentel, G. C., and A. L. McClellan, "Hydrogen Bonding" in *Ann. Rev. Phys. Chem.,* H. Eyring, C. J. Christensen, and H. S. Johnston, Eds., **22**, 350, (1971).

Popovich, A. T., R. E. Jervis, and O. Trass, *Chem. Eng. Sci.*, **19**, 357–365, (1964).

Poutanen, A. A., and A. I. Johnson, *Can. J. Chem. Eng.*, **38**, 93, (1960).

Pyle, C., A. P. Colburn, and H. R. Duffey, *Ind. Eng. Chem.*, **42**, 1042–1047, (1950).

Rao, E. V. L., R. Kumar, and N. R. Kuloor, *Chem. Eng. Sci.*, **21**, 867–880, (1966).

Rose, P. M., and R. C. Kintner, *A. I. Ch. E. J.*, **12**, 530–534, (1966).

Row, S. B., J. H. Koffolt, and J. R. Withrow, *Trans. A. I. Ch. E.*, **37**, 559–595, (1941).

Ruby, C. L., and J. C. Elgin, *Chem. Eng. Prog. Symp. Ser.*, **51**, No. 16, 17–29, (1955).

Sawistowski, H., and G. E. Goltz, *Trans. Inst. Chem. Eng. (Lond.,)* **41**, 174, (1963).

Scheele, G. F., and B. J. Meister, *A. I. Ch. E. J.*, **14**, 9–19, (1968).

Siemes, W., *Chem.-Ing.-Tech.*, **28**, 727–731, (1956).

Skelland, A. H. P., and C. L. Caenepeel, *A. I. Ch. E. J.*, **18**, 1154–1163, (1972).

Skelland, A. H. P., and W. L. Conger, *Ind. Eng. Chem., Process Des. and Dev.*, **12**, 448–454, (1973).

Skelland, A. H. P., and A. R. H. Cornish, *A. I. Ch. E. J.*, **9**, 73–76, (1963).

Skelland, A. H. P., and A. R. H. Cornish, *Can. J. Chem. Eng.*, **43**, 302–305, (1965).

Skelland, A. H. P., and C. L. Hemler, unpublished work at the University of Notre Dame, (1969).

Skelland, A. H. P., and S. S. Minhas, *A. I. Ch. E. J.*, **17**, 1316–1324, (1971).

Skelland, A. H. P., and V. K. Raval, *Can. J. Chem. Eng.*, **50**, 41–44, (1972).

Skelland, A. H. P., and R. M. Wellek, *A. I. Ch. E. J.*, **10**, 491–496, 789, (1964).

Smith, B. D., *Design of Equilibrium Stage Processes,* McGraw-Hill, New York, (1963), Chapter 16.

Thornton, J. D., *Chem. Eng. Sci.*, **5**, 201, (1956).

Treybal, R. E., *Liquid Extraction,* 2nd ed., McGraw-Hill, New York, (1963), pp. 470, 480.

Treybal, R. E., *Mass Transfer Operations,* 2nd ed., McGraw-Hill, New York, (1968).

Treybal, R. E., and F. E. Dumoulin, *Ind. Eng. Chem.*, **34**, 709–713, (1942).

Ueyama, K., *Kagaku Kogaku.*, **21**, 766, (1957).

Vermeulen, T., *Ind. Eng. Chem.*, **45**, 1664, (1953).

Weaver, R. E. C., L. Lapidus, and J. C. Elgin, *A. I. Ch. E. J.*, **5**, 533, (1959).

Wellek, R. M., A. K. Agrawal, and A. H. P. Skelland, *A. I. Ch. E. J.*, **12**, 854–862, (1966).

Wellek, R. M., and A. H. P. Skelland, *A. I. Ch. E. J.*, **11**, 557, (1965).

Zenz, F. A., *Pet. Refiner*, **36**, No. 8, 147, (1957).

Zheleznyak, A. S., *J. Appl. Chem. U.S.S.R.*, **40**, 834, (1967).

9

Design of Cooling Towers From Rate Equations

The terms "humidification" and "dehumidification" refer to a variety of engineering processes involving the exchange of mass and energy between a pure liquid and an effectively insoluble gas. The material constituting the liquid phase represents the mass transferred, either by evaporation or condensation. The presence of only one liquid component means that concentration gradients and resistance to the transfer of mass are both absent from the liquid phase, but these processes are nevertheless complicated because of the need for quantitative treatment of simultaneous heat and mass transfer between the gas and the liquid.

Operations aimed primarily at modifying the gas phase include cooling, humidifying, and dehumidifying the gas. A hot liquid may be cooled by a combination of sensible heat transfer and evaporation into a gas when direct contact is made between the phases.

An extremely widespread and important application occurs in the cooling of water with air. Water which has absorbed heat while flowing through condensers or heat exchangers may be cooled for reuse by suitable contact with atmospheric air. Of the equipment used in this process—spray ponds, cooling ponds, crossflow towers, natural-draft cooling towers, and mechanical-draft cooling towers, only the last are considered in detail here.

Definitions

The material undergoing transfer will be designated component A and will be referred to as vapor when in the nonliquid state. The insoluble, nontransferring gas will be designated component B. Whereas mass or molal units and concentrations expressed in mass or mole fractions are convenient in some areas of mass transfer, the humidification literature is predominantly in terms of mass units relative to a fixed mass of the inert component B. Most available charts and tabulated data are in these terms, which are accordingly adopted in this chapter. Many of the definitions in common use were given by Grosvenor (1908).

Humidity, sometimes called "specific humidity" or "humidity ratio," is defined as the number of pounds of vaporized component A per pound of gas B. Thus if the gas phase total pressure is P, the humidity is

$$\mathcal{H}=\frac{M_A p_A}{M_B(P-p_A)} \tag{9.1}$$

where P is assumed low enough for the gas-vapor mixture to be regarded as ideal. At saturation under the prevailing temperature, $p_A=P_A$, and

$$\mathcal{H}_S=\frac{M_A P_A}{M_B(P-P_A)} \tag{9.2}$$

Relative humidity, usually expressed as a percentage, is a meteorological term given by

$$\mathcal{H}_R=100\frac{p_A}{P_A} \tag{9.3}$$

Percentage humidity or percentage absolute humidity is defined to be the ratio of equation 9.1 to equation 9.2, expressed as a percentage:

$$\mathcal{H}_A=100\frac{p_A(P-P_A)}{P_A(P-p_A)} \tag{9.4}$$

Comparison between equations 9.3 and 9.4 shows that the relation between the relative and percentage or absolute humidities is

$$\mathcal{H}_A=\mathcal{H}_R\frac{(P-P_A)}{(P-p_A)} \tag{9.5}$$

Humid heat is the heat needed to increase the temperature of one pound of gas B plus its associated vapor A by 1°F. In algebraic form,

$$c_s = c_{pB} + \mathcal{H} c_{pA} \tag{9.6}$$

For the air-water-vapor system at normal temperatures the average specific heats are

$$c_{p\,H_2O} = c_{pA} = 0.45\,\text{Btu}/(\text{lb})(°\text{F})$$

$$c_{p\,\text{air}} = c_{pB} = 0.24\,\text{Btu}/(\text{lb})(°\text{F})$$

Thus, with equation 9.6, for this system,

$$c_s = (0.24 + 0.45\,\mathcal{H})\ \text{Btu}/(\text{lb dry air})(°\text{F}) \tag{9.7}$$

Enthalpy or humid enthalpy of a gas-vapor mixture is taken to be the enthalpy of the mixture per pound of the insoluble carrier gas B. Thus

$$H = H_B + \mathcal{H} H_A \tag{9.8}$$

where

H = humid enthalpy, Btu/lb of B,

H_B = enthalpy of B, Btu/lb of B,

H_A = enthalpy of A, Btu/lb of A,

$\mathcal{H}$ = humidity, lb of A/lb of B.

The standard reference state for humid-enthalpy calculations is indicated by stating the temperature and pressure and with the vapor (A) condensed to liquid form. Different reference temperatures are often used for the two components, so that ideally,

$$H = H_B + \mathcal{H} H_A = \int_{T_-}^{T_+} c_{pB}\, dT + \mathcal{H}\left(\int_{T_0}^{T_+} c_{pA}\, dT + \lambda_{A0}\right) \tag{9.9}$$

where λ_{A0} is the latent heat of vaporization of component A at T_0, in Btu/lb. When T_- equals T_0,

$$H = \int_{T_-}^{T_+} (c_{pB} + \mathcal{H} c_{pA})\, dT + \mathcal{H}\lambda_{A-}$$

$$= \int_{T_-}^{T_+} c_s\, dT + \mathcal{H}\lambda_{A-}$$

The usual reference state for water is liquid at 32°F and 1 atm, and for dry air it is 0°F and 1 atm. The enthalpy of an air–water-vapor mixture at t°F is then given by equation 9.9 as

$$H = 0.24(t°\text{F} - 0) + \mathcal{H}[0.45(t°\text{F} - 32) + 1075.2]$$

in which 1075.2 is the latent heat of vaporization of water at 32°F in Btu/lb. Simplifying,

$$H = c_s t + 1060.8\mathcal{H} \text{ Btu/lb of dry air} \tag{9.10}$$

Humid volume of a mixture of gas B and vapor A is the total volume, measured in cubic feet, of one pound of dry gas plus its associated vapor at any specified temperature and pressure. Assuming ideal behavior, therefore, at t°F and P atm,

$$v_H = \frac{359(t+460)}{492P}\left(\frac{1}{M_B} + \frac{\mathcal{H}}{M_A}\right) \text{ ft}^3\text{/lb of dry air.} \tag{9.11}$$

The saturated volume is given by equation 9.11 with $\mathcal{H}$ replaced by $\mathcal{H}_S$ from equation 9.2.

Dry-bulb temperature. This is the temperature measured when a thermometer is located within a mixture of vapor and gas in the usual way.

Dew point—when a specimen of humid gas is cooled at constant pressure and humidity, a temperature will be reached at which the gas is saturated with moisture. The temperature at which this occurs is the dew point.

The Wet-Bulb Temperature

Suppose that a large amount of a mixture of vapor A and gas B at a temperature above its dew point is brought into adiabatic contact with a very small amount of liquid A having a temperature initially about the same as that of the gas. Under such circumstances changes in the condition of the liquid do not significantly affect the temperature or humidity of the gas with change in time.

Because the gas is unsaturated, the liquid begins to evaporate under the influence of the partial pressure difference $P_{Aw} - p_A$, where P_{Aw} is the vapor pressure of A at the surface temperature of the liquid and p_A is the partial pressure of A in the bulk of the gas. This removal of latent heat by the evaporation process will cool the liquid—eventually below the temperature of the bulk of the gas, T_G. Heat will consequently flow from the gas to the liquid, which will reach a steady-state temperature T_w, the wet-bulb temperature.

Measurement of the wet-bulb temperature is effected by using a thermometer with a bulb covered with a wick which is thoroughly wetted with the liquid in question. Suspension of this thermometer in a fast-moving stream of the gas under investigation enables T_w to be read when steady-state conditions are reached.

The process may be represented schematically in Figure 9.1, which shows a small liquid droplet surrounded by a large amount of gas at a dry-bulb temperature T_G. At steady-state conditions the heat transferred to the liquid from the gas is just sufficient to provide the latent heat of vaporization and the sensible heat needed to raise the temperature of the evaporated material from T_w to T_G:

$$q_H = M_A N_A A[\lambda_{Aw} + c_{pA}(T_G - T_w)] \tag{9.12}$$

where λ_{Aw} is the latent heat of vaporization of A at T_w in Btu/lb. In the absence of significant radiation from the surroundings, and neglecting the effect of mass transfer on h_G, the rate of heat transfer is also given by

$$q_H = h_G A(T_G - T_w) \tag{9.13}$$

where A is the liquid surface and h_G is the heat-transfer coefficient in the gas. The rate of mass transfer may be expressed as

$$M_A N_A A = M_A k_G A(P_{Aw} - p_A) \tag{9.14}$$

Thus combining equation 9.12 and the steady-state equations 9.13 and 9.14,

$$P_{Aw} - p_A = \frac{h_G(T_G - T_w)}{M_A k_G[\lambda_{Aw} + c_{pA}(T_G - T_w)]} \tag{9.15}$$

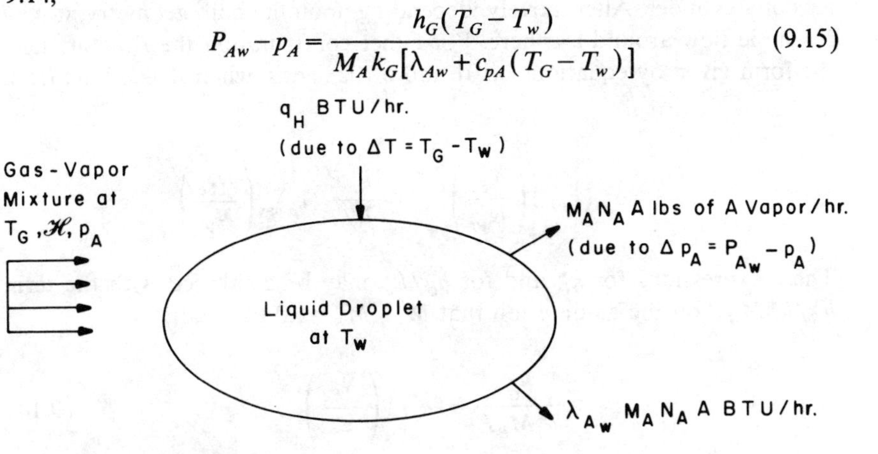

Figure 9.1. Steady-state heat and mass transfer between a small amount of liquid and a large amount of gas.

When $p_A \ll P$ and $P_{Aw} \ll P$, equation 9.1 may be solved for P_{Aw} and p_A and the results inserted in equation 9.15 to obtain

$$\mathcal{H}_w - \mathcal{H} = \left(\frac{h_G}{k_G^* M_B P} \right) \left[\frac{T_G - T_w}{\lambda_{Aw} + c_{pA}(T_G - T_w)} \right] \tag{9.16}$$

where $\mathcal{H}_w$ is the saturation humidity at the wet-bulb temperature T_w. When $c_{pA}(T_G - T_w)$ may be neglected in comparison with λ_{Aw},

$$\mathcal{H}_w - \mathcal{H} = \left(\frac{h_G}{k_G^* M_B P} \right) \left(\frac{T_G - T_w}{\lambda_{Aw}} \right) \tag{9.17}$$

It is evident that measurement of the wet- and dry-bulb temperatures of a gas specimen will enable its humidity $\mathcal{H}$ to be calculated from equation 9.17, provided that the term $h_G / k_G^* M_B P$ can be evaluated. In this regard it may be noted that for ideal gases,

$$N_A = k_G^* \Delta p_A = k_c^* \Delta c_A = k_c^* \frac{\Delta p_A}{RT}$$

so

$$k_G^* = \frac{k_c^*}{RT}$$

The gas flow around a thermometer may resemble flow normal to the axis of a cylinder. Alternatively, depending upon the bulb geometry, it may resemble flow around a sphere. For either configuration the j factors take the form given by equations 6.101 and 6.102. Thus, when A is dilute, from $j_H = j_D$,

$$\frac{h_G}{k_c^*} = \rho_{\text{av}}(c_p)_{\text{av}} \left(\frac{N_{\text{Sc}}}{N_{\text{Pr}}} \right)_{\text{av}}^{2/3} = \frac{M_{\text{av}} P}{RT} (c_p)_{\text{av}} \left(\frac{N_{\text{Sc}}}{N_{\text{Pr}}} \right)_{\text{av}}^{2/3}$$

These expressions for k_G^* and for h_G / k_c^* may be combined with the term $h_G / k_G^* M_B P$ on the assumption that $M_{\text{av}} / M_B \doteq 1.0$ to obtain

$$\frac{h_G}{k_G^* M_B P} \doteq (c_p)_{\text{av}} \left(\frac{N_{\text{Sc}}}{N_{\text{Pr}}} \right)_{\text{av}}^{2/3} \tag{9.18}$$

In the case of the air–water-vapor system at a pressure of 1 atm, the equality in equation 9.18 assumes a value of about 0.26. For many organic vapors in air the value is greater by a factor of around 2, partly because of changes in the relative magnitudes of Schmidt and Prandtl numbers.

In view of the constancy of the term $h_G/k_G^* M_B P$ for a given system, as shown by equation 9.18, it follows from equation 9.17 that the wet-bulb temperature is determined only by the temperature and humidity of the gas when radiation effects are negligible.

The Adiabatic Saturation Temperature

In contrast with the physical situation considered in developing relationships for the wet-bulb temperature, attention is now directed to the case of a limited amount of unsaturated gas in adiabatic contact with a recirculated body of liquid large enough for its temperature to stay constant at T_s. The gas is both humidified and cooled during this process, and with sufficiently intimate contact, it will leave at temperature T_s in a condition of saturation. It will be supposed that the liquid lost by evaporation is made up with liquid which is also at T_s, the adiabatic saturation temperature.

It should be noted that, unlike the situation considered in Figure 9.1, the quantities T_G and p_A are no longer constant, so that steady-state equations analogous to 9.13 and 9.14 cannot be written for the present conditions, shown schematically in Figure 9.2.

An enthalpy balance may be written for the process with T_s as datum and using inlet values of T_G and $\mathcal{H}$:

$$(c_{pB} + \mathcal{H} c_{pA})(T_G - T_s) + \mathcal{H}\lambda_{As} = \mathcal{H}_{T_s}\lambda_{As}$$

or, with equation 9.6,

$$\mathcal{H}_{T_s} - \mathcal{H} = c_s\left(\frac{T_G - T_s}{\lambda_{As}}\right) \tag{9.19}$$

Relation between the Wet-Bulb and Adiabatic Saturation Temperatures

Comparison between equations 9.17 and 9.19 shows that the two expressions become identical—and hence T_w becomes equal to T_s—when $h_G/k_G^* M_B P$ equals the humid heat c_s. Equation 9.18 indicates that this occurs when $N_{Sc} = N_{Pr}$ and $c_s = (c_p)_{av}$. For most systems these requirements do not hold, but, fortuitously, the conditions are approximately satisfied in the important case of the air–water-vapor system when the humidity is not very high. The approximation $h_G/k_G^* M_B P \doteq c_s$ for the air–water-vapor system is known as the Lewis relation (Lewis, 1922) and the ratio N_{Sc}/N_{Pr}

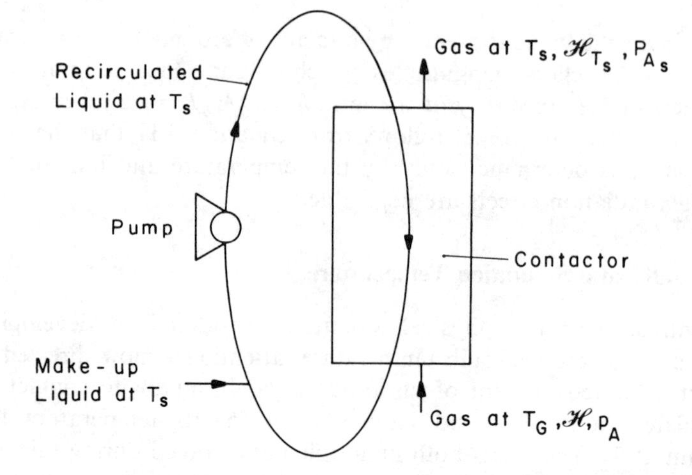

Figure 9.2. Steady-state heat and mass transfer between a large amount of liquid and a limited amount of gas.

is called the Lewis number, N_{Le}. As a result, then, of the approximate validity of the Lewis relation, the wet-bulb and adiabatic saturation temperatures are effectively equal for air–water-vapor mixtures. In most other systems the adiabatic saturation temperature is lower than the wet-bulb temperature, as shown experimentally by Sherwood and Comings (1932).

The Psychrometric Chart

The properties of mixtures of any given gas-vapor system, as defined earlier in this chapter, are conveniently represented on the psychrometric chart. An example of such a chart for the air–water-vapor system at a total pressure of 1 atm appears in Figure 9.3.

Vapor-pressure data as a function of temperature have been used with equation 9.2 to prepare the 100-percent-saturation curve, which relates to temperature as abscissa and humidity as right-hand ordinate in the figure. The definition of percentage humidity shows that the curves for various constant percentage humidities were established by interpolating linearly between the 100-percent-saturation curve and the abscissa.

The plot of humid heat versus humidity is a representation of equation 9.7 and uses axes at the top and the right side of the chart.

Equation 9.11 has been used to obtain the lines showing specific volume of dry air and saturated volume as functions of temperature. The volume scale appears at the left in the figure. The humid volume of air with a given percentage humidity is found by linear interpolation between the two lines at the appropriate temperature.

The adiabatic saturation lines are plots of equation 9.19, each line corresponding to a constant value of T_s represented by the abscissa of its intersection with the 100-percent-saturation curve. Equation 9.19 shows that for a given T_s it yields a plot of $\mathcal{H}$ versus T_G with a slope of $-c_s/\lambda_{As}$. According to equation 9.6, c_s varies with humidity, and of course λ_{As} is also somewhat dependent upon T_s. It follows, therefore, that the adiabatic saturation lines are neither strictly linear nor parallel in rectangular coordinates. Some presentations of the chart compensate for this by distorting the axes so as to render the lines both linear and parallel in order to facilitate interpolation. This refinement appears unnecessary in Figure 9.3, in view of the fact that, although two adiabatic saturation lines remote from each other—say, at the top and bottom of the chart—are not parallel, any two consecutive or immediately adjacent lines are sufficiently close to being linear and parallel that the error in interpolation is negligible. Indeed, if greater accuracy is desired, equation 9.19 itself may be applied to the problem at hand.

Psychrometric charts for the air–water-vapor system at atmospheric pressure but covering much wider ranges of temperature are given in Perry (1950, pp. 759, 765, 766). Page 767 of the same reference gives a chart for water vapor and air mixed with combustion gases, and pp. 813–816 give charts for the binary systems of air mixed, respectively, with vapors of carbon tetrachloride, benzene, toluene, and *ortho*-xylene. Psychrometric charts may of course be prepared for other systems at any specified pressure. Inspection of the relevant equations among 9.1 to 9.19 shows that the information needed for such a construction consists of the molecular weights and specific heats of components A and B, and the latent heat of vaporization and vapor pressure of component A, both as functions of temperature over the range of interest. The use of the psychrometric chart will be illustrated with the following example.

Illustration 9.1.

The air supplied to a processing unit at a total pressure of 1 atm is found to have a dry-bulb temperature of 80°F and a wet-bulb temperature of 60°F. Use the psychrometric chart and relevant equations to evaluate the properties of this humid air.

SOLUTION. For the air–water-vapor system, T_w and T_s are effectively equal. The condition of the air is therefore represented by the intersection of the vertical at 80°F with that adiabatic saturation line which terminates on the 100-percent-saturation curve at a temperature of 60°F. The humidity of the air is accordingly read from Figure 9.3 as 0.0065 lb of water vapor per pound of dry air.

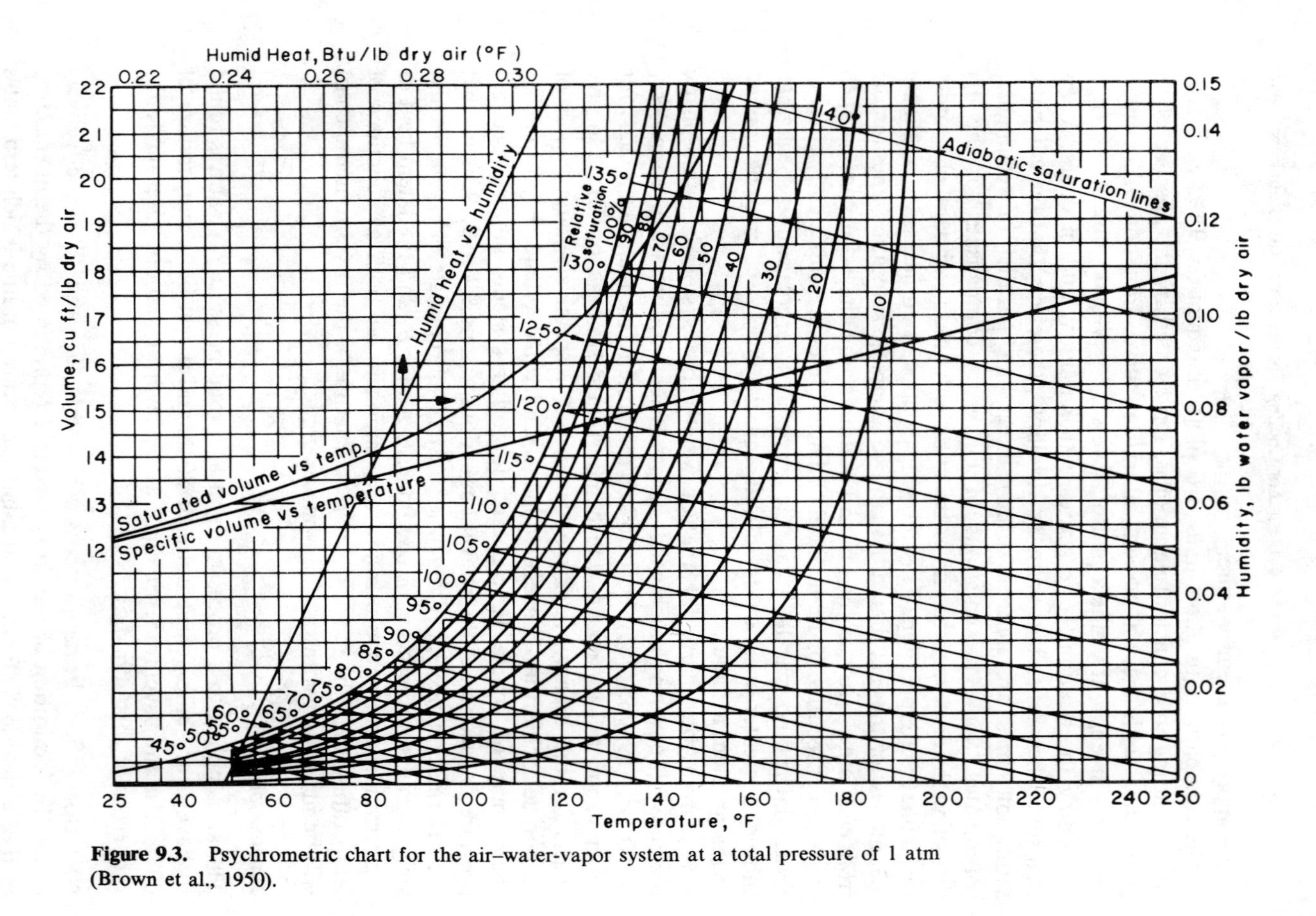

Figure 9.3. Psychrometric chart for the air–water-vapor system at a total pressure of 1 atm (Brown et al., 1950).

The intersection of the vertical at 80°F with the 100-percent-saturation curve shows that this air would hold 0.0222 lb of water vapor per pound of dry air when saturated. The percentage humidity of the air supply is therefore

$$\mathcal{H}_A = 100\left(\frac{0.0065}{0.0222}\right) = 29.3 \text{ percent}$$

The partial pressure of water vapor in the air is obtained from equation 9.1 as

$$p_A = \frac{\mathcal{H}P}{\dfrac{M_A}{M_B} + \mathcal{H}}$$

In the present example, $P = 1$ atm, $M_A = 18$, $M_B = 29$, and $\mathcal{H} = 0.0065$, so that

$$p_A = \frac{0.0065(1)}{\frac{18}{29} + 0.0065} = 0.01036 \text{ atm.}$$

From steam tables, P_A at $80°\text{F} = 0.5067/14.7 = 0.0345$ atm. The relative humidity is therefore

$$\mathcal{H}_R = 100\left(\frac{0.01036}{0.0345}\right) = 30.1 \text{ percent}$$

The humid heat is read at the top of Figure 9.3 as the abscissa of the intersection of the humid-heat line with the horizontal through $\mathcal{H} = 0.0065$. The value found is $c_s = 0.241$ Btu/(°F)(lb dry air), which is within 1 percent of that calculated from equation 9.7.

Equation 9.10 shows the humid enthalpy of the air to be

$$H = 0.241(80) + 1060.8(0.0065) = 26.18 \text{ Btu/lb dry air}$$

The specific volume of dry air at 80°F is found from the psychrometric chart to be 13.6 ft^3/lb dry air, and the saturated volume at the same temperature is 14.1 ft^3/lb dry air. The humid volume is therefore

$$v_H = 13.6 + (14.1 - 13.6)0.293 = 13.75 \text{ ft}^3\text{/lb dry air}$$

The dew point is read as the abscissa of the intersection of the 100-percent-saturation curve with the horizontal through $\mathcal{H} = 0.0065$. A dew point of 46°F is obtained by this procedure.

The definitions and relationships presented so far in this chapter are of value in a most important operation, namely, the cooling of water by contact with air. In this application water which has absorbed heat while flowing through condensers or heat exchangers is cooled before reuse by contact with atmospheric air in some suitable equipment. Mechanical-draft cooling towers are among the most commonly used devices for this purpose, and their design is considered in some detail here. The treatment utilizes an enthalpy driving force as first put forward by Merkel (1925).

In mechanical-draft towers, air flows upwards under the influence of a fan located at either the top or the bottom of the tower. Water is distributed near the top and flows countercurrently to thc air over a system of wooden grids or baffles intended to promote intimate contact between the gas and the liquid. Design procedures involve the formulation of an operating line, rate equations, and NTU and HTU relationships in terms which express the simultaneous transfer of heat and mass, so as to establish tower dimensions and operating conditions which will achieve a specified water-cooling load.

The Operating Line

A relationship connecting some appropriate bulk characteristic of each of the two phases at a given section of the tower is obtained by a suitable balance and is called the operating line. Consider the water-cooling tower sketched in Figure 9.4, which also defines much of the terminology to be used. Note that the primes on G'_B and L' denote flow rates in mass units instead of the molal units mostly used in Chapter 7. This facilitates the use of published psychrometric data for the air–water-vapor system, which is commonly in these terms.

The water is to be cooled from a specified T_{L1} to T_{L2}, and it is desired to find the necessary tower height Z. The quantities which are known or deducible from wet- and dry-bulb temperature measurements as shown in Illustration 9.1 are T_{L1}, T_{L2}, T_{G2}, G'_{B2}, L'_1, H_2, and $\mathcal{H}_2$. A value is also selected for the column cross section S. An enthalpy balance over the differential volume $S\,dZ$ gives

$$d(L'h) = G'_B\,dH \tag{9.20}$$

Consideration of the magnitudes of the specific heat c_L and latent heat of vaporization of water shows that only a small fraction of the water is evaporated during the cooling process. In fact, for each 10°F cooling, about 1 percent of the water is evaporated (Denman, 1961). Then, noting that the variation in c_L is small,

$$d(L'h) = L'c_L\,dT_L = G'_B\,dH \tag{9.21}$$

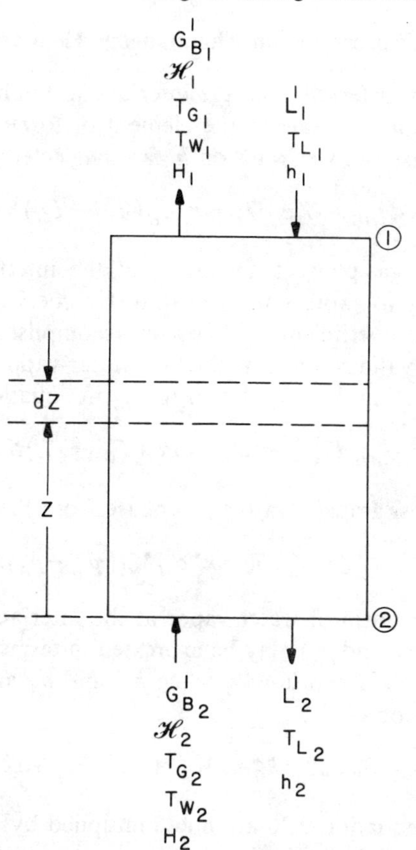

Figure 9.4. Terminology for a water cooling tower. a_H, a_M: interfacial areas for heat and mass transfer per unit packed volume, ft^2/ft^3; G'_B, L': flow rate of dry air in the air-water vapor mixture, and water flow rate, lb/hr; Z: tower height; S: cross-sectional area of empty column; H, h: gas and liquid enthalpies.

Integration over the tower gives

$$H_1 - H_2 = \frac{L' c_L}{G'_B}(T_{L1} - T_{L2}) \tag{9.22}$$

Equation 9.22 defines the operating line for the tower in the H-T_L coordinate system. The plot will be essentially linear because of the effective constancy of the term $L' c_L / G'_B$.

Transfer-Unit Expressions for Simultaneous Heat and Mass Transfer

Sensible heat is transferred from the interface to the bulk of the gas-vapor mixture at the following rate in the element of tower volume $S\,dZ$, if the effects of low mass-transfer rates on h_G are neglected:

$$dq_{HG} = G'_B c_s\, dT_G = h_G a_H (T_{Gi} - T_G) S\, dZ \tag{9.23}$$

where T_{Gi} is the gas-phase temperature at the interface. It will later be found convenient to express the heat-transfer coefficient, h_G, in terms of the mass-transfer coefficient k_G^*. This is accomplished by means of the Lewis relation, valid only for the air–water-vapor system, giving $h_G \doteq k_G^* M_B P c_s$. Inserting this expression for h_G in equation 9.23 results in

$$G'_B c_s\, dT_G = k_G^* a_H M_B P c_s (T_{Gi} - T_G) S\, dZ \tag{9.24}$$

The rate of mass transfer in the gas phase from the interface is

$$d\mathfrak{N}_A M_A = G'_B d\,\mathcal{H} = k_G^* a_M M_A (p_{Ai} - p_A) S\, dZ \tag{9.25}$$

The partial pressures of water vapor at the interface and in the bulk of the gas stream (p_{Ai} and p_A) may be expressed in terms of humidities by the use of equation 9.1. Accordingly, when p_{Ai} and p_A are much less than P, equation 9.25 becomes

$$G'_B d\,\mathcal{H} = k_G^* a_M M_B P (\mathcal{H}_i - \mathcal{H}) S\, dZ \tag{9.26}$$

Both sides of equation 9.26 are now multiplied by 1060.8 and the result added to equation 9.24 to obtain the following expression for simultaneous heat and mass transfer:

$$G'_B [1060.8\, d\mathcal{H} + c_s\, dT_G] = 1060.8 k_G^* a_M M_B P (\mathcal{H}_i - \mathcal{H}) S\, dZ$$
$$+ k_G^* a_H M_B P c_s (T_{Gi} - T_G) S\, dZ \tag{9.27}$$

Equation 9.10 shows that the contents of the square brackets in equation 9.27 are equal to dH. Then, assuming that $a_H = a_M = a$, as found experimentally at high flow rates of both phases (McAdams et al., 1949; Hensel and Treybal, 1952),

$$G'_B\, dH = k_G^* a M_B P [1060.8(\mathcal{H}_i - \mathcal{H}) + c_s (T_{Gi} - T_G)] S\, dZ \tag{9.28}$$

Combining equations 9.10 and 9.28,

$$G'_B\, dH = k_G^* a M_B P (H_i - H) S\, dZ \tag{9.29}$$

Equation 9.29 is merely a rearrangement of equation 9.27 and therefore represents simultaneous heat and mass transfer, showing that an enthalpy difference is the appropriate driving force. It is restricted to the air–water-vapor system because the derivation involved the use of the Lewis relation. The expression may be rearranged for integration, assuming $k_G^* a / G'_B$ is constant, as follows:

$$(\mathrm{NTU})_G = \int_{H_2}^{H_1} \frac{dH}{H_i - H} = \frac{k_G^* a M_B P S}{G'_B} \int_0^Z dZ = \frac{k_G^* a M_B P S Z}{G'_B} \tag{9.30}$$

The first integral defines the number of individual gas-enthalpy transfer units. Physically, this may be regarded as the number obtained by dividing the average enthalpy driving force, $(H_i - H)_{\mathrm{av}}$, into the total change in gas-phase enthalpy achieved in the tower, $H_1 - H_2$. The corresponding height of an individual gas-enthalpy transfer unit is given by

$$(\mathrm{HTU})_G = \frac{Z}{(\mathrm{NTU})_G} = \frac{G'_B}{k_G^* a M_B P S} \tag{9.31}$$

The application of equation 9.30 to the determination of the $(\mathrm{NTU})_G$ requires a knowledge of the interfacial gas-phase enthalpy H_i corresponding to the enthalpy of the bulk of the gas, H, at any given cross section in the tower. This may be obtained in terms of temperatures in the liquid and appropriate constants by considering the rate at which heat is transferred to the interface from the main body of the liquid. Thus

$$dq_{HL} = L' c_L dT_L = h_L a (T_L - T_{Li}) S\, dZ \tag{9.32}$$

A combination of equations 9.21, 9.29, and 9.32 gives

$$h_L a (T_L - T_{Li})\, dZ = k_G^* a M_B P (H_i - H)\, dZ$$

If it is assumed that resistance to heat and mass transfer is negligible at the interface, then T_{Li} equals T_{Gi}, and H_i is the saturation enthalpy corresponding to T_{Gi}, so that

$$-\frac{h_L a}{k_G^* a M_B P} = \frac{H_i - H}{T_{Gi} - T_L} \tag{9.33}$$

Equations 9.30 and 9.33 are used to evaluate $(\mathrm{NTU})_G$ and Z by a graphical procedure, using a plot of H versus T_L for saturated air, which also constitutes a plot of H_i versus T_{Gi}.

In systems other than air–water vapor, for which the Lewis relation is invalid, or when a_H does not equal a_M, equation 9.29 will not hold. Reference should then be made to Lewis and White (1953) and Olander (1960, 1961).

Evaluation of Number of Transfer Units

The integral in equation 9.30 may be evaluated numerically by graphical integration. Information for this procedure is obtained from the plot of the saturation or equilibrium curve and the operating line in (H, T_L) coordinates, as sketched in Figure 9.5. (Reference should also be made to Figure 9.4.)

Quantities which are known or deducible from wet- and dry-bulb temperature measurements include T_{L1}, T_{L2}, T_{G2}, G'_{B2}, L'_1, H_2, $\mathcal{H}_2$, and c_L. Now $G'_{B2} = G'_{B1} = G'_B$ and $L'_2 \doteq L'_1 \doteq L'$, so that substitution into equation 9.22 enables H_1 to be evaluated and the operating line to be plotted in Figure 9.5. The equilibrium curve for saturated air could be obtained from water vapor-pressure data and equations 9.2, 9.7, and 9.10 or from the tabulated data in Perry (1963, pp. **15**-6–7).

For any intermediate point in the tower, such as (H, T_L) on the operating line, the corresponding value of H_i is located at the intersection with the equilibrium curve of the line through (H, T_L) of slope $-h_L a / k_G^* a M_B P$, in accordance with equation 9.33. Assuming, therefore, that values of $h_L a / k_G^* a$ are available (see later) this procedure enables evaluation of a series of $H_i - H$ values between H_1 and H_2. A plot of $1/(H_i - H)$ versus H is then prepared, and the $(\mathrm{NTU})_G$ is given by the area under the curve

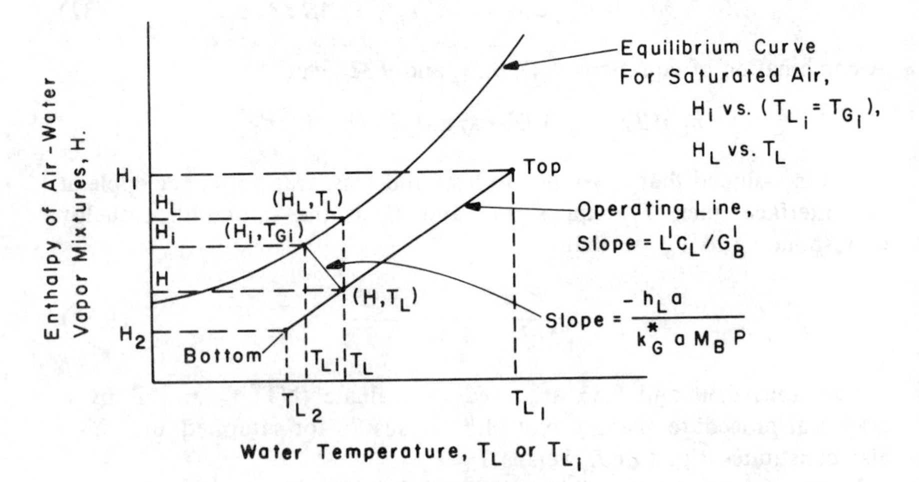

Figure 9.5. Evaluation of components in expressions for NTU—water cooling.

between the limits of integration. The tower height is next obtained from the right side of equation 9.30.

As a first approximation, in the absence of information on the ratio $h_L a/k_G^* a$, it is sometimes assumed that the temperature drop through the liquid is negligible, in which case $T_L = T_{Li} = T_{Gi}$. Equation 9.33 shows that $-h_L a/k_G^* a M_B P$ is then infinite, so that H_i is estimated as the ordinate of a point on the equilibrium curve vertically above the corresponding point (H, T_L) on the operating line. This approximation, in which H_i is estimated as H_L, leads to replacement of the *individual* driving force $H_i - H$ by an *overall* driving force $H_L - H$, because H_L is the enthalpy of air saturated at the bulk temperature of the liquid, T_L, and H is the existing enthalpy of the air at the point in question. However, such an overall driving force is more properly used with an overall coefficient K_G^*, defined with reference to equation 9.29 as follows:

$$G'_B dH = k_G^* a M_B P(H_i - H) S\,dZ = K_G^* a M_B P(H_L - H) S\,dZ \quad (9.34)$$

Rearrangement yields the following expressions for the number and height of overall gas-enthalpy transfer units:

$$(\mathrm{NTU})_{OG} = \int_{H_2}^{H_1} \frac{dH}{H_L - H} = \frac{K_G^* a M_B P S Z}{G'_B} \quad (9.35)$$

$$(\mathrm{HTU})_{OG} = \frac{G'_B}{K_G^* a M_B P S} \quad (9.36)$$

It is important to note, as pointed out in Chapter 4, that the use of constant overall coefficients is justified only when the equilibrium curve is effectively linear over the range of interest, or alternatively, in the present case, when the resistance to heat transfer in the liquid phase is negligible, so that $T_L = T_{Li} = T_{Gi}$. Evidence regarding the validity of the latter alternative is currently inconclusive.

The availability of individual or, when appropriate, overall coefficients or HTU values from pilot-plant or full-scale experimental runs enables the required tower height to be calculated as

$$(\mathrm{NTU})_G(\mathrm{HTU})_G = (\mathrm{NTU})_{OG}(\mathrm{HTU})_{OG} = Z \quad (9.37)$$

As noted in Chapter 7, it is generally preferable to design in terms of transfer units, rather than by direct use of transfer coefficients with integrated forms of equation 9.34. This is because the HTU varies less with flow rate than coefficients do, thereby providing a more stable basis for design.

The Axial Temperature-Enthalpy Profile for the Air–Water-Vapor Mixture

A knowledge of the variation in temperature and enthalpy of the air–water-vapor mixture along the tower is of interest; location of the T_G-H profile beneath the equilibrium curve corresponds to air that is below saturation and incapable of fog formation. The information may be used in estimating the individual transfer coefficients $h_G a$, $h_L a$, and $k_G^* a$ from an experimental run on a pilot or full-scale unit. The procedure for establishing the T_G-H profile will accordingly be outlined here, following Mickley (1949).

Equation 9.29 is divided by equation 9.23 to obtain

$$\frac{G_B' \, dH}{G_B' c_s \, dT_G} = \frac{k_G^* a M_B P (H_i - H) S \, dZ}{h_G a (T_{Gi} - T_G) S \, dZ}$$

or

$$\frac{dH}{dT_G} = \frac{H_i - H}{T_{Gi} - T_G} \tag{9.38}$$

after cancellation of the term $h_G / k_G^* M_B P$ with c_s, in accordance with the Lewis relation. Equation 9.38 shows that the rate of change of air enthalpy with air temperature equals the slope of the line joining the point representing the bulk conditions of the air, (H, T_G), to the point representing the corresponding interfacial conditions of the air, (H_i, T_{Gi}). The latter point, of course, lies on the equilibrium curve.

The steps in the construction of the T_G-H profile will now be itemized with reference to Figures 9.4 and 9.6, starting at the bottom of the tower.

Step 1. Draw line *ab*, of slope $-h_L a / k_G^* a M_B P$, through point (H_2, T_{L2}). This construction follows equation 9.33 and locates the interfacial conditions (point *b*) at the bottom of the tower.

Step 2. Locate point *c*, having the coordinates (H_2, T_{G2}), from the specified state of the inlet air.

Step 3. Join points *c* and *b*, and in accordance with equation 9.38, the air condition will follow line *cb*.

Step 4. At point *d* the air conditions are sufficiently modified to require a new direction for the T_G-H profile. Accordingly, from point *e* (at the intersection of the operating line with the horizontal through *d*) draw a line of slope $-h_L a / k_G^* a M_B P$ to locate the new interfacial conditions at point *f*.

Step 5. Join point *d* to point *f* to obtain the new path for the air condition. Continue along this path until point *g* is reached; then locate the new interfacial conditions (point *i*) from point *h*.

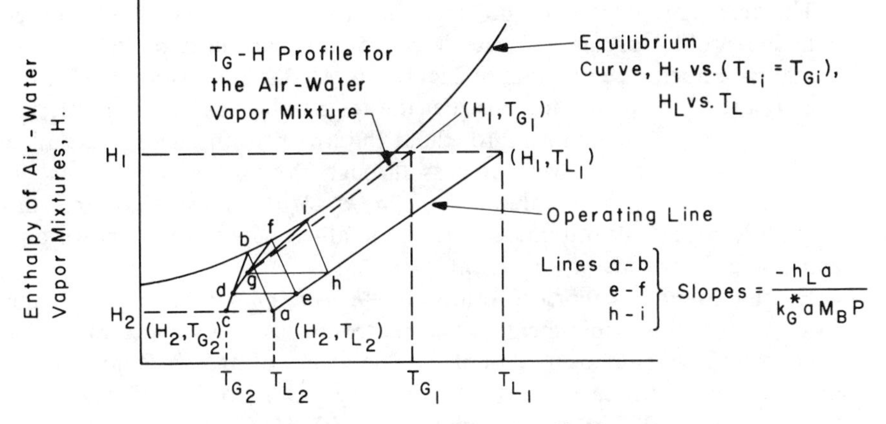

Figure 9.6. Construction of the T_G-H profile for the air–water-vapor mixture in a water cooler.

Step 6. In this manner construct the curve through points $c, d, g, \ldots$, finally reaching the point (H_1, T_{G1}), the condition of the air leaving the contactor. This curve is the desired enthalpy-temperature profile for the air–water-vapor mixture in the tower.

The vertical intervals between lines $ac, de, gh, \ldots$ are arbitrary and contribute to the approximate nature of the method. An additional factor causing inaccuracy is the frequently close approach to saturation of the air near the outlet.

Experimental Evaluation of Individual Rate Coefficients

The three individual rate coefficients, $h_G a$, $h_L a$, and $k_G^* a$, used in equations 9.23, 9.25, and 9.32, and subsequent manipulations, may be determined by a graphical technique with data obtained in a single pilot-plant run using the packing and operating conditions to be encountered on the full scale. It should be noted, however, that rate coefficients on the small scale frequently exceed those found on the large scale, perhaps because of liquid channeling and poorer gas-liquid contact in larger equipment.

The inlet and outlet temperatures of the water and the inlet and outlet wet- and dry-bulb temperatures of the gas are measured, together with the water flow rate, in the single experimental run. This permits the calculation of the enthalpy and humidity of the gas stream at each end of the tower by the methods of Illustration 9.1.

The terminal points on the operating line and on the temperature-enthalpy profile for the air–water-vapor mixture are next plotted on the enthalpy-temperature diagram of Figure 9.6. A constant, *assumed* value for $-h_L a/k_G^* a M_B P$ is used to construct the temperature-enthalpy profile for the air–water-vapor mixture throughout the tower, in the manner detailed earlier. If this profile does not pass through the known upper terminal point (H_1, T_{G1}), a new value of $-h_L a/k_G^* a M_B P$ is assumed and the procedure repeated until coincidence is obtained between the known and constructed locations of (H_1, T_{G1}).

Having found the correct value of $-h_L a/k_G^* a M_B P$ in this way, the $(\text{NTU})_G$ for this experiment is evaluated by graphical integration of equation 9.30 as previously described. The values of M_B, P, Z, and G_B'/S used in the run are known, allowing $k_G^* a$ to be obtained as $G_B'(\text{NTU})_G/M_B PSZ$. Since $k_G^* a$ and $-h_L a/k_G^* a M_B P$ are now known, $h_L a$ can be calculated. The quantity $h_G a$ is finally obtained from the Lewis relation for the air–water-vapor system, as $h_G a \doteq k_G^* a M_B P c_{s,\text{av}}$, where $c_{s,\text{av}}$ is the average of the humid heats at the top and bottom of the tower.

This experimental determination assumes that the process is adiabatic and that the interfacial areas for heat and mass transfer are the same $(a_H = a_M)$. Although the latter assumption is approximately true at high gas and liquid flow rates, at lower rates a_H is greater than a_M. This is because not all of the packing is wetted at lower flow rates, but both wet and dry surfaces contribute to heat transfer, whereas only the wetted surface is involved in mass transfer. A rather high degree of accuracy is needed in collecting the data because point (H_1, T_{G1}) is often near the equilibrium curve. Attempts to avoid this problem by a reduction in tower height may cause contributions from end effects to become excessive. (See Problem 9.4 at the end of this chapter.)

Some experimental data on mass-transfer coefficients in this type of equipment are given by McKelvey and Brooke (1959) and by Norman (1961). These same references also provide extensive practical information on the construction and operation of cooling towers.

The Minimum Air-to-Water Ratio

The terminology of Figure 9.4 is used in Figure 9.7 which shows operating lines for two different sets of conditions. Any location where the operating line touches (point 1) or becomes tangential (point 2) to the equilibrium curve results in zero driving force at that point. When $H_i - H$ or $H_L - H$ becomes zero, the column height becomes infinite, in accordance with equations 9.30 and 9.35. These conditions therefore represent limiting values for the slope of the operating line $(L'c_L/G_B')$ from equation 9.22, and correspond to minimum values of G'_B for given sets of L', T_{L1}, T_{L2}, and H_2. Operable conditions require G'_B greater than the minimum.

It is of interest to consider the variety of air conditions at the foot of the tower that are possible in successful operation. When the Lewis relation holds and $a_H = a_M$, equations 9.29 and 9.34 demonstrate that an enthalpy difference is the relevant driving force for simultaneous transfer of heat and mass. Thus it is only necessary that the enthalpy H_2 of the entering air be below H_{L2}, the latter being the saturation enthalpy of the bulk of the air at the bulk water temperature T_{L2}. The development of equation 9.19 shows the adiabatic saturation process to be one of essentially constant enthalpy, so that the enthalpy of the air is determined by its adiabatic saturation temperature T_s—which effectively equals the wet-bulb temperature for the air–water-vapor system. In consequence, it is quite feasible for T_{G2} to be above T_{L2}, provided that T_{w2} is below T_{L2}. The tower will also function using inlet air with a humidity of 100 percent, provided that the air temperature is less than T_{L2}.

Cooling-tower design usually calls for $T_{L2} - T_{w2}$, the so-called "approach," to be of the order of 10°F, using a value of T_{w2} high enough to be anticipated only, say, 2.5 to 5 percent of the time during the summer.

Illustration 9.2

A small induced-draft cooling tower is to be constructed to cool water

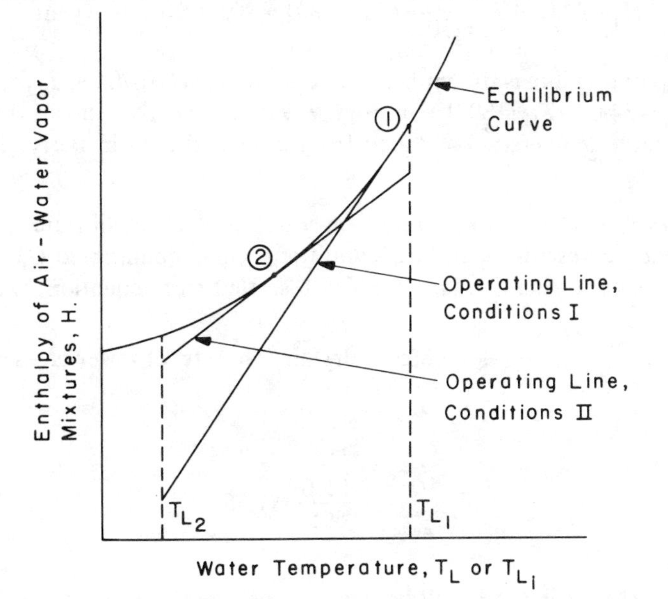

Figure 9.7. Two sets of conditions illustrating the minimum air/water ratio.

from 110 to 80°F by means of air at 1 atm in countercurrent flow. The dry-bulb temperature of the inlet air will be 78°F and, for design purposes, the wet-bulb temperature appropriate to the region will be taken as 70°F. The mass flow rates of the water and dry-air streams will be 1500 and 1250 lb/(hr)(ft^2 of tower cross section) respectively. Determine

(a) The factor by which the air flow rate exceeds the minimum value.
(b) The height of packing required when allowance is made for the heat-transfer resistance in the water phase.
(c) The packed height indicated when the water-phase resistance to heat transfer is neglected.

SOLUTION. Refer to Figures 9.4, 9.5, and 9.7. The enthalpy of any air–water-vapor mixture will be effectively that corresponding to saturation at its wet-bulb temperature. This is because the adiabatic saturation process is one of essentially constant enthalpy, H_{T_S}, where H_{T_S} is the enthalpy of saturated gas at T_s; also, for this system, $T_s \doteq T_w$. Thus, from Perry (1963, p. **15**-6),

$$H_2 = 34.09 \text{ Btu/lb dry air}$$

Substituting in equation 9.22,

$$H_1 = 34.09 + \frac{1500(1)}{1250}(110-80) = 70.09 \text{ Btu/lb dry air}$$

The operating line is drawn between the points ($H_1 = 70.09$, $T_{L1} = 110$°F) and ($H_2 = 34.09$, $T_{L2} = 80$°F) in Figure 9.8, which also shows the equilibrium curve, H_i versus T_{Gi}, taken from tabulated data in Perry (1963, p. **15**-6–7).

SOLUTION (a). The broken line tangential to the equilibrium curve in Figure 9.8 represents operating conditions corresponding to $G'_{B\,\min}$. The upper terminal of the line is (90, 110°F), so that from equation 9.22,

$$\frac{G'_{B\min}}{S} = \frac{1500(1)(110-80)}{90-34.09} = 805 \text{ lb dry air/(hr)(ft}^2\text{ of tower cross section)}$$

Therefore

$$\frac{G'_B}{G'_{B\min}} = \frac{1250}{805} = 1.55$$

SOLUTION (b). The tower will be packed with wooden slats in grid form in the manner used in the experiments of W. J. Thomas and P. Houston [*Brit.*

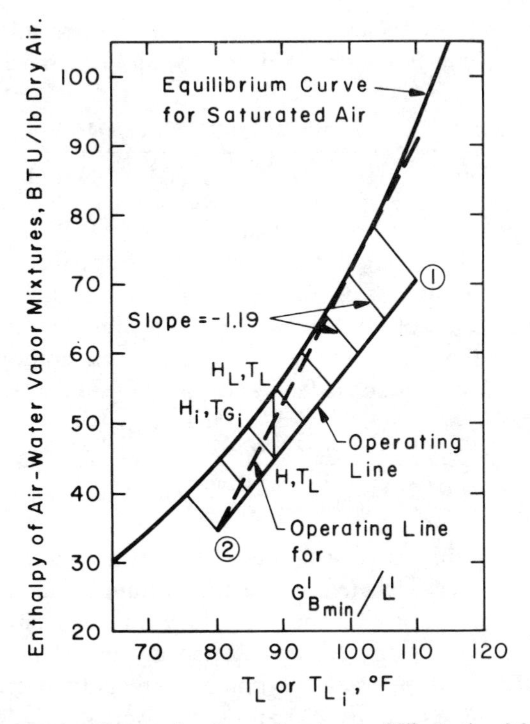

Figure 9.8. The operating diagram for the cooling tower of Illustration 9.2.

Chem. Eng., **4**, 160–163, 217–222, (1959)]. These investigators used a tower 1 ft square in cross section and obtained the following correlations of their measurements for packed heights greater than 0.25 ft, after correction for end effects:

$$h_G a = 1.16 c_s \left(\frac{L'}{S}\right)^{0.26} \left(\frac{G'_B}{S}\right)^{0.72} \tag{a}$$

$$h_L a = 0.03 \left(\frac{L'}{S}\right)^{0.51} \left(\frac{G'_B}{S}\right)^{1.00} \tag{b}$$

$$k_G^* a = 0.04 \left(\frac{L'}{S}\right)^{0.26} \left(\frac{G'_B}{S}\right)^{0.72} \tag{c}$$

for $865 \leqslant G'_B/S \leqslant 1680$, $1000 \leqslant L'/S \leqslant 2000$.

The flow rates for the tower under consideration are within the ranges for which the above expressions were established. Since the packings are identical and both towers are small, these relationships will be adopted for the present problem. Equations 9.33 and (b) and (c) above give

$$-\frac{h_L a}{k_G^* a M_B P} = -\frac{0.03(1500)^{0.51}(1250)}{0.04(1500)^{0.26}(1250)^{0.72}(29)(1)} = -1.19$$

The construction shown schematically in Figure 9.5 is carried out for the present case in Figure 9.8. Some of the values obtained appear in Table 9.1. (Corresponding values of H_L are also listed for use in part c.)

The fourth column of this table is plotted against the first in Figure 9.9. The area under the curve between the limits of integration gives $(\mathrm{NTU})_G = 6.92$, so that from equations 9.30 and (c) above,

$$Z = \frac{1250(6.92)}{0.04(1500)^{0.26}(1250)^{0.72}(29)(1)} = 6.54 \text{ ft}$$

SOLUTION (c). If the water-phase resistance to heat transfer is neglected, H_i is estimated as $H_{L'}$, located on the equilibrium curve vertically above a given point (H, T_L) on the operating line. In such circumstances $k_G^* a \doteq K_G^* a$. Values of $1/(H_L - H)$ are plotted against H in Figure 9.9. The area under the curve between the limits of integration gives $(\mathrm{NTU})_{OG} = 3.14$, so that from equations 9.35 and (c) above,

$$Z = \frac{1250(3.14)}{0.04(1500)^{0.26}(1250)^{0.72}(29)(1)} = 2.97 \text{ ft}$$

These results indicate that the liquid-film resistance to heat transfer is not negligible, in accordance with the findings of Thomas and Houston, who cite four references in support of this contention. In contrast, G. Cribb (*Brit. Chem. Eng.*, **4**, 264–266, 1959) cites evidence and references

Table 9.1. Values obtained for the cooling tower of Illustration 9.2.

H	H_i	H_L	$\frac{1}{H_i - H}$	$\frac{1}{H_L - H}$
34.09	39.5	43.7	0.185	0.104
40.00	44.7	49.2	0.213	0.109
50.00	54.7	60.0	0.213	0.100
60.00	65.5	73.8	0.182	0.072
70.09	78.5	92.3	0.119	0.045

indicating the reverse conclusion for practical design purposes. The point remains somewhat controversial at present.

Part b, for example, could alternatively be solved with a digital computer. The equilibrium curve (H_i versus T_{Gi}) is always the same for the air–water-vapor system and may be empirically represented by an equation such as $H_i = a + bT_{Gi} + cT_{Gi}^2$, where a, b, and c are known constants for a given T_{Gi}-range. This could be used with equations 9.22, 9.30, and 9.33 to program the form of analytical solution shown by Foust et al. (1960, pp. 308–9).

Illustration 9.3

The following observations have been made during the operation at 1 atm of a countercurrent water cooling tower located in a hot and arid region.

Water measurements:	Air measurements:
Flow rate: 1500 lb/(hr)(ft^2)	Inlet dry bulb temperature: 122°F
Inlet temperature: 115°F	Inlet wet bulb temperature: 65°F
Outlet temperature: 93°F	Outlet dry bulb temperature: 100.5°F
	Outlet wet bulb temperature: 93°F

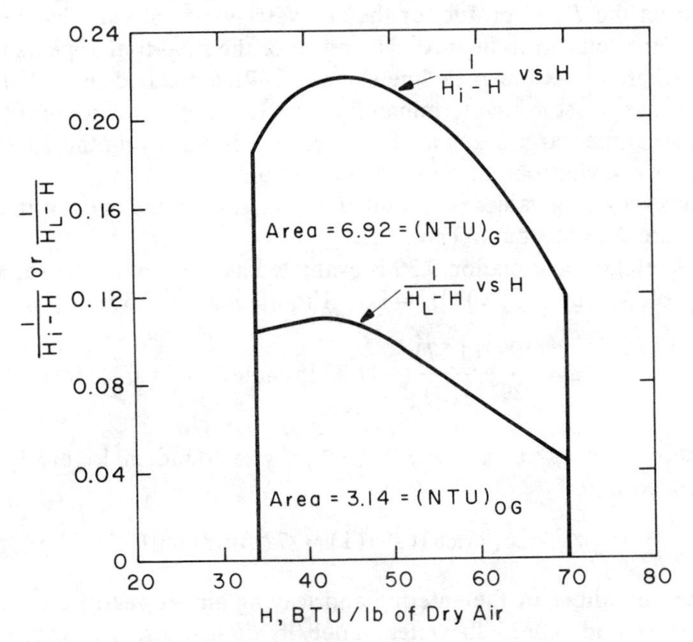

Figure 9.9. Graphical integration to obtain NTUs for Illustration 9.2.

If the packed height of the mechanical-draft tower is 5 ft, determine:

(a) The air flow rate.

(b) The individual coefficients $h_G a$, $h_L a$, and $k_G^* a$ under these operating conditions.

SOLUTION (a). Refer to Figures 9.4 and 9.6. From the wet-bulb temperatures of the inlet and outlet air streams and the tabulation in Perry (1963, pp. **15**-6–7),

$$H_1 = 60.25 \text{ Btu/lb dry air}$$

$$H_2 = 30.06 \text{ Btu/lb dry air}$$

and from equation 9.22,

$$\frac{G'_B}{S} = \frac{1500(1)(115-93)}{60.25-30.06} = 1095 \text{ lb/(hr)(ft}^2\text{)}$$

SOLUTION (b). The equilibrium curve is plotted in Figure 9.10, along with the terminal points on the operating line $[(H_1, T_{L1}), (H_2, T_{L2})]$ and on the T_G-H profile for the air–water-vapor mixture $[(H_1, T_{G1}), (H_2, T_{G2})]$. A constant, assumed value of the tie-line slope, $-h_L a/k_G^* a M_B P$, is used to construct the T_G-H profile for the air–water-vapor mixture by the six-step procedure detailed in the text. The result of the final trial appears in Figure 9.10, where an assumed tie-line slope of -2.26 resulted in a T_G-H profile for the gas phase which terminated correctly on the known point (H_1, T_{G1}). Only six rather large segments have been used in drawing the T_G-H profile, in order to avoid obscuring the construction.

Corresponding values of H and H_i are read from the ends of the tie lines in Figure 9.10 to obtain Table 9.2.

The integral in equation 9.30 is evaluated as the sum of the values in the final column, giving $(\text{NTU})_G = 1.57$. From equation 9.30,

$$k_G^* a = \frac{1095(1.57)}{29(1)(5)} = 11.88 \text{ lb-mole/(ft}^3\text{)(hr)(atm)}$$

The correct value of $-h_L a/k_G^* a M_B P$ was found in Figure 9.10 to be -2.26, so that

$$h_L a = 2.26(11.88)(29)(1) = 777 \text{ Btu/(hr)(ft}^3\text{)(°F)}$$

The humidities of the entering and leaving air are read from Figure 9.3 as 0.0007 and 0.0325 lb water vapor/lb of dry air. The corresponding

Table 9.2. Values obtained for the cooling tower of Illustration 9.3.

H	H_i	$\frac{1}{H_i - H}$	ΔH	$\left(\frac{1}{H_i - H}\right)_{av}$	$\Delta H\left(\frac{1}{H_i - H}\right)_{av}$
30.06	48.8	0.0534			
			4.94	0.0537	0.2653
35	53.5	0.0541			
			5	0.0541	0.2705
40	58.5	0.0541			
			5	0.0538	0.2690
45	63.7	0.0535			
			5	0.0524	0.2620
50	69.5	0.0513			
			5	0.0500	0.2500
55	75.5	0.0488			
			5.25	0.0475	0.2494
60.25	81.9	0.0462			
					Sum = 1.5662

humid heats are obtained from the same figure as 0.240 and 0.253 Btu/(lb dry air)(°F). Thus $c_{s_1,av} = 0.5(0.240 + 0.253) = 0.246$, and from the Lewis relation,

$$h_G a = (11.88)(29)(1)(0.246) = 84.6 \text{ Btu/(hr)(ft}^3\text{)(°F)}$$

Additional runs using various heights of packing would enable at least partial corrections to be made for end effects in the manner of W. J. Thomas and P. Houston [*Brit. Chem. Eng.*, **4**, 160–163, 217–222, (1959)].

Illustration 9.4

To what temperature will the water be cooled in the tower of Illustration 9.3 if

(a) The air flow rate is increased (without flooding) by 100 percent?
(b) The inlet wet-bulb temperature of the air increases to 75°F?

These changes are to be considered separately and in turn.

SOLUTION (a). The slope of the operating line corresponding to the increased air rate is

$$\frac{L' c_L}{G'_B} = \frac{1500(1)}{2(1095)} = 0.685$$

For the conditions of Illustration 9.3 the correct tie line slope, $-h_L a / k_G^* a M_B P$, was found to be -2.26. The slope under the new conditions of air flow depends upon the way in which $h_L a$ and $k_G^* a$ vary with G'_B/S. This, in turn, depends on the type of packing used; thus, in a

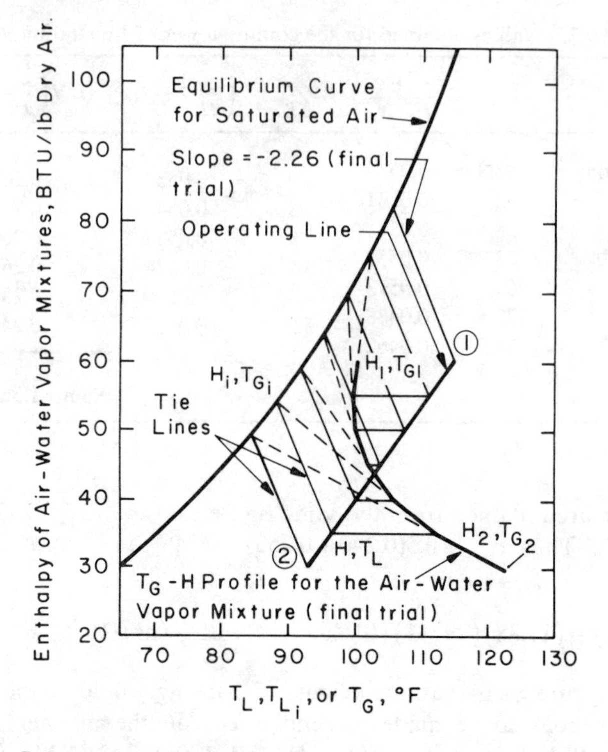

Figure 9.10. Construction of the T_G-H profile for the gas phase in the cooling tower of Illustration 9.3.

4-ft-square tower, H. J. Lowe and D. G. Christie [*Inst. Mech. Eng. Heat Transfer Symp.*, Paper 113, 933, (1962)] found that

$$k_G^* a \propto \left(\frac{L'}{S}\right)^{1-n} \left(\frac{G_B'}{S}\right)^n$$

where n ranged from 0.4 to 0.8 with variations in packing. For illustration purposes in the present case, it is supposed that $h_L a$ and $k_G^* a$ vary with G_B'/S raised to the respective exponents 1.0 and 0.72, as in equations (b) and (c) of Illustration 9.2. The new tie-line slope is therefore

$$-2.26 \frac{2^{1.0}}{2^{0.72}} = -2.75$$

In Illustration 9.3,

$$(\text{NTU})_G = \frac{k_G^* a M_B P S Z}{G_B'} = 1.57$$

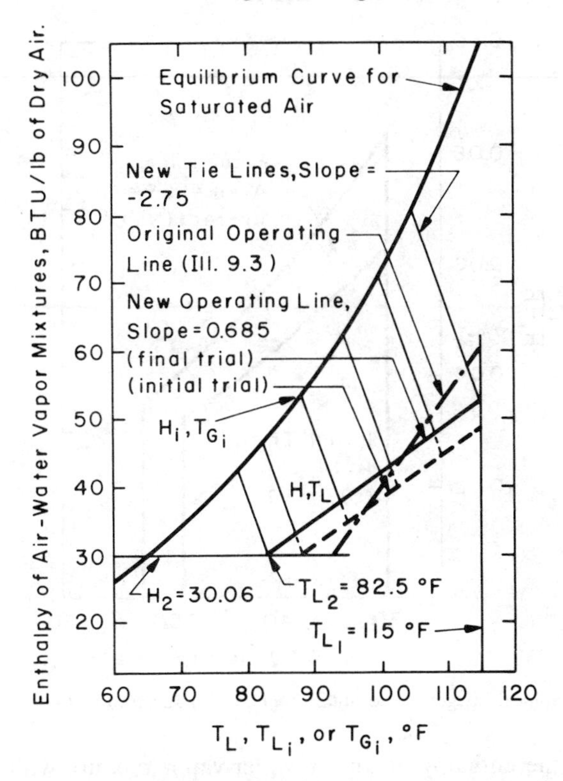

Figure 9.11. Location of the operating line for the cooling tower of Illustration 9.4*a*.

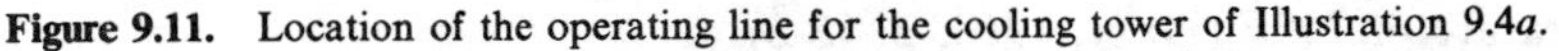

so that under the new conditions of part (a),

$$(\text{NTU})_G = 1.57\frac{2^{0.72}}{2} = 1.29$$

With the new slopes of the operating line and tie lines now known, the position of the operating line is found by trial and error, such that $(\text{NTU})_G = 1.29$. The upper terminal of the operating line lies on the vertical through $T_{L1} = 115°\text{F}$, and the lower terminal will be on the horizontal at $H_2 = 30.06$ Btu/lb dry air. The results of the initial and final trials are shown in Figures 9.11 and 9.12, the latter being constructed in a manner analogous to that in Illustration 9.2. The water is evidently cooled to 82.5°F, as indicated by the abscissa of the lower terminal of the operating line.

SOLUTION (b). The solution for this case is similar in principle to that for part (a). The slope of the operating line is the same as for Illustration 9.3, because the air and water flow rates are now unchanged. The upper terminal of the operating line will lie on the vertical at $T_{L1} = 115°\text{F}$, and the lower terminal will be on the horizontal at $H_2 = 38.61$ Btu/lb dry air.

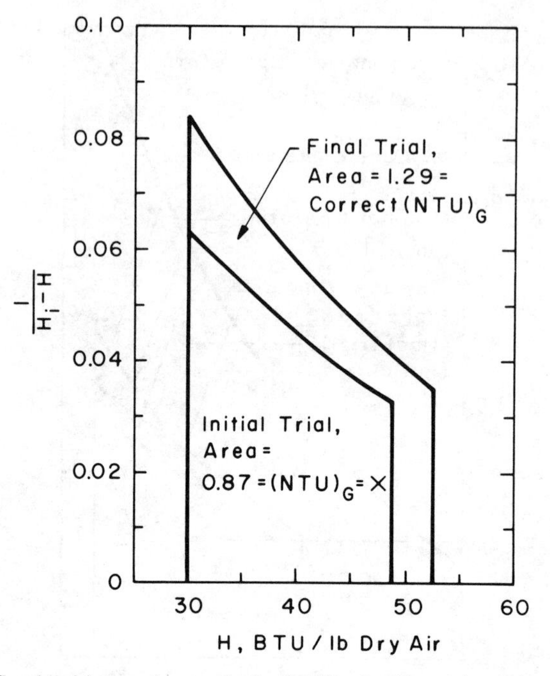

Figure 9.12. Graphical integration to obtain $(NTU)_G$ for Illustration 9.4*a*.

The latter is the enthalpy of an air–water-vapor mixture with a wet-bulb temperature of 75°F (Perry, 1963, p. **15**-6).

The slope of the tie lines and the $(NTU)_G$ will also remain the same as in Illustration 9.3. This follows because $h_L a$ and $k_G^* a$ remain constant in view of the unaltered flow rates (see, e.g., equations b and c in Illustration 9.2). In addition, the variation in physical properties with humidity will not be sufficient to change $k_G^* a$ significantly (see, e.g., Illustration 5.7).

The operating line is accordingly located by trial and error, so as to satisfy all the above conditions. The correct position is found when the subsequent integration (analogous to Figure 9.12) gives $(NTU)_G = 1.57$. The operating line resulting from the final trial in Figure 9.13 fulfills all these requirements, and the abscissa of its lower terminal shows that the water is cooled to 96°F.

Cocurrent Flow

Cocurrent flow of gas and liquid is not widely used, except in certain spray chambers and in some laboratory studies such as experiments on a wetted-wall column. In these cases, when the liquid is being cooled in downward flow, the combination is one of hot liquid with gas of low enthalpy at the top of the column, and cold liquid with highly humid gas

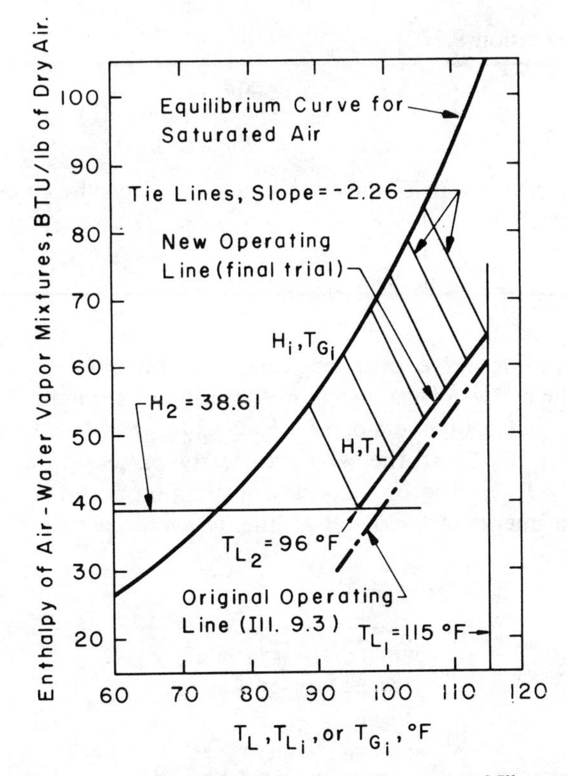

Figure 9.13. Location of the operating line for the cooling tower of Illustration 9.4*b*.

of high enthalpy at the bottom of the column. The operating line is accordingly beneath the equilibrium curve but with negative slope, in contrast to the positive slope shown in Figure 9.5.

Illustration 9.5

Water enters an experimental spray tower at 100°F and is cooled to 81°F by a cocurrent stream of air. The air enters with wet- and dry-bulb temperatures of 45 and 72°F, and leaves with wet- and dry-bulb temperatures of 63 and 76°F.

If the water flow rate is 500 lb/(hr)(ft^2), the tower height is 4.5 ft, and the operation takes place at atmospheric pressure, evaluate the air flow rate and the individual coefficients $h_G a$, $h_L a$, and $k_G^* a$ under these conditions.

SOLUTION. From the wet-bulb temperatures and the tabulation in Perry (1963, p. 15-6),

$$H_1 = 17.65 \text{ Btu/lb dry air}$$

$$H_2 = 28.57 \text{ Btu/lb dry air}$$

and from equation 9.22,

$$17.65-28.57=-\frac{500(1)}{G'_B/S}(100-81)$$

where the negative slope of the operating line reflects the co-current nature of the process.

$$\frac{G'_B}{S}=868\ \text{lb}/(\text{hr})(\text{ft}^2)$$

The equilibrium curve, operating line, and terminal points on the T_G-H profile for the air–water-vapor mixture are plotted in Figure 9.14. Trial and error shows that a tie-line slope $(-h_L a/k_G^* a M_B P)$ of -3.2 enables construction of a T_G-H profile which correctly passes through the known terminal (H_2, T_{G2}). (The broken-line construction is shown for only three of the five segments of the T_G-H profile, to avoid obscuring the diagram.)

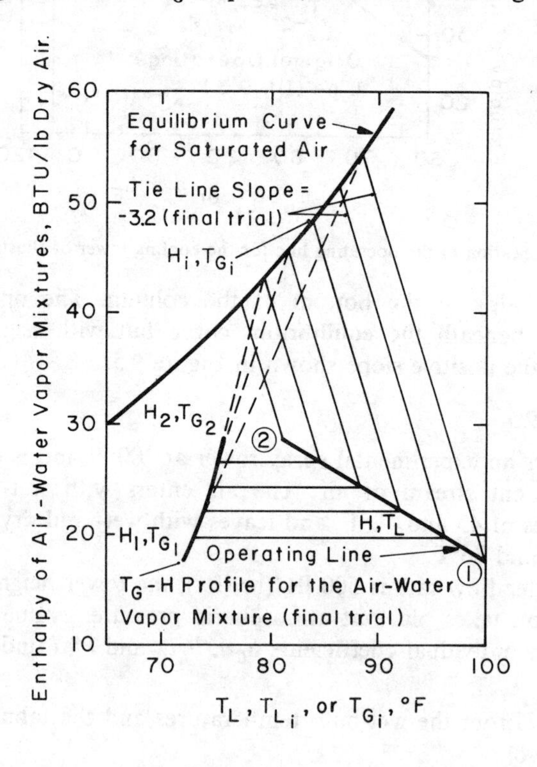

Figure 9.14. Construction of the T_G-H profile for the gas phase in the spray tower of Illustration 9.5.

Corresponding values of H and H_i are read from the ends of tie lines and used to evaluate equation 9.30 as

$$(\mathrm{NTU})_G = -\int_{H_2}^{H_1} \frac{dH}{H_i - H} = 0.5 = \frac{k_G^* a(29)(1)(4.5)}{868}$$

where the minus sign before the integral arises from the cocurrent nature of the operation.

$$k_G^* a = 3.32\ \text{lb-mole}/(\text{ft}^3)(\text{hr})(\text{atm})$$

$$h_L a = 3.2(3.32)(29)(1) = 308\ \text{Btu}/(\text{hr})(\text{ft})^3(°\text{F})$$

From equations 9.7 and 9.10,

$$\mathcal{H}_1 = \frac{H_1 - 0.24t_1}{0.45t_1 + 1060.8} = \frac{17.65 - 0.24(72)}{0.45(72) + 1060.8} \qquad (t_1 \text{ in } °\text{F})$$

$$= 0.00034\ \text{lb water vapor/lb dry air}$$

$$\mathcal{H}_2 = \frac{28.57 - 0.24(76)}{0.45(76) + 1060.8} = 0.0094\ \text{lb water vapor/lb dry air}$$

Substituting in equation 9.7,

$$c_{s1} = 0.24 + 0.45(0.00034) = 0.240\ \text{Btu}/(\text{lb dry air})(°\text{F})$$

$$c_{s2} = 0.24 + 0.45(0.0094) = 0.244\ \text{Btu}/(\text{lb dry air})(°\text{F})$$

$$c_{s,\text{av}} = 0.242\ \text{Btu}/(\text{lb dry air})(°\text{F})$$

The Lewis relation gives

$$h_G a = (3.32)(29)(1)(0.242) = 23.3\ \text{Btu}/(\text{hr})(\text{ft}^3)(°\text{F})$$

Dehumidification

The dehumidification of a warm gas-vapor mixture may be effected by countercurrent contact with a cold liquid having the same composition as the vapor. Part of the vapor is condensed, and cooling of the gas is accompanied by warming of the liquid. In the air–water-vapor system, where the Lewis relation applies, and when $a_H = a_M$, the graphical construction procedures are as described for water cooling, except that the operating line is now located above the equilibrium curve with positive slope, and the individual and overall driving forces are reversed. Thus equations 9.30 and 9.35 are used with the respective driving forces $H-H_i$ and $H-H_L$.

Illustration 9.6

Warm, moist air is being cooled and dehumidified in a packed tower at 1 atm by countercurrent contact with cold water. The air enters at a rate of 1400 lb dry air/(hr)(ft^2) and with wet- and dry-bulb temperatures of 115 and 135°F. It leaves with a wet-bulb temperature of 85°F. The water enters the top of the column at a temperature of 65°F and a flow rate of 1.35 times the minimum operable value. Measurements at these flow rates on reduced heights of the same packing (such that the outlet air was unsaturated) permitted the evaluation of $-h_L a/k_G^* a M_B P$ as -2.5.

Calculate

(a) The water flow rate.
(b) The water-outlet temperature.
(c) The outlet dry-bulb temperature and humidity of the air.
(d) The water removed from the air stream per 1000 lb of dry air.
(e) The $(\mathrm{NTU})_G$.

SOLUTION (a). Refer to Figure 9.4. The hot and cold ends of the column are reversed in comparison with the case for water cooling. The top of the column is now represented by the *lower* terminal of the operating line, point 1, on the H-T diagram of Figure 9.15, located by the following coordinates:

$$T_{w1}=85°\mathrm{F}, \text{ so } H_1=49.43 \text{ Btu/lb dry air}$$

$$T_{L1}=65°\mathrm{F}$$

The upper terminal of the operating line will lie on the horizontal through $H_2=104.98$ Btu/lb dry air. This is the enthalpy of an air–water-vapor mixture with $T_{w2}=115°$F (Perry, 1963, p. **15**-7).

The broken line from point 1 to the intersection of the equilibrium curve with the horizontal at H_2 represents operating conditions corresponding to $L'_{\min}$. The upper terminal of the line is (104.98, 115°F), so that from equation 9.22,

$$\frac{L'_{\min}}{S}=\frac{1400(104.98-49.43)}{(1)(115-65)}=1555 \text{ lb water/(hr)(ft}^2)$$

$$\frac{L'}{S}=1.35(1555)=2100 \text{ lb water/(hr)(ft}^2)$$

SOLUTION (b). From equation 9.22,

$$T_{L2}=65+\frac{1400}{2100(1)}(104.98-49.43)=102°\text{F}$$

SOLUTION (c). The operating line and the inlet air condition (H_2, T_{G2}) are plotted in Figure 9.15. Tie-line slopes of -2.5 are used to construct the T_G-H profile for the air–water-vapor mixture by the six-step procedure detailed in the text. To avoid overcrowding, the broken-line construction is shown for only the top four segments of the profile, which crosses the equilibrium curve at point *P*. Fog would be expected to appear in the tower beyond this point, although the subsequent path of the T_G-H profile is unclear. As stated by Thomas and Houston [*Brit. Chem. Eng.*, **4**, 220 (1959)],

> The alternatives are that the air becomes super-saturated, indicated by the T_G-H profile crossing the equilibrium curve, or that the profile follows the line of the equilibrium curve.

In the present case it is probable that the air will emerge saturated at 85°F. Figure 9.3 shows that its humidity will then be 0.0264 lb water vapor/lb dry air.

SOLUTION (d).

$$\text{Water removed/1000 lb dry air}=1000(\mathcal{H}_2-\mathcal{H}_1)$$

$$=1000(0.0628-0.0264)=36.4 \text{ lb water}$$

SOLUTION (e). The direction of heat and mass transfer here is opposite to that for water cooling. The left side of equation 9.30 therefore becomes

$$(\text{NTU})_G=\int_{H_1}^{H_2}\frac{dH}{H-H_i}$$

Corresponding values of H and H_i are read as the ordinates of the upper and lower terminals of tie lines in Figure 9.15 and used to evaluate the integral, as in Figures 9.9 and 9.12 or as in Illustration 9.3. The result is $(\text{NTU})_G=3.12$.

The water temperature is below the dew point of the air stream at all locations in the tower.

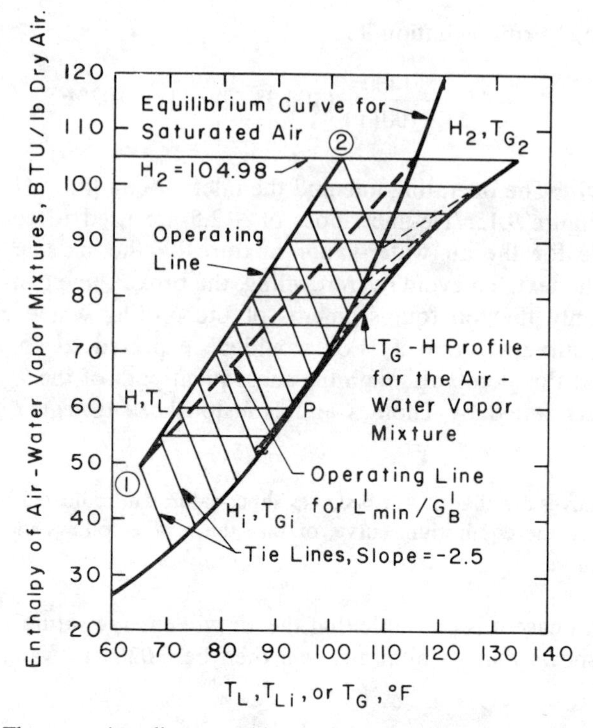

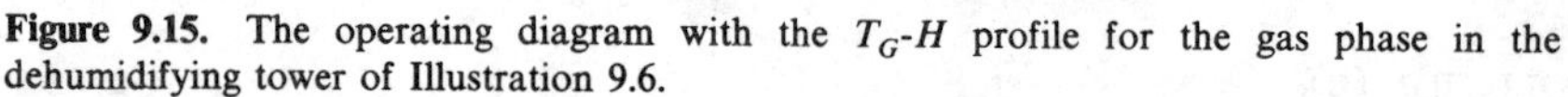
Figure 9.15. The operating diagram with the T_G-H profile for the gas phase in the dehumidifying tower of Illustration 9.6.

Humidification and Gas Cooling with Liquid at the Adiabatic Saturation Temperature

Consider the situation shown in Figure 9.2, in which a continuous stream of gas is brought into contact with recirculated liquid. The temperature throughout the liquid attains the adiabatic saturation temperature of the inlet gas, T_s, and the degree of saturation of the exit gas depends on the intimacy and duration of contact between the two phases. Normally the rate of addition of make-up liquid is low enough that small departures of its temperature from T_s have a negligible effect upon the conditions of the main body of liquid.

In this operation $T_{L1} = T_{L2} = T_{Li} = T_s$, so that the operating line on the H-T_L diagram shrinks to a point. The gas cools and humidifies along that adiabatic saturation line that terminates at T_s on the psychrometric chart. But the adiabatic saturation process is one of essentially constant enthalpy, H_{T_S}, where H_{T_S} is the enthalpy of saturated gas at T_s. There is consequently no enthalpy driving force in this process, and the "operating point" just referred to lies *on* the equilibrium curve in the H-T_L diagram. This diagram therefore cannot be used; instead, design may be achieved in terms of either the temperature or humidity driving force, both of which are confined to the gas phase. These terms will not involve the Lewis relation, so that the treatment applies to any gas-liquid combination. The assumption $a_H = a_M$ may also be relaxed.

Relationships in terms of heat transfer are obtained from equation 9.23, which is written as follows for this process:

$$G'_B c_s \, dT_G = h_G a_H (T_s - T_G) S \, dZ \tag{9.39}$$

where T_{Gi} throughout the tower has the constant, known value T_s. This, of course, contrasts with the situation in water cooling, where T_{Gi} varies throughout the contactor. Rearranging and integrating,

$$(\text{NTU})'_G = \ln \frac{T_{G2} - T_s}{T_{G1} - T_s} = \frac{h_G a_H S Z}{G'_B c_{s,\text{av}}} \tag{9.40}$$

$$(\text{HTU})'_G = \frac{G'_B c_{s,\text{av}}}{h_G a_H S} \tag{9.41}$$

Alternative relationships in terms of mass transfer are obtainable from equation 9.26, which takes the following form in the present case:

$$G'_B \, d\mathcal{H} = k^*_G a_M M_B P (\mathcal{H}_{T_s} - \mathcal{H}) S \, dZ \tag{9.42}$$

where $\mathcal{H}_i$ throughout the tower has the constant, known value $\mathcal{H}_{T_s}$, namely, the saturation humidity at T_s. Rearranging and integrating,

$$(\text{NTU})''_G = \ln \frac{\mathcal{H}_{T_s} - \mathcal{H}_2}{\mathcal{H}_{T_s} - \mathcal{H}_1} = \frac{k^*_G a_M M_B P S Z}{G'_B} \tag{9.43}$$

$$(\text{HTU})''_G = \frac{G'_B}{k^*_G a_M M_B P S} \tag{9.44}$$

The ratio of equation 9.44 to 9.41 shows that

$$\frac{(\text{HTU})''_G}{(\text{HTU})'_G} = \frac{h_G a_H}{k^*_G a_M M_B P c_{s,\text{av}}} \tag{9.45}$$

In the special case of the air–water-vapor system, the Lewis relation will prevail, so that when $a_H = a_M$, the right side of equation 9.45 becomes unity and $(\text{HTU})''_G = (\text{HTU})'_G$. In consequence, since $Z = (\text{NTU})'_G(\text{HTU})'_G$

$=(\text{NTU})''_G(\text{HTU})''_G$, it follows that $(\text{NTU})'_G = (\text{NTU})''_G$ for this particular situation.

Illustration 9.7

Air is being cooled and humidified adiabatically at 1 atm by passage through a bed of 1-in. ceramic Raschig rings which are randomly packed to a depth of 1 ft. Water is continuously recirculated countercurrently through the bed from a small reservoir, so that its temperature attains a single, steady value at all locations in the system. If the wet- and dry-bulb temperatures of the entering air are 95 and 235°F and the flow rates of the ingoing water and air are respectively 1500 lb/(hr)(ft^2) and 5300 ft^3/(hr)(ft^2), determine

(a) The temperature and humidity of the air leaving the bed.
(b) The water flow rates which will give air with a humidity of 0.034 lb water vapor/lb dry air, for dry-air rates of 200 and 500 lb/(hr)(ft^2).
(c) The condition of the outlet air if the bed height is increased by 67 percent with the original flow rates.
(d) The ratio a_H/a_M.

SOLUTION (a). Under adiabatic conditions the recirculating water will acquire the adiabatic saturation temperature of the entering air; thus the water temperature is $T_s \doteq T_{w2} = 95°\text{F}$. Figure 9.3 shows that $\mathcal{H}_2 = 0.005$ lb water vapor/lb dry air, so that from equation 9.11,

$$v_{H_2} = \frac{359(235+460)}{492(1)}\left(\frac{1}{29} + \frac{0.005}{18}\right) = 17.65 \text{ ft}^3 \text{ humid air/lb dry air}$$

$$\frac{G'_B}{S} = \frac{5300}{17.65} = 300 \text{ lb dry air/(hr)(ft}^2\text{)}$$

The application of equation 9.40 or 9.43 will require a numerical value of $h_G a_H$ or $k_G^* a_M$. Correlations of these coefficients with air and water flow rates in countercurrent operation through beds of ceramic Raschig rings were obtained by F. Yoshida and T. Tanaka [*Ind. Eng. Chem.*, **43**, 1467–1473, (1951)]. Their bed was 10 in. in diameter and 12.5 in. deep, and the water temperatures were constant at the adiabatic saturation values. The apparatus was constructed to minimize end effects. The resulting correlations were

$$h_G a_H = 0.117\left(\frac{G'_B}{S}\right)\left(\frac{L'}{S}\right)^{0.2} \tag{A}$$

$$k_G^* a_M M_B P = 0.45\left(\frac{G'_B}{S}\right)\left(\frac{L'}{S}\right)^{0.2} \tag{B}$$

for the ranges $137 \leqslant G'_B/S \leqslant 586$, $201 \leqslant L'/S \leqslant 4160$, $158°F \leqslant T_{G2} \leqslant 302°F$; and for 0.6-, 1.0-, and 1.4-in. Raschig rings in random packing. It will be supposed that the air and water enter and leave the bed in the same manner as in the work of Yoshida and Tanaka. Conditions should then permit the adoption of equations A and B above for the present case.

For illustration purposes, Solutions a and b are carried out using equation 9.43 with equation B, whereas Solution c is performed using equation 9.40 and equation A.

From Figure 9.3, when $T_{w2} = 95°F$, $\mathcal{H}_{T_s} = 0.037$, so that equations 9.43 and B give

$$(\text{NTU})''_G = \ln \frac{0.037 - 0.005}{0.037 - \mathcal{H}_1} = \frac{0.45(300)(1500)^{0.2}(1)}{300} = 1.943$$

$$\mathcal{H}_1 = 0.032 \text{ lb water vapor/lb dry air}$$

The air condition through the bed follows the adiabatic saturation line which terminates at 95°F in Figure 9.3. The abscissa of a point on this line for which $\mathcal{H}_1 = 0.032$ gives T_{G1} as 117°F.

SOLUTION (b). In this assembly the same result will be obtained for a variety of air flow rates, assuming only that

$$137 \leqslant G'_B/S \leqslant 586$$

$$(\text{NTU})''_G = \ln \frac{0.037 - 0.005}{0.037 - 0.034} = \frac{0.45(G'_B/S)(L'/S)^{0.2}(1)}{G'_B/S}$$

$$L'/S = 3000 \text{ lb water/(hr)(ft}^2)$$

Figure 9.3 shows that $T_{G1} = 107°F$.

SOLUTION (c). The term $c_{s,\text{av}}$ will initially be approximated by c_{s2} and adjustment made after the first estimate of $\mathcal{H}_1$. From equation 9.7,

$$c_{s2} = 0.24 + 0.45(0.005) = 0.242 \text{ Btu/(lb dry air)(°F)}$$

Equations 9.40 and A above give

$$(\text{NTU})'_G = \ln \frac{235 - 95}{T_{G1} - 95} = \frac{0.117(300)(1500)^{0.2}(1.67)}{300(0.242)} = 3.49$$

$$T_{G1} = 99.3°F$$

The ordinate of a point with this abscissa on the 95°F adiabatic saturation line of Figure 9.3 shows $\mathcal{H}_1 = 0.0357$ lb water vapor/lb dry air. This is used to calculate c_{s1} as 0.256, so that $c_{s,\text{av}}$ becomes 0.249. The revised result is then

$$(\text{NTU})'_G = 3.49\left(\frac{0.242}{0.249}\right) = 3.39$$

$$T_{G1} = 99.7°\text{F}$$

$$\mathcal{H}_1 = 0.0356 \text{ lb water vapor/lb dry air}$$

SOLUTION (d). The ratio of equation A to equation B gives

$$\frac{h_G a_H}{k_G^* a_M M_B P} = \frac{0.117}{0.45} = 0.26$$

Combination with the Lewis relation shows that

$$\frac{a_H}{a_M} = \frac{0.26}{c_s} \doteq 1.0$$

Evidently, in the experiments from which equations A and B were developed, $a_H \doteq a_M$.

It should be noted that results different from equations A and B have been obtained, for example, by W. H. McAdams, J. B. Pohlenz, and R. C. St. John [*Chem. Eng. Prog.*, **45**, 241, (1949)] for adiabatic humidification using 1-in. carbon rings as packing. Further differences were found for the same operation by S. L. Hensel and R. E. Treybal [*Chem. Eng. Prog.*, **48**, 362, (1952)], whose packing consisted of 1.5-in. Berl saddles.

NOMENCLATURE

A, B	Components A and B (for the air–water-vapor system, A is water vapor and B is air).
A	Area; total interfacial area, ft^2.
a, a_H, a_M	Interfacial area per unit packed volume; for heat transfer; for mass transfer, ft^2/ft^3.
c_A	Concentration of component A, lb-mole/ft^3.
c_L	Specific heat of liquid, Btu/(lb)(°F).

c_{pA}, c_{pair}, $(c_p)_{av}$, c_{pB}, c_{pH_2O} — Specific heat at constant pressure for component A; for air; average for a mixture; for component B: for water vapor, Btu/(lb)(°F).

c_s — Humid heat defined by and above equation 9.6, Btu/(lb B)(°F).

$c_{s,av}$ — Average of c_s at top and bottom of tower, Btu/(lb B)(°F).

G'_B — Mass flow rate of dry air in the air–water-vapor mixture, lb-mass/hr.

H — Enthalpy or humid enthalpy (defined by and above equations 9.8 and 9.9) in the bulk of the gas, Btu/lb B.

H_A, H_B — Enthalpy of A; of B, Btu/lb A or B.

H_i — Local value of H for the gas phase at the interface, Btu/lb B.

H_L — Enthalpy or humid enthalpy of air saturated at the bulk temperature T_L of the liquid, Btu/lb air (B).

H_{T_s} — Enthalpy or humid enthalpy of saturated gas at T_s, Btu/lb B.

$\mathcal{H}$ — Humidity or specific humidity (defined by and above equation 9.1) in the bulk of the gas, lb A/lb B.

$\mathcal{H}_A$ — Percentage humidity or percentage absolute humidity, defined by and above equation 9.4.

$\mathcal{H}_i$ — Local value of $\mathcal{H}$ at the interface, lb A/lb B.

$\mathcal{H}_R$ — Relative humidity, defined by equation 9.3.

$\mathcal{H}_S$ — Saturation value of $\mathcal{H}$, defined by equation 9.2, lb A/lb B.

$\mathcal{H}_{T_s}$ — The saturation value of $\mathcal{H}$ at T_s, lb A/lb B.

$\mathcal{H}_w$ — The saturation value of $\mathcal{H}$ at T_w, lb A/lb B.

$(HTU)_G$, $(HTU)'_G$, $(HTU)''_G$ — Heights of individual G-phase transfer units defined by equations 9.31, 9.41, and 9.44, respectively, ft.

$(HTU)_{OG}$ — Height of overall G-phase transfer unit defined by equation 9.36, ft.

h — Enthalpy of liquid, Btu/lb.

h_G, h_L — Heat-transfer coefficients in the gas and liquid phases, Btu/(hr)(ft^2)(°F).

$h_G a$, $h_L a$	Capacity coefficients of heat transfer in the gas and liquid phases, Btu/(hr)(ft^3)(°F).
j_D, j_H	The j factors for mass and heat transfer, equations 6.101 and 6.102.
K_G^*	Overall mass-transfer coefficient for low transfer rates, defined by equation 9.34, lb-mole/(ft^2)(hr)(atm).
k_c^*	Individual mass-transfer coefficient, $N_A/\Delta c_A$, ft/hr.
k_G, k_G^*	Individual mass-transfer coefficient; at low transfer rates, $N_A/\Delta p_A$, lb-mole/(ft^2)(hr)(atm).
$k_G^* a$	Capacity coefficient of mass transfer at low transfer rates, lb-mole/(ft^3)(hr)(atm).
L'	Mass flow rate of liquid water, lb-mass/(hr).
M_A, M_{av}, M_B	Molecular weights: of A; average; of B.
N_A	Molal flux of component A relative to stationary coordinates, lb-mole/(ft^2)(hr).
$\mathfrak{N}_A$	Transfer rate of A in entire column, lb-mole/hr.
N_{Le}	Lewis number, N_{Sc}/N_{Pr}.
N_{Pr}	Prandtl number, defined in Chapter 6.
N_{Sc}	Schmidt number, defined in Chapter 6.
$(NTU)_G$, $(NTU)'_G$, $(NTU)''_G$	Number of individual G-phase transfer units, defined by equations 9.30, 9.40, and 9.43, respectively.
$(NTU)_{OG}$	Number of overall G-phase transfer units, defined by equation 9.35.
P	Total pressure, atm.
P_A	Vapor pressure of component A, atm.
P_{As}, P_{Aw}	Values of P_A at temperatures T_s and T_w, atm.
p_A	Partial pressure of component A in the bulk of the gas, atm.
p_{Ai}	Local value of p_A at the interface, atm.
q_H	Rate of heat transfer, Btu/hr.
q_{HG}	Rate of sensible heat transfer from the interface to the bulk of the gas-vapor mixture in the entire column, Btu/hr.
q_{HL}	Rate of heat transfer to the interface from the bulk of the liquid in the entire column, Btu/hr.
R	Gas constant, 0.73 (atm)(ft^3)/(lb-mole)(°R).
S	Cross-sectional area of empty column, ft^2.

T, T_G, T_{Gi}, T_L, T_{Li}, T_s, T_w	Absolute temperature; temperature in the bulk of the gas; in the gas at the interface; in the bulk of the liquid; in the liquid at the interface; adiabatic saturation temperature; wet-bulb temperature, °R or °F.
T_-, T_+, T_0	Temperatures in equation 9.9, °R.
t	Temperature, °F.
v_H	Humid volume, defined by and above equation 9.11, ft^3/lb dry air.
Z	Tower height, ft.
λ_{As}, λ_{Aw}, λ_{A-}, λ_{A0}	Latent heat of vaporization of component A at T_s; at T_w; at T_-; at T_0, Btu/lb.
ρ_{av}	Average density, lb-mass/ft^3.
Subscripts:	
1 and 2	refer to conditions at the top and bottom of the tower, respectively.

PROBLEMS

9.1 The air supplied to a dryer consists of 10,000 lb dry air/hr at 145°F and 20 percent humidity. The supply is prepared from air initially at 70°F and 30 percent humidity by heating to 235°F, humidifying adiabatically at 1 atm to the required moisture content, and then reheating to 145°F. Determine

(a) The temperature of the air leaving the humidifier.
(b) The adiabatic saturation temperature in the humidifier.
(c) The heat loads of the heater and reheater.
(d) The temperature to which the air would be raised before entering the humidifier and the corresponding adiabatic saturation temperature if all heating took place in the initial single stage.
(e) The humidity, relative humidity, humid heat, humid enthalpy, humid volume, wet-bulb temperature, and dew point of the final conditioned air.

9.2 Water is cooled from 115 to 85°F by countercurrent contact with air in an induced-draft cooling tower at 1 atm. The air enters with a dry-bulb temperature of 85°F, and the design wet-bulb temperature for the location is 65°F. The water flow rate is 1250 lb/(hr)(ft^2 of tower cross section), and the dry-air rate is 1.5 times the minimum. The mass-transfer coefficient k_G^*a is estimated to be 30 lb-mole/(ft^3)(hr)(atm).

Determine the dry-air flow rate, the wet-bulb temperature of the air leaving the tower, and the packed height required if $h_L a/k_G^* a$ is 60.

9.3 The performance of a 1-ft-square countercurrent water cooling tower packed with wooden slats was studied by Thomas and Houston [*Brit. Chem. Eng.*, **4**, 160–163, 217–222, (1959)]. The following data are from their experiment PP4, performed at 1 atm with a packed height of $3\frac{3}{8}$ inches.

Water rate	1000 lb/hr ft^2
Inlet water temperature	110.2°F
Outlet water temperature	96.7°F
Inlet air temperature (dry bulb)	73.8°F
Inlet air temperature (wet bulb)	62.6°F
Outlet air temperature (dry bulb)	82.9°F
Outlet air temperature (wet bulb)	80.5°F

For these conditions,

(a) Determine the $(\text{NTU})_G$, $h_G a$, $h_L a$, and $k_G^* a$ (all uncorrected for end effects).

(b) Compare these results with values calculated from the correlations a, b, and c cited in Illustration 9.2, which are all corrected for end effects.

9.4 A procedure for eliminating end effects from measurements on a countercurrent, packed water cooling tower was described by Thomas and Houston [*Brit. Chem. Eng.*, **4**, 217, (1959)]. Provided the exit air is unsaturated, a plot of $(\text{NTU})_G$ as ordinate versus packed height was found to be linear at fixed air and water rates. Extrapolation to zero $(\text{NTU})_G$ gives the packed height equivalent to the end effects. Alternatively, extrapolation to zero packed height gives the $(\text{NTU})_{G0}$ corresponding to the end effects. The $(\text{NTU})_{GP}$ equivalent to the packing equals $(\text{NTU})_G-(\text{NTU})_{G0}$ and is used to calculate values of the individual transfer coefficients which are free from end effects.

Using the method of Illustration 9.3, Thomas and Houston obtained the following uncorrected values of $(\text{NTU})_G$: 0.682, 0.907, 1.149, and 1.421, corresponding to packed heights of $3\frac{3}{8}$, $6\frac{3}{4}$, $10\frac{1}{8}$, and $13\frac{1}{2}$ in. These results were obtained at air and water rates of 1680 and 1000 lb/(hr)(ft^2), respectively, in a 1-ft-square experimental tower at 1 atm. Determine:

(a) The height of packing equivalent to the end effects in this case.

(b) The $(\text{NTU})_{G0}$ corresponding to the end effects.

(c) The individual transfer coefficients after correction for end effects under these operating conditions. Compare the results with values calculated from equations a, b, and c in Illustration 9.2.

9.5 A wetted-wall column has a length of 4 ft and an internal diameter of 1 in., and is being used for co-current contact between water and air at atmospheric pressure. The water flow rate is 128 lb/hr with inlet and outlet temperatures of 125 and 119°F, respectively. The air enters at a rate of 12 lb/hr with wet- and dry-bulb temperatures of 66 and 124°F. If the ratio h_L/k_G^* is 600, evaluate the $(\mathrm{NTU})_G$, $(\mathrm{HTU})_G$, and k_G^*a for these conditions.

9.6 Plot the axial temperature-enthalpy profile for the air–water-vapor mixture in the column of Problem 9.5, and obtain the wet- and dry-bulb temperatures and humidity of the air leaving the unit. Prepare qualitative sketches showing temperature distributions on either side of the interface between air and water and the directions of transfer of sensible heat in each phase in the upper and lower parts of the column.

9.7 A warm air stream removes 250 lb/hr of water vapor from a dryer. The air leaves the dryer with wet- and dry-bulb temperatures of 93 and 105°F, and enters a countercurrent, packed dehumidifying tower at 1 atm for removal of the water vapor acquired in the dryer. The air leaves the packed tower with wet- and dry-bulb temperatures of 80 and 82°F, and is then returned to the dryer after reheating. Water enters the packed tower at 70°F and leaves at 75°F. The tower has a cross section of 20 ft^2, and it is known that $h_La/k_G^*a = 435$ (Btu)(atm)/(°F)(lb-mole) for this unit when the dry-air and water flow rates are, respectively, about 1250 and 4000 lb/(hr)(ft^2 of tower cross section).

Determine whether the value cited for h_La/k_G^*a is relevant to the present operation. Estimate the $(\mathrm{NTU})_G$ under these conditions and locate the T_G-H profile for the air–water-vapor mixture in the tower.

9.8 The two adiabatic humidifiers in Problem 9.1 consist of randomly packed beds of 1.4-in. ceramic Raschig rings. Countercurrent contact with continuously recirculated water is used in both cases, with the two phases entering and leaving the beds in the same way as in the work of Yoshida and Tanaka [*Ind. Eng. Chem.*, **43**, 1467–1473, (1951)]. If the water and dry air enter each humidifier at the respective rates of 1250 and 580 lb/(hr)(ft^2), estimate the bed diameter and packed heights necessary to obtain air with the desired conditions in the two cases. For the humidifier used in part d of Problem 9.1, what water rate would be required if the packed height is 1 ft?

REFERENCES

Brown, G. G., et al., *Unit Operations,* Wiley, New York, (1950), p. 545.

Denman, W. W., *Ind. Eng. Chem.*, **53**, 817, (1961).

Foust, A. S., L. A. Wenzel, C. W. Clump, L. Maus, and L. B. Andersen, *Principles of Unit Operations*, Wiley, New York, (1960).

Grosvenor, W. M., *Trans. A. I. Ch. E.,* **1**, 184–202, (1908).

Hensel, S. L., and R. E. Treybal, *Chem. Eng. Prog.*, **48**, 362–370, (1952).

Lewis, W. K., *Trans. Am. Soc. Mech. Eng.*, **44**, 325, (1922).

Lewis, J. G., and R. R. White, *Ind. Eng. Chem.*, **45**, 486, (1953).

McAdams, W. H., J. B. Pohlenz, and R. C. St. John, *Chem. Eng. Prog.*, **45**, 241–252, (1949).

McKelvey, K. K., and M. Brooke, *The Industrial Cooling Tower*, Van Nostrand, Princeton, N. J., (1959).

Merkel, F., *Mitt. Forschungsarb. Heft.* 275, (1925).

Mickley, H. S., *Chem. Eng. Prog.*, **45**, 739–745, (1949).

Norman, W. S., *Absorption, Distillation, and Cooling Towers*, Longmans, Green, London, (1961).

Olander, D. R., *A. I. Ch. E. J.*, **6**, 346, (1960).

Olander, D. R., *Ind. Eng. Chem.*, **53**, 121, (1961).

Perry, J. H., Ed., *Chemical Engineers' Handbook*, 3rd ed., McGraw-Hill, New York, (1950); 4th ed., R. H. Perry, C. H. Chilton, and S. D. Kirkpatrick, Eds., (1963).

Sherwood, T. K., and E. W. Comings, *Trans. A. I. Ch. E.*, **28**, 88 (1932).

Appendix

Table 1. Lennard-Jones force constants as determined from viscosity data.[a]

Molecule	Compound	σ, Å	ϵ/k_B, °K
A	Argon	3.542	93.3
He	Helium	2.551[b]	10.22
Kr	Krypton	3.655	178.9
Ne	Neon	2.820	32.8
Xe	Xenon	4.047	231.0
Air	Air	3.711	78.6
AsH_3	Arsine	4.145	259.8
BCl_3	Boron chloride	5.127	337.7
BF_3	Boron fluoride	4.198	186.3
$B(OCH_3)_3$	Methyl borate	5.503	396.7
Br_2	Bromine	4.296	507.9
CCl_4	Carbon tetrachloride	5.947	322.7
CF_4	Carbon tetrafluoride	4.662	134.0
$CHCl_3$	Chloroform	5.389	340.2
CH_2Cl_2	Methylene chloride	4.898	356.3
CH_3Br	Methyl bromide	4.118	449.2
CH_3Cl	Methyl chloride	4.182	350
CH_3OH	Methanol	3.626	481.8
CH_4	Methane	3.758	148.6
CO	Carbon monoxide	3.690	91.7

Table 1. Continued

Molecule	Compound	σ, Å	ϵ/k_B, °K
COS	Carbonyl sulfide	4.130	336.0
CO_2	Carbon dioxide	3.941	195.2
CS_2	Carbon disulfide	4.483	467
C_2H_2	Acetylene	4.033	231.8
C_2H_4	Ethylene	4.163	224.7
C_2H_6	Ethane	4.443	215.7
C_2H_5Cl	Ethyl chloride	4.898	300
C_2H_5OH	Ethanol	4.530	362.6
C_2N_2	Cyanogen	4.361	348.6
CH_3OCH_3	Methyl ether	4.307	395.0
CH_2CHCH_3	Propylene	4.678	298.9
CH_3CCH	Methylacetylene	4.761	251.8
C_3H_6	Cyclopropane	4.807	248.9
C_3H_8	Propane	5.118	237.1
n-C_3H_7OH	n-Propyl alcohol	4.549	576.7
CH_3COCH_3	Acetone	4.600	560.2
CH_3COOCH_3	Methyl acetate	4.936	469.8
n-C_4H_{10}	n-Butane	4.687	531.4
iso-C_4H_{10}	Isobutane	5.278	330.1
$C_2H_5OC_2H_5$	Ethyl ether	5.678	313.8
$CH_3COOC_2H_5$	Ethyl acetate	5.205	521.3
n-C_5H_{12}	n-Pentane	5.784	341.1
$C(CH_3)_4$	2, 2-Dimethylpropane	6.464	193.4
C_6H_6	Benzene	5.349	412.3
C_6H_{12}	Cyclohexane	6.182	297.1
n-C_6H_{14}	n-Hexane	5.949	399.3
Cl_2	Chlorine	4.217	316.0
F_2	Fluorine	3.357	112.6
HBr	Hydrogen bromide	3.353	449
HCN	Hydrogen cyanide	3.630	569.1
HCl	Hydrogen chloride	3.339	344.7
HF	Hydrogen fluoride	3.148	330
HI	Hydrogen iodide	4.211	288.7

Table 1. Continued

Molecule	Compound	σ, Å	ϵ/k_B, °K
H_2	Hydrogen	2.827	59.7
H_2O	Water	2.641	809.1
H_2O_2	Hydrogen peroxide	4.196	289.3
H_2S	Hydrogen sulfide	3.623	301.1
Hg	Mercury	2.969	750
$HgBr_2$	Mercuric bromide	5.080	686.2
$HgCl_2$	Mercuric chloride	4.550	750
HgI_2	Mercuric iodide	5.625	695.6
I_2	Iodine	5.160	474.2
NH_3	Ammonia	2.900	558.3
NO	Nitric oxide	3.492	116.7
NOCl	Nitrosyl chloride	4.112	395.3
N_2	Nitrogen	3.798	71.4
N_2O	Nitrous oxide	3.828	232.4
O_2	Oxygen	3.467	106.7
PH_3	Phosphine	3.981	251.5
SF_6	Sulfur hexafluoride	5.128	222.1
SO_2	Sulfur dioxide	4.112	335.4
SiF_4	Silicon tetrafluoride	4.880	171.9
SiH_4	Silicon hydride	4.084	207.6
$SnBr_4$	Stannic bromide	6.388	563.7
UF_6	Uranium hexafluoride	5.967	236.8

[a]R. A. Svehla, *NASA Tech. Rept. R-132*, Lewis Research Center, Cleveland, Ohio, 1962. For estimation of unlisted force constants use

$$\frac{\epsilon}{k_B} = 0.75\, T_c; \quad \sigma = \tfrac{5}{6} V_c^{1/3}$$

where the symbols are defined in Chapter 3. Taken from R. C. Reid and T. K. Sherwood, *The Properties of Gases and Liquids*, 2nd ed., McGraw-Hill, New York, 1966, pp. 632–633.

[b]The potential σ was determined by quantum-mechanical formulas.

Table 2. Values of the collision integral $\Omega_{D,AB}$ based on the Lennard-Jones potential.[a]

k_BT/ϵ_{AB} [b]	$\Omega_{D,AB}$ [b]	k_BT/ϵ_{AB}	$\Omega_{D,AB}$	k_BT/ϵ_{AB}	$\Omega_{D,AB}$
0.30	2.662	1.65	1.153	4.0	0.8836
0.35	2.476	1.70	1.140	4.1	0.8788
0.40	2.318	1.75	1.128	4.2	0.8740
0.45	2.184	1.80	1.116	4.3	0.8694
0.50	2.066	1.85	1.105	4.4	0.8652
0.55	1.966	1.90	1.094	4.5	0.8610
0.60	1.877	1.95	1.084	4.6	0.8568
0.65	1.798	2.00	1.075	4.7	0.8530
0.70	1.729	2.1	1.057	4.8	0.8492
0.75	1.667	2.2	1.041	4.9	0.8456
0.80	1.612	2.3	1.026	5.0	0.8422
0.85	1.562	2.4	1.012	6	0.8124
0.90	1.517	2.5	0.9996	7	0.7896
0.95	1.476	2.6	0.9878	8	0.7712
1.00	1.439	2.7	0.9770	9	0.7556
1.05	1.406	2.8	0.9672	10	0.7424
1.10	1.375	2.9	0.9576	20	0.6640
1.15	1.346	3.0	0.9490	30	0.6232
1.20	1.320	3.1	0.9406	40	0.5960
1.25	1.296	3.2	0.9328	50	0.5756
1.30	1.273	3.3	0.9256	60	0.5596
1.35	1.253	3.4	0.9186	70	0.5464
1.40	1.233	3.5	0.9120	80	0.5352
1.45	1.215	3.6	0.9058	90	0.5256
1.50	1.198	3.7	0.8998	100	0.5130
1.55	1.182	3.8	0.8942	200	0.4644
1.60	1.167	3.9	0.8888	400	0.4170

[a] From J. O. Hirschfelder, C. F. Curtiss, and R. B. Bird, *Molecular Theory of Gases and Liquids*, Wiley, New York, 1954.
[b] Hirschfelder uses the symbols T^* for k_BT/ϵ_{AB} and $\Omega^{(1,1)\bigstar}$ in place of $\Omega_{D,AB}$.

Table 3. Atomic diffusion volumes for use in equation 3.5.[a]

	Atomic and Structural Diffusion Volume Increments, v		
C	16.5	(Cl)	19.5
H	1.98	(S)	17.0
O	5.48	Aromatic or Hetero-	
(N)	5.69	cyclic rings	−20.2
	Diffusion Volumes of Simple Molecules, $\sum v$		
H_2	7.07	CO_2	26.9
D_2	6.70	N_2O	35.9
He	2.88	NH_3	14.9
N_2	17.9	H_2O	12.7
O_2	16.6	(CCl_2F_2)	114.8
Air	20.1	(SF_6)	69.7
Ne	5.59		
Ar	16.1	(Cl_2)	37.7
Kr	22.8	(Br_2)	67.2
(Xe)	37.9	(SO_2)	41.1
CO	18.9		

[a]From E. N. Fuller, P. D. Schettler, and J. C. Giddings, *Ind. Eng. Chem.*, **58** (5), 19–27, (May, 1966). Parentheses indicate a value based on only a few data.

Table 4. Eigenvalues and derivatives relating to equations 5.131, 5.133, 5.142, and 5.144-5.146 for mass transfer in laminar flow through tubes.[a]

j	β_j	$(\partial\phi/\partial\beta)_{j,r_+=1}$	$(d\phi_j/dr_+)_{r_+=1}$
1	2.70436 44199	−0.50089 91914	−1.01430 04587
2	6.67903 14493	0.37146 22734	1.34924 16221
3	10.67337 95381	−0.31826 44696	−1.57231 93392
4	14.67107 84627	0.28648 21001	1.74600 43350
5	18.66987 18645	−0.26449 06034	−1.89085 71240
6	22.66914 33588	0.24799 44920	2.01646 66530
7	26.66866 19960	−0.23496 76067	−2.12816 47501
8	30.66832 33409	0.22430 62663	2.22925 54182
9	34.66807 38224	−0.21534 85062	−2.32194 33391
10	38.66788 33469	0.20766 87724	2.40778 11647
11	42.66773 38055	−0.20097 87384	−2.48790 82547

[a]From G. M. Brown, *A. I. Ch. E. J.*, **6**, 179–183, (1960).

Table 5. Eigenfunctions relating to equation 5.131 for mass transfer in laminar flow through tubes.[a]

r_+	ϕ_1	ϕ_2	ϕ_3	ϕ_4	ϕ_5	ϕ_6	r_+
0.00	1.00000000	1.00000000	1.00000000	1.00000000	1.00000000	1.00000000	0.00
0.05	0.99543708	0.97232998	0.93009952	0.87000981	0.79384934	0.70387285	0.05
0.10	0.98184469	0.89180935	0.73545009	0.53108099	0.30228888	0.07488082	0.10
0.15	0.95950842	0.76560457	0.45728832	0.11310661	−0.18271103	−0.36382498	0.15
0.20	0.92889268	0.60469973	0.15247311	−0.23303152	−0.40260123	−0.32121955	0.20
0.25	0.89062392	0.42260986	−0.12055081	−0.39912334	−0.29328114	0.03117856	0.25
0.30	0.84546827	0.23385711	−0.31521322	−0.35914102	0.00054293	0.28981991	0.30
0.35	0.79430462	0.05242653	−0.40611068	−0.16904247	0.24974768	0.22529155	0.35
0.40	0.73809441	−0.10959274	−0.39208452	0.06793183	0.29907438	−0.04765772	0.40
0.45	0.67784945	−0.24301831	−0.29305410	0.24985073	0.14844781	−0.24623845	0.45
0.50	0.61459912	−0.34214076	−0.14234190	0.31507157	−0.07973259	−0.20531795	0.50
0.55	0.54935825	−0.40482321	0.02279222	0.25703015	−0.24057455	0.00683579	0.55
0.60	0.48309693	−0.43218156	0.16968455	0.11416883	−0.25522963	0.19749710	0.60
0.65	0.41671327	−0.42794545	0.27593866	−0.05472615	−0.13791706	0.22889512	0.65
0.70	0.35100978	−0.39762949	0.33148788	−0.19604286	0.03610027	0.10372100	0.70
0.75	0.28667401	−0.34764861	0.33742641	−0.27757775	0.18482648	−0.07719162	0.75
0.80	0.22426362	−0.28449432	0.30272281	−0.29224076	0.25918362	−0.20893147	0.80
0.85	0.16419583	−0.21405616	0.24015828	−0.25200258	0.25273850	−0.24408867	0.85
0.90	0.10674088	−0.14113350	0.16262482	−0.17762079	0.18817264	−0.19521652	0.90
0.95	0.05201900	−0.06914371	0.08046102	−0.08916322	0.09629593	−0.10234372	0.95
1.00	0.0000000	0.00000000	0.00000000	0.00000000	0.00000000	0.00000000	1.00

[a]From G. M. Brown, *A. I. Ch. E. J.*, **6**, 179–183, (1960).

Table 6. Eigenvalues and derivatives relating to equations 5.180–5.183 for mass transfer in laminar flow between flat, parallel plates.[a]

j	γ_j	$(\partial\psi/\partial\gamma)_{j,h_+=1}$	$(d\psi_j/dh_+)_{h_+=1}$
1	1.68159 53222	−0.99043 69608	−1.42915 55060
2	5.66985 73459	1.17910 73461	3.80707 01070
3	9.66824 24625	−1.28624 87056	−5.92023 79188
4	13.66766 14426	1.36201 96175	7.89253 51208
5	17.66737 35653	−1.42132 56612	−9.77094 42849
6	21.66720 53243	1.47040 11597	11.57980 87072
7	25.66709 64863	−1.51246 03349	−13.33387 89738
8	29.66702 10447	1.5493860066	15.04298 83445
9	33.66696 60687	−1.58238 01630	−16.71412 93950
10	37.66692 44563	1.61225 92197	18.35251 24063

[a]From G. M. Brown, *A. I. Ch. E. J.*, **6**, 179–183, (1960). Higher eigenvalues and related constants may be estimated from the following relationships relevant to equations 5.181–5.183:

$$\gamma_j = 4(j-1) + \tfrac{5}{3}$$

$$-G_j(d\psi_j/dh_+)_{h_+=1} = 2.02557\gamma_j^{-1/3}$$

Table 7. Eigenfunctions relating to equation 5.180 for mass transfer in laminar flow between flat, parallel plates.[a]

h_+	ψ_1	ψ_2	ψ_3	ψ_4	ψ_5	ψ_6	h_+
0.00	1.00000000	1.00000000	1.00000000	1.00000000	1.00000000	1.00000000	0.00
0.05	0.99646885	0.96010074	0.88546018	0.77552799	0.63468541	0.46854394	0.05
0.10	0.98591788	0.84377192	0.56853419	0.20356398	−0.19367528	−0.56064196	0.10
0.15	0.96847352	0.66076030	0.12249535	−0.45940625	−0.88289598	−1.00084872	0.15
0.20	0.94434300	0.42615789	−0.35125799	−0.92127048	−0.94139939	−0.39858805	0.20
0.25	0.91380933	0.15883318	−0.74804895	−0.98588465	−0.33711948	0.61543949	0.25
0.30	0.87722437	−0.12047066	−0.98430775	−0.63425823	0.50149201	1.01611220	0.30
0.35	0.83500029	−0.39113244	−1.01518005	−0.02144606	1.00745915	0.42451165	0.35
0.40	0.78759965	−0.63450596	−0.84140636	0.60158620	0.86317657	−0.57373954	0.40
0.45	0.73552430	−0.83542525	−0.50482523	0.99796998	0.18131026	−1.05774223	0.45
0.50	0.67930340	−0.98321739	−0.07498014	1.03501041	−0.61211784	−0.62716953	0.50
0.55	0.61948097	−1.07215385	0.36866934	0.71960900	−1.06882651	0.32311352	0.55
0.60	0.55660303	−1.10134778	0.75396988	0.17433415	−0.97130705	1.03602948	0.60
0.65	0.49120483	−1.07417239	1.02875272	−0.42445889	−0.40612604	1.02291537	0.65
0.70	0.42379830	−0.99732402	1.16687578	−0.91420090	0.34092183	0.35052804	0.70
0.75	0.35485987	−0.87967811	1.16760186	−1.19276885	0.96322186	−0.53240219	0.75
0.80	0.28481910	−0.73108692	1.04990111	−1.23289457	1.27030674	−1.16513731	0.80
0.85	0.21404783	−0.56124735	0.84407970	−1.06843948	1.23140209	−1.32944153	0.85
0.90	0.14285022	−0.37873136	0.58311581	−0.76562278	0.92849439	−1.07197987	0.90
0.95	0.07145363	−0.19022795	0.29544440	−0.39311565	0.48542365	−0.57342785	0.95
1.00	0.00000000	0.00000000	0.00000000	0.00000000	0.00000000	0.00000000	1.00

[a] G. M. Brown, *A. I. Ch. E. J.*, **6**, 179–183, (1960).

Table 8. Eigenvalues and constants in equation 5.184.[a]

j	η_j	F_j	$\chi_j(1)$
1	4.287224	0.175024	−1.26970
2	8.30372	−0.051725	1.4022
3	12.3114	0.02506	−1.4911

[a] R. D. Cess and E. C. Shaffer, *Appl. Sci. Res., A*, **8**, 339–344, (1959). Eigenvalues and constants for $j > 3$ may be estimated from

$$\eta_j = 4j + \tfrac{1}{3}$$

$$F_j = (-1)^{j+1} 2.4727 \eta_j^{-11/6}$$

$$\chi_j(1) = (-1)^j 0.97103 \eta_j^{1/6}$$

Author Index

Subject Index